LONDON MATHEMATICAL SOCIETY LECTURE NOTE SERIES

Managing Editor: Professor J.W.S. Cassels, Department of Pure Mathematics and Mathematical Statistics, University of Cambridge, 16 Mill Lane, Cambridge CB2 1SB, England

The books in the series listed below are available from booksellers, or, in case of difficulty, from Cambridge University Press.

34 Representation theory of Lie groups, M.F. ATIYAH *et al*
36 Homological group theory, C.T.C. WALL (ed)
39 Affine sets and affine groups, D.G. NORTHCOTT
40 Introduction to H_p spaces, P.J. KOOSIS
46 p-adic analysis: a short course on recent work, N. KOBLITZ
49 Finite geometries and designs, P. CAMERON, J.W.P. HIRSCHFELD & D.R. HUGHES (eds)
50 Commutator calculus and groups of homotopy classes, H.J. BAUES
57 Techniques of geometric topology, R.A. FENN
59 Applicable differential geometry, M. CRAMPIN & F.A.E. PIRANI
66 Several complex variables and complex manifolds II, M.J. FIELD
69 Representation theory, I.M. GELFAND *et al*
74 Symmetric designs: an algebraic approach, E.S. LANDER
76 Spectral theory of linear differential operators and comparison algebras, H.O. CORDES
77 Isolated singular points on complete intersections, E.J.N. LOOIJENGA
79 Probability, statistics and analysis, J.F.C. KINGMAN & G.E.H. REUTER (eds)
80 Introduction to the representation theory of compact and locally compact groups, A. ROBERT
81 Skew fields, P.K. DRAXL
82 Surveys in combinatorics, E.K. LLOYD (ed)
83 Homogeneous structures on Riemannian manifolds, F. TRICERRI & L. VANHECKE
86 Topological topics, I.M. JAMES (ed)
87 Surveys in set theory, A.R.D. MATHIAS (ed)
88 FPF ring theory, C. FAITH & S. PAGE
89 An F-space sampler, N.J. KALTON, N.T. PECK & J.W. ROBERTS
90 Polytopes and symmetry, S.A. ROBERTSON
91 Classgroups of group rings, M.J. TAYLOR
92 Representation of rings over skew fields, A.H. SCHOFIELD
93 Aspects of topology, I.M. JAMES & E.H. KRONHEIMER (eds)
94 Representations of general linear groups, G.D. JAMES
95 Low-dimensional topology 1982, R.A. FENN (ed)
96 Diophantine equations over function fields, R.C. MASON
97 Varieties of constructive mathematics, D.S. BRIDGES & F. RICHMAN
98 Localization in Noetherian rings, A.V. JATEGAONKAR
99 Methods of differential geometry in algebraic topology, M. KAROUBI & C. LERUSTE
100 Stopping time techniques for analysts and probabilists, L. EGGHE
101 Groups and geometry, ROGER C. LYNDON
103 Surveys in combinatorics 1985, I. ANDERSON (ed)
104 Elliptic structures on 3-manifolds, C.B. THOMAS
105 A local spectral theory for closed operators, I. ERDELYI & WANG SHENGWANG
106 Syzygies, E.G. EVANS & P. GRIFFITH
107 Compactification of Siegel moduli schemes, C-L. CHAI
108 Some topics in graph theory, H.P. YAP
109 Diophantine analysis, J. LOXTON & A. VAN DER POORTEN (eds)
110 An introduction to surreal numbers, H. GONSHOR
111 Analytical and geometric aspects of hyperbolic space, D.B.A. EPSTEIN (ed)
113 Lectures on the asymptotic theory of ideals, D. REES
114 Lectures on Bochner-Riesz means, K.M. DAVIS & Y-C. CHANG
115 An introduction to independence for analysts, H.G. DALES & W.H. WOODIN
116 Representations of algebras, P.J. WEBB (ed)

117 Homotopy theory, E. REES & J.D.S. JONES (eds)
118 Skew linear groups, M. SHIRVANI & B. WEHRFRITZ
119 Triangulated categories in the representation theory of finite-dimensional algebras, D. HAPPEL
121 Proceedings of *Groups - St Andrews 1985*, E. ROBERTSON & C. CAMPBELL (eds)
122 Non-classical continuum mechanics, R.J. KNOPS & A.A. LACEY (eds)
124 Lie groupoids and Lie algebroids in differential geometry, K. MACKENZIE
125 Commutator theory for congruence modular varieties, R. FREESE & R. MCKENZIE
126 Van der Corput's method of exponential sums, S.W. GRAHAM & G. KOLESNIK
127 New directions in dynamical systems, T.J. BEDFORD & J.W. SWIFT (eds)
128 Descriptive set theory and the structure of sets of uniqueness, A.S. KECHRIS & A. LOUVEAU
129 The subgroup structure of the finite classical groups, P.B. KLEIDMAN & M.W.LIEBECK
130 Model theory and modules, M. PREST
131 Algebraic, extremal & metric combinatorics, M-M. DEZA, P. FRANKL & I.G. ROSENBERG (eds)
132 Whitehead groups of finite groups, ROBERT OLIVER
133 Linear algebraic monoids, MOHAN S. PUTCHA
134 Number theory and dynamical systems, M. DODSON & J. VICKERS (eds)
135 Operator algebras and applications, 1, D. EVANS & M. TAKESAKI (eds)
136 Operator algebras and applications, 2, D. EVANS & M. TAKESAKI (eds)
137 Analysis at Urbana, I, E. BERKSON, T. PECK, & J. UHL (eds)
138 Analysis at Urbana, II, E. BERKSON, T. PECK, & J. UHL (eds)
139 Advances in homotopy theory, S. SALAMON, B. STEER & W. SUTHERLAND (eds)
140 Geometric aspects of Banach spaces, E.M. PEINADOR and A. RODES (eds)
141 Surveys in combinatorics 1989, J. SIEMONS (ed)
142 The geometry of jet bundles, D.J. SAUNDERS
143 The ergodic theory of discrete groups, PETER J. NICHOLLS
144 Introduction to uniform spaces, I.M. JAMES
145 Homological questions in local algebra, JAN R. STROOKER
146 Cohen-Macaulay modules over Cohen-Macaulay rings, Y. YOSHINO
147 Continuous and discrete modules, S.H. MOHAMED & B.J. MÜLLER
148 Helices and vector bundles, A.N. RUDAKOV et al
149 Solitons, nonlinear evolution equations and inverse scattering, M.J. ABLOWITZ & P.A. CLARKSON
150 Geometry of low-dimensional manifolds 1, S. DONALDSON & C.B. THOMAS (eds)
151 Geometry of low-dimensional manifolds 2, S. DONALDSON & C.B. THOMAS (eds)
152 Oligomorphic permutation groups, P. CAMERON
153 L-functions and arithmetic, J. COATES & M.J. TAYLOR (eds)
154 Number theory and cryptography, J. LOXTON (ed)
155 Classification theories of polarized varieties, TAKAO FUJITA
156 Twistors in mathematics and physics, T.N. BAILEY & R.J. BASTON (eds)
157 Analytic pro-*p* groups, J.D. DIXON, M.P.F. DU SAUTOY, A. MANN & D. SEGAL
158 Geometry of Banach spaces, P.F.X. MÜLLER & W. SCHACHERMAYER (eds)
159 Groups St Andrews 1989 volume 1, C.M. CAMPBELL & E.F. ROBERTSON (eds)
160 Groups St Andrews 1989 volume 2, C.M. CAMPBELL & E.F. ROBERTSON (eds)
161 Lectures on block theory, BURKHARD KÜLSHAMMER
162 Harmonic analysis and representation theory for groups acting on homogeneous trees, A. FIGA-TALAMANCA & C. NEBBIA
163 Topics in varieties of group representations, S.M. VOVSI
164 Quasi-symmetric designs, M.S. SHRIKANDE & S.S. SANE
166 Surveys in combinatorics, 1991, A.D. KEEDWELL (ed)
167 Stochastic analysis, M.T. BARLOW & N.H. BINGHAM (eds)
168 Representations of algebras, H. TACHIKAWA & S. BRENNER (eds)
174 Lectures on mechanics, J.E. MARSDEN
175 Adams memorial symposium on algebraic topology 1, N. RAY & G. WALKER (eds)
176 Adams memorial symposium on algebraic topology 2, N. RAY & G. WALKER (eds)
177 Applications of categories in computer science, M.P. FOURMAN, P.T. JOHNSTONE, & A.M. PITTS (eds)
178 Lower K- and L-theory, A. RANICKI
179 Complex projective geometry, G. ELLINGSRUD, C. PESKINE, G. SACCHIERO & S.A. STROMME (eds)

London Mathematical Society Lecture Note Series. 179

Complex Projective Geometry

Edited by
G. Ellingsrud
University of Bergen
C. Peskine
Université Paris VI
G. Sacchiero
Universita di Trieste
S.A. Stromme
University of Bergen

CAMBRIDGE
UNIVERSITY PRESS

Published by the Press Syndicate of the University of Cambridge
The Pitt Building, Trumpington Street, Cambridge CB2 1RP
40 West 20th Street, New York, NY 10011-4211, USA
10 Stamford Road, Oakleigh, Victoria 3166, Australia

First published 1992

Library of Congress cataloguing in publication data available

British Library cataloguing in publication data available

ISBN 0 521 43352 5

Transferred to digital printing 2004

TABLE OF CONTENTS

INTRODUCTION

In the recent years, one has seen the developements of new methods in the study of projective complex varieties. Among others, techniques involving algebraic vector bundles and deformation theory have had an increasingly important role. Surprisingly or not, these developements often stem from problems in traditional Projective Geometry. They were therefore followed by a renewed interest in the works and ideas of the old masters of the subject.

These new trends were studied during two conferences organized, in collaboration, by the Universities of Trieste ("Projective Varieties", 19-24 June 1989) and Bergen ("Vector Bundles and Special Projective Embeddings", 3-16 July 1989). We publish here selected papers from these meetings, hoping they will reflect the main subjects of interest as well as the pleasant working atmosphere of this month of Projective Complex Geometry.

Acknowledgements.

The conference "Projective Varieties" in Trieste was made possible by the supports of the "Consiglio Nazionale delle Ricerche" (C.N.R.), the regional authorities of "Friuli-Venezia Giulia" and the University of Trieste.

The conference "Vector Bundles and Special Projective Embeddings" in Bergen was made possible by the support of the Norwegian Research Council for Sciences and Humanities" (N.A.V.F.). The support given by the French authorities through "le Conseiller Scientifique" at the French Embassy in Norway allowed for a particularly important French participation. The University in Bergen provided an excellent working environment.

Geir Ellingsrud
Christian Peskine
Gianni Sacchiero
Stein Arild Strømme

PROJECTIVE VARIETIES

LIST OF TALKS

Albano A.	Deformations of lines on the Fermat quintic threefold and the infinitesimal Hodge conjecture
Andreatta M.	On the adjunction process over a projective surface in positive characteristic
Arrondo E.	Clifford bounds in higher rank
Ballico E.	Geometry in characteristic p
Beauville A.	Jacobians of spectral curves and completely integrable Hamiltonian systems
Bolondi G.	Cohomological stratas of Hilbert schemes of curves
Brodmann M.	A priori bounds of Castelnuovo type for cohomological Hilbert functions
Catanese F.	Manifolds with an irrational pencil and classification of surfaces with $q=p_g=1$
Ciliberto C.	Gaussian maps and cones over canonical curves
Clemens H.	D-modules and deformations
Conte A.	A tribute to G.Fano
Ellia Ph.	Surfaces lisses de $\mathbf{P}^4$ reglées en coniques
Ellingsrud G.	Numerical equivalence for generic surfaces (I)
Ghione F.	Hommage à Corrado Segre
Green M.	Higher obstructions to deforming the cohomology of line bundles (II)
Gruson L.	An involution on the variety M(0,2) of instanton bundle on $\mathbf{P}^3$ with Chern classes $c_1=0$, $c_2=2$
Hirschowitz A.	Voie est ou voie ouest?
Iarrobino A.	Rational curves on scrolls and the restricted tangent bundle
Kaji H.	On the Gauss maps of space curves in characteristic p
Kleiman S.	The double-point cycle of a map of codimension one
Laksov D.	Explicit formulas for characteristic numbers of complete quadrics under the linear maps
Lazarsfeld R.	Higher obstructions to deforming the cohomology of line bundles (I)
Mukai S.	Fano 3-folds
Müller-Stach S.	Compactifications of $\mathbf{C}^3$
Ottaviani O.	A linear bound on the t-normality of codimension two subvarieties of $\mathbf{P}^n$
Peskine C.	Numerical equivalence for generic surfaces (II)
Ramanan S.	Hyperelliptic curves and spin varieties
Ran Z.	Generic projection
Serrano F.	Algebraic quasi-bundles
Strano R.	Generalized Laudal's lemma
Strømme S.A.	On the cohomology of a geometric quotient
Vainsencher I.	On the Hilbert scheme component of elliptic quartic curves
Voisin C.	On the Noether-Lefschetz theorem
Wahl J.	Gaussian maps on algebraic curves
Walter C.	Spectral sequences and Massey products for space curves conjectures
Xambo S.	Enumerative rings of symmetric varieties

VECTOR BUNDLES AND SPECIAL PROJECTIVE EMBEDDINGS

LIST OF TALKS

Alexander J.	Special rational surfaces of degree 9 in $\mathbf{P}^4$
Arrondo E.	Vector bundles and linkage of congruences II
Aure A.	Surfaces on quintic threefolds associated to the Horrocks-Mumford bundle
Beauville A.	Projective embeddings of moduli spaces of vector bundles on curves
Beltrametti M.	Threefolds of degree 9 and 10 in $\mathbf{P}^5$
Braun R.	On the normal bundle of surfaces in $\mathbf{P}^4$
Catanese F.	Irregular manifolds and surfaces of general type with p_g=q=1
Ciliberto C.	Surfaces of general type with p_g=q=1
Clemens H.	The quartic double solid revisited
D'Almeida J.	Régularité des courbes de $\mathbf{P}^r$
Debarre O.	Smooth images of simple abelian surfaces
Decker W.	Sections of the Horrocks bundle on $\mathbf{P}^5$
Ein L.	Cohen-Macaulay modules and geometry of Severi varieties
Ein L.	Subvarieties of generic complete intersections
Ellia	Problèmes d'existence et de construction de courbes de $\mathbf{P}^3$
Ellingsrud G.	Numerical equivalence on generic surfaces II
Floystad G.	Construction of space curves with good properties
Ghione F.	Le morphisme d'Abel-Jacobi en rang supérieur
Gruson L.	About the work of G. Halphen in algebraic geometry
Horrocks G.	Algebraic deformations of bundles on a punctured spectrum
Hulek K.	Degeneration of non-principally polarized abelian surfaces
Johnsen T.	Secant plane varieties
Koelblen L.	Rational surfaces on a quartic solid with isolated singularities in $\mathbf{P}^4$
Kreuzer M.	Vector bundles with good sections
Laksov D.	Weierstrass point on integral varieties
Lange H.	Cubic equation for abelian varieties
Le Potier J.	Fibré déterminant et courbes de droites de saut
Martin Deschamps M.	Classification of space curves
Mestrano N.	Line bundles on the universal curve of genus g
Mezzetti E.	Smooth nonspecial surfaces in $\mathbf{P}^4$
Mukai S.	Vector bundles and special Grassmannian embeddings of K3-surfaces
Pedrini C.	A moving lemma for singular varieties
Peskine C.	Numerical equivalence on generic surfaces I
Ramanan S.	Remarks on Hirzebruch's proof of the Riemann-Roch theorem
Ramanan S.	Vector bundles and special projective embeddings of curves
Ran Z.	Deformation of maps
Ranestad K.	Surfaces of degree 10 in $\mathbf{P}^4$
Roberts J.	A recenr result of Zak on secant varieties
Sacchiero G.	Surfaces lisses de $\mathbf{P}^4$ reglées en coniques
Sankaran G.	Choosing compactifications of moduli spaces
Schneider M.	Concernng two conjectures of Hartshorne's
Schreyer F.O.	An approach to rank 2 vector bundles on $\mathbf{P}^4$
Sols I.	Vector bundles and linkage of congruences I
Sommese A.	Recent results in adjunction theory
Stromme S.A.	Deformation of unstable vector bundles
Stromme S.A.	On the cohomology of a geometric quotient
Trautmann G.	Instanton bundles on odd dimensional projective spaces and Poncelet varieties
Trautmann G.	Picard group of the compactification of M(0,2)
Vainsencher I.	Conical sextuples in $\mathbf{P}^2$
Van de Ven A.	An exemple concerning Chern classes
Verra A.	Fano models of Enriques surfaces and trisecant lines
Weinfurtner R.	Classification of singular surfaces of Bordiga type

LIST OF PARTICIPANTS

PROJECTIVE VARIETIES

Trieste 19-24 June 1989

A. Albano (Torino), J. Alexander (Angers), A. Alzati (Milano), E. Ambrogio (Torino), M. Andreatta (Milano), E. Arrondo (Madrid), E. Ballico (Trento), B. Basili (Paris), A. Beauville (Paris), G. Beccari (Torino), G. Bolondi (Trento), F. Bottacin (Trieste), C M. Brodmann (Zurich), M. Brundu (Trieste), A. Buraggina (Trieste), G. Carrà Ferro, G. Casnati (Ferrara), F. Catanese (Pisa), G. Ceresa (Roma), F. Chersi (Trieste), A. Chiandit (Trieste), C. Ciliberto (Roma), M.G. Cinquegrani (Catania), H. Clemens (Salt Lake City), E. Colombo (Pavia), A. Conte (Torino), P. Cragnolini (Pisa), V. Cristante (Padova), J. D'Almeida (Caen), A. Del Centina (Ferrara), V. Di Gennaro (Napoli), A. Dolcetti (Pisa), F. Ducrot (Angers), Ph. Ellia (Nice), G. Ellingsrud (Bergen), B. Fantechi (Pisa), G. Ferrarese (Torino), G. Fløystad (Bergen), E. Frigerio (Milano), R. Gattazzo (Padova), F. Ghione (Roma), A. Gimigliano (Genova), S. Giuffrida (Catania), M. Green (Los Angeles), L. Gruson (Lille), A. Hirschowitz (Nice), A. Iarrobino (Boston), M. Idà (Trieste), T. Johnsen (Tromsø), H. Kaji (Tokio), S. Kleiman (Boston), D. Laksov (Stockholm), A. Lascu (Ferrara), F. Laytimi (Lille), R. Lazarsfeld (Los Angeles), H. Le Ming (Trieste), A. Logar (Trieste), B. Loo (Trieste), A. Lopez (Riverside), R. Maggioni (Catania), M. Manaresi (Bologna), P. Maroscia (Potenza), M. Mendes Lopes (Lisboa), J.Y. Mérindol (Strasbourg), P. Merisi (Trieste), E. Mezzetti (Trieste), R. Mirò Roig (Barcelona), I. Morrison (Paris), S. Mukai (Nagoya), S. Müller-Stach (Bayreuth), J.C. Naranjo (Barcelona), C. Oliva (Milano), P. Oliverio (Cosenza), C.E. Olmos (Trieste), G. Ottaviani (Firenze), R. Pardini (Pisa), Ch. Peskine (Paris), L. Picco Botta (Salerno), N. Pla Garcia (Barcelona), A. Predonzan (Trieste), S. Ramanan (Bombay), Z. Ran (Riverside), I. Raspanti (Trieste), R. Re (Catania), M. Roggero (Torino), F. Rossi (Trieste), G. Sacchiero (Trieste), D. Schaub (Angers), E. Sernesi (Roma), F. Serrano (Barcelona), R. Smith (Athens), E. Stagnaro (Padova), R. Strano (Catania), S.A. Strømme (Bergen), G. Tedeschi (Torino), A. Tortora (Roma), F. Tovena (Pisa), I. Vainsencher (Recife), R. Vila Freyer (Trieste), C. Voisin (Paris), J. Wahl (Chapel Hill), C. Walter (New Brunswick), S. Xambò (Barcelona).

VECTOR BUNDLES
AND
SPECIAL PROJECTIVE EMBEDDINGS

Bergen 3-16 July 1989

J. Alexander (Angers), E. Arrondo (Madrid), A. Aure (Trondheim), B. Basili (Paris), A. Beauville (Paris), M. Beltrametti (Genova), R. Braun (Bayreuth), F. Catanese (Pisa), C. Ciliberto (Roma), H. Clemens (Salt Lake City), J. D'Almeida (Caen), O. Debarre (Paris), W. Decker (Kaiserslautern), M. Deschamps (Paris), L. Ein (Chicago), Ph. Ellia (Nice), G. Ellingsrud (Bergen), G. Fløystad (Bergen), F. Ghione (Roma), L. Gruson (Lille), A. Holme (Bergen), G. Horrocks (Newcastle), K. Hulek (Bayreuth), M. Idà (Trieste), T. Johnsen (Tromsø), L. Koelblen (Paris), M. Kreuzer (Regensburg), D. Laksov (Stockholm), H. Lange (Erlangen), J. Le Potier (Paris),P. Lelong (Paris), N. Mestrano (Nice), E. Mezzetti (Trieste), S. Mukai (Nagoya), C. Pedrini (Genova), Ch. Peskine (Paris), S. Ramanan (Bombay), Z. Ran (Riverside), K. Ranestad (Oslo), J. Roberts (Minneapolis), G. Sacchiero (Trieste), G. Sankaran (Cambridge), M. Schneider (Bayreuth),F.O. Schreyer (Bayreuth), M. Skiti (Nancy), I. Sols (Madrid), A. Sommese (Notre Dame), S.A. Strømme (Bergen), G. Trautmann (Kaiserslautern), I. Vainsencher (Recife), A. Van de Ven (Leiden), A. Verra (Genova), C. Victoria (Barcelona), R. Weinfurtner (Bayreuth).

SPECIALITY ONE RATIONAL SURFACES IN $\mathbb{P}^4$

James Alexander.

Dept. de mathématiques
Faculté des sciences
Université de Angers
2, boulevard Lavoisier
49045 ANGERS CEDEX.

Introduction: We work over an algebraically closed field **k** of characteristic zero, except in section (3) where the characteristic is arbitry. By a surface we will mean a smooth projective surface and a curve will be any effective divisor on a surface. We recall that in [A], the speciality of a rational surface X in $\mathbb{P}^n$ is defined to be the number $q(1)=h^1(\mathcal{O}_X(H))$, where H is a hyperplane section of X. We say that X is special or non-special in accordance with $q(1)>0$ or $q(1)=0$.

In [A], a complete classification of non-special rational surfaces in $\mathbb{P}^4$ was given, showing that the linearly normal ones form, for each degree $3\leq d\leq 9$, a single irreducible family. Recently in [E-P] it was shown that there are only a finite number of irreducible components of the Hilbert scheme of $\mathbb{P}^4$ containing rational surfaces; in particular the degrees of such surfaces is bounded. The results which we present here are a contribution to the eventual determination of all such components and contributes to the classification of surfaces in $\mathbb{P}^4$ of small degree [A], [A-R], [R], [Ro],and varieties with small invariants $[I_1,I_2,I_3]$.

We will be concerned with rational surfaces of speciality one in $\mathbb{P}^4$. By $[O_1,O_2,O_3]$ these have degree eight or more and a simple argument shows that their degree is at most eleven (prop.(1.1)). By $[O_3]$, those in degree eight form a single irreducible family and in degree ten we have the following theorem from [R]:

Theorem:(Ranestad) *Let X be a speciality one rational surface of degree ten in $\mathbb{P}^4$, then X is a blowing up of $\mathbb{P}^2$ in thirteen points $x_1,\ldots,x_{13}$, embedded in $\mathbb{P}^4$ by the linear system*

$$|D|=|\pi^*14L-6E_1-4E_2-\ldots-4E_{10}-2E_{11}-E_{12}-E_{13}|$$

where $\pi:X\to\mathbb{P}^2$ is the blowing up in the x_i's, L is a line in $\mathbb{P}^2$ and E_i is the fiber of π over x_i. /

In §(5) we will prove the following theorem:

Theorem(1): *The Hilbert scheme of speciality one, degree ten, rational surfaces in $\mathbb{P}^4$ is irreducible of dimension 38. /*

In degree nine we have the following:

Theorem(2): *(a) Let X be a rational surface of degree nine in $\mathbb{P}^4$ with speciality $q(1)>0$, then $q(1)=1$ and the first and second adjunctions of X give a canonical sequence of birational morphisms of rational surfaces*

$$X \xrightarrow{f_1} X_1 \xrightarrow{f_2} X_2$$

where X_2 is canonically a cubic surface in $\mathbb{P}^3$. The morphism f_2 blows up in three distinct closed points x_7, x_8, x_9, while f_1 blows up in six distinct closed points $x_{10}, \ldots, x_{15}$. Letting K_1 and K_2 be the inverse images on X of the canonical divisors on X_1 and X_2 respectively, the linear system of hyperplane sections of X is given by

$$|H| = |-K-K_1-K_2|$$

where K is the canonical divisor on X.

(b) The Hilbert scheme of special rational surfaces of degree nine in $\mathbb{P}^4$ is irreducible of dimension 42. /

Remark: If we let $f_3: X_2 \rightarrow \mathbb{P}^2$ be one of the finitely many expressions for X_2 as $\mathbb{P}^2$ blown up in six points $x_1, \ldots, x_6$, we obtain an expression for X as $\mathbb{P}^2$ blown up in fifteen points $x_1, \ldots, x_{15}$, with

$$|H| = |\pi^*9L-3E_1-\ldots-3E_6-2E_7-2E_8-2E_9-E_{10}-\ldots-E_{15}|. /$$

The difficulty which one confronts in trying to show that the Hilbert scheme is irreducible, arises from the speciality of the surface. As was explained in [A] the speciality of the surface X is reflected in the special position of the blown up points of $\mathbb{P}^2$. As in the non-special case we construct a parameter family for special rational surfaces, as a universal family of blowings up of $\mathbb{P}^2$. In the non-special case the irreducibility is an automatic consequence of this construction,since the universal variety of ordered blowings up of $\mathbb{P}^2$ is itself irreducible. However in the special case, we are working over a closed subset of this variety which is not apriori irreducible. To treat this phenomena, we are obliged to establish certain properties of the linear systems determining the special position of the points, for all the surfaces in the family. In the case of degree nine, our objective is to show that every configuration of fifteen points, giving rise to a surface in $\mathbb{P}^4$ via the linear system $|H|$ of theorem(2), is a specialisation of the following generic configuration:

Proposition(3): *Let $x_1,...,x_9$, be nine generic closed points of $\mathbb{P}^2$ and let $\mathcal{P}$ be the generic pencil of plane sextics which are singular at $x_1,...,x_6$ and tangent at x_7,x_8,x_9. Then the base of $\mathcal{P}$ contains six further distinct closed points $x_{10},...,x_{15}$. The generic special rational surface of degree nine in $\mathbb{P}^4$ is then obtained by blowing up the x_i's and embedding by the linear system $|H|$ given in theorem(2).* /

This leaves open the question of speciality one, rational surfaces of degree eleven in $\mathbb{P}^4$ as much for the existence as for the irreducibility of the family.

SECTION 1: Uniqueness.

In this section we use freely the general theory on the adjunction mapping (see [S],[So],[V]).

Proof : (of theorem (2)(a)).

Firstly we will show that if q(1)=1, then X is of the form indicated in the theorem. Let H be the general hyperplane section of X in $\mathbb{P}^4$ and let K be the canonical divisor on X. Then by [A] section (5), we have

$$H^2 = 9 \ , \ H \, . \, K = 3 \ , \ K^2 = -6 \ , g = 7$$

where g is the genus of H. By [So] (1.5) and (3.1), the linear system $|H+K|$ is without base points and the resulting morphism $\phi : X \to \mathbb{P}^6$ induces a birational morphism $f_1 : X \to X_1$ of X to its smooth image X_1 in $\mathbb{P}^6$, so that f_1 blows up X_1 in a finite number s_1 of distinct closed points.

Let H_1 be the general hyperplane section of X_1 in $\mathbb{P}^6$. Then we have

$$H_1{}^2 = 9 \ ; \ H_1 \, . \, K_1 = -3 ; \ K_1{}^2 = -6 + s_1 \ ; \ g_1 = 4 \ ,$$

where g_1 is the genus of H_1. Once again by [So] (1.5), $|H_1+K_1|$ is without base points and induces a morphism $\phi_1 : X_1 \to \mathbb{P}^3$.

By the general theory of the adjunction mapping, we have several cases to consider. Firstly, since $|H_1+K_1|$ has no base points we have

$$0 \leq |H_1+K_1|^2 = -3+s_1$$

Elimination of the case $s_1 = 3$:

If $s_1 = 3$, then by [So] (2.1), the image of X_1 by ϕ_1 is a smooth twisted cubic in $\mathbb{P}^3$. Letting $g : X_1 \to \mathbb{P}^1$ be the induced morphism of X_1 to its image, the singular fibers of g are of the form $P_1 + P_2$ with $P_i \cong \mathbb{P}^1$ (i = 1, 2) and $P_1^2 = -1 = P_2^2 = -P_1 . P_2$.

Blowing down P_1 in each singular fiber we obtain a factorisation

$$X_1 \xrightarrow{g_1} F_e \xrightarrow{g_2} \mathbb{P}^1 \qquad (e \geq 0)$$

of g, where g_1 expresses X_1 as the blowing up of a Hirzebruch surface F_e ([H],(v)) in eleven points and g_2 is the canonical map expressing F_e as a fibered projective line over $\mathbb{P}^1$.

Now note that we can suppose e = 1. This is clear if e = 0 or 1. If $e \geq 2$, let C_0 be the unique section of g_2 with negative self intersection ($C_0^2 = -e$) and let λ be the number of blown up points which lie on C_0. Let C_0' be the strict transform of C_0 on X_1, so that $C_0^2 = - e - \lambda$.Then using the facts

$$H_1 + K_1 = g_1^*(3f) \qquad (\text{f one fiber of } g_2)$$

$$K_1 = g_1^*(-2C_0 - (e + 2)f) - E_1 - \dots - E_{11}$$

where E_i (i = 1, ..., 11) are the blown down curves of the singular fibers, we conclude that $0 < H_1 . C_0 = 5 - e - \lambda$, since H_1 is very ample and C_0 is effective.

This shows that $\lambda \leq 4 - e$ and we can suppose that $E_1, \dots, E_{e-1}$ don't meet C_0. Finally by exchanging E_i for its complementary component in the singular fiber of g, for i = 1, ..., e-1, we obtain the desired factorisation through F_1. In fact the complementary components are the strict transforms of fibers of g_2, so they meet C_0 in just one point. After blowing down we obtain a section over $\mathbb{P}^1$ with self intersection -1.

Now we can project F_1 to $\mathbb{P}^2$ to obtain a birational morphism $p : X_1 \to \mathbb{P}^2$ which blows up $\mathbb{P}^2$ in twelve points $x_1, \dots, x_{12}$ so ordered that H_1 takes the form $H_1 = p^*6L - 4E_1 - E_2 - \dots - E_{12}$ .Letting $\pi = p \circ f_1$, we have

$$H = \pi^*9L - 5E_1 - 2E_2 - \dots - 2E_{12} - E_{13} - E_{14} - E_{15}.$$

Now we will show that H is not very ample. There is at least one curve C in the linear system $|\pi^*4L - 2E_1 - E_2 - \dots - E_{12}|$ and for such C we have $H . C = 4$, $p_a(C) = 2$. However there are no projective curves, reducible or not of arithmetic genus two and degree four.

We have now shown that $s_1 > 3$. In this case $(H_1 + K_1)^2 > 0$ so that by [So] (2.3), the image X_2 of X_1 by ϕ_1 is a hypersurface of $\mathbb{P}^3$ of degree two or more. By [So] (2.3) and (3.1), except for the case $K_1^2 = 1$ ($s_1 = 7$) corresponding to the Bertini Involution [So] (2.5.2), the image X_2 of X_1 is a linearly normal rational surface in $\mathbb{P}^3$. In the latter case there are only two possibilities ; X_2 is a smooth quadric ($s_1 = 5$) or a smooth cubic surface ($s_1 = 6$).

Elimination of the case $s_1 = 7$.

In this case, the induced morphism $f_2 : X_1 \to X_2$ of X_1 to its image in $\mathbb{P}^3$, is the Bertini involution ; so we have a birational morphism [D] (p.66) $p : X_1 \to \mathbb{P}^2$ expressing X_1 as $\mathbb{P}^2$ blown up in eight points $x_1, \ldots, x_8$, H_1 takes the form $H_1 = p^*6L-2E_1-\ldots-2E_8$. Letting $\pi = p \circ f_1$ we find

$$H = \pi^*9L-3E_1- \ldots -3E_8-E_9- \ldots -E_{15} .$$

Now in the linear system $| \pi^*3L-E_1-\ldots-E_9 |$ there is a curve C with H.C =2 and p_a (C) = 1, showing that H is not very ample.

Elimination of the case where $s_1 = 5$

In this case, the induced morphism $f_2 : X_1 \to X_2$ of X_1 to its image in $\mathbb{P}^3$ is the blowing up of the smooth quadric X_2, in nine points. Blowing up a point on X_2 and blowing down to $\mathbb{P}^2$, we obtain a birational morphism $p : X_1 \to \mathbb{P}^2$ expressing X_1 as $\mathbb{P}^2$ blown up in ten points $x_1, \ldots, x_{10}$ and H_1 takes the form $H_1 = p^*5L-2E_1-2E_2-E_3-\ldots- E_{10}$. Letting $\pi = p \circ f_1$, we have $H = \pi^*8L-3E_1-3E_2-2E_3-\ldots- 2E_{10}-E_{11}-\ldots-E_{15}$. Now let C be a curve in the linear system $| \pi^*4L-2E_1-2 E_2-E_3- \ldots -E_{10} |$ .Then we have H . C = 4 and $p_a(C) = 1$, so that C is contained in a hyperplane of $\mathbb{P}^4$. This shows that the linear system $| H - C | = | \pi^*4L - E_1 - \ldots - E_{15} |$contains a curve D with H . D = 5 and p_a (D) = 3. Such a curve is necessarily the union of a plane quartic D_1 and a line Δ, meeting D_1 in a point. We then have D_1 in a linear system $| \pi^*dL - r_1E_1 - \ldots - r_{15} E_{15} |$ with $d \leq 4$. Since p_a $(D_1) = 3$, we have $d = 4$ and $r_i \leq 1$, but this implies that Δ is not effective.

The case $s_1 = 6$

In this case, the induced morphism $f_2 : X_1 \to X_2$ of X_1 to its image in $\mathbb{P}^3$, is the blowing up of the smooth cubic surface $X_2 \subset \mathbb{P}^3$. The hyperplane section of X_2 is the anticanonical system giving $| H | = | - K - K_1 - K_2 |$ in the notation of the theorem. Now it is well known that every smooth cubic surface in $\mathbb{P}^3$ is a blowing of $\mathbb{P}^2$ in six distinct closed points in one of finitely many ways [H] (v).

The fact that there are no rational surfaces of degree nine in $\mathbb{P}^4$ with q(1)>1, follows from ([A-R] §1). /

Proposition(1.1): *There are no speciality one rational surfaces in $\mathbb{P}^4$ of degree twelve or more.*

preuve: In fact in this case we would have |-K| effective with H.(-K)<0 which is impossible.

SECTION 2: RESULTS ON LINEAR SYSTEMS.

In this section we will establish numerous results about certain linear systems on a special rational surface of degree nine in $\mathbb{P}^4$, and a number of results on very ampleness. The reader is advised to consult this section as a reference for the proofs of theorems in later sections.

We let X be a special rational surface of degree nine in $\mathbb{P}^4$, then, as indicated in theorem (2)(a) and the remark which follows it, we have a birational morphism $\pi : X \to \mathbb{P}^2$ expressing X as $\mathbb{P}^2$ blown up in fifteen points $x_1, ..., x_{15}$ by three successive blowings up. We can thus write

$$K = \pi^*(-3L)+E_1+ ... +E_{15}$$

$$K_1 = \pi^*(-3L)+ E_1+ ... +E_9$$

$$K_2 = \pi^*(-3L)+E_1+ ... + E_6$$

As before, H is a general hyperplane section of X.

The important lemmas (2.4), (2.7), (2.8) allow us to stratify the parameter scheme $\mathcal{H}_1$ which will be constructed in §4. The first strate $\mathcal{H}_2 \subset \mathcal{H}_1$ is given by the equivalent conditions of lemma (2.7) and the second $\mathcal{H}_3 \subset \mathcal{H}_2$ is given by the conditions of lemma (2.8) . We show that the open subset $\mathcal{H}_2 - \mathcal{H}_3$ of $\mathcal{H}_2$ is irreducible and that $\mathcal{H}_3$ is the union of three disjoint irreducible components W'(i) (i=7,8,9,) corresponding to the cases i=7,8,9 of lemma (2.8). We then show by an indirect argument from deformation theory , that the generic members of $\mathcal{H}_2$ et W(i) (i=7,8,9) are degenerations of the generic member of $\mathcal{H}_1$ described in proposition (3).

Lemma (2.1): *We have $h^\circ(\mathcal{O}(-K_1)) = 1$ and $h^\circ(\mathcal{O}(-K-K_2)) = 2$.*

Proof : It is clear that $h^\circ(\mathcal{O}(-K_1)) \geq 1$. Let C be an effective divisor in $|-K_1|$, then we have the canonical exact sequence :

$$o \to \mathcal{O}(-K-K_2) \to \mathcal{O}(H) \to \mathcal{O}_C(H) \to o$$

and, since H.C = 3 and $p_a(C)=1$, it follows that the image of the canonical map $H^\circ(\mathcal{O}(H)) \to H^\circ(\mathcal{O}_C(H))$ is three dimensional. Since $h^\circ(\mathcal{O}(H)) = 5$, this gives $h^\circ(\mathcal{O}(-K-K_2)) =2$.

Now let D be an effective divisor in $|-K-K_2|$, then, since H.D =6 and p_a (D) = 4, it follows that the image of the canonical map $H^\circ(\mathcal{O}(H)) \to H^\circ(\mathcal{O}_D(H))$ is at least four dimensional. This gives $h^\circ(\mathcal{O}(-K_1)) = 1$.

Definition (2.2) : From now on, we let C be the unique curve in $|-K_1|$ and we let D be the generic curve in $|-K-K_2|$. The following lemma shows that C has no multiple components.

Lemma (2.3) : *Let Δ be an effective divisor on X such that $H.\Delta = 3$ and $p_a(\Delta)$*

= 1, then Δ is a plane cubic in $\mathbb{P}^4$ without multiple components.

Proof : It is immediate that Δ is a plane cubic and since no plane curve on a smooth projective surface can have a multiple component, this proves the lemma. /

We now investigate what happens if C is reducible.

Lemma (2.4) : *Suppose C is a reducible curve, C = A + B, where A is a line and B is a conic. The following two cases are the only ones possible :*

(i) $A^2 = -2$, $B^2 = -2$

(ii) $A^2 = -3$, $B^2 = -1$

In case (ii), B is smooth and B is one of the three curves E_7, E_8, E_9.

Proof : (The first part is trivial). By the two connexity of $|H|$ we conclude that $A^2 \leq -1$.If $A^2 = -1$,then A is one of the curves $E_{10},\dots, E_{15}$. The latter would imply that B = C - A had arithmetic genus one contradicting the fact that B is a conic. So we have $A^2 \leq -2$. If $B = B_1 + B_2$ is reducible, then the same argument shows that $B_1{}^2 \leq -2$ and $B_2{}^2 \leq -2$ implying that $B^2 = B_1{}^2 + B_2{}^2 + 2 \leq -2$. If B is smooth, then B is birational to its image by the first adjunction on X so $0<(H+K).B = 2+B.K$ and by the adjunction formula we conclude that $B^2 \leq -1$. Now, noting that $C^2 = 0$ and $A.B = 2$, we have $A^2 + B^2 = -4$ which gives the two cases and shows that B is smooth in case (ii).

Now suppose that B is smooth and that $B^2 = -1$. Then $(H + K).B = 1$ and since, as seen above, B does not meet $E_{10}, \dots, E_{15}$ we conclude that by the first adjunction on X, the image of B is a line on X_1 with self intersection -1. This shows that B is one of the curves E_7, E_8, E_9.

Lemma (2.5) : *The linear system $|D|$ is a complete pencil and the generic curve D in $|D|$ is smooth of genus four. The base of $|D|$ is smooth of degree $D^2 = 3$. Equally, we have $\mathcal{O}_D(H) = \omega_D$ the canonical sheaf on D and*

$$\mathcal{O}_D(D) = \mathcal{O}_D(E_7 + E_8 + E_9)$$

so that the base of the pencil $|D|$ is formed by the three closed points $y_i = D \cap E_i$; $i = 7, 8, 9$.

Proof : We note firstly that if $|D|$ has no fixed components, then since $D^2 = 3$, no two curves of $|D|$ have a singularity at a point in the base of $|D|$ and using Bertini's theorem in characteristic zero we can conclude that the generic curve in $|D|$ is smooth and this will prove the first part of the lemma. We note firstly that any fixed component of $|D|$ is contained in the plane of C.

Step (i) : No fixed component of $|D|$ is a component of C.

If C was a fixed component of $|D|$, then $|H - 2C| = |D-C|$ would be a complete pencil and 2C would be a plane curve of degree $H.(2C) = 6$ and arithmetic genus $p_a(2C) = 1$, but this is impossible. If C = A + B and A is a fixed component of $|D|$, then, as above, $C + A = 2A + B$ is a plane curve, but as in the proof of lemma (2.3), $2A$ is not a plane curve.

Step (ii) : $|D|$ has no fixed components that are not components of C.

Let Δ be the fixed curve of $|D|$. Since Δ and C are in a common plane and have no components in common we have $\Delta.C \geq 3$. Since $|D - \Delta|$ has no fixed components we have $3-\Delta.C = (D-\Delta).C \geq 0$. This gives $\Delta . C = 3$ and Δ is a line in $\mathbb{P}^4$.

Now since the complete pencil $|D - \Delta|$ has no fixed components, it follows that the generic curve D_0 in $|D - \Delta|$ is integral and the fact $D_0 . C = 0$ implies that there are disconnected curves in $|D_0 + C| = |H - \Delta|$. By [So](0.10.1), $|H - \Delta|$ is without base points and we conclude that $(H - \Delta)^2 = 0$. A simple calculation then gives $\Delta^2 = -7$, $\Delta . D = -2$, $p_a(D_0) = 0$.

Finally, as in the proof of [So] (1.1.1) , the morphism $X \to \mathbb{P}^2$ induced by the linear system $|H - \Delta|$ factorises as

$$X \xrightarrow{r} \mathbb{P}^1 \xrightarrow{s} \mathbb{P}^2$$

where s embeds $\mathbb{P}^1$ as a conic and r has connected fibers. Since the connected curves C and D_0 lie in separate fibers of the flat morphism r we should have $p_a(D_0) = p_a(C) = 1$, which is a contradiction.

We have now shown that the generic curve D in $|D|$ is smooth and since $H . D = 6$ and $p_a(D) = 4$, it follows that D is a canonical curve of degree six and genus 4 in a hyperplane of $\mathbb{P}^4$. This shows that $\mathcal{O}_D(H) = \omega_D$ on D. Noting that $\omega_D = \mathcal{O}_D(D + K)$, a simple manipulation gives $\mathcal{O}_D(D) = \mathcal{O}_D(E_7 + E_8 + E_9)$ and, since x_7, x_8, x_9 are distinct closed points of X_1 (notation of theorem (2)), the curves E_i ($i = 7, 8, 9$) have no points in common. Noting that $D . E_i = 1$ ($i = 7, 8, 9$) this completes the proof. /

Lemma (2.6) : *For $i = 7, 8, 9$, we have $h^\circ(O(D-E_i)) = 1$, so that the linear system $|D-E_i|$ contains a unique curve D_i.*

Proof : In fact $y_i = D \cap E_i$ is, by lemma (2.5), a fixed point of the complete pencil $|D|$, so considering the cohomology of the followingcanonical exact sequence

$$o \to \mathcal{O}(D-E_i) \to \mathcal{O}(D) \to \mathcal{O}_{E_i}(D) \to o$$

we see that the canonical map, $H^\circ(\mathcal{O}(D)) \to H^\circ(\mathcal{O}_{E_i}(D))$ has a one dimensional image. This gives $h^\circ(\mathcal{O}(D-E_i)) = 1$. /

Lemma (2.7) : *Let Δ be the base of the pencil $|D|$ and let $G = H - 2C$, then the following conditions are equivalent :*

(i) the canonical map , $H^\circ(\mathcal{O}(H + K)) \to H^\circ(\mathcal{O}_\Delta(H + K))$ is not

surjective.

(ii) the canonical map , $H^{\circ}(\mathcal{O}_D(H + K)) \to H^{\circ}(\mathcal{O}_{\Delta}(H + K))$ is not surjective.

(iii) $\mathcal{O}_D(E_{10} + ... + E_{15}) = \omega_D$

(iv) $\mathcal{O}_D(D) = \mathcal{O}_D(C)$

(v) $\mathcal{O}_D(G) = \mathcal{O}_D$

(vi) The divisor G on X is effective.

what is more, if the above conditions are verified then $h^{\circ}(O(2C)) = 2$.

Proof :(i)⇔(ii). This results from a consideration of the cohomology of the canonical exact sequence

$$0 \to \mathcal{O}(E_{10} + ... + E_{15}) \to \mathcal{O}(H + K) \to \mathcal{O}_D(H + K) \to 0$$

noting that $h^1(\mathcal{O}(E_{10} + ... + E_{15})) = 0$.

(ii)⇔(iii). The ideal sheaf of Δ, as a divisor on D, is $\mathcal{O}_D(-D)$, so we have the following canonical exact sequence

$$0 \to \mathcal{O}_D(E_{10} + ... + E_{15}) \to \mathcal{O}_D(H + K) \to \mathcal{O}_{\Delta}(H + K) \to 0$$

Since $(H + K) . D = 9$, we conclude that $H + K$ is non-special on D. This shows that we have (ii) if and only if $h^1(\mathcal{O}_D(E_{10} + ... + E_{15})) > 0$ and by Clifford's theorem, this is equivalent to (iii), since in any case $h^{\circ}(\mathcal{O}_D(E_{10} + ... + E_{15})) \geq 3$.

(iii) ⇔ (iv) ⇔ (v) These equivalences result by simple manipulation noting

(v) ⇔ (vi) We have the exact sequence $0 \to \mathcal{O}(-C) \to \mathcal{O}(G) \to \mathcal{O}_D(G) \to 0$ with $h^1(\mathcal{O}(-C)) = h^1(\mathcal{O}(C + K)) = h^1(\mathcal{O}(E_{10} + ... + E_{15})) = 0$ and $h^{\circ}(\mathcal{O}(-C)) = 0$. This shows that $H^{\circ}(\mathcal{O}(G)) \approx H^{\circ}(\mathcal{O}_D(G))$.

Since $G . D = 0$, we have $h^{\circ}(\mathcal{O}_D(G)) > 0$ if and only if $\mathcal{O}_D(G) \approx \mathcal{O}_D$ and this gives the desired equivalence. Now suppose that G is effective, then from the exact sequence $0 \to \mathcal{O}(2C) \to \mathcal{O}(H) \to \mathcal{O}_G(H) \to 0$ and the facts $H . G = 3$, $p_a(G) = 1$, we conclude that the image of the canonical map $H^{\circ}(\mathcal{O}(H)) \to H^{\circ}(\mathcal{O}_G(H))$ is three dimensional. This gives $h^{\circ}(\mathcal{O}(2C)) = 2$. /

Lemma (2.8) : *If the conditions of lemma (2.7) are verified and the complete pencil $|2C|$ is not base point free, then $C = A + B$ where $H . A = 1$, $A^2 = -3$ and B is the fixed curve of $|2C|$ with $H . B = 2$; $B^2 = -1$. In particular B is one of the curves E_i ($i = 7, 8, 9$) and A is smooth and irreducible.*

Proof : Since $C^2 = 0$, if $|2C|$ is not base point free, it has a fixed component whose integral components are components of C. Since $h^{\circ}(\mathcal{O}(C)) = 1$ and $h^{\circ}(\mathcal{O}(2C)) = 2$ it follows that C is not a fixed component of $|2C|$.

As such we can write $C = A + B$ where A is a component of the fixed curve of $|2C|$ and B has no components in common with the fixed curve of $|2C|$. As such,

$$|2C - A| = |A + 2B|$$

is a complete pencil and $0 \le (A + 2B) . B = 2 + 2B^2$, so that $B^2 \ge -1$. Now by lemma (2.4) we conclude that $B^2 = -1$ and that B is a smooth conic in $\mathbb{P}^4$ with $B = E_i$ (i = 7, 8, 9). Finally | 2C | cannot have 2A as fixed component, otherwise |2B| would be a pencil without fixed components and $(2B)^2 \ge 0$, but this contradicts $B^2 = -1$. /

Lemma (2.9) : *Let Δ be the image on X_1 of the pencil | D | on X, let A be the base of Δ and let B be the base of | D |. Then*

(i) The induced morphism $B \to A$ is a closed immersion

(ii) A contains the closed subscheme $\{ x_{10} , \dots , x_{15} \}$

(iii) A is curvelinear.

Let $f_0 : X_0 \to X_1$ be the blowing up of X_1 in the image of B and let Δ_0 bethe strict transform on X_0 of the pencil Δ. Then the base of Δ_0 is canonically isomorphic to Δ.

Proof :This results from the fact that $x_{10} , \dots , x_{15}$ are distinct closed points of X_1 and any one of the points y_i (of lemma (2.5)) is infinitely closed to at most one of the points $x_{10} , \dots , x_{15}$. /

Proposition (2.10): *Let X be a surface and let |H| be a complete linear system on X. We suppose that there is a curve C and a pencil |D| on X, such that C+|D| $\subset$ |H|. If the following conditions are verified*

(1) The canonical map, $H^0(\mathcal{O}(H)) \to H^0(\mathcal{O}_C(H))$ is surjective

(2) The canonical map, $H^0(\mathcal{O}(H)) \to H^0(\mathcal{O}_D(H))$ is surjective for every curve D in |D|,

(3) $\mathcal{O}(H)$ is very ample on C and on every curve in |D|, then the linear system |H| is very ample on X.

Proof: Let Z be some finite degree two closed subscheme of X. If Z is contained in C or one of the curves in |D|, then |H| separates Z by (1) and (2). We can thus suppose that Z meets C, or a curve in |D|, in at most a closed point. In particular we can suppose that Z does not meet the base of the pencil |D|. If Z meets C, then C+D separates Z for a general choice of D in |D|. If Z does not meet C then some curve D' in |D| meets Z in a closed point and C+D' separates Z. /

Proposition (2.11) : *Let $C = C_1 + C_2$ be the sum of two effective divisors on a projective surface X and let $\mathcal{L}$ be an invertible sheaf on C. We suppose that the following three conditions are verified:*

(a) The canonical map, $H^\circ(\mathcal{L}) \to H^\circ(\mathcal{L} \otimes_C \mathcal{O}_{C_i})$ (i = 1,2) is

surjective.

(b) $\mathcal{L} \otimes_C \mathcal{O}_{C_i}$ *is very ample on* C_i *(i=1,2).*

(c) $\mathcal{L} \otimes_C \mathcal{O}_{C_2}(-C_1)$ *has no base points on* C_2.

Then $\mathcal{L}$ *is very ample on C.*

Proof : Writing the residual exact diagram [A] (1.4) of C_2 with respect to C_1 we obtain the exact sequence on C

$$0 \to \mathcal{O}_{C_2}(-C_1) \to \mathcal{O}_C \to \mathcal{O}_{C_1} \to 0$$

Now apply the same technique as used in [A] (2.7). /

We finish this section with a lemma on the Geyser involution [D]. This is the natural involution $\sigma : X \to X$ defined on the blowing up $\pi : X \to \mathbb{P}^2$ of the projective plane in seven points $x_1, \ldots, x_7$ almost general position (loc. cit.). The quotient of X by the involution σ is a finite morphism of degree two, $p : X \to \mathbb{P}^2$, induced by the anticanonical system $|-K|$ on X. The orbits of σ or, what is essentially the same, of p, are the finite degree two subschemes of X which are not separated by the anticanonical system. In termes of the morphism π, for any choice of an eigth point x_8, the pencil of curves through x_8, has a base of degree two formed by x_8 and a point x_9 possibly infinitely near to x_8. The following lemma results directly from [D] (p. 67-68).

Lemma (2.12) : With the above notation we let $x_1,\ldots,x_7$ be generic in $\mathbb{P}^2$. Let $D \in |-2K|$ be the generic curve. Then D is non hyperelliptic. /

Section (3): THE GENERIC SURFACE IN DEGREE NINE .

In this section we will work over an algebraically closed field of arbitry characteristic. We construct a speciality one rational surface of degree nine in $\mathbb{P}^4$, which in characteristic zero is just the generic one.

Proposition(3.1): *Let* $x_1,\ldots,x_9$ *be generic closed points of* $\mathbb{P}^2$ *and for i= 7,8,9, let* y_i *, be the generic point infinitely close to* x_i*. Let*

$$X_0 \to X_1 \to X_2 \to \mathbb{P}^2$$

be the successive blowings up in $x_1,\ldots,x_6$*, then* x_7,x_8,x_9*, then* y_7,y_8,y_9*.. Let* $p: X_0 \to \mathbb{P}^4$ *be the composite morphism and let* E_i *(resp.* F_i *) be the fiber over* x_i *(resp.* y_i*). Then we have:*

(a) The generic curve D' in the linear system

$$|D'| = |p^*6L-2E_1-\ldots-2E_6-E_7-E_8-E_9-F_7-F_8-F_9|$$

is smooth and the base Δ *of the generic pencil in* $|D'|$ *is smooth formed by six closed points* $x_{10},\ldots,x_{15}$ *any two of which form a generic family with* $x_1,\ldots,x_9$*..*

(b) If $X_0 \to Y_0$ *is the blowing down of* F_7 *and* $p_0 : Y_0 \to \mathbb{P}^2$ *is the morphism to* $\mathbb{P}^2$*, then the generic curve in the linear system*

$$|D'_7| = |p_0{}^*6L-2E_1-\dots-2E_7-E_8-E_9-F_8-F_9|$$

is smooth and we have

$$h^0(\mathcal{O}(D'_7)) = 3\ ,\quad h^0(\mathcal{O}(D'_7-E_i)) = 0 \quad (i=8,9).$$

Proof: The generic curves in $|D'|$ and $|D'_7|$ are smooth by construction since they are the strict transforms of generic plane nodal curves with six and seven nodes respectivly. To prove the remaining part of *(a)* , it will be enough to show that:

(i) $h^0(\mathcal{O}(D')) = 4$

(ii) $|D'|$ has no base points and is very ample on an open subset of the generic curve D'.

In fact, this shows that Δ contains two generic closed points of X_0 (condition (i)) and that D' is birational to its image in $\mathbb{P}^2$ under the morphism induced by the linear system $|D'|$ (condition (ii)). We then deduce from Bertini's theorem and the two transitive action of the algebraic fundamental group on Δ, that Δ is smooth and its points are two by two generic on X_0.

Since conditions (i) and (ii) are open properties we can take the points x_i, y_i in special position and show that these conditions hold there. So let T be the generic plane quintic with six nodes $x_1,\dots,x_6$ and let x_i, y_i (i=7,8,9) be in the smooth part of T (y_i being infinitely close to x_i). Let $p:X_0\to\mathbb{P}^2$, be the corresponding blowing up and let T_0 be the strict transform of T on X_0. Since $h^1(\mathcal{O}(p^*L))=0$ the exact sequence

$$0 \to \mathcal{O}(p^*L) \to \mathcal{O}(D') \to \mathcal{O}_{T_0}(D') \to 0$$

is exact on global sections and since $T_0.D'=0$ with $T_0\approx\mathbb{P}^1$, we conclude that $|D'|$ has no base points in T_0. Finally, since $h^0(\mathcal{O}(p^*L))=3$ and $|p^*L|$ has no base points, we conclude that $h^0(\mathcal{O}(D')) = 4$ and that $|D'|$ has no base points. In fact, $|p^*L|$ is very ample outside the total transform of T so that the linear series with fixed component, $T+|p^*L| \subset |D'|$, is very ample outside of the total transform of T. Since the canonical map $H^0(\mathcal{O}(D')) \to H^0(\mathcal{O}_{D'}(D'))$ is surjective we conclude that $|D'|$ is very ample on an open subset of D'.

To finish the proof of *(b)*, we return to the generic configuration of the points x_i, y_i given in the proposition. We let C' be the unique (smooth elliptic) curve in $|p_0{}^*3L-E_1-\dots-E_9|$. Since y_8 is the generic tangent direction at x_8, we have $h^0(\mathcal{O}(C'-F_8))=0$. Considering the two exact sequences

$$0 \to \mathcal{O}(C'-\Sigma_{i=8,9}(E_i-F_i)) \to \mathcal{O}(D'_7) \to \mathcal{O}_{C'}(D'_7) \to 0$$

$$0 \to \mathcal{O}(C'-F_8+(E_9-F_9)) \to \mathcal{O}(D'_7-E_8) \to \mathcal{O}(D'_7-E_8) \to 0$$

both of which are exact on global sections, we find using $C'.D'_7=3$ and $C'.(D'_7-E_8)=2$ that $h^0(\mathcal{O}(D'_7))=3$, $h^0(\mathcal{O}(D'_7-E_8))=2$ as required. /

Proposition(3.2): *Let $x_1,\dots,x_{15}$ be the points given by propersition (3.1) and let*

$$X \to X_1 \to X_2 \to \mathbb{P}^2$$

be the successive blowings up in $x_1,\dots,x_6$, then $x_7,\dots,x_9$, then $x_{10},\dots,x_{15}$. We let $\pi : X \to \mathbb{P}^2$ be the composite morphism and we set

$$|C| = |\pi^* 3L-E_1-\dots-E_9|$$

$$|D| = |\pi^* 6L-2E_1-\dots-2E_6-E_7-\dots-E_{15}|$$

where E_i is the fiber over x_i and L is a line of $\mathbb{P}^2$. Then $|H|=|C+D|$ is very ample on X of projective dimension four.

Proof: Setting $|D_i|=|D-E_i|$ $(i=7,8,9)$, we have by proposition (3.1),

$$h^0(\mathcal{O}(C))=1, \quad h^1(\mathcal{O}(C))=0 \tag{3.2.1}$$

$$h^0(\mathcal{O}(D))=2, \quad h^1(\mathcal{O}(D))=0$$

$$h^0(\mathcal{O}(D_i))=1, \quad h^1(\mathcal{O}(D_i))=1 \qquad (i=7,8,9)$$

and the unique curves in $|C|$ and $|D_i|$ $(i=7,8,9)$ are smooth, as is the generic curve D in $|D|$.

As well, the base of the complete pencil $|D|$ is formed by the three closed points $y_i = E_i \cap D$ $(i=7,8,9)$.

To prove that $|H|$ is very ample we apply lemma (2.11) to the pencil with fixed component $C+|D| \subset |H|$.

step(1): If $D \in |D|$ is not of the form D_i+E_i $(i=7,8,9)$ then $\mathcal{O}_D(H)=\omega_D$ (the dualizing sheaf on D) and ω_D is very ample on D.

In fact, if D is such a curve, then D meets E_i transversally in y_i $(i=7,8,9)$, so that $\mathcal{O}_D(D)=\mathcal{O}_D(E_7+E_8+E_9)$ which is equivalent to $\mathcal{O}_D(H)=\mathcal{O}_D(H+K)=\omega_D$. What is more, since all such curves meet E_i $(i=10,\dots,15)$ transversally by construction, it follows that any such curve is isomorphic to its image in the linear system $|\pi^*6L-2E_1-\dots-2E_6|$ on X_2. However it is well known that the dualizing sheaf of every curve in this linear system is very ample for a sufficiently general choice of $x_1,\dots x_6$ (consider X_2 as a cubic surface in $\mathbb{P}^3$).

step(2): $h^0(\mathcal{O}(H))=5$, and the canonical map $H^0(\mathcal{O}(H)) \to H^0(\mathcal{O}_D(H))$ is surjective for all D in $|D|$.

In fact we have the exact sequence

$$0 \to \mathcal{O}(C) \to \mathcal{O}(H) \to \mathcal{O}_D(H) \to 0$$

with $h^0(\mathcal{O}(C))=1$, $h^1(\mathcal{O}(C))=0$ and $\mathcal{O}_D(H)=\omega_D$ for generic D in $|D|$.

step(3): $\mathcal{O}_C(H)$ is very ample on C and the canonical map $H^0(\mathcal{O}(H)) \to H^0(\mathcal{O}_C(H))$ is surjective.

In fact, since C is smooth and elliptic with $H.C=3$, we have $h^0(\mathcal{O}_C(H))=3$, $h^1(\mathcal{O}_C(H))=0$. The rest follows fom the exact sequence

$$0 \to \mathcal{O}(D) \to \mathcal{O}(H) \to \mathcal{O}_C(H) \to 0$$

noting that $h^1(\mathcal{O}(D))=1= h^1(\mathcal{O}(H))$.

step(4): $\mathcal{O}(H)$ is very ample on D_i+E_i $(i=7,8,9)$.

By symmetry we can set i=7. We will use lemma(2.10). By proposition (3.1) (b), D_7 meets E_8 and E_9 transversally in y_8 and y_9 respectively, so that $\mathcal{O}_{D7}(D) = \mathcal{O}_{D7}(E_8+E_9)$ and, by a simple manipulation we find that, $\mathcal{O}_{D7}(H) = \mathcal{O}_{D7}(D_7+K)$ is the dualizing sheaf on D_7. Again by (3.1) (b), D_7 is smooth and as in step one, the curves C and D_7 are isomorphic to their images in the linear systems $|-K_Y|$ and $|-2K_Y|$ respectively, on the blowing up Y of $\mathbb{P}^2$ in $x_1,\ldots x_7$. Now by lemma (2.12), we conclude that the dualizing sheaf on D_7 is very ample. Noting that $h^1(\mathcal{O}(C+E_7))=0$ on X, we conclude that the following exact sequence

$$0 \to \mathcal{O}(C+E_7) \to \mathcal{O}(H) \to \mathcal{O}_{D7}(H) \to 0$$

is exact on global sections, showing that $\mathcal{O}(H)$ separates in D_7.

To see that $\mathcal{O}(H)$ separates in E_7, we consider the two exact sequences

(3.2.2) $$0 \to \mathcal{O}(H-E_7) \to \mathcal{O}(H) \to \mathcal{O}_{E7}(H) \to 0$$

$$0 \to \mathcal{O}(C) \to \mathcal{O}(H-E_7) \to \mathcal{O}_{D7}(H-E_7) \to 0$$

Since $\mathcal{O}_{D7}(H)= \omega_{D7}$ and $E_7 \cap D_7$ is an effective divisor of degree two on D_7, it follows that, since ω_{D7} is very ample on D_7, we have $h^i(\mathcal{O}_{D7}(H-E_7))=0$ for i=0,1. Combining this with (3.2.1), we find $h^1(\mathcal{O}(H-E_7))=1= h^1(\mathcal{O}(H))$, so that (3.2.2) is exact on global sections. Noting that $H.E_7=2$ and $E_7 \approx \mathbb{P}^1$,we conclude that $\mathcal{O}(H)$ separates in E_7. Finally, noting that $(H-D_7).E_7=0$, we conclude that $h^1(\mathcal{O}_{E7}(H-E_7))=0$ and that $\mathcal{O}_{E7}(H-E_7)$ is without base points on E_7. This shows that $\mathcal{O}(H)$ separates in D_7+E_7 by (2.10) completing the proof of the proposition. /

SECTION 4: PROOF OF THEOREM(2)(b).

Let T be the **k**-scheme which parametises blowings up of $\mathbb{P}^2$ by three successive blowings up in six closed points $x_1,\ldots,x_6$,then three closed points x_7,x_8,x_9,then three closed points y_7,y_8,y_9 with y_i infinitely close to x_i (i =7,8,9). Then T is a smooth irreducible algebraic variety. (See [A] (3) for this type of construction).Now let $X_0 \to X_1 \to X_2 \to \mathbb{P}_T^2$ be the universal blowing up over T and let E_i (resp. F_i) be the inverse image on X_0 of the blown up section x_i (resp. y_i). Let $p = f_3 \circ f_2 \circ f_0$ and define

$$\mathcal{C}' = \mathcal{O}(p^*3L-E_1-\ldots-E_9)$$

$$\mathcal{D}' = \mathcal{O}(p^*6L-2E_1-\ldots-2E_6-E_7-E_8-E_9-F_7-F_8-F_9)$$

We let T_0 be the open subset of T where

(*) (i) $h^\circ(\mathcal{T}_t') = 1$

(ii) $h^\circ(\mathcal{D}_t') \leq 5$

(iii) $h^\circ(\mathcal{T}_t' \otimes \mathcal{T}_t') \leq 2$

and let S_0 be the open subset S_0 of the relative grassmaninan S of pencils of Cartier divisors associated to the invertible sheaf D' on $X_0 \times_T T_0$ where

(**) (iv) the pencil contains a smooth curve

(v) the base $\Delta \to X_0$ of the pencil is smooth

(vi) the induced morphism $\Delta \to X_1$ is a closed immersion.

Now over S_0 we have (with an obvious abuse of notation) the universal blowing up

$$X \xrightarrow{f_1} X_1 \xrightarrow{f_2} X_2 \xrightarrow{f_3} \mathbb{P}_S{}^2$$

where $f_1 : X \to X_1$ is the blowing up of X_1 in Δ.

Let $\pi = f_3 \circ f_2 \circ f_1$, let E_i be the fiber of π over x_i (i = 1, ..., 9), let E be the inverse image on X of $\Delta \subset X_1$, and finally set

$$\mathcal{T} = \mathcal{O}(\pi^*3L-E_1-\ldots-E_9)$$

$$\mathcal{D} = \mathcal{O}(\pi^*6L-2E_1-\ldots-2E_6-E_7-E_8-E_9-E)$$

$$\mathcal{L} = \mathcal{T} \otimes \mathcal{D}$$

The following lemma resumes the properties of these invertible sheaves on X, viewed fiber by fiber.

Lemma(4.1): With the above notation we have at each point $s \in S_0$,

(i) $h^\circ(\mathcal{T}_s) = 1$

(ii) $h^\circ(\mathcal{D}_s) = 2$

(iii) the generic curve D in $|H^\circ(\mathcal{D}_s)|$ is smooth

(iv) $\mathcal{L}_s \otimes \mathcal{O}_D$ is the canonical sheaf on D

(v) $h^\circ(\mathcal{L}_s) = 5$

Proof : Parts (i), (ii), (iii) resulte directily from (*) and (**). By the construction we have $\mathcal{O}_D(D) = \mathcal{O}_D(E_7+E_8+E_9)$since the y_i (i = 7, 8, 9) are chosen in the fibers E_i (i = 7, 8, 9). This gives $\mathcal{O}_D(D+K) = \mathcal{O}_D \otimes \mathcal{L}_s = \omega_D$ which is the canonical divisor on D. We then have the exacte sequence :

$$0 \to \mathcal{T}_s \to \mathcal{L}_s \to \omega_D \to 0$$

with $h^1(\mathcal{T}_s) = 0$ and $h^\circ(\omega_D) = 4$, so that $h^\circ(\mathcal{L}_s) = 5$. /

Lemma(4.2) : *The set of points $s \in S_0$ where $\mathcal{L}_s$ is very ample with $h^\circ(\mathcal{L}_s) = 5$ is an open subset $\mathcal{H} \subset S_0$.*

Proof : Since at all points $s \in S_0$ we have $h^\circ(\mathcal{L}_s)=5$ by lemma (4.1), it follows by [A] (3.8) that the subset $\mathcal{H} \subset S_0$ formed by those points $s \in S_0$ such that $\mathcal{L}_S$ is very ample on X_S is an open subset of S_0. /

Let $T_1 \subset T_0$ be the open subset formed by those points $t \in T_0$ such that $h^\circ(\mathcal{D}_t') = 4$ (note that $h^\circ(\mathcal{D}_t') \geq 4$ by Riemman-Roch). Then over T_1, the grassmanian Sx_TT_1 of pencils of Cartier divisors associated to $\mathcal{D}'$, is smooth with irreducible fibers. Since T is irreducible, so is T_1 and it follows that $\mathcal{H} x_{T_0} T_1$ is irreducible. We let $\mathcal{H}_1$ be the closure in $\mathcal{H}$ of the open subset $\mathcal{H} x_{T_0} T_1$.

Now let $T_2 = T_0 - T_1$. Then by (*) (ii) we have $h^\circ(\mathcal{D}'_t) = 5$ at all points $t \in T_2$. It follows that over T_2, the grassmanian Sx_TT_2 is smooth with irreducible fibers, so that $\mu : \mathcal{H} x_{T_0} T_2 \rightarrow T_2$ is smooth with irreducible fibers as well.

Let $T_2' = \mu(Hx_{T_0}T_2)$, which is an open subset of T_2. By lemma (2.7) we have $h^\circ(\mathcal{C}_t' \otimes \mathcal{C}_t') = 2$ at all points $t \in T_2'$. Let T_2'' be the open subset of T_2' formed by those points $t \in T_2'$ such that $| H^\circ(\mathcal{C}_t' \otimes \mathcal{C}_t') |$ contains a smooth curve and let $\mathcal{H}_2$ be the closure in $\mathcal{H}$ of the subset $T_2'' x_{T_0} \mathcal{H} \subset \mathcal{H}$. It is not difficult to see that T_2'' est irreducible . In fact it is dominated by an open subset of the variety of plane irreducible ellyptic sextics.

Now let $T_3 = T'_2 - T''_2$. By lemma (2.8) at each point $t \in T_3$, the curve C in $|H^\circ(\mathcal{C}_t)|$ is reducible of the form $(C-E_i)+E_i$ (i=7,8,9). Let C_0 be the integral plane cubic with a singularity at x_i whose strict transform is $(C-E_i)$. In view of lemma (2.8),it is clear that T_3 is the disjoint union of the closed subsets W(i), where C_0 is singular at x_i (i = 7, 8, 9). It is easy to see that W(i) is irreducible. Let $\mathcal{H}_3$ be the closure of $T_3 x_T \mathcal{H}$ in $\mathcal{H}$.

Now T_2'' has dimension 17 and T_3 has dimension 16. As such the subschemes $\mathcal{H}_2$, $\mathcal{H}_3$ of $\mathcal{H}$ have dimensions 24 and 23 respectively. These give rise to corresponding subschemes of the Hilbert scheme of special rational surfaces of degree nine in $\mathbb{P}^4$, of dimensions 40 and 39 respectively. However if X is a speciality one rational surface of degree nine in $\mathbb{P}^4$, then since $h^2(\mathcal{N}_{X/\mathbb{P}^4})=0$, where $\mathcal{N}_{X/\mathbb{P}^4}$ is the normal bundle of X in $\mathbb{P}^4$, we have

$$h^0(\mathcal{N}_{X/\mathbb{P}^4}) - h^1(\mathcal{N}_{X/\mathbb{P}^4}) = \chi(\mathcal{N}_{X/\mathbb{P}^4}) = 42$$

which is well known to be the minimum possible value for the dimension of the component of the Hilbert scheme containing X. It follows that the "images" of $\mathcal{H}_2$ and $\mathcal{H}_3$, lie in the closure of the "image" of $\mathcal{H}_1$ where, by "image", we mean the image of the canonical morphisme

$$\rho: \underline{\mathrm{Isom}}_{\mathcal{H}}(\pi_* \mathcal{L}, \mathcal{O}_{\mathcal{H}}^5) \rightarrow \underline{\mathrm{Hilb}}(\mathbb{P}^4/k).$$

Since $\mathcal{H}_1$ is irreducible of dimension 26, its "image" is irreducible of dimension 42. By (2.5), (2.9) and the fact that |H+K| is very ample on X_1,the conditions of (*) and (**) are verified for every speciality one rational surface of degree nine in $\mathbb{P}^4$, showing that the image of ρ is the whole of the Hilbert scheme of speciality one degree nine rational surfaces in $\mathbb{P}^4$. This completes the proof of theorem(2)(b). /

Remark: Using a raffinement of the arguments emploied in §3 one can show that the varieties $\mathcal{H}_2$ and W(i) i=1,2,3 are non empty.

SECTION 5: SPECIALITY ONE IN DEGREE TEN

In this section we will prove theorem(1). We consider speciality one rational surfaces of degree ten in $\mathbb{P}^4$. In [R], it was shown that all such surfaces X arise in the following way: The first and second adjunctions on X give a canonical sequence of blowings up

$$X \xrightarrow{f_1} X_1 \xrightarrow{f_2} X_2 \approx \mathbb{P}^2(x_1,\dots,x_{10})$$

where X_2 is an ordered blowing up of $\mathbb{P}^2$ in ten points $x_1,\dots,x_{10}$. The morphism f_2 blows up in one closed point x_{11} while f_1 blows up in two distinct closed points x_{12}, x_{13}. Letting E_i be the exceptional fiber over x_i (i=1,...,13) and π: X $\to \mathbb{P}^2$ be the composit morphism, we can express the hyperplane section of X in $\mathbb{P}^4$ as the linear system:

$$|H| = |\pi^*14L - 6E_1 - 4E_2 - \dots - 4E_{10} - 2E_{11} - E_{12} - E_{13}|$$

where L is a general line in $\mathbb{P}^2$.

Before proving theorem(1), we will establish a number of results concerning certain linear systems on X. The following notation is that of [R]. We let

$$|C| = |\pi^*7L - 3E_1 - 2E_2 - \dots - 2E_{10}|$$

$$|D| = |\pi^*10L - 4E_1 - 3E_2 - \dots - 3E_{10}|$$

$$|C_1| = |C - E_{11}|, \qquad |C_0| = |C_1 - E_{12} - E_{13}|,$$

$$|D_1| = |D - E_{11}|, \qquad |D_0| = |D_1 - E_{12} - E_{13}|.$$

We will use K_1 and K_2 to denote the canonical divisors on X_1 and X_2 as well as their inverse images on X, while K will be the canonical divisorin X. The information we need to prove the theorem is given by proposition (5.2) below. First we need two lemmas.

(5.1) **Lemma:** *In the above notation, for every speciality one rational surface of degree ten in $\mathbb{P}^4$, we have*

(i) $h^0(\mathcal{O}(C-2E_i)) = 0$;i = 11,12,13

(ii) $h^0(\mathcal{O}(C)) = 3$ $h^0(\mathcal{O}(C_1)) = 2$,$h^0(\mathcal{O}(C_0)) = 1$

(iii) C_0 is a plane quartic in $\mathbb{P}^4$

(iv) the generic curve in $|C|$ is smooth.

Proof: We have $H.(C-2E_i)=4$ with $p_a(C-2E_i)=2$, but no projective curve has these invariants, so this gives (i). Now in any case $h^0(\mathcal{O}(C))\geq 3$, so this gives $h^0(\mathcal{O}(C))=3$, $h^0(\mathcal{O}(C_1))=2$. Since $H.C_1=6$ with $p_a(C_1)=3$, it follows that no curve in $|C_1|$ is contained in a plane and, since $C_0=H-C_1$ we conclude that $h^0(\mathcal{O}(C_0))\leq 1$. Now considering an effective divisor in $|C_1-E_{12}|$ we see that it has invariants $H.(C_1-E_{12})=5$, $p_a(C_1-E_{12})=3$. Such a divisor is reducible of the form $\Delta+\Gamma$, where Γ is a plane quartic and Δ is a line meeting Γ in one point. We will show that $\Delta=E_{13}$ so that Γ is the unique curve in $|C_0|$. This will finish the proof of *(ii)* and *(iii)*.

In fact, by the Hodge index theorem, we have $\Gamma^2\leq 1$ and by the two connexity of $|H|$, we have $\Delta^2\leq -1$. However

$$\Gamma^2+\Delta^2=(C_1-E_{12})^2-2=0$$

showing that $\Gamma^2=1$, and $\Delta^2=-1$. Since the only lines on X with self intersection -1 are E_{12} and E_{13}, we conclude from *(i)* that $\Delta=E_{13}$.

To prove *(iv)* we will show that the generic curve in $|C_1|$ is smooth. Since by *(i)*, $|C|$ has no base points in E_{11}, this will give the desired result. We begin by writing $|C_1|=T+|C_1-T|$ where T is the fixed part of $|C_1|$. Since by *(i)* and *(ii)*, neither E_{12} nor E_{13} is a fixed component of $|C_1|$, it follows that T is a component of C_0. We write $C_0=T+S$.

Now $|C_1-C_0|=|E_{12}+E_{13}|$ so that T is not the whole of C_0 and, since no plane curve on X can have a multiple component, it follows that S and T have no component in common. Noting that each of E_{12} and E_{13} meet C_0 in one point, we have $0\leq E_i.S\leq 1$ (i=12,13) and since $|C_1-T|=|S-E_{12}-E_{13}|$ has no fixed components, we have

$$(S-E_{12}-E_{13}).E_i\geq 0;\ i=12,13$$

$$(S-E_{12}-E_{13}).S\geq 0$$

From the first inequality we get $S.(E_{12}+E_{13})=2$ and from the second we get $S^2\geq -2$. However $H.C_0=4$, so that

$$2(S^2-S.T-1)=H.S\leq 3$$

with $S.T=3$, contradicting the fact that $S^2=-2$. This shows that $|C_1|$ has no fixed components and, since $C_1{}^2=3$ we conclude by Bertini's theorem in characteristic zero that the generic curve in $|C_1|$ is smooth. /

(5.2) **Lemma:** *For any speciality one rational surface of degree ten in* $\mathbb{P}^4$ *we have*

(a) $h^0(\mathcal{O}(-K_2))=0$

(b) $h^0(\mathcal{O}(D_0))=1$

(c) $h^0(\mathcal{O}(D)) = 2$

(d) C_0 and D_0 have no common component and each meets $E_i (i=11,12,13)$ transversally.

(e) $\mathcal{O}_D(H) = \omega_D$ the dualising sheaf on D_0

(f) $\mathcal{O}_{Do}(C_0) = \mathcal{O}_{Do}(E_{11})$ and $h^0(\mathcal{O}_{Do}(C_0)) = 1$

(g) $D_0 \cap C_0 = D_0 \cap E_{11} = C_0 \cap E_{11} = y_{11}$ is a closed point

(h) $h^0(\mathcal{O}(H-D_0)) = 2$.

Proof: *(a)* Since $H.K_2=0$.

(b) We have $h^0(\mathcal{O}(H-D_0)) \geq 2$ with $p_a(H-D_0)=2$ so that no effective divisor in $|H-D_0|$ is a plane curve. This shows that $h^0(\mathcal{O}(D_0)) \leq 1$. Since C_0 is a plane quartic, we have $\mathcal{O}_C(H) = \mathcal{O}_{Co}(C_0+K)$ giving $\mathcal{O}_{Co}(D_0) = \mathcal{O}_{Co}(E_{11})$. Since E_{11} meets C_0 in a closed point we conclude from the exact sequence

$$0 \rightarrow \mathcal{O}(-K_2) \rightarrow \mathcal{O}(D_0) \rightarrow \mathcal{O}_{Co}(D_0) \rightarrow 0$$

and *(a)*, that $1 \geq h^0(\mathcal{O}(D_0)) \geq h^0(\mathcal{O}_{Co}(D_0)) \geq 1$.

(c) Let C be a smooth curve in $|C|$ (lemma(5.2)). From the exact sequence

$$0 \rightarrow \mathcal{O}(-K_2) \rightarrow \mathcal{O}(D) \rightarrow \mathcal{O}_C(D) \rightarrow 0$$

we find $h^0(\mathcal{O}(D)) = h^0(\mathcal{O}_C(D)) \geq 2$, since $D.C=4$ and $p_a(C)=3$. Now by Clifford's theorem $h^0(\mathcal{O}_C(D)) \geq 3$ if and only if $\mathcal{O}_C(D) = \mathcal{O}_C(C+K)$, but this would imply $\mathcal{O}_C(-2K_2) = \mathcal{O}_C$. Considering the exact sequence

$$0 \rightarrow \mathcal{O}(-\pi^*L+E_1) \rightarrow \mathcal{O}(-K_2) \rightarrow \mathcal{O}_C(-K_2) \rightarrow 0$$

and noting that all the cohomology of the first two terms is zero, we find $h^0(\mathcal{O}_C(-K_2))=0$.

(d) Both C_0 and D_0 are plane curves in $\mathbb{P}^4$ and their union is not contained in a hyperplane. As such the plane of C_0 meets that of D_0 in one point. We have already seen that $h^0(\mathcal{O}(C-2E_i))=0$ for $i=11,12,13$ (lemma(5.2)) so that C_0 meets E_i transversally. The same argument applies to $h^0(\mathcal{O}(D-2E_i))$ to conclude that D_0 meets E_i transversally for $i=11,12,13$.

(e) Since $H.D_0=4$, $p_a(D_0)=3$ we have $\mathcal{O}_{Do}(H)=\omega_{Do}$.

(f) From *(e)* we find $\mathcal{O}_{Do}(C_0) = \mathcal{O}_{Do}(E_{11})$ and writing $C_0=D_0+K-E_{12}-E_{13}$ we find $\mathcal{O}_{Do}(C_0) = \omega_{Do}(-E_{12}-E_{13})$. Since ω_{Do} is very ample and D_0 meets each of E_{12} and E_{13} transversally, we have $h^0(\mathcal{O}_{Do}(C_0)) = h^0(\omega_{Do})-2=1$.

(g) follows directly from *(e)* and *(f)*.

(h) since D_0 is a plane quartic. /

We now formulate the information we need in the:

(5.3) **Proposition:** *For every speciality one rational surface of degree ten in $\mathbb{P}^4$, we have the following properties:*

(1) $h^0(\mathcal{O}(D)) = 2$, $h^0(\mathcal{O}(D+K_2)) = 3$, $h^0(\mathcal{O}(K_2)) = 0$

(2) *D_0 is isomorphic to its image on X_2 and the dualizing sheaf on*

(3) *The image of D_0 on X_2 is smooth at x_{11}.*

(4) *C_0 is isomorphic to its image on X_2 and the images of C_0 and D_0 on X_1 contain x_{12} and x_{13} in their intersection. What is more, if $f:X_0 \to X_1$ is the blowing up of X_1 in the image y'_{11}' of $y_{11} = C_0 \cap D_0$, then the strict transformes C'_0, D'_0 of C_0 and D_0 on X_0 intersect in two closed points dominating x_{12} and x_{13}.*

(5) The image of C_0 on X_2 is smooth at x_{11}.

(6) $h^0(\mathcal{O}(H-D_0))=2$.

Proof: Noting that $D+K_2=C$, *(1)* follows from (5.2) *(ii)*, (5.3) *(c)* and the fact that $H.K_2=0$. Parts *(2)*,*(3)* and *(5)* follow from (5.3) *(d)* and *(e)* as does the first part of *(4)*. The part *(6)* is just (5.3) *(h)*.

The remaining part of (4) results from the fact that x_{12} and x_{13} are distinct points on the smooth part of C_0 and D_0. /

Construction of an irreducible parameter scheme:

Let $\mathfrak{F}$ be the functor which to a **k**-scheme T associates the set of triples (X_2, D, x_{11}) where

(1) $p:X_2 \to \mathbb{P}^2{}_T$ is an ordered blowing up of $\mathbb{P}_T{}^2$ in ten sections $(x_1,\dots,x_{10})$ over T such that, if we let E_i be the inverse image of x_i on X_2 and

$$\mathcal{O}(D) = \mathcal{O}(p^*10\ L-4E_1-3E_2-\dots-3E_{10})$$

then $h^0(\mathcal{O}(D_t))=2$, $h^0(\mathcal{O}(D_t+K_{2,t}))=3$, for all points $t \in T$, where $\mathcal{O}(K_2)$ is the relative canonical sheaf of X_2 over T.

(2) D is a relative Cartier divisor on X_2 associated to the invertible sheaf $\mathcal{O}(D)$ such that the dualizing sheaf on D is very ample.

(3) $x_{11}:T \to D$ is a section of D over T such that D is smooth at x_{11}.

Then it is easily verified that $\mathfrak{F}$ is representable by a smooth irreducible quasi-projective variety V of dimension 22.

Over V we have the universal triple

$$V \xrightarrow{x_{11}} D \to X_2 \xrightarrow{p} \mathbb{P}^2{}_V$$

Letting $f_2: X_1 \to X_2$ be the blowing up of x_{11}, the strict transform on X_1

meets the inverse image E_{11} of x_{11} in a section $y'_{11}=D' \cap E_{11}$ of D over V, because D is smooth at x_{11}. Consider the blowing up f_0: $X_0 \to X_1$ of X_1 in y'_{11} and let D'_0 be the strict transform of D on X_0. We put

$$\mathcal{O}(C'_0) = \mathcal{O}(D'_0+K_0-E_{11}-F_{11})$$

where F_{11} is the inverse image of y'_{11} and $\mathcal{O}(K_0)$ is the relative dualizing sheaf of X_0 over V.

(5.4) **Lemma:** *$h^0(\mathcal{O}(C'_0))=1$ at all points of V.*

Proof: We have the exact sequence on X_0

$$0 \to \mathcal{O}(f^*K_2) \to \mathcal{O}(C'_0) \to \mathcal{O}_{D_0'}(C'_0) \to 0$$

where $f=f_0 \circ f_2$. Since $\omega_{D'o}(-E_{11}-E_{12}) = \mathcal{O}_{D'o}(C'_O)$ on $D'_0 \approx D$ and ω_D is very ample on D we have $h^0(\mathcal{O}_{D'o}(C'_0)) = h^0(\omega_D)-2 =1$ at all points of V, where ω_D and $\omega_{D'o}$ are the relative dualizing sheaves of D and D'_0 over V. Since $h^i(\mathcal{O}(K_2)) = 0$ (i=0,1) at all points of V, this gives the result. /

Now since $h^0(\mathcal{O}(C'_0))=0$ at all points of V, we have a well defined Cartier divisor C'_0 on X_0. Let V' be the open subset of V where the following two conditions are verified:

(4) $B = C'_0 \cap D'_0$ is finite and smooth, and the induced morphism $f_{0,B}$: $B \to X_1$ is a closed immersion.

(5) $h^0(\mathcal{O}(C'_0-E_{11}))=0$.

Now over V', let $f_1:X \to X_1$ be the blowing up of $B_1=f_0(B)$ and let E be the inverse image of B_1 on X. Letting $C"_0=f_0(C'_0)$ and putting $D"_0=D=f_0(D'_0)$ we obtain the strict transformes

$$C_0 \in |f^*C"_0-E|, \quad D_0 \in |f_1{}^*D"_0-E|$$

with $C"_0 \approx C_0$ and $D"_0 \approx D_0$. Equally we have by construction that

$$C_0 \cap D_0 = C_0 \cap E_{11} = D_0 \cap E_{11} = y_{11}$$

is a section of X over V' dominating y'_{11}. As such at all points of V' we have $\mathcal{O}_{Do}(C_0) \approx \mathcal{O}_{Do}(E_{11})$ and consequently $\mathcal{O}_{Do}(H) \approx \mathcal{O}_{Do}(D_0+K)$, where $\mathcal{O}(H) = \mathcal{O}(2C_0+E)$ and $\mathcal{O}(K)$ is the relative dualizing sheaf of X over V'.

Now considering the following exact sequence on X

$$0 \to \mathcal{O}(H-D_0) \to \mathcal{O}(H) \to \mathcal{O}_{Do}(H) \to 0$$

with $\omega_{Do}= \mathcal{O}_{Do}(H)$ and $h^0(\omega_{Do})=3$ at all points of V', we conclude that

(6) $h^0(\mathcal{O}(H-D_0))=2$, $h^1(\mathcal{O}(H-D_0))=0$

on an open subset V" of V'. As such $h^0(\mathcal{O}(H_t))=5$ at all points $t \in V"$ and $\mathcal{O}(H_t)$ is very ample on an open subset $V_0 \subset V"$ (see [A]).

Let $\pi : X \to \mathbb{P}_{Vo}{}^2$ be the canonical morphism and let W=

$\underline{\text{Isom}}(\pi_* \mathcal{O}(H), \mathcal{O}_{V_0}{}^5)$. Since by proposition (5.4) any rational surface of speciality one and degree ten in $\mathbb{P}^4$ is isomorphic to one of the surfaces in the family V_0, it follows that the canonical map from the Hilbert scheme $\mathcal{H}$ of speciality one, degree ten rational surfaces in $\mathbb{P}^4$ is dominant. Since V_0 is irreducible so are W and $\mathcal{H}$. Now a standard calculation gives the dimension of $\mathcal{H}$ as $\dim V_0 + 16 = 38$. /

BIBLIOGRAPHY:

[A] Alexander,J — *Surfaces rationnelles non spéciales dans $\mathbb{P}^4$*, Math . Z **200**, 87-110 (1988).

[Au] Aure, A — *On surfaces in $\mathbb{P}^4$*, Thesis, University of Oslo, Oslo (1987).

[A-R] Aure,A Ranestad,K — *On smooth surfaces of degree nine in $\mathbb{P}^4$*, Preprint, University of Oslo,Oslo (1990).

[Ar] Arbarello,E et al. — *Geometry of algebraic curves, Vol.1*, Grundlehren der Mathematischen Wissenschaften **267**, Springer Verlag, New York.

[D] Demazure,M — *Surfaces de Del Pezzo*, Seminaire sur les singularités, **LNM 777**, 67-68, 1976-77.

[E-P] Ellinsgrud,G Peskine,Ch — *Sur les surfaces lisse de $\mathbb{P}^4$*, Inv. Math., vol.**95**, Fas. 1, 1-12, (1989).

[H] Hartshorne,R — *Algebraic Geometry*, Graduate Texts in Math. **52**, New York, Springer (1977).

[I_1] Ionescu,P — *Embedded projective varieties of small invariants*, Proceedings of the week of Algebraic Geometry, Bucharest 1982. Springer Lecture Notes in Math., Vol **1056**,(1984).

[I_2] Ionescu,P — *Embedded projective varieties of small invariants <u>II</u>*, Revue Roumaine de Mathematiques Pure et Appliquées, Tome **XXXI**, №**6** (1986).

[I_3] Ionescu,P — *Embedded projective varieties of small invariants <u>III</u>* Preprint Series in Math. ,№**59**, Instit.Nat.Pentra Creatie Stiintifica si Technica, Inst. de Math., (1988).

[I_4] Ionescu,P — *Generalised adjunction and applications*,Math. Proc. Camb. Phil. Soc. **99**, p.457, (1986).

[K] Katz,S — *Hodge numbers of linked surfaces in $\mathbb{P}^4$*. Preprint Dept. of math., Univ. of Oklahoma, Norman, OK 73019.

[O_1] Okonek,C — *Moduli reflexive Garben und Flachen von kleinem Grad in $\mathbb{P}^4$*. Math. Z. **184**, 549-572 (1983).

[O_2] Okonek,C *Uber 2-codimensionale Utermannigfaltigkeiten vom Grad 7 im $\mathbb{P}^4$ und $\mathbb{P}^5$.* Math. Z. **187**, 209-219, (1984).

[O_3] Okonek,C *Flachen vom Grad 8 im $\mathbb{P}^4$.* Math. Gotting. Sond. Geom Anal., Heft **8**, 1-42 (1985).

[R] Ranestad,K *On smooth surfaces of degree ten in $\mathbb{P}^4$.* Thesis, Univ. of Oslo (1989).

[Ro] Roth,L *On the projective classification of surfaces.* Proc. of London Math. Soc. **42**, 142-170 (1937).

[S] Serrano,F *The adjunction mapping and hyperellyptic divisors on a surface.* J. fur die reine ange. Math. **381**, 90-109 (1987).

[So] Sommese, A *Hyperplane sections of projective surfaces: I, the adjunction mapping.* Duke Math. J. **46**,377-401 (1979).

[V] Van de Ven, A *On the two connectedness of very ample divisors on a surface.* Duke Math. J. **46**, 403-407 (1979).

BOUNDING SECTIONS OF BUNDLES ON CURVES

By *Enrique Arrondo** and *Ignacio Sols**

Departamento de Algebra
Facultad de Ciencias Matemáticas
Universidad Complutense
28040 Madrid, Spain
E-mail:w258 @ EMDUCM11 (Bitnet)

1. Introduction.

Let C be a smooth irreducible curve of genus g and E a rank two vector bundle of degree d on it. Using the notations of [H] Ch.V §2, we write $-e$ for the minimum degree of a twist $E \otimes L^{-1}$ having sections (for any line bundle L on C). This is in fact an invariant of the ruled surface $\mathbf{P}(E)$. Therefore, there can be sections for both E and $\check{E} \otimes \omega_C$ only if $-e \leq d \leq 4g-4+e$. Assume $\mathbf{P}(E)$ is not the trivial ruled surface. If $d=-e$ then $h^0(E)=1$, and if $d=4g-4+e$ then $h^0(\check{E}\otimes\omega_C)=1$, hence $h^0(E)=d-2g+3$. We propose the following:

Conjecture. *In the above notations, if* $-e \leq d \leq 4g-4+e$ *and* $\mathbf{P}(E)$ *is not* $C\times\mathbf{P}^1$, *then* $h^0(E) \leq \frac{d+e}{2}+1$.

Observe that, in geometric terms, our conjecture is claiming that for a ruled surface of $\mathbf{P}^r$ of degree d the minimal degree of a directrix (which is $\frac{d-e}{2}$) is at most $d-r$. Francesco Severi proposes the problem of bounding this minimal degree in [S] page 13. After proving this is at most $d-\frac{r}{2}$, he says that: *"La questione dunque delle direttrice minima resta aperta"*.

In this paper we show this conjecture to be true for hyperelliptic curves or for non-semistable vector bundles. We also show that this bound is reached for any values of d, e.

For semistable vector bundles we prove the bound

$$h^0(E) \leq \frac{d}{2}+2$$

with equality if and only if $E \cong \mathcal{O}_C \oplus \mathcal{O}_C$, $E \cong \omega_C \oplus \omega_C$ or C is hyperelliptic and $E \cong L \oplus L$, L being a multiple of the g^1_2.

This bound is weaker than the one conjectured, although it is sharp as a bound depending only on d. It sharpens the former bound $h^0(E) \leq \frac{d}{2}+g$ obtained in [P] by a direct generalization to higher rank of the proof of Clifford lemma in [A-C-G-H].

* Partially supported by CAICYT grant No. PB86-0036

We also prove the bound

$$h^0(E) \leq \frac{d}{2} + R$$

for semistable bundles E of arbitrary rank R which are generically generated by global sections. We also yield from here an upper bound of the dimension of the component of E in the variety $\mathcal{W}^r_{R,d}(C)$ of semistable bundles on C of degree d and rank R having at least $r+1$ sections, thus generalizing Martens' theorem.

This bound is all should be needed for a theory of higher rank divisors (in the sense suggested by Ghione in [G])

$$0 \to \mathcal{O}_C^R \to E \to T \to 0$$

(rk$E = R$, T=torsion sheaf on C), since E being generically generated by global sections is just the condition of existence of such "divisors" associated to E. We, however conjecture that the above bound holds for any semistable bundle E with no further conditions.

We thank the referee for pointing us several improvements, specially for the proof of Prop. 4.

2. Proof of the bounds in rank two.

We need first some more notations and elementary facts. Keeping still the notations of [H], we write $\mathcal{E} = E \otimes L^{-1}$ for the minimum twist of E having global sections (notice that $\mathcal{E}$ is not necessarily unique). We also set $\mathfrak{e} = \wedge^2 \mathcal{E}$, so that we have an exact sequence:

$$0 \to \mathcal{O}_C \to \mathcal{E} \to \mathfrak{e} \to 0$$

which, tensored with L, yields another exact sequence

$$(*) \quad 0 \to L \to E \to L \otimes \mathfrak{e} \to 0$$

presenting E as an extension of two line bundles of degrees $\frac{d+e}{2}$ and $\frac{d-e}{2}$.

We observe immediately that $\mathcal{E}$ has two sections if and only if it is the trivial bundle $\mathcal{O}_C \oplus \mathcal{O}_C$ and it has only one section otherwise. We recall that E is stable (resp. semistable) if and only if $e < 0$ (resp. $e \leq 0$). We can prove now our conjecture for non-semistable vector bundles.

Proposition 1. *If E is not semistable and $-e \leq d \leq 4g - 4 + e$, then $h^0(E) \leq \frac{d+e}{2} + 1$.*

Proof. We see from (*) that $h^0(E) \leq h^0(L) + h^0(L \otimes \mathfrak{e})$ and we distinguish four cases:

-If $h^1(L) = h^1(L \otimes \mathfrak{e}) = 0$, then $h^1(E) = 0$ and from the Riemann-Roch theorem we get $h^0(E) = d - 2g + 2$, which is bounded by $\frac{d+e}{2}$ from our hypothesis $d \leq 4g - 4 + e$

-If $h^1(L) = 0$ and $h^1(L \otimes \mathfrak{e}) \neq 0$, we apply Riemann-Roch to L and Clifford lemma to $L \otimes \mathfrak{e}$ and get $h^0(E) \leq \frac{d+e}{2} - g + 1 + \frac{d-e}{4} + 1 \leq \frac{d+e}{2} + 1$, again from $d \leq 4g - 4 + e$.

-If $h^1(L) \neq 0$ and $h^1(L \otimes \mathfrak{e}) = 0$, using again Clifford and Riemann-Roch theorems we get $h^0(E) \leq \frac{d+e}{4} + 1 + \frac{d-e}{2} - g + 1 \leq \frac{d}{2} + 1 \leq \frac{d+e}{2} + 1$ since $e > 0$.

If both L and $L \otimes \mathfrak{e}$ are special, by applying Clifford lemma to them we obtain $h^0(E) \leq \frac{d+e}{4} + 1 + \frac{d-e}{4} + 1 = \frac{d}{2} + 2 \leq \frac{d+e}{2} + 1$ (except if $e = 1$, but in this case the first inequality is strict, since $\frac{d+e}{4}$ and $\frac{d-e}{4}$ are not both integer). ∎

We take care now of the semistable case in the following:

Proposition 2. *If C is not hyperelliptic, E is semistable and $-e \leq d \leq 4g-4+e$, then $h^0(E) \leq \frac{d}{2}+1$ unless E is either $\mathcal{O}_C \oplus \mathcal{O}_C$ or $\omega_C \oplus \omega_C$.*

Proof. We consider again the exact sequence (*) and the bound $h^0(E) \leq h^0(L) + h^0(L \otimes \mathfrak{e})$. Now we see that from our hypothesis $e \leq 0$ and $-e \leq d \leq 4g-4+e$ we have $0 \leq \frac{d+e}{2} \leq \frac{d-e}{2} \leq 2g-2$, so that we can apply Clifford theorem to L and $L \otimes \mathfrak{e}$. If neither L nor $L \otimes \mathfrak{e}$ is $\mathcal{O}_C$ or ω_C we obtain, since C is not hyperelliptic, that $h^0(E) \leq \frac{d+e}{4} + \frac{1}{2} + \frac{d-e}{4} + \frac{1}{2} = \frac{d}{2} + 1$. Hence, we are left with four cases:

-If $L = \mathcal{O}_C$ then $E = \mathcal{E}$ and the result is trivial.

-If $L \otimes \mathfrak{e} = \mathcal{O}_C$ then L has degree zero since E is stable and $d \geq 0$. Therefore the result is clear in this case.

The remaining cases are just dual, replacing E with $\check{E} \otimes \omega_C$. ∎

We prove now the conjecture in the hyperelliptic case.

Proposition 3. *If C is hyperelliptic and $-e \leq d \leq 4g-4+e$, then $h^0(E) - h^0(\mathcal{E}) \leq \frac{d+e}{2}$ and for any values of d and e such that $d \equiv e - 2(\mathrm{mod} 4)$ there exists a vector bundle E achieving the bound.*

Proof. First, notice that such a bound is equivalent to the inequality $h^0(E) + h^0(\check{E} \otimes \omega_C) \leq h^0(\mathcal{E}) + h^0(\check{\mathcal{E}} \otimes \omega_C)$ (just applying Riemann-Roch). Also, the bound is trivially true for $d = -e-2$. Hence, it will suffice to prove the bound for $-e-2 \leq d \leq 2g-2$. Since we know it holds for $d = -e-2$, $-e$, we can assume $d \geq -e+2$ and use induction on d.

Consider in $\mathrm{Pic}^1 C$ the following subschemes:

$$X = \{M \in \mathrm{Pic}^1 C \mid h^0(L \otimes M) > 0\}$$
$$Y = \{M \in \mathrm{Pic}^1 C \mid h^0(\omega_C \otimes L^{-1} \otimes \mathfrak{e}^{-1} \otimes M) > 0\}$$

In the Chow ring of $\mathrm{Pic}^1 C$ they are positive multiples of powers of the theta divisor, which is ample, so that they will have non-empty intersection if $\dim X + \dim Y \geq g$. This is true since

$$\dim X = \min\{\tfrac{d+e}{2} + 1, g\}$$
$$\dim Y = \min\{2g-2-\tfrac{d-e}{2} + 1, g\}$$

and $e \geq -g$ (cf. [N]).

Hence, there exist positive divisors A and B defined as zero sections of $L \otimes M$ and $\omega_C \otimes L^{-1} \otimes \mathfrak{e}^{-1} \otimes M$. Let A' be the greatest common divisor of A and B and $B' = A+B-A'$. Then, we have an exact sequence

$$0 \to \mathcal{O}_C(A') \to \mathcal{O}_C(A) \oplus \mathcal{O}_C(B) \to \mathcal{O}_C(B') \to 0$$

If we write $L' = \mathcal{O}_C(A') \otimes M^{-1}$ we obtain another exact sequence

$$0 \to L' \to L \oplus (\omega_C \otimes L^{-1} \otimes \mathfrak{e}^{-1}) \to \omega_C \otimes L'^{-1} \otimes \mathfrak{e}^{-1} \to 0$$

which, tensored with $\mathcal{E}$ and writing $E' = \mathcal{E} \otimes L'$, yields

$$0 \to E' \to E \oplus (\check{E} \otimes \omega_C) \to \check{E}' \otimes \omega_C \to 0$$

Hence, $h^0(E) + h^0(\check{E} \otimes \omega_C) \leq h^0(E') + h^0(\check{E}' \otimes \omega_C)$ and since $\deg E' \geq -e+2$, it will suffice to prove $\deg E' < d$ (from our induction hypothesis). Since $d \leq 2g-2$, this is equivalent to prove that the divisor A is not contained in B. We will distinguish two cases:

-If $H^0(\omega_C \otimes L^{-2} \otimes \mathfrak{e}^{-1}) = 0$, then $|B - A| = \emptyset$ and we are done.

-If there exist a positive divisor D defined by a section of $\omega_C \otimes L^{-2} \otimes \mathfrak{e}^{-1}$, then we use our hypothesis that C is hyperelliptic to conclude that we can choose our divisor A so that $|A|$ has dimension at least one. Therefore, we can find an equivalent divisor A_1 not contained in $A + D$. Using A_1 and $A + D$ in the above argument, we complete the proof of the bound.

We show now that the bound can be achieved for any value of d and $e \leq 0$ (we exclude the case $E = L \oplus L$ since it is clear). Indeed, after $-e + 2$ general natural transformations of the trivial ruled surface one obtains a ruled surface $\mathbf{P}(\mathcal{F})$ where $\mathcal{F}$ is a vector bundle on C having index of stability e and appearing as kernel in

$$0 \to \mathcal{F} \to \mathcal{O}_C^2 \to T \to 0$$

(T being a torsion sheaf of length $-e+2$). Tensoring with a line bundle M of degree $\frac{d-e+2}{2}$ having exactly $\frac{d-e+2}{4} + 1$ independent sections we obtain $E = \mathcal{F} \otimes M$ of degree d having

$$h^0(E) \geq 2h^0(M) - (-e + 2) = \frac{d + e}{2}$$

and hence exactly $\frac{d+e}{2}$ independent sections. ■

Having proved this result, we can complete now the statement of Prop. 2:

Proposition 2'. *If E is semistable and $-e \leq d \leq 4g - 4 + e$, then $h^0(E) - h^0(\mathcal{E}) \leq \frac{d}{2}$.*

3. Bounds for arbitrary rank.

Proposition 4. *Let E be a rank R vector bundle on C of degree d such that both E and $\check{E} \otimes \omega_C$ are generically generated by global sections. Then,*

$$h^0(E) \leq \frac{d}{2} + R.$$

Proof. Let E' be the image of the evaluation map

$$\begin{array}{ccccccc} H^0(E) \otimes \mathcal{O}_C & & \xrightarrow{\epsilon} & & E & \to & \operatorname{coker}\epsilon & \to & 0 \\ & \searrow & & \nearrow & & & & & \\ & & E' & & & & & & \\ & \nearrow & & \searrow & & & & & \\ 0 & & & & 0 & & & & \end{array}$$

so that E' has the same rank R and $H^0(E) = H^0(E')$. Denote by $d' = \deg E'$.

Let $U \subseteq \operatorname{Hom}(\mathcal{O}_C^R, E') \times \operatorname{Hom}(E, \omega_C^R) = \operatorname{Hom}(\mathcal{O}_C^R, E) \times \operatorname{Hom}(E, \omega_C^R)$ be the dense open subset consisting of those pairs of monomorphisms $\alpha : \mathcal{O}_C^R \hookrightarrow \alpha' E' \hookrightarrow E, \beta : E \hookrightarrow \omega_C^R$ such that coker(α') is the structure sheave of a reduced divisor D' whose support is disjoint with the support of cokerϵ and cokerβ (non-emptiness of U follows from E' being generated by global sections: after a monomorphism β is chosen, the generic monomorphism α' will do).

Therefore, their composition $\gamma = \beta \circ \alpha : \mathcal{O}_C^R \to \omega_C^R$ has cokernel coker$\gamma = \mathcal{O}_{D'} \oplus T_{\gamma, D'}$, where $T_{\gamma, D'}$ is a torsion sheaf containing cokerϵ, the supports of these direct factors being mutually disjoint.

Let $V \subseteq \mathrm{Hom}(\mathcal{O}_C^R, \omega_C^R)$ be the image of U by the obvious composition map

$$\mathrm{Hom}(\mathcal{O}_C^R, E) \times \mathrm{Hom}(E, \omega_C^R) \to \mathrm{Hom}(\mathcal{O}_C^R, \omega_C^R)$$

It is clear that U is invariant under the action of $G = \mathrm{Aut}(E)$ $\varphi(\alpha, \beta) = (\varphi \circ \alpha, \varphi \circ \beta)$ (for $\varphi \in G$) and that the above map and thus its restriction $f : U \to V$ * is constant over the orbits of this action, so that the fiber at each

$$0 \to \mathcal{O}_C^R \xrightarrow{\gamma} \omega_C^R \to T_\gamma \to 0$$

in V is union of G-orbits, in fact as we shall see a finite number of orbits:

Let $p_1, \ldots, p_n$ (with $n \geq d'$) be the points of $\mathrm{Supp}T_\gamma$ such that $\dim(T_\gamma \otimes \mathbf{C}(p_i)) = 1$ and $p_i \notin \mathrm{SuppCoker}\epsilon$. Write

$$T_\gamma = \mathbf{C}(p_1) \oplus \ldots \oplus \mathbf{C}(p_n) \oplus \tilde{T}_\gamma$$

the direct factors being thus of mutually disjoint supports.

For each one of the $\binom{n}{d'}$ subset D' of $\{p_1, \ldots, p_n\}$ of cardinality d' there is exactly one G-orbit of pairs (α, β) applying by f to γ, namely the one given by the display

$$\begin{array}{ccccccccc}
 & & & & 0 & & 0 & & \\
 & & & & \downarrow & & \downarrow & & \\
0 & \to & \mathcal{O}_C^R & \to & E & \to & T_\alpha = T_{\alpha'} \oplus \mathrm{coker}\epsilon = \mathcal{O}_{D'} \oplus T & \to & 0 \\
 & & \| & & \downarrow & & \downarrow & & \\
0 & \to & \mathcal{O}_C^R & \to & \omega_C^R & \to & T_\gamma & \to & 0 \\
 & & & & \downarrow & & \downarrow & & \\
 & & & & T_\beta & = & T'_{\gamma,D'} & & \\
 & & & & \downarrow & & \downarrow & & \\
 & & & & 0 & & 0 & &
\end{array}$$

where $T'_{\gamma,D'} = \oplus_{i \notin D'} \mathbf{C}(p_i) \oplus \tilde{T}_\gamma$. It is clear from the construction that all elements of the fiber of f at γ appear in this way, so this fiber is the finite union of $\binom{n}{d'}$ orbits of G. The dimension of such orbits is at most $\dim G = R^2$. Therefore,

$$\dim U - R^2 = Rh^0(E) + Rh^0(\check{E} \otimes \omega_C) - R^2 \leq \dim\mathrm{Hom}(\mathcal{O}_C^R, \omega_C^R) = R^2 g$$

i.e., $h^0(E) + h^0(\check{E} \otimes \omega_C) \leq Rg + R$. Combining this bound with the Riemann-Roch equality

$$h^0(E) - h^0(\check{E} \otimes \omega_C) = d + R(1 - g)$$

we obtain that

$$h^0(E) \leq \frac{d}{2} + R$$

as wanted. ∎

Corollary 5. *Let E be a semistable vector bundle of degree d and rank R on a smooth connected curve C. Assume $\check{E} \otimes \omega_C$ is generically generated by global sections. Then*

$$h^0(E) \leq \frac{d}{2} + R$$

Proof. We assume $h^0(E) \neq 0$ since otherwise the result is trivial. Let E' be the image of the evaluation map $H^0(E) \otimes \mathcal{O}_C \to E$ and denote by R' and d' the rank and degree of E'. Then E' is generated by global sections. Furthermore, $\check{E}' \otimes \omega_C$ is generically generated by global sections since the composite map

$$H^0(\check{E} \otimes \omega_C) \otimes \mathcal{O}_C \to \check{E} \otimes \omega_C \to \check{E}' \otimes \omega_C$$

is generically surjective as the first is so and the second is surjective.

We conclude, from the former proposition applied to E' and from the stability of E, that

$$h^0(E) = h^0(E') \leq \frac{d'}{2} + R' \leq \frac{d}{2} + R \quad \blacksquare$$

Corollary 6. *Let E be a semistable vector bundle which is generically generated by global sections and assume that $h^0(\check{E} \otimes \omega_C) \neq 0$. Then*

$$h^0(E) \leq \frac{d}{2} + R$$

Proof. Since $\check{E} \otimes \omega_C$ satisfies the hypothesis of corollary 5,

$$h^0(E) = h^0(\check{E} \otimes \omega_C) + d + R - Rg \leq \frac{-d + R(2g-2)}{2} + R + d + R - Rg = \frac{d}{2} + R \quad \blacksquare$$

We have thus shown that $h^0(E) \leq \frac{d}{2} + R$ whenever two of the following three conditions are satisfied:

1) E is semistable.
2) E is generically generated by global sections.
3) $\check{E} \otimes \omega_C$ is generically generated by global sections.

We thus

Conjecture. *Let E be a semistable vector bundle of rank R and degree d on a smooth connected curve C. Then*

$$h^0(E) \leq \frac{d}{2} + R$$

In working with ruled varieties $\mathbf{P}(E)$ on a curve C (for E a vector bundle of rank R) the rank R morphisms

$$0 \to \mathcal{O}_C^R \to E \to T \to 0$$

(T being a torsion sheaf) seem to be the right generalization of divisor

$$0 \to \mathcal{O}_C \to L \to T \to 0$$

(L being a line bundle) on a curve. Call them (as Ghione in [G]) effective divisors of rank R, and call them semistable if the bundle E is. A complete series of rank R divisors, i.e.

all those associated to a same vector bundle E, is thus an open set $|E|$ in $Gr(R, H^0(E))$ (and is nonempty if and only if E is generically by global sections). The above corollary 6 can be then rephrased in the following way:

Theorem 7. *Let $|E|$ be a complete series of semistable divisors of rank R and degree d on a smooth connected curve of genus g.*

If $d \geq R(2g-2)$, then $\dim|E| = d - Rg + R - 1$

If $d \leq R(2g-2)$, then $\dim|E| \leq \frac{d}{2} + R - 1$

Taking L a line bundle on a hyperelliptic curve achieving Clifford bound we see that the above bound is achieved by $E = L \oplus \ldots \oplus L$.

4. On Martens' theorem in higher rank.

Just as in the case of line bundles (cf. [A-C-G-H] Ch. IV §5), Clifford bound can be re-interpreted as a bound for the dimension of certain subschemes of the moduli space of (semistable) vector spaces. First, we need a generalization to higher rank of the base-point-free pencil trick (cf. [A-C-G-H] page 126).

Lemma 8. *Let E be a rank R vector bundle on C that is spanned by a space V of $R+1$ global sections. Then, for any bundle F on C, the kernel of the natural map $V \otimes H^0(F) \to H^0(E \otimes F)$ is isomorphic to $H^0(F \otimes L^{-1})$, where $L = c_1(E)$.*

Proof. From our hypothesis, there is a natural epimorphism $\mathcal{O}_C^{R+1} \to E$ whose kernel, which is a line bundle, must be L^{-1}, by just computing first Chern classes. Tensoring with F the exact sequence $0 \to L^{-1} \to \mathcal{O}_C^{R+1} \to E \to 0$ and taking cohomology, we get our result. ∎

Let $\mathcal{W}_{d,R}$ be the moduli space of semistable rank R vector bundles on C of degree d. Denote with $\mathcal{W}_{d,R}^r(C)$ the subscheme parameterizing those bundles E having at least $r+1$ global sections.

Proposition 9. *Let Y be a component of $\mathcal{W}_{d,R}^r(C)$ such that the bundle E corresponding to its generic point is spanned by its sections, has not automorphisms different from the identity and $E \otimes \wedge^R E$ is strongly special. Then, $\dim Y \leq (R+1)(\frac{d}{2} - r)$.*

Proof. The annihilator of the cotangent space to Y at E (which we take general) turns out to be the image of the natural map

$$\mu : H^0(E) \otimes H^0(\check{E} \otimes \omega_C) \to H^0(E \otimes \check{E} \otimes \omega_C)$$

Hence,

$$\dim Y \leq h^0(E \otimes \check{E} \otimes \omega_C) - \dim \mathrm{Im}\mu = h^0(E \otimes \check{E} \otimes \omega_C) - h^0(E)h^1(E) + \dim \mathrm{Ker}\mu$$

Consider a chain $V_{R+1} \subseteq V_{R+2} \subseteq \ldots \subseteq V_{r+1} = H^0(E)$ of sub-vector spaces V_i of dimension i (we can assume $h^0(E) = r+1$) and write μ_i for the restriction of μ to $V_i \otimes H^0(\check{E} \otimes \omega_C)$. It is easy to check that $\dim \mathrm{Ker}\mu_i - \dim \mathrm{Ker}\mu_{i-1} \leq h^1(E)$ for all i. Therefore

$$\dim \mathrm{Ker}\mu \leq (r-R)h^1(E) + \dim \mathrm{Ker}\mu_{R+1}$$

From lemma 8, the kernel of μ_{R+1} is $H^0(\check{E} \otimes \omega_C \otimes \wedge^R \check{E})$. We can apply Corollary 6 to $E \otimes \wedge^R E$ so that $h^0(E \otimes c_1 E) \leq \frac{(R+1)d}{2} + R$ and get

$$\dim\mathrm{Ker}\mu_{R+1} \leq \frac{(R+1)d}{2} + R - (R+1)d + R(g-1) = Rg - (R+1)\frac{d}{2}$$

Summing up, and using the equalities $h^0(E \otimes \check{E} \otimes \omega_C) = R^2(g-1)+1$ (from Riemann-Roch and the fact that the only automorphisms of E are the constants), $h^0(E) = r+1$ and $h^1(E) = r+1+R(g-1)-d$, we get

$$\dim Y \leq R^2(g-1)+1-(r+1)(r+1+R(g-1)-d)+(r-R)(r+1+R(g-1)-d)+Rg-(R+1)\frac{d}{2}$$

which is the wanted inequality. ∎

ADDED IN PROOF: Recently, Tan has proved that any plane quartic gives a counterexample to our first conjecture about rank two vector bundles. In fact, his method generalizes to show that any smooth plane curve is a counterexample.

References.

[A-C-G-H] *Arbarello, E. - Cornalba, M. - Griffiths, P. A. - Harris, J.*, The geometry of algebraic curves, Springer GmW **267** (1985).

[G] *Ghione, F.*, Le morphism de Abel-Jacobi en rang supérieur, lecture in the conference "Vector bundles and special projective invariants" held in Bergen July 1989.

[H] *Hartshorne, R.* Algebraic Geometry, Springer GTM **52** (1977).

[N] *Nagata, M.*, On self-intersection number of a section on a ruled surface, Nagoya Math. J. **37** (1970), 191-196.

[P] *Pedreira, M.*, Preprint University of Santiago (1988).

[S] *Severi, F.*, Sulla clasificazione delle rigate algebriche, Rend. Mat. **2** (1941), 1-32.

The Smooth Surfaces of Degree 9 in $\mathbf{P}^4$

by Alf Bjørn Aure *) and Kristian Ranestad **)

§0 Introduction

The classification of smooth surfaces of low degree d in $\mathbf{P}^4$ goes back to the Italian geometers at the turn of the century. They treated the cases $d \leq 6$. More recently, Ionescu and Okonek have treated the cases $d = 7$ and $d = 8$ ([Io],[O1,O2]). Their work were complemented by a result of Alexander to give a classification ([A1]). In this paper we find the possible numerical invariants of surfaces of degree 9, and describe for each set of invariants the family of surfaces with the given invariants. Some of the results are mentioned in [Ra], which deals with the case $d = 10$.

We work over an algebraically closed field of characteristic 0.

The first result is the following

(0.1) Theorem. *Let S be a smooth nondegenerate surface of degree 9 in $\mathbf{P}^4$ with sectional genus π, Euler-Poincaré characteristic χ and canonical class K. Then S is a regular surface with $K^2 = 6\chi - 5\pi + 23$, where*

$\pi = 6$ and $\chi = 1$ and S is rational or S is the projection of an Enriques surface of degree 10 in $\mathbf{P}^5$ with center of projection on the surface, or

$\pi = 7$ and $\chi = 1$ and S is a rational surface, or $\chi = 2$ and S is a minimal properly elliptic surface, or

$\pi = 8$ and $\chi = 2$ and S is a $K3$-surface with five (-1)-lines, or $\chi = 3$ and S is a minimal surface of general type, or

$\pi = 9$ and $\chi = 4$ and S is linked $(3,4)$ to a cubic scroll (possibly singular/reducible), or

$\pi = 10$ and $\chi = 5$ and S is a complete intersection $(3,3)$, or

$\pi = 12$ and $\chi = 9$ and S is linked $(2,5)$ to a plane.

Furthermore if $\pi \geq 7$ then S is contained in at least two quartic hypersurfaces.

The theorem gives the possible candidates for a smooth surface of degree 9 in $\mathbf{P}^4$ and a description of a general surface with $\pi \geq 9$. Its proof is found in §2. It leaves us with the question of existence and of a description of a general surface when $\pi \leq 8$. For the rational surfaces this question is rigorously answered by Alexander (cf. [A2 Theorem 2]). He shows

(0.2) Theorem (Alexander). *Smooth rational surfaces S of degree 9 in $\mathbf{P}^4$ exist and belong to two families. Any surface of the first family ($\pi = 6$) is the blowing up of $\mathbf{P}^2$ in 10 points in sufficiently general position, embedded by the linear system $|13l - \sum_{i=1}^{10} 4E_i|$, where l is the total transform of a line in $\mathbf{P}^2$, and the E_i are the exceptional curves of the blowing up map. Any surface S of the second family ($\pi = 7$) is the blowing up of $\mathbf{P}^2$ in 15 points, embedded by the linear system $|9l - \sum_{i=1}^{6} 3E_i - \sum_{i=7}^{9} 2E_i - \sum_{i=10}^{15} E_i|$ which satisfies the condition that there is a pencil of curves $|6l - \sum_{i=1}^{6} 2E_i - \sum_{i=7}^{15} E_i|$ with base points on the exceptional curves E_7, E_8, E_9.*

In §3 we give general constructions of surfaces with each of the remaining sets of invariants. The existence of an Enriques surface with $\pi = 6$ is easily shown using the knowledge of the embeddings of minimal Enriques surfaces of degree 10 in $\mathbf{P}^5$ (cf. [Co Theorem 5.5] and [CV Corollary 2.13]). So we will concentrate on showing the existence of the surfaces with $\pi = 7$ and $\pi = 8$, including another construction of the rational surfaces with $\pi = 7$. This can be done in a uniform manner. By (0.1) these surfaces are contained in at least two quartic hypersurfaces. We construct reducible locally Cohen-Macaulay surfaces of degree 7 which are linked $(4,4)$ to smooth surfaces S of degree 9 with the desired invariants. Furthermore we show that all the surfaces can be obtained in this way. In the case of the $K3$-surface, the result is not new; a smooth surface of degree 7 and sectional genus 4 will do the job, so we describe the other cases. Let V_4 and V_4' be quartic hypersurfaces containing S. By Bezout's theorem, the union of the 5-secant lines, say S_5, is contained in $V_4 \cap V_4'$. The expected dimension of the family of 5-secant lines is one, and in our case one can show that the expected dimension equals the actual dimension. Thus S_5 consists of scrolls and planes, and the formulae of LeBarz (cf. [LB p. 798]) give an upper bound for its degree. Let $V_4 \cap V_4' = S \cup S_5 \cup S'$ for some surface S'. We describe S' and S_5 in each of the four cases, and use them to prove the existence of S. Our second result is summarized in

(0.3) Theorem. a) *Any smooth rational surface S of degree 9 with sectional genus 7 in $\mathbf{P}^4$ is linked $(4,4)$ to a reducible surface $T = P_0 \cup P_1 \cup P_2 \cup P_3 \cup T_1$, where P_1, P_2 and P_3 are three planes that meet pairwise in points, P_0 is the plane spanned by the points of their pairwise intersection and T_1 is a degenerate cubic surface, which meets P_1, P_2 and P_3 in lines and P_0 in three points. The union of the three planes P_1, P_2 and P_3 is the union of the 5-secants to S.*

b) *Any smooth elliptic surface S of degree 9 and sectional genus 7 in $\mathbf{P}^4$ is linked $(4,4)$ to a reducible surface $T = T_0 \cup P$ which lies on a cubic hypersurface V with a double plane along P. V is the projection of the Segre embedding of $\mathbf{P}^2 \times \mathbf{P}^1$ into $\mathbf{P}^5$, and T_0 is the image in V of a product $C \times \mathbf{P}^1$, where C is a plane cubic curve. T_0 is the union of the 5-secants to S.*

c) *Any smooth nonminimal $K3$-surface S of degree 9 with sectional genus 8 is linked $(4,4)$ to a smooth rational surface T of degree 7 and sectional genus 4. S has no 5-secants.*

d) *Any smooth surface of general type S of degree 9 with sectional genus 8 is linked $(4,4)$ to a reducible surface $T = T_0 \cup P_0$, where T_0 is linked $(2,4)$ to two planes which meet in a point p, and P_0 is a plane which meets T_0 in two lines through p. P_0 is the union of the 5-secants to S.*

§1 Preliminaries

In this paragraph we will give the basic definitions and results that are needed. For a smooth surface S in a projective space we denote by H and K the class of a hyperplane and the canonical divisor respectively. The structure sheaf is $\mathcal{O}_S$, and the Euler-Poincaré characteristic is $\chi = \chi(\mathcal{O}_S)$. The degree is d and the sectional genus π is the genus of a general hyperplane section. For short we write $\mathcal{O}_S(n) = \mathcal{O}_S(nH)$.

For smooth surfaces in $\mathbf{P}^4$ with normal bundle N_S there is the relation [H,p. 434],

$$d^2 - c_2(N_S) = d^2 - 10d - 5H \cdot K - 2K^2 + 12\chi(S) = 0,$$

which expresses the fact that S has no double points. This relation together with the adjunction formula [H,p. 361] yields for smooth surfaces of degree 9 in $\mathbf{P}^4$

$$\text{(1.1)} \qquad H \cdot K = 2\pi - 11 \quad \text{and} \quad K^2 = 6\chi - 5\pi + 23.$$

Since $H \cdot (K - \frac{1}{9}(H \cdot K)H) = 0$, the Hodge Index Theorem implies that

$$\text{(1.2)} \qquad K^2 \leq \frac{1}{9}(H \cdot K)^2$$

The first and only major theorem on smooth surfaces in $\mathbf{P}^4$, which we use over and over, is the

(1.3) Theorem (Severi). *All smooth surfaces in $\mathbf{P}^4$, except for the Veronese surfaces, are linearly normal.*

Proof. See [Se].□

The adjunction mapping is defined by the linear system $|H+K|$. By Riemann-Roch and the Kodaira Vanishing Theorem,

$$h^0(\mathcal{O}_S(H+K)) = \chi + \pi - 1. \tag{1.4}$$

(1.5) Theorem (Sommese, Van de Ven). *Let S be a smooth surface with a very ample divisor H and a canonical divisor K. Then $|H+K|$ is base-point free unless S is a scroll or a Veronese surface or a plane. Assuming $|H+K|$ has no base points, S is ruled in conics if and only if $(H+K)^2 = 0$, and if S is not rational and $(H+K)^2 > 0$, then the adjunction mapping is the composition of the contraction of (-1)-lines on S and an embedding, unless $S \cong \mathbf{P}(E)$ where E is an indecomposable rank 2 bundle on an elliptic curve, and $H \equiv 3B$ where B is an effective divisor with $B^2 = 1$ on S, in which case the degree of the adjunction mapping is 3.*

Proof. See [SV].□

In $\mathbf{P}^4$ there is the following additional result.

(1.6) Theorem. *Any smooth scroll in $\mathbf{P}^4$ has degree at most five.*

Proof. See [Au 1] and [La].□

(1.7) Linkage.(cf. [PS]) Two surfaces S and S' are said to be linked (m,n) if there exist hypersurfaces V and V' of degree n and m respectively such that $V \cap V' = S \cup S'$. There are the standard sequences of linkage, namely

$$0 \longrightarrow \mathcal{O}_S(K) \longrightarrow \mathcal{O}_{S\cup S'}(m+n-5) \longrightarrow \mathcal{O}_{S'}(m+n-5) \longrightarrow 0$$

$$0 \longrightarrow \mathcal{O}_S(K) \longrightarrow \mathcal{O}_S(m+n-5) \longrightarrow \mathcal{O}_{S\cap S'}(m+n-5) \longrightarrow 0$$

The first sequence yields the relation between the Euler-Poincaré characteristics

$$\chi(S') = \chi(V \cap V') - \chi(\mathcal{O}_S(m+n-5)). \tag{1.8}$$

The corresponding sequence for linkage of curves in $\mathbf{P}^3$ yields the following relation between the sectional genera.

$$\pi(S) - \pi(S') = \frac{1}{2}(m+n-4)(d(S) - d(S')). \tag{1.9}$$

To determine the surfaces to which our surfaces are linked we will use:

(1.10) Proposition. *If S and T are linked, then S is locally Cohen-Macaulay if and only if T is locally Cohen-Macaulay.*

For a proof see [PS Proposition 1.3].

For the constructions we will use the following

(1.11) Proposition. *If T is a local complete intersection surface in $\mathbf{P}^4$, which scheme-theoretically is cut out by hypersurfaces of degree d, then T is linked to a smooth surface S in the complete intersection of two hypersurfaces of degree d.*

For a proof see [PS Proposition 4.1].

(1.12) Remark. (Peskine, private communication). A slight modification of the conditions of this proposition is allowable, without changing the conclusion. Namely, at a finite set of points T need not be a local complete intersection. It suffices that it is locally Cohen-Macaulay, and that the tangent cone at that point is linked to a plane in a complete intersection. The proof is an application of the proof of (1.11) to the strict transform of T in the blow-up of $\mathbf{P}^4$ in the points where T is not a local complete intersection.

§2 The Invariants
(The proof of Theorem 0.1)

Let us finish the easy cases when the sectional genus is large:

(2.1) Lemma. *If S is a smooth surface of degree 9 in $\mathbf{P}^4$ and $\pi \geq 10$, then either S is a complete intersection $(3,3)$ with $\pi = 10$ and $\chi = 5$ or S is linked (2,5) to a plane and $\pi = 12$ and $\chi = 9$.*

Proof. It is wellknown that any surface of odd degree on a quadric hypersurface is linked to a plane. By (1.9) one finds $\pi = 12$. If S is not contained in a quadric, then (see for instance [Au 2, Prop.1.7]) $\pi \leq 10$, with equality if and only if S is a complete intersection $(3,3)$. The values for χ follows from (1.8).□

(2.2) Lemma. *If S is a smooth surface of degree 9 in $\mathbf{P}^4$, then $\pi \geq 6$.*

Proof. Suppose $\pi < 6$. By (1.1), $H \cdot K \leq -1 < 0$; hence (see [B,p.112]) S is birationally ruled and since $S \neq \mathbf{P}^2$ (Severi's Theorem), $K^2 \leq 8\chi$ and $\chi \leq 1$. Furthermore, (1.5) and (1.6) give $(H+K)^2 \geq 0$ which implies that $K^2 \geq -H^2 - 2H \cdot K \geq -9+2 = -7$, so $\chi \in \{0,1\}$. If $\chi = 0$, then $K^2 \leq 0$ and by (1.1), $K^2 = 23 - 5\pi$; hence $\pi \leq 4$ is impossible and $K^2 = -2$ if $\pi = 5$. If $\chi = 1$, then $K^2 \leq 8$ and by (1.1), $K^2 = 29 - 5\pi$; again $\pi \leq 4$ is impossible and $K^2 = 4$ if $\pi = 5$.

By (1.2), $K^2 \leq 0$ when $\pi = 5$; this leaves us with $\chi = 0$. But then $(H+K)^2 = 5$ and $h^0(\mathcal{O}_S(H+K)) = 4$, a contradiction by (1.5) since a smooth quintic surface in $\mathbf{P}^3$ has positive χ.□

(2.3) Lemma. *A smooth surface of degree 9 in $\mathbf{P}^4$ has Euler-Poincaré characteristic $\chi \geq 1$.*

Proof. Suppose $\chi \leq 0$. If S is birationally ruled in conics, then $(H+K)^2 = 0$; hence $K^2 = 13 - 4\pi$. Combined with (1.1) we find $\pi = 6\chi + 10$: so $\chi \in \{-1, 0\}$. But this is impossible by (2.1) and (2.2).

Now S is mapped to a surface of degree 5 or more by the adjunction mapping, because a smooth surface of degree 4 or less has $\chi > 0$; hence $5 \leq (H+K)^2 = 6\chi + 10 - \pi \leq 10 - \pi$. But then $\pi \leq 5$, contradicting (2.2).□

We can now find the possible smooth surfaces S with $6 \leq \pi \leq 9$.

(2.4) The case $\pi = 6$:
The relations (1.1) and (1.2) read $H \cdot K = 1$, $K^2 = 6\chi - 7$ and $K^2 \leq 0$; hence by (2.3): $\chi = 1$ and $K^2 = -1$.

Suppose S is irregular, then $|K| \neq \emptyset$ since $\chi = 1$. Now, $(H+K) \cdot K = 0$ and $H \cdot K = 1$, so the canonical divisor is an exceptional line. Blowing down this line gives a minimal surface with trivial canonical class and $\chi = 1$; this is impossible by the Enriques classification. Consequently $|K| = \emptyset$ and S is regular.

If $|2K| = \emptyset$, then Castelnuovo's criterion [BPV,p.190] implies that S is rational. This is a surface of the first family in (0.2).

If $|2K| \neq \emptyset$, then $2K = 2E$ where E is an exceptional line, because $(H+K) \cdot 2K = 0$ and $H \cdot K = 1$. Thus S is an Enriques surface blown up in one point. The blowing down map is a factor of the adjunction mapping, which embeds the blown down surface as a surface of degree 10 in $\mathbf{P}^5$, since $(H+K)^2 = 10$ and $h^0(\mathcal{O}_S(H+K)) = 6$. Therefore S is the projection of an Enriques surface of degree 10 in $\mathbf{P}^5$ with a center of projection on the surface.

(2.5) The case $\pi = 7$:
The relations (1.1) and (1.2) read $H \cdot K = 3$, $K^2 = 6\chi - 12$ and $K^2 \leq 1$, so $\chi \in \{1, 2\}$.
If $\chi = 1$, then $K^2 = -6$. Furthermore, $(H+K) \cdot nK < 0$ when $n \geq 1$, so $|K| = |2K| = \emptyset$. Therefore S is regular, and S is a rational surface by Castelnuovo's criterion. This is a surface of the second family in (0.2).
If $\chi = 2$, then $|K| \neq \emptyset$ and $K^2 = 0$. Thus any $C \in |K|$ has arithmetic genus 1 and degree 3, so it is a plane cubic curve. It follows that S is minimal: Otherwise we could write $C = K_0 + E$, where E is a (-1)-curve. But $E \cdot C = E \cdot K = -1 = E^2$ which implies that $K_0 \cdot E = 0$; this is impossible since C is a plane curve. Since $K^2 = 0$ and $12K \neq 0$, it follows from the Enriques classification that S is an elliptic surface. S is regular: Otherwise K moves in a pencil, which means by Severi's Theorem that the residual curve $H - K$ is a plane sextic curve. But $H - K$ has arithmetic genus 4, so this is impossible.

(2.6) Remark. a) A curve of degree 9 and genus 7 in $\mathbf{P}^4$ is a Castelnuovo curve [GH,p.527]. So if C is a curve of degree 9 and genus 8 or more in a projective space, then $h^0(\mathcal{O}_C(1)) \leq 4$.
b) Any surface of degree 9 in $\mathbf{P}^4$ with $\pi \leq 9$ has, by Serre duality, $h^2(\mathcal{O}_S(n)) = h^0(\mathcal{O}_S(K - nH)) = 0$ when $n \geq 1$, because $(K - nH) \cdot H < 0$. Similarly $h^1(\mathcal{O}_H(n)) = 0$ when $n \geq 2$ and H is integral.
c) Severi's Theorem and Riemann-Roch together with b), yield the speciality $h^1(\mathcal{O}_S(1)) = 5 - \chi(\mathcal{O}_S(1)) = \pi - \chi - 5$.

(2.7) By (2.6c) the surfaces with $\pi = 7$ have speciality 1 and 0 respectively.

The following lemma is an application of Porteous' formula.

(2.8) Lemma. *Let S be a smooth surface in $\mathbf{P}^4$, which is not a Veronese surface. If $a = h^1(\mathcal{O}_S(n))$ and $b = h^1(\mathcal{O}_S(n+1))$, or $a = h^1(\mathcal{I}_S(n+1))$ and $b = h^1(\mathcal{I}_S(n))$, and $a \geq b > 0$, then there is a subscheme $V \subset \check{\mathbf{P}}^4$ of dimension at least 3-a+b parametrizing hyperplane sections for which the natural map $H^1(\mathcal{O}_S(n)) \to H^1(\mathcal{O}_S(n+1))$ is not surjective (resp. the map $H^0(\mathcal{I}_S(n+1)) \to H^0(\mathcal{I}_H(n+1))$ is not surjective).*

Proof. Consider the cohomology of the natural exact sequences

$$0 \longrightarrow \mathcal{O}_S(n) \longrightarrow \mathcal{O}_S(n+1) \longrightarrow \mathcal{O}_H(n+1) \longrightarrow 0$$

resp.

$$0 \longrightarrow \mathcal{I}_S(n) \longrightarrow \mathcal{I}_S(n+1) \longrightarrow \mathcal{I}_H(n+1) \longrightarrow 0$$

when H varies. V is defined by the maximal minors of an $a \times b$ matrix with entries in $H^0(\mathcal{O}_S(1))$, and since S is linearly normal, the entries correspond to the points of $\check{\mathbf{P}}^4$.□

(2.9) Lemma. *Let S be a smooth surface in $\mathbf{P}^4$. If $H \cdot K \leq d - 4$ and $h^1(\mathcal{O}_S(1)) \leq 2$, then $h^1(\mathcal{O}_S(2)) = 0$.*

Proof. Assume that $h^1(\mathcal{O}_S(2)) > 0$. Now, $H \cdot (H+K) < 2d$ so $h^1(\mathcal{O}_H(2)) = 0$ for any integral H. Thus $h^1(\mathcal{O}_S(2)) \leq 2$. By (2.8) and its proof the variety $V \subset \check{\mathbf{P}}^4$ parametrizing hyperplane sections for which $h^1(\mathcal{O}_H(2)) > 0$ contains a plane, so there is a line $L \subset \mathbf{P}^4$ contained in a net of hyperplanes for which $h^1(\mathcal{O}_H(2)) > 0$. The line L lies on S; otherwise the general hyperplane through L cuts out an integral curve, impossible. If the general section $C \in |H - L|$ is integral with arithmetic genus p, then it follows from the exact sequence

$$0 \longrightarrow \mathcal{O}_C(2H - L) \longrightarrow \mathcal{O}_H(2H) \longrightarrow \mathcal{O}_L(2H) \longrightarrow 0$$

that

$$2p-2 \geq (H-L)\cdot(2H-L) = 2d-3+L^2,$$

while the genus formula says that

$$2\pi - 2 = d + H\cdot K = -2+2p-2+2(H-L)\cdot L = 2p-2-2L^2,$$

so

$$L^2 \geq d - H\cdot K - 3 \geq 1,$$

impossible. If the general section $C \in |H-L|$ is not integral, then, by Bertini's theorem, $|H-L|$ is composed with a pencil of plane curves. It has no basepoint since a basepoint would be a singular point on S. Thus $(H-L)^2 = 0$ and hence $L^2 = 2-d$ and $\mathcal{O}_C(C) \cong \mathcal{O}_C$. It follows that

$$\mathcal{O}_C(2H-L) \cong \mathcal{O}_C(H)$$

with $h^1(\mathcal{O}_C(H)) > 0$. Therefore C consists of plane curves of degree at least four. But

$$C^2 + C\cdot K = C\cdot K = H\cdot K - L\cdot K \leq d-4-d+4 = 0,$$

impossible. □

(2.10) By (2.5) and (2.8) and (2.9) we get that $h^1(\mathcal{O}_S(k)) = 0$ when $k \geq 2$ for surfaces with $\pi = 7$. By Riemann-Roch $h^0(\mathcal{O}_S(4)) = \chi(\mathcal{O}_S(4)) = \chi + 66$. Since $h^0(\mathcal{O}_{\mathbf{P}^4}(4)) = 70$, it follows that the rational surface is contained in at least 3 quartics and the elliptic surface in at least 2 quartics. On the other hand neither of them is contained in a cubic hypersurface: They cannot lie on a quadric (cf. proof of (2.1)), so since $\pi = 7$ they lie on at most one cubic hypersurface. Now $h^1(\mathcal{O}_S(2)) = 0$, so $h^1(\mathcal{I}_S(2)) = 1$ (resp. 2), and $h^1(\mathcal{I}_S(3)) = 3$ (resp. 4). (2.8) now yields $h^0(\mathcal{I}_H(3)) = 2$ for a family of hyperplane sections H. But this is easily seen to be impossible knowing the arithmetic genus π.

(2.11) The case $\pi = 8$:
The relations (1.1) and (1.2), read $H\cdot K = 5$, $K^2 = 6\chi - 17$ and $K^2 \leq 2$ so $\chi \in \{1,2,3\}$. If $\chi = 1$, then $K^2 = -11$. As in the case $\pi = 7$ and $\chi = 1$ we find that S must be rational. We will show that this is impossible: By (2.6c), $h^1(\mathcal{O}_S(1)) = 2$, so (2.9) tells that $h^1(\mathcal{O}_S(2)) = 0$. But by Riemann-Roch $h^0(\mathcal{O}_S(2)) = \chi(\mathcal{O}_S(2)) = 14 = h^0(\mathcal{O}_{\mathbf{P}^4}(2)) - 1$. Consequently S is on a quadric; this is impossible when $\pi = 8$ (see proof of (2.1)).

If $\chi = 2$, then $K^2 = -5$ and $|K| \neq \emptyset$. Furthermore, $(H+K)\cdot K = 0$; hence $K = \sum_{i=1}^{5} E_i$, where the E_i are five skew exceptional lines. Blowing down these lines yields a surface with trivial canonical class which is regular with $\chi = 2$; this is a $K3$ surface. Now (1.3) and (2.9) imply that $h^1(\mathcal{O}_S(1)) = 1$ and $h^1(\mathcal{O}_S(n)) = 0$, for $n > 1$.

If $\chi = 3$, then $K^2 = 1$ and $h^1(\mathcal{O}_S(1)) = 0$ by (2.6c). From the natural exact sequence

$$0 \longrightarrow \mathcal{O}_S(n) \longrightarrow \mathcal{O}_S(n+1) \longrightarrow \mathcal{O}_H(n+1) \longrightarrow 0$$

and (2.6a) it follows that $h^1(\mathcal{O}_S(n)) = 0$, when $n \geq 0$, in particular S is regular. Since $K^2 > 0$ and $|K| \neq \emptyset$, S is a surface of general type. To see that S is minimal, assume it is not. Then $K = K_0 + E$, where E is a (-1)-curve. But $K_0^2 = 2$ and $H\cdot K_0 \leq 4$ which contradicts (1.2) with K_0 in the place of K.

(2.12) The two types of surfaces with $\pi = 8$ have $h^0(\mathcal{O}_S(4)) = \chi(\mathcal{O}_S(4)) = \chi + 62$. So the two types of surfaces are contained in at least 6 or 5 quartic hypersurfaces respectively. As in (2.10) one can show that neither of these surfaces are contained in a cubic.

(2.13) The case $\pi = 9$:
The relations (1.1) and (1.2) read $H\cdot K = 7$, $K^2 = 6\chi - 22$ and $K^2 \leq 5$ so $\chi \in \{1,2,3,4\}$.

Since $h^1(\mathcal{O}_S(1)) = 4 - \chi$, by Severi, and $\mathcal{O}_H(2H)$ is nonspecial for a smooth H,

$$h^1(\mathcal{O}_S(n)) \leq 4 - \chi$$

when $n \geq 2$. Hence

$$h^0(\mathcal{O}_S(3)) = \chi(\mathcal{O}_S(3)) + h^1(\mathcal{O}_S(3)) = \chi + 30 + h^1(\mathcal{O}_S(3)) \leq 34$$

and S is contained in a cubic hypersurface. Similarly $h^0(\mathcal{O}_S(4)) \leq 62$, so S is contained in a quartic which is not a multiple of the cubic. Therefore S is linked $(3,4)$ to a cubic surface (possibly singular/reducible). By (1.9), the hyperplane section of S is linked to a cubic curve of genus 0. By linkage, this curve is locally Cohen-Macaulay. Such a curve is projectively Cohen-Macaulay ([E, Ex.1, p. 430]). Again, by linkage, a hyperplane section of S and therefore S itself has this property, and we can lift the linkage between the curves to a linkage of S and a cubic scroll (possibly singular/reducible). (1.8) implies $\chi(S) = 4$, excluding the possibilities $\chi \in \{1, 2, 3\}$.

§3 Constructions

(3.1) A nonminimal Enriques surface with $\pi = 6$ is the projection of a minimal smooth Enriques surface of degree 10 in $\mathbf{P}^5$ from a general point on the surface. Smooth Enriques surfaces of degree 10 in $\mathbf{P}^5$ are well understood, in fact any Enriques surface has a linear system of degree 10 and projective dimension 5 without base points (see Cossec [Co Theorem 5.5]), for very ampleness it suffices to require that any elliptic curve on the surface has degree at least 3 with respect to the linear system, and that there are no rational curves on it. These surfaces have exactly 20 plane cubic curves whose planes form the union of the trisecants to the surface (cf. [CV Theorem 2.12 and Corollary 2.13]). Thus projecting the surface from a point on the surface away from these 20 planes, we get a smooth surface of degree 9 in $\mathbf{P}^4$.

(3.2) A rational surface with $\pi = 6$ is the blowing-up of $\mathbf{P}^2$ in 10 points in general position embedded in $\mathbf{P}^4$ by the linear system

$$|H| = |13l - \sum_{i=1}^{10} 4E_i|$$

where l is the pullback of a line in $\mathbf{P}^2$ and E_i are the exceptional curves under the blowing-up map. (See Alexander [A1]).

The surfaces S of degree 9 with $\pi(S) > 6$ all lie on at least two quartic hypersurfaces (see (2.9) and (2.11)). Therefore it is natural to ask for surfaces T of degree 7 to which they are linked. A first suggestion is given by the 5-secant formula of Le Barz (see [LB p. 798]). If S has a one-dimensional family of 5-secants and S_5 is their union, then the formula gives an upper bound for the degree of S_5, which must be a component of T. For the constructions we will start with surfaces T of degree 7 and use the Bertini type result on linkage (1.11) to get smooth surfaces of degree 9. From the linkage formula (1.9) one gets that if S is linked to T in the complete intersection V of two quartic hypersurfaces, then

$$\pi(T) = \pi(S) - 4.$$

The first exact sequence of linkage (see (1.7)) now reads

$$0 \longrightarrow \mathcal{O}_S(K_S) \longrightarrow \mathcal{O}_V(3) \longrightarrow \mathcal{O}_T(3) \longrightarrow 0,$$

which shows that

$$\textbf{(3.3)} \qquad \chi(S) = \chi(\mathcal{I}_T(3)) + 1 \quad \text{and} \quad h^0(\mathcal{O}_S(K)) = h^0(\mathcal{I}_T(3)).$$

Proof of Theorem (0.3a):

(3.4) If S is a smooth rational surface with $\pi = 7$, then we know from (2.9) that S does not lie on a cubic hypersurface and that it lies on a net of quartic hypersurfaces. Therefore S is linked $(4,4)$ to a surface T of degree 7 and sectional genus 3. We use the intrinsic description of the linear system embedding S in $\mathbf{P}^4$ which is due to Alexander (cf. [A2 Proposition 3]), to describe T. It says that S is the blow-up of $\mathbf{P}^2$ in 15 points and

$$H \equiv 9l - \sum_{i=1}^{6} 3E_i - \sum_{j=7}^{9} 2E_j - \sum_{k=10}^{15} E_k,$$

where l is the pullback of a line in $\mathbf{P}^2$ and the E_i, $i = 1,\dots,15$ are the exceptional curves of the blow-up map. The 15 points are chosen so that there is a pencil of curves

$$D \equiv 6l - \sum_{i=1}^{6} 2E_i - \sum_{j=7}^{9} E_j - \sum_{k=10}^{15} E_k$$

with base points on E_7, E_8 and E_9.

Residual to the pencil of curves D there is a plane cubic curve

$$C \equiv H - D \equiv 3l - \sum_{i=1}^{9} E_i.$$

Let P_0 be the plane of C. Since the base points of the pencil $|D|$ are on the curves E_i, $i = 7, 8, 9$, there are curves

$$D_m \equiv 6l - \sum_{i=1}^{6} 2E_i - \sum_{j=7}^{9} E_j - \sum_{k=10}^{15} E_k - E_{m+6},$$

where $m = 1, 2, 3$, on S. They are plane quartic curves in $\mathbf{P}^4$. Let P_1, P_2, P_3 be the planes of these curves. Since $(H - D_m)^2 = 1$, the pencils $|H - D_m|$ have a base point p_m in the planes P_m, and any line in P_m through p_m is a 5-secant to S. So $S \cup P_m$ is not Cohen-Macaulay at this point, and any quartic containing S must also contain P_m. Since each P_m meets P_0 along a line, P_0 is also contained in any quartic containing S. Therefore $T = T_0 \cup P_0 \cup P_1 \cup P_2 \cup P_3$, for some surface T_0 of degree 3. By linkage, $T_0 \cap S$ is a hyperplane section H of S. On the other hand since $P_0 \cup P_1 \cup P_2 \cup P_3$ is locally Cohen-Macaulay so is $S \cup T_0$, therefore T_0 is a degenerate cubic surface.

This is the uniqueness part of Theorem (0.3a).

(3.5) For the existence we start with T and show that it gives rise to a smooth rational surface S. Let T_0 be a Del Pezzo cubic surface in a hyperplane H_0 of $\mathbf{P}^4$. Let L_1, L_2 and L_3 be three skew lines on T_0 and let L_0 be a line not contained in T_0 which meets all three L_i. Let P_0 be a plane through L_0 not contained in H_0, and let p_1, p_2 and p_3 be three noncolinear points in P_0 away from L_0. The lines L_i and the points p_i span three planes which we denote by P_i, $i = 1, 2, 3$. Let

$$T = T_0 \cup P_0 \cup P_1 \cup P_2 \cup P_3.$$

(3.6) Lemma. *$\chi(\mathcal{I}_T(3)) = 0$, T does not lie on a cubic hypersurface and is cut out by quartic hypersurfaces.*

Proof. Let $T_{123} = P_0 \cup P_1 \cup P_2 \cup P_3$, and $T_{12} = P_0 \cup P_1 \cup P_2$ and consider the exact sequences of sheaves of ideals,

$$0 \longrightarrow \mathcal{I}_{T_{123}}(k) \longrightarrow \mathcal{I}_T(k+1) \longrightarrow \mathcal{I}_{T\cap H_0}(k+1) \longrightarrow 0 \qquad k = 2,3$$

$$0 \longrightarrow \mathcal{I}_{T_{12}}(k-1) \longrightarrow \mathcal{I}_{T_{123}}(k) \longrightarrow \mathcal{I}_{T_{123}\cap H_3}(k) \longrightarrow 0 \qquad k = 2,3$$

where H_0 is the hyperplane of T_0, while H_3 is a general hyperplane containing P_3. Now T_{12} is a degenerate cubic scroll, while $T_{123} \cap H_3$ is the union of a plane and two skew lines, so that

$$h^0(\mathcal{I}_{T_{123}}(2)) = h^1(\mathcal{I}_{T_{123}}(2)) = 0,$$

while $\mathcal{I}_{T_{123}}(3)$ is generated by global sections. On the other hand $T \cap H_0$ is the union of a Del Pezzo cubic surface and a line, so it is not contained in any cubic while it is cut out by quartic surfaces. The lemma follows taking global sections of the first sequence.□

By (1.11), T is linked to a smooth surface S in the complete intersection of two quartic hypersurfaces. Since $\chi(S) = \chi(\mathcal{I}_T(3)) + 1 = 1$ and $\pi(S) = \pi(T) + 4 = 7$ the double point formula implies that $K_S^2 = -6$. Now $H \cdot K_S = 3$, so S must be rational.

From the argument in (3.4) we see that the union of the planes P_1, P_2 and P_3 is the union of the 5-secants to S, and that

$$P_0 \cup P_1 \cup P_2 \cup P_3$$

is contained in any quartic hypersurface which contains S.

This concludes the proof of Theorem (0.3a).

(3.7) Remark. A straightforward count shows that the dimension of the family of surfaces constructed is 41, which equals the Euler characteristic of the normal bundle to S in $\mathbf{P}^4$.

Proof of Theorem (0.3b):

(3.8) Any smooth elliptic surface S with $\pi = 7$ is linked $(4,4)$ to a surface T of degree 7 and with $\pi(T) = 3$. We first give a description of T, as formulated in Theorem (0.3b). Secondly, we show that it gives rise to a smooth elliptic surface of degree 9.

(3.9) Lemma. *The family of 5-secants to S is one-dimensional.*

Proof. From Le Barz results the family of 5-secants is at least one-dimensional. If it is two-dimensional or more, then the 5-secants either fill out any irreducible quartic hypersurface containing S, or S contains a plane curve of degree at least 5. The first case is impossible since, by (2.9), S is contained in a pencil of irreducible quartics. In the second case the plane curve C meets the canonical curve K in at most 3 points, i.e. $C \cdot K \leq 3$, while the Hodge Index Theorem implies that $C^2 \leq \frac{1}{9}(H \cdot C)^2$. From the adjunction formula

$$H \cdot C(H \cdot C - 3) = C^2 + C \cdot K \leq \frac{1}{9}(H \cdot C)^2 + 3$$

which is impossible when $H \cdot C \geq 5$.□

(3.10) Remark. The proof extends to show that S contains no plane quartic curve C, since the inequalities above in that case imply that $C^2 = 1$ and that $C \cdot K = 3$. If D is the residual curve $H - C - K$, then $D^2 + D \cdot K = 2$ while $D \cdot H = 2$ which is absurd.

Thus the union of the 5-secants fill out a surface, which we denote by S_5. This surface is a component of the surface T which is linked $(4,4)$ to S. Let P be the plane of the canonical curve on S.

(3.11) Lemma. *P is a component of T, but if $T = T_0 \cup P$, then P is not a component of T_0.*

Proof. First we show that $T = T_0 \cup P$, where P is the plane of the canonical curve K: Consider the pencil $|H-K|$ of curves on S. Any fix component of this curve lies in P. Since $(H-K)\cdot K = 3$ it must be a line L with $L^2 = -2-L\cdot K = -5$, but then $K\cdot(H-K-L) = 0$, which means that the pencil $|H-K-L|$ is part of the bicanonical pencil on S, which is absurd. Thus $|H-K|$ has only isolated base points, in fact since $(H-K)^2 = 3$, the base-point scheme B is a subscheme of length 3 in the plane P and the general member of $|H-K|$ is smooth. Note that, by (1.10), $S \cup P$ is not locally Cohen-Macaulay along B. Consider $S \cap P$ as a subcheme of P. If it is contained in a quartic plane curve, then B is contained in a line. The general member C of $|H-K|$ is a smooth complete intersection of a quadric and a cubic, therefore if B is contained in a line, $|K_{|C}|$ must be a trigonal series. But considering the global sections of the exact sequence

$$0 \longrightarrow \mathcal{O}_S \longrightarrow \mathcal{O}_S(H-K) \longrightarrow \mathcal{O}_C(H-K) \longrightarrow 0,$$

the surface S must be irregular which is absurd. Thus $K \cup B$ is not contained in a quartic curve, which means that P is a component of T.

P is clearly not a component of S_5, so S_5 must be a component of T_0. Since, by the 5-secant formula, there are six 5-secants to S meeting a general plane, $S_5 = T_0$ unless some component S_C of S_5 contributes to the 5-secant formula more than the degree of this component. If S_C is reduced, then S_C must contain several pencils of 5-secant lines to S, i.e. S_C is a plane or a quadric and S meets S_C in a plane quartic and in several points outside this quartic (resp. S meets the quadric in a curve of type (n,m), where $n \geq m \geq 5$). The latter is impossible since S has degree 9, while the former is impossible by (3.10). If S_C is not reduced we have no general argument, but for the lemma the only possible cases are again that $S_{C,red}$ is a plane or a quadric which lead to contradictions as above.□

Now $h^0(\mathcal{I}_T(3)) = 1$, by (3.3).

(3.12) Lemma. *The cubic hypersurface V which contains T is singular along the plane P of the canonical curve of S.*

Proof. Consider T_0, it is linked $(4,4)$ to $P \cup S$ so, by the linkage formulas, T_0 has sectional genus 1. Now, any hyperplane containing P will meet T_0 along a curve with an embedded component of degree 3 along the base locus B. Let C_0 be the one dimensional part of this curve. It has arithmetic genus at least 4, and is of degree 6. By linkage, C_0 has a component in the plane P which is a cubic curve, so the residual curve must also be a plane cubic curve, and C_0 must be a complete intersection. The planes of these cubic curves must clearly all lie in V, so V contains a pencil of planes. But the only cubics with a pencil of planes are those with a double plane, so V must be singular along P.□

(3.13) Now any cubic hypersurface with a double plane is a projection from $\mathbf{P}^5$. Let V_0 be a cubic threefold in $\mathbf{P}^5$ which projects to V in $\mathbf{P}^4$. Then V_0 is the birational image of a $\mathbf{P}^2$-bundle over $\mathbf{P}^1$, which we denote by V_1. Then the Picard group of V_1 is of rank 2 and is generated by the pullback of a hyperplane section, which we denote by H, and a plane of the ruling which we denote by F. If T_1 is the pullback of T_0 on V_1, then $T_1 \equiv 3H - 3F$ since T_0 meets each plane in a cubic curve and is of degree 6. We use this to show:

(3.14) Lemma. *V is the projection of a smooth cubic threefold in $\mathbf{P}^5$.*

Proof. If V_0 is singular, then it is the cone over a rational cubic scroll in $\mathbf{P}^4$, or it is the cone over a twisted cubic curve with vertex a line.

If V_0 is a cone over a twisted cubic curve, then the preimage of the vertex line in V_1 has to be a component of T_1. This means that the image of this line in $\mathbf{P}^4$ is an embedded component of T_0. But this is impossible since, by (1.10), T_0 is locally Cohen-Macaulay everywhere except along B which is a scheme of finite length.

If V_0 is the cone over a rational cubic scroll, then the pencil of planes in V all have a common point in P. Now, the union $P \cup T_0$ is locally Cohen-Macaulay and lies on some quartic hypersurface independent of V, since $h^0(\mathcal{I}_{P\cup T_0}(4)) = h^0(\mathcal{O}_S(H+K)) + 2 = 10$. Therefore $P \cup T_0$ is linked $(3,4)$ to some surface U which, by (1.10), is locally Cohen-Macaulay. Pulling back to V_1, the class of $U + P$ is $H + 3F$. The plane P is clearly a component of U, and the class of the double point locus on V_1 is $H - F$, so residual to P the surface U consists of 4 planes in V. But if they all have a common point then U cannot be Cohen-Macaulay at this point.□

Since V_0 is smooth, $V_0 \cong V_1$. Thus V_0 is the Segre embedding of $\mathbf{P}^2 \times \mathbf{P}^1$ into $\mathbf{P}^5$, and any surface in the class $3H - 3F$ is the product of a plane cubic curve and $\mathbf{P}^1$. This concludes the uniqueness part of Theorem (0.3b).

For the proof of Theorem (0.3b), we are left to show that one can actually get a smooth surface S starting with the plane P and the scroll T_0. For this we assume, for simplicity, that T_1 is smooth and show:

(3.15) Lemma. *T_0 lies on a pencil of cubic hypersurfaces, and is linked $(3,3)$ to three planes in the pencil of planes on V.*

Proof. Let p be a general point in $\mathbf{P}^5$, and let $\pi_p : T_1 \to \mathbf{P}^4$ be the projection of T_1 from p. Now, any plane in V_1 is residual to a net of quadric surfaces in the net of hyperplanes through the plane. Each quadric spans a $\mathbf{P}^3$ and they meet T_1 along three skew lines. Thus the net of quadrics gives rise to a net of $\mathbf{P}^3$s, one of which contains the point p. Let Q_p be the quadric of this $\mathbf{P}^3$. The three lines $T_1 \cap Q_p$ are mapped into a plane by π_p. The plane spanned by each one of the lines and the point p, meet the quadric Q_p in another line, which in turn is the intersection of the quadric with one of the planes of U_1. Projecting to $\mathbf{P}^4$ we get three planes each through one of the three lines. If $T_0 = \pi_0(T_1)$ and P_1, P_2, P_3 are these three planes, then it remains to show that $T_0 \cup P_1 \cup P_2 \cup P_3$ is a complete intersection $(3,3)$. But this is easy since a general hyperplane section is the union of an elliptic curve of degree 6 and three foursecants. Using nonspeciality and Riemann-Roch, the elliptic curve lies on two cubics so by Bezout we are done. □

(3.16) Starting with three planes we may now go backwards for the construction of S. So let P_1, P_2 and P_3 be three planes in $\mathbf{P}^4$ which meet pairwise in points, and let P be the plane spanned by the three points of their pairwise intersection. Let U_1 and U_2 be general cubic hypersurfaces which contain the union $P_1 \cup P_2 \cup P_3$ such that U_1 is singular along P while U_2 does not contain P. Let T_0 be the surface linked to $P_1 \cup P_2 \cup P_3$ in the complete intersection $U_1 \cap U_2$.

(3.17) Proposition. *Let $T = T_0 \cup P$, then T is linked $(4,4)$ to a regular and minimal smooth elliptic surface of degree 9 with $\chi = 2$.*

Proof. First we show that T_0 has good properties.

(3.18) Lemma. *$P_1 \cup P_2 \cup P_3$ is cut out by cubic hypersurfaces.*

Proof. Let H_3 be a general hyperplane containing P_3, then $H_3 \cap (P_1 \cup P_2 \cup P_3)$ is the union of P_3 and the skew lines $H_3 \cap P_i$ $i = 1, 2$. Consider the cohomology associated to the exact sequence

$$0 \longrightarrow \mathcal{I}_{P_1\cup P_2}(2) \longrightarrow \mathcal{I}_{P_1\cup P_2\cup P_3}(3) \longrightarrow \mathcal{I}_{H_3\cap(P_1\cup P_2\cup P_3)}(3) \longrightarrow 0$$

Now $\mathcal{I}_{H_3\cap(P_1\cup P_2\cup P_3)}(3)$ and $\mathcal{I}_{P_1\cup P_2}(2)$ are clearly generated by global sections. Since $h^1(\mathcal{I}_{P_1\cup P_2}(2)) = 0$ the lemma follows.□

(3.18) and (1.11) implies that T_0 is smooth outside P. Let $N_i = P_i \cap P$ $i = 1,2,3$. Since U_2 does not contain P, the intersection $U_2 \cap P = (N_1 \cup N_2 \cup N_3)$. Now U_1 is singular along P, so the union $T_0 \cup (P_1 \cup P_2 \cup P_3)$ must be singular along the three lines N_i. Thus T_0 meets P along $N_1 \cup N_2 \cup N_3$ and is singular at the points of intersection $N_i \cap N_j$.

(3.19) Lemma. *T_0 is an elliptic scroll with three improper double points, and meets each of the planes P_1, P_2 and P_3 along a quartic curve.*

Proof. First, $\pi(T_0) = 1$ from (1.9), secondly (1.7) implies that the one-dimensional part of $T_0 \cap P_i, i = 1,2,3$ is a quartic curve. We need to show that there is no intersection outside this curve, which means that the pencil of curves D_i residual to the plane quartic in the hyperplanes containing P_i has no base points. The residual intersection to P_i in $U_1 \cap U_2$ meets each of these hyperplanes in the complete intersection of two quadric surfaces. The planes P_j and P_k, $\{i,j,k\} = \{1,2,3\}$ meet these hyperplanes in two skew lines, so each curve D_i must be two skew lines. Thus T_0 is a scroll. Its normalization is an elliptic ruled surface since the sectional genus is one, so the pencil of curves D_i cannot have base points.□

(3.20) Lemma. *T lies on one cubic hypersurface and is cut out by quartic hypersurfaces.*

Proof. First of all, T clearly lies on U_1. Using the pencil of cubic hypersurfaces generated by U_1 and U_2 and the pencil of hyperplanes through P, we furthermore get that any base points of the system of $H^0(\mathcal{I}_T(4))$ outside T lie in the planes P_i, $i = 1,2,3$. But since T meets each of these planes along quartic curves, any such quartic hypersurface would then contain the P_i by Bezout. Thus we will argue by counting dimensions that some quartic contains T but not P_i, so that T is cut out by quartics. In fact since T_0 is an elliptic scroll of degree 6, by Riemann-Roch, $\chi(\mathcal{I}_{T_0}(3)) = -1$, so $h^0(\mathcal{I}_{T_0}(3)) = 2$ yields $h^1(\mathcal{I}_{T_0}(3)) = 3$. Furthermore, if H_i is the special hyperplane spanned by P and P_i, then $h^0(\mathcal{I}_{T_0 \cap H_i}(4)) = 13$. Thus from the cohomology of the exact sequence

$$0 \longrightarrow \mathcal{I}_{T_0}(3) \longrightarrow \mathcal{I}_{T_0}(4) \longrightarrow \mathcal{I}_{T_0 \cap H_i}(4) \longrightarrow 0$$

$h^0(\mathcal{I}_{T_0}(4)) \geq 12$. But $h^0(\mathcal{I}_{T_0 \cup P_1 \cup P_2 \cup P_3}(4)) = 10$, so there is some quartic hypersurface which contains T_0 but not $P_1 \cup P_2 \cup P_3$. By symmetry that means that there is some quartic containing T_0 which does not contain any of the planes P_i. □

From (1.11) T is linked to a smooth surface S in the complete intersection of two quartic hypersurfaces. Now $\pi(S) = \pi(T) + 4 = 7$, and $h^0(\mathcal{O}_S(K)) = h^0(\mathcal{I}_T(3)) = 1$, so the invariants follows from (2.5). This concludes the proof of (3.17).□

(3.21) Remark. By linkage $K_S + (S \cap T)_S \equiv 3H_S$, so it follows that the canonical curve on S is residual to $S \cap T$ in $S \cap U_1$. Therefore K_S is the curve of the intersection $P \cap S$, which is a plane cubic curve. Since U_1 is singular along P,

$$2K_S + (S \cap T_0)_S \equiv 3H_S$$

so the pencil of cubic hypersurfaces containing T_0 cuts out a pencil of elliptic curves on S residual to the intersection $S \cap T_0$. This pencil has a double fiber $2K_S$.

The curve $S \cap T_0$ meets the general fibre of the ruling on T_0 in 5 points, so from the construction T_0 is the union of the 5-secants to S.

This concludes the proof of Theorem (0.3b).

(3.22) Remark. A straightforward count shows that the dimension of the family of surfaces constructed is 43, which equals the Euler characteristic of the normal bundle to S in $\mathbf{P}^4$.

Proof of Theorem (0.3c):

From (2.11) we know that any smooth $K3$-surface S with $\pi = 8$ is linked $(4,4)$ to a surface T of degree 7 and with $\pi(T) = 4$.

(3.23) Lemma. *For a general pencil of quartic hypersurfaces containing S, T is a smooth surface.*

Proof. We will in fact show that S is cut out by quartic hypersurfaces. By (1.11) the lemma then follows.

Let Z be the intersection of the quartics containing S. If $S \neq Z$, then we can find some line L in P_4 such that $L \cap S \neq L \cap Z$ scheme-theoretically. Let P be a general plane through L, then we may assume that $Z \cap P$ is a finite scheme of length at least ten. Consider the exact sequences of sheaves of ideals

(3.24) $$0 \longrightarrow \mathcal{I}_S(k-1) \longrightarrow \mathcal{I}_S(k) \longrightarrow \mathcal{I}_{H\cap S}(k) \longrightarrow 0 \quad k = 3,4$$

(3.25) $$0 \longrightarrow \mathcal{I}_{H\cap S}(3) \longrightarrow \mathcal{I}_{H\cap S}(4) \longrightarrow \mathcal{I}_{P\cap S}(4) \longrightarrow 0,$$

where H is a hyperplane containing P. From (2.11) and (2.12) and Riemann-Roch, $h^1(\mathcal{I}_S(3)) = 0$, and the long exact sequence of cohomology associated to (3.24) yields $h^i(\mathcal{I}_{H\cap S}(k)) = 0$ for $i > 0, k > 2$. Applying this again to (3.25), gives $h^1(\mathcal{I}_{P\cap S}(k)) = 0$ for $k > 2$. Therefore the restriction mapping

$$H^0(\mathcal{I}_S(4)) \rightarrow H^0(\mathcal{I}_{P\cap S}(4))$$

is surjective and the sections of $H^0(\mathcal{I}_{P\cap S}(4))$ define a linear system of quartic curves with a baselocus of length at least ten and projective dimension at least 5. But this is only possible if the linear system has a fixed curve, which contradicts our choice of P. □

The smooth surface T has invariants $\chi_T = 1$ and $K^2 = -2$, from (3.3) and (1.1). Since $H \cdot K = -1$, the surface T must be rational (see [O2]).

(3.26) Going the other way we can construct S (see [Ka p. 94]): A smooth surface T of degree 7 and $\pi = 4$ is scheme-theoretically cut out by quartic hypersurfaces and lies on exactly one cubic hypersurface. Thus T is linked to a smooth surface S of degree 9 and $\pi = 8$. Calculating the invariants one can see that S is the blow up of a $K3$-surface in 5 points.

This concludes the proof of Theorem (0.3c).

(3.27) Remark. For this nonminimal $K3$-surface there is a classical construction with the grassmannian G of lines in $\mathbf{P}^5$: Let V be the threefold in $\mathbf{P}^5$ which is the union of the lines corresponding to a general member of the equivalence class of H^6, where H is a Plücker divisor. Then V is known to be smooth, its general hyperplane section is a $K3$-surface of degree 9 with five (-1)-lines. Using the above one can show that any $K3$-surface of degree 9 in $\mathbf{P}^4$ can be constructed this way.

(3.28) Remark. The dimension of the family of the surfaces constructed is 48, which equals the Euler characteristic of the normal bundle to S in $\mathbf{P}^4$.

Proof of Theorem (0.3d):

(3.29) Any smooth surface S of general type with $\pi = 8$ is linked $(4,4)$ to a surface T of degree 7 and with $\pi(T) = 4$. We first give a description of T, as formulated in Theorem (0.3c). Secondly, we show that it gives rise to a smooth surface of general type of degree 9.

On S the canonical curves K move in a pencil residual to a plane quartic curve $C \equiv H - K$. Since $K^2 = 1$, the pencil of canonical curves has a base point p in the plane P of C, and $S \cup P$ is not Cohen-Macaulay at p. Therefore P lies in any quartic hypersurface containing S, which means that $T = P \cup T_0$ for some surface T_0 of degree 6. From the linkage formulas T_0 has sectional genus 3 and is locally Cohen-Macaulay except at p. Furthermore we may assume that T_0 is irreducible: From (2.11) we know that $S \cup P$ lies on 5 quartic hypersurfaces. Restricting to a general plane these form a linear system of plane quartic curves with ten assigned base points. Since a general plane section of $S \cup P$ cannot be contained in a conic, this linear system has no further base points, which means by Bertini that for a general pencil of quartic hypersurfaces T_0 is irreducible.

(3.30) Lemma. *T_0 is linked $(2,4)$ to two skew planes.*

Proof. Now from (3.3) we know that T lies on a pencil of cubic hypersurfaces. If they were irreducible, then, by (1.10), T would be linked $(3,3)$ to a locally Cohen-Macaulay surface U of degree 2 with sectional genus -1. But this is impossible, so the cubic hypersurfaces containing T are reducible. Since T_0 is irreducible, T_0 must be contained in a quadric hypersurface Q, which is singular at p. If Q is singular along a line through p, then T_0 would have an embedded component along this line since $\pi(T_0) = 3$. But this contradicts the above, so Q is a cone over a smooth quadric surface, and T_0 is linked to two skew planes in the complete intersection of Q and a quartic hypersurface.□

Note that from linkage $P \cap T_0$ is a plane conic which is singular at p. This concludes the uniqueness part of Theorem (0.3d).

(3.31) We are left to show that starting from P and T_0 we can get a smooth surface S. In fact let P_0, P_1, P_2 be three planes through a point p such that any two of them span $\mathbf{P}^4$, and let V_4 be a general quartic hypersurface which contains these three planes. Then V_4 has a quadratic singularity at p. Let V_2 be a general quadric which contains P_1 and P_2. Then $P_1 \cup P_2$ is linked to a surface T_0 in the complete intersection $V_2 \cap V_4$. The surface T_0 is, by (1.11), smooth outside the point p, while the tangent cone at p consists of two planes spanning $\mathbf{P}^4$. The quadric V_2 has two pencils of planes through p. The planes P_1 and P_2 belong to one pencil, whose planes meet T_0 along curves of degree four by (1.7). The planes of the other pencil therefore meet T_0 along conics, which shows that T_0 is rational, ruled in conics. By (1.9), $\pi(T_0) = 3$. So by the natural exact sequence

$$0 \longrightarrow \mathcal{O}_{T_0}(n) \longrightarrow \mathcal{O}_{T_0}(n+1) \longrightarrow \mathcal{O}_H(n+1) \longrightarrow 0$$

it follows that

$$h^2(\mathcal{I}_{T_0}(n)) = h^1(\mathcal{O}_{T_0}(n)) = 0 \qquad \text{for} \quad n \geq 0. \tag{3.32}$$

The plane P_0 meets T_0 along $P_0 \cap V_2$, which is two lines through p, so if

$$T = T_0 \cup P_0,$$

then the tangent cone of T at p is linked to a plane in two quadric hypersurfaces, while T is a locally Cohen-Macaulay elsewhere.

(3.33) Lemma. *T is cut out by quartic hypersurfaces.*

Proof. It follows from the construction that $T \cup P_1 \cup P_2$ is cut out by quartic hypersurfaces, namely V_4 and the unions of V_2 with quadrics containing P_0. Now, from Riemann-Roch and (3.32), $h^0(\mathcal{I}_{T_0}(4)) \geq 17$, so from a dimension count alone $h^0(\mathcal{I}_T(4)) \geq 11$ since $T_0 \cap P_0$ is a conic. On the other hand, the space of quartics in $H^0(\mathcal{I}_T(4))$ which has V_2 as a factor has dimension $15 - 6 = 9$, so there is some quartic U_4 independent of V_2 and V_4 in $H^0(\mathcal{I}_T(4))$, i.e. U_4 does not contain P_1 or (by symmetry) P_2. Since $T_0 \cap P_i, i = 1,2$ are plane quartic curves, U_4 does not meet $P_i, i = 1,2$ outside T_0. We conclude that T is cut out by quartic hypersurfaces.□

Thus (1.11) applies, and T is linked to a smooth surface S in a pencil of quartic hypersurfaces. Furthermore, S meets P_0 along a quartic curve C, and at p. So P_0 will be a component of the union of 5-secants to S, in fact since T_0 is not a scroll P_0 is the union of the 5-secants. On the other hand, any cubic hypersurface containing S has V_2 as a component, and $V_2 \cap S = T_0 \cap S$, so by linkage, the canonical curves on S are the curves residual to C in the hyperplanes through P_0. Thus the canonical curves have no fixed components, and S is minimal. Finally, $\pi(S) = 8$ and $\chi(S) = \chi(\mathcal{I}_T(3)) + 1 = 3$ from linkage, and $K_S^2 = 1$ by the double point formula. This concludes the proof of Theorem (0.3d).

(3.34) Remark. A straightforward count shows that the dimension of the family of the surfaces constructed is 50, which equals the Euler characteristic of the normal bundle to S in $\mathbf{P}^4$.

Acknowledgement. We would like to thank the referee for valuable comments and remarks.

References

[A1] Alexander, J.: Surfaces rationelles non-speciales dans $\mathbf{P}^4$. Math. Z. **200**, 87-110 (1988).

[A2] Alexander, J.: Speciality one rational surfaces in $\mathbf{P}^4$. to appear in these proceedings.

[Au 1] Aure, A.B.: On surfaces in projective 4-space. Thesis, Oslo (1987).

[Au 2] Aure, A.B.: The smooth surfaces in $\mathbf{P}^4$ without apparent triple points. Duke Math.J. **57**, 423-430, (1988).

[BPV] Barth, W., Peters, C., Van de Ven, A.: Compact Complex Surfaces. Springer. (1984).

[B] Beauville, A.; Surfaces algébriques complexes. Astérisque **54**, Soc. Math. France, (1978).

[CV] Conte A. Verra A.: Reye Constructions for Nodal Enriques Surfaces. preprint (1990).

[Co] Cossec, F.R.: On the Picard Group of Enriques Surfaces. Math. Ann. **271**, 577-600 (1985).

[E] Ellingsrud, G.:Sur le schéma de Hilbert des variétés de codimension 2 dans $\mathbf{P}^e$ à cône de Cohen-Macaulay, Ann. Scient. Ec. Norm. Sup. **8**, 423-432 (1975),

[EP] Ellingsrud, G., Peskine, C.: Sur les surfaces lisses de $\mathbf{P}^4$. Invent. Math. **95**, 1-11 (1989).

[F] Fulton,W.:Intersection Theory. Springer. (1984)

[GH] Griffiths,P., Harris,J.: Principles of Algebraic Geometry. Wiley-Interscience.(1978).

[Ha] Hartshorne, R.: Algebraic Geometry. Springer. (1977).

[Io] Ionescu, P.: Embedded projective varieties of small invariants (i,ii,iii) in S.L.N. **1056** 142-186 (1984), Rev.Roum. Math. Pures. Appl. **31** 539-544 (1986), S.L.N. **1417**, 138-154 (1990)

[Ka] Katz, S.: Hodge numbers of linked surfaces in $\mathbf{P}^4$. Duke Math. J. **55**, Vol 1, - (1987).

[La] Lanteri, A.: On the existence of scrolls in $\mathbf{P}^4$. Lincei-Rend. Sc. fis.mat.e nat. - **Vol.LXIX**- nov. 1980, 223-227.

[LB] Le Barz, P.: Formules pour les multisecantes des surfaces. C. R. Acad. Sc. Paris, t.**292** Serie I, 797-799 (1981).

[O1] Okonek, C.: Über 2-codimensionale Untermannigfaltigkeiten vom grad 7 in $\mathbf{P}^4$ und $\mathbf{P}^5$. Math. Z. **187**, 209-219 (1984).

[O2] Okonek, C.: Flächen vom grad 8 in $\mathbf{P}^4$. Math. Z. **191**, 207-223 (1986).

[PS] Peskine, C., Szpiro, L.: Liaison des variétés algébriques I. Invent. Math. **26**, 271-302 (1974).

[Ra] Ranestad, K.: On smooth surfaces of degree ten in the projective fourspace. Thesis, Univ. of Oslo (1988)

[Ro] Roth, L.: On the projective classification of surfaces. Proc. of London. Math. Soc., **42**, 142-170 (1937).

[SV] Sommese, A.J., Van de Ven, A.: On the adjunction mapping. Math. Ann. **278**, 593-603 (1987).

*) Institutt for matematiske fag, Norges tekniske høgskole, N-7034 Trondheim.

**) Matematisk Institutt, Universitetet i Oslo, N-0316 Oslo 3.

COMPACTIFYING THE SPACE OF ELLIPTIC QUARTIC CURVES

Israel Vainsencher & Dan Avritzer

INTRODUCTION.

An elliptic quartic curve (Γ_4^1 for short) is the complete intersection of a (unique) pencil of quadric surfaces. We obtain an explicit description of the irreducible component **H** of the Hilbert scheme parametrizing all specializations of Γ_4^1's.

Global descriptions of complete families of curves in projective space are seldom found in the literature. The first non trivial case, concerning the family of twisted cubics in projective 3-space, was treated by R. Piene and M. Schlessinger [PS] (cf. also [EPS] and [V],[V']). Later, a partial compactification of this family was considered by S. Kleiman, A. Strφmme and S. Xambó [KSX] in order to compute characteristic numbers, and in particular to verify the number, found by Schubert, of twisted cubics tangent to 12 quadric surfaces in general position.

There is a natural rational map from the grassmannian **G** of pencils of quadrics to **H**, assigning to a pencil π its base locus $\beta(\pi)$. The map β is not defined along the subvariety **B** of **G** consisting of pencils with a fixed component.

Let $\mathbf{G}'$ denote the blowup with center **B**. Let $\beta' : \mathbf{G}' \cdots\longrightarrow \mathbf{H}$ be the induced rational map; it improves the situation in the sense that the image of β' covers now all Cohen-Macaulay curves (e.g., planar cubic union a unisecant line).

A pleasant and surprising aspect of the geometry of $\mathbf{G}'$ is the appearence of the locus of indeterminacy of β as the variety of flags of doublets in a plane, denoted $\mathbf{C}'$. The normal bundle of $\mathbf{C}'$ in $\mathbf{G}'$ is identified to the bundle of planar quartic forms singular along a doublet. Blowing up $\mathbf{G}'$ along $\mathbf{C}'$ replaces $\mathbf{C}'$ by the parameter space of the family of subschemes consisting of singular plane quartics with an imbedded component of length 2 supported at a doublet contained in the singular locus. Now our main result may be stated thus:

THEOREM. H is isomorphic to the blowup of $\mathbf{G}'$ along $\mathbf{C}'$.

This enables us to obtain the Chow (or cohomology) ring of **H**. We have, in particular, the following.

COROLLARY. *Pic* $\mathbf{H} \cong \mathbb{Z}^3$.

The picture in the next page summarizes the description of **H** (cf.§5 for details).

As an application we compute the number, 6,383,765,416 of Γ_4^1 incident to 16 lines in general position. Other applications will appear elsewhere.

The main tool employed in the proof of the theorem are the Fitting ideals of the natural map of bundles over **G** induced by multiplication of quadratic forms, cf. §4 below. We've made extensive use of versions of REDUCE and MACAULAY for PC in order to handle some of the calculations. Sort of a rehearsal for the technics employed here appeared in [AV].

1. NOTATION.

Let $\mathcal{F}$ be the vector space of linear forms in the variables x_1, x_2, x_3, x_4; $\mathbf{P}^{3\vee}$ denotes the dual projective space with tautological sequence

$$\mathcal{L} \rightarrowtail \mathcal{F} \twoheadrightarrow \mathcal{H} \qquad (1.1)$$

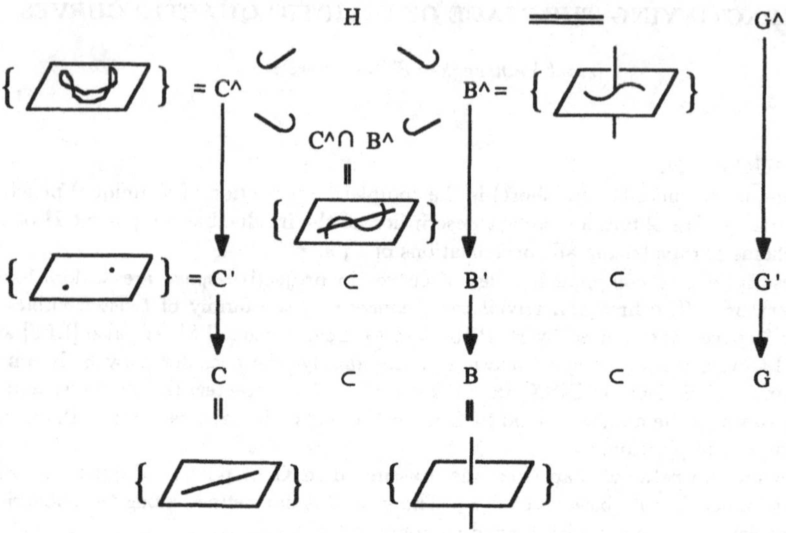

where rank $\mathcal{L} = 1$.

Let $\mathbf{G}(2,4)$ be the grassmannian of lines in $\mathbb{P}^3$, with tautological sequence,

$$\mathcal{K} \rightarrowtail \mathcal{F} \twoheadrightarrow \mathcal{Q}. \tag{1.2}$$

Write $\mathbf{G} = \mathbf{G}(2,10) = \mathbf{G}(2, S_2\mathcal{F})$ for the grassmannian of pencils of quadrics, with tautological sequence

$$\mathcal{A} \rightarrowtail S_2\mathcal{F} \twoheadrightarrow \mathcal{B} \qquad (\text{rank}\mathcal{A} = 2) \tag{1.3}$$

The variety consisting of pairs (plane,line) in $\mathbb{P}^3$, $\mathbf{B} = \mathbb{P}^{3\vee} \times \mathbf{G}(2,4)$ imbeds in $\mathbf{G}$ as the locus of pencils with a fixed component; pictorially:

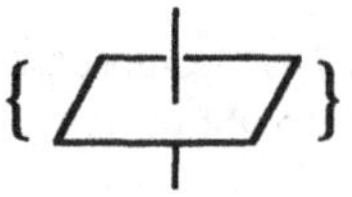

On $\mathbf{B}$ we have the diagram of maps of bundles,

$$\begin{array}{ccccc} & & \mathcal{L} & & \\ & & \downarrow & & \\ \mathcal{K} & \rightarrowtail & \mathcal{F} & \twoheadrightarrow & \mathcal{Q} \\ & & \downarrow & & \\ & & \mathcal{H} & & \end{array} \tag{1.4}$$

We set for later reference,

$$\mathcal{M} = coker(\mathcal{L} \longrightarrow \mathcal{Q}) = coker(\mathcal{K} \longrightarrow \mathcal{H}) = \mathcal{F}/(\mathcal{K} + \mathcal{L}). \tag{1.5}$$

Restriction of the tautological sequence of $\mathbf{G}$ over $\mathbf{B}$ yields the diagram,

$$\begin{array}{ccccc} \mathcal{LK} & \rightarrowtail & \mathcal{L}\,\mathcal{F} & \twoheadrightarrow & \mathcal{L}\,\mathcal{Q} \\ \| & & \downarrow & & \downarrow \\ \mathcal{A} & \rightarrowtail & S_2\mathcal{F} & \twoheadrightarrow & \mathcal{B} \\ & & \downarrow & & \downarrow \\ & & S_2\mathcal{H} & = & S_2\mathcal{H} \end{array} \tag{1.6}$$

Let $\mathbf{C} = \mathbb{P}(\mathcal{K}) = \mathbb{P}(\mathcal{H})$ denote the closed orbit of $\mathbf{B}$, consisting of pencils such that the distinguished line, axis of the residual pencil of planes, falls inside the fixed component.

Notice $\mathbf{C}$ is the scheme of zeros of $\mathcal{L} \longrightarrow \mathcal{Q}(1.4)$; it carries the diagram of maps of bundles,

$$\begin{array}{ccccc} \mathcal{L} & = & \mathcal{L} & & \\ \downarrow & & \downarrow & & \\ \mathcal{K} & \rightarrowtail & \mathcal{F} & \twoheadrightarrow & \mathcal{Q} \\ \downarrow & & \downarrow & & \| \\ \mathcal{R} & \rightarrowtail & \mathcal{H} & \twoheadrightarrow & \mathcal{Q} \end{array} \qquad (1.7)$$

For a vector bundle $\mathcal{W}$, the tautological line subbundle of $\mathcal{W}_{|\mathbb{P}(\mathcal{W})}$ is denoted by $\mathcal{O}_{\mathcal{W}}(-1)$. Thus, we write, $\mathcal{L} = \mathcal{O}_{\mathcal{K}}(-1)$, $\mathcal{R} = \mathcal{O}_{\mathcal{H}}(-1)$. The dual of $\mathcal{W}$ is denoted $\mathcal{W}^\vee$. For sheaves of modules $\mathcal{E}, \mathcal{F}$ we often write $\mathcal{E}\mathcal{F} = \mathcal{E} \otimes \mathcal{F}$ and omit pullbacks for short.

2. NORMAL BUNDLE OF B IN G.

Think of a pencil $\mathbf{b} \in \mathbf{B}$ as a limit along a 1 parameter family of pencils corresponding to honest proper intersection curves; these converge to a well defined (flat) limit curve, one component of which is the distinguished line of $\mathbf{b}$. There arises a residual cubic curve in the fixed plane of $\mathbf{b}$ passing through the intersection with the distinguished line.

It turns out that for $\mathbf{b}$ in $\mathbf{B} - \mathbf{C}$, the limit cubic depends only on the normal direction to $\mathbf{B}$ at $\mathbf{b}$ defined by the 1 parameter family.

Computing the tangent complex of $\mathbf{B}$ in $\mathbf{G}$ we find the following.

2.1 PROPOSITION. Let $\mathcal{N} = \mathcal{N}_{\mathbf{B}/\mathbf{G}}$ be the normal bundle of $\mathbf{B}$ in $\mathbf{G}$. There is a natural map of bundles, $\varphi : \mathcal{N} \longrightarrow \mathcal{L}^\vee \Lambda^2 \mathcal{K}^\vee S_3\mathcal{H}$ such that

(1) off $\mathbf{C}, \varphi$ is an isomorphism onto the subbundle $\mathcal{L}^\vee\Lambda^2\mathcal{K}^\vee\, ker(S_3\mathcal{H} \twoheadrightarrow S_3\mathcal{M})$.

Thus, off $\mathbf{C}$, the bundle $\mathbb{P}(\mathcal{N})$ parametrizes the configurations consisting of a plane cubic union an incident line.

(2) on $\mathbf{C}, \varphi$ drops rank, fitting into the diagram with exact sequence,

$$\begin{array}{ccccc} \Lambda^2\mathcal{K}^\vee S_2\mathcal{Q} & \rightarrowtail & \mathcal{N} & \twoheadrightarrow & \mathcal{L}^{\vee 2}\mathcal{K}^\vee\mathcal{R}S_2\mathcal{H} \\ & & & & \downarrow \\ & & & & \mathcal{L}^\vee\Lambda^2\mathcal{K}S_3\mathcal{H}. \end{array}$$

Proof. We consider the tangent / normal bundle sequence and plug in natural maps,

$$\begin{array}{ccccccc} \mathcal{T}\mathbf{B} = \mathcal{L}^\vee\mathcal{H} \oplus \mathcal{K}^\vee\mathcal{Q} & \rightarrowtail & \mathcal{T}\mathbf{G}_{|\mathbf{B}} & = & \mathcal{L}^\vee\mathcal{K}^\vee\mathcal{B} & \longrightarrow & \mathcal{N} \\ & & & & \downarrow & & \\ & & & & \mathcal{L}^\vee\mathcal{K}^\vee S_2\mathcal{H} & & \\ & & & & \| & & \searrow \varphi \\ & & & & \mathcal{L}^\vee(\Lambda^2\mathcal{K}^\vee)\mathcal{K}S_2\mathcal{H} & & \\ & & & & a\downarrow & & \\ & & & & \mathcal{L}^\vee(\Lambda^2\mathcal{K}^\vee)\mathcal{R}S_2\mathcal{K} & \xrightarrow{b} & \mathcal{L}^\vee(\Lambda^2\mathcal{K}^\vee)S_3\mathcal{H} \end{array}$$

One checks that the composition from top left to bottom right, $\mathcal{T}\mathbf{B} \longrightarrow \mathcal{L}^\vee(\Lambda^2\mathcal{K}^\vee)S_3\mathcal{H}$ is zero whence φ is well defined. The bottom vertical map a is split injective off $\mathbf{C}$ and its cokernel fits into the commutative diagram,

$$\begin{array}{ccc} \mathcal{L}^\vee(\Lambda^2\mathcal{K}^\vee)\mathcal{R}S_2\mathcal{H} & \xrightarrow{b} & \mathcal{L}^\vee(\Lambda^2\mathcal{K}^\vee)S_3\mathcal{H} \\ \downarrow & & \downarrow e \\ coker(a)=\mathcal{L}^\vee(\Lambda^2\mathcal{K}^\vee)\mathcal{M}\, S_2\mathcal{H} & \longrightarrow & \mathcal{L}^\vee(\Lambda^2\mathcal{K}^\vee)S_3\mathcal{M}. \end{array}$$

with $\mathcal{M}$ as in (1.5) and the right vertical map e being defined at a fibre, say over (l,h), by restriction of forms to the intersection of the line l and the plane h. One sees at once that the image of ba is equal to $ker(e)$ off $\mathbf{C}$. This proves the first assertion. Restrict now over $\mathbf{C}$. In view of (1.7), the map a factors as

$$\mathcal{L}^\vee(\Lambda^2\mathcal{K}^\vee)\mathcal{K}S_2\mathcal{R} \twoheadrightarrow \mathcal{L}^\vee(\Lambda^2\mathcal{K}^\vee)\mathcal{R}S_2\mathcal{R} \rightarrowtail \mathcal{L}^\vee(\Lambda^2\mathcal{K}^\vee)\mathcal{R}S_2\mathcal{R},$$

Standard diagram chase obtains the epimorphism, $\gamma : \mathcal{N}_{|C} \twoheadrightarrow \mathcal{L}^\vee(\Lambda^2\mathcal{K}^\vee)\mathcal{R}S_2\mathcal{K}$. It remains to show that $ker\gamma = \Lambda^2\mathcal{K}^\vee S_2\mathcal{Q}$. This follows by studying the commutative diagram of maps over $\mathbf{C}$,

$$\begin{array}{ccccc} \mathcal{L}^\vee\mathcal{H}\oplus\mathcal{K}^\vee\mathcal{Q} & \rightarrowtail & \mathcal{R}^\vee\mathcal{L}^\vee\mathcal{K}\mathcal{Q}\oplus\mathcal{R}^\vee\mathcal{L}^\vee\mathcal{B}/\mathcal{R}^\vee\mathcal{Q} & \twoheadrightarrow & ker\,\gamma \\ \| & & \downarrow & & \downarrow \\ \mathcal{T}\mathbf{B} & \rightarrowtail & \mathcal{T}\mathcal{G}=\mathcal{L}^\vee\mathcal{K}^\vee\mathcal{B} & \twoheadrightarrow & \mathcal{N} \\ & & \downarrow & & \downarrow \\ & & \mathcal{L}^{-2}S_2\mathcal{H} & = & \mathcal{L}^{-2}S_2\mathcal{H} \end{array}$$

Details are straightforward and left for the reader. ∎

3. QUARTICS SINGULAR ALONG A DOUBLET.

Set $\mathcal{V} = \Lambda^2\mathcal{K}^\vee S_2\mathcal{Q}$, and put $C' = \mathbb{P}(\mathcal{V})\times_{G(2,4)}\mathbf{C}$, the projective bundle parametrizing the configurations consisting of a doublet in a plane in $\mathbf{P}^3$, as pictured below.

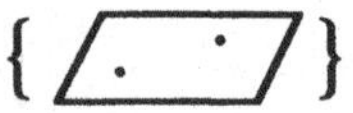

Let $\mathcal{E}$ denote the rank 9 vector subbundle of $S_4\mathcal{H}_{|C'}$ consisting of plane quartic forms singular at a doublet. The fibre of $\mathcal{E}$ at a doublet given by the homogeneous ideal $(x_1(x_1 - x_4), x_2, x_3)$ (resp.,(x_1^2, x_2, x_3)) and the plane $x_3 = 0$ may be identified with the subspace of quartic forms in the variables x_1, x_2, x_4 contained in the ideal $(x_1^2(x_1-x_4)^2, x_2^2, x_1(x_1 - x_4)x_2)$ (resp. $(x_1^4, x_2^2, x_1^2x_2)$).

3.1 LEMMA. The class of $\mathcal{E}$ in the Grothendieck ring of $\mathbf{C}'$ is given by,

$$\mathcal{E} = \mathcal{R}^{\otimes 2}(S_2\mathcal{H} + \mathcal{L}\mathcal{O}_{\mathcal{V}}(-1)\mathcal{Q} + \mathcal{L}^{\otimes\epsilon}\mathcal{O}_{\mathcal{V}}(-2)).$$

Proof. This is a special case of the de Jonquières-type formula considered in [V"]. For the reader's convenience we outline the procedure. We construct the appropriate incidence correspondence of triples (q,d,h) where d stands for a doublet in the plane h and q denotes a quartic curve in that plane with equation lying in the square of the ideal of d in h. More precisely, write the diagram, $\mathbf{D} \subseteq \mathbb{P}(S_2\mathcal{Q})\times_{\mathbf{G(2,4)}}\mathbb{P}(\mathcal{Q}^\vee)$ where $\mathbf{D}$ denotes the universal divisor of degree 2 in the family of lines in $\mathbf{P}^3$. We have

$$\mathcal{O}(\mathbf{D}) = \mathcal{O}_{\mathcal{Q}^\vee}(2)\otimes\mathcal{O}_{S_2\mathcal{Q}}(1). \qquad (3.2)$$

Pulling back to $\mathbb{P}(\mathcal{K})$ we obtain the imbedding $\mathbb{P}(\mathcal{Q}^\vee) \hookrightarrow \mathbb{P}(\mathcal{H}^\vee)$ with normal bundle $\mathcal{R}^\vee(1)$ (see (1.7)) hence the imbeddings,

$$\mathbf{D} \hookrightarrow \mathbb{P}(S_2\mathcal{Q}) \times_{\mathbb{P}(\mathcal{K})} \mathbb{P}(\mathcal{Q}^\vee) \hookrightarrow \mathbb{P}(S_2\mathcal{Q}) \times_{\mathbb{P}(\mathcal{K})} \mathbb{P}(\mathcal{H}^\vee).$$

We write $2\mathbf{D}$ for the subscheme defined by the square of the ideal of $\mathbf{D}$ in $\mathbb{P}(S_2\mathcal{Q}) \times_{\mathbb{P}(\mathcal{K})} \mathbb{P}(\mathcal{H}^\vee)$.
(Warning: not as a divisor in $\mathbb{P}(S_2\mathcal{Q}) \times_{\mathbb{P}(\mathcal{K})} \mathbb{P}(\mathcal{Q}^\vee)$*!!!)*
We have the exact sequences,

$$\mathcal{N}^\vee_{\mathbf{D}/\mathbb{P}(\mathcal{H}^\vee)} \rightarrowtail \mathcal{O}_{2\mathbf{D}} \twoheadrightarrow \mathcal{O}_{\mathbf{D}}, \tag{3.3}$$

and

$$\mathcal{O}_{\mathbf{D}}(\mathbf{D}) \rightarrowtail \mathcal{N}^\vee_{\mathbf{D}/\mathbb{P}(\mathcal{H}^\vee)} \twoheadrightarrow \mathcal{R}^\vee \otimes \mathcal{O}_{\mathcal{Q}^\vee}(1)_{|\mathbf{D}}. \tag{3.4}$$

On the other hand, bring into the picture the universal plane quartic, namely, the relative divisor $\mathbf{Q} \subseteq \mathbb{P}(S_4\mathcal{H}) \times_{\mathbb{P}(\mathcal{K})} \mathbb{P}(\mathcal{H}^\vee)$, with $\mathcal{O}(\mathbf{Q}) = \mathcal{O}_{S_4\mathcal{H}}(1) \otimes \mathcal{O}_{\mathcal{H}^\vee}(4)$. We consider now the incidence correspondence defined by the requirement "$\mathbf{Q} \supseteq 2\mathbf{D}$". More precisely, let

$$\pi : \mathbb{P}(S_4\mathcal{H}) \times_{\mathbb{P}(\mathcal{K})} \mathbb{P}(S_2\mathcal{Q}) \times_{\mathbb{P}(\mathcal{K})} \mathbb{P}(\mathcal{H}^\vee) \longrightarrow \mathbb{P}(S_4\mathcal{H}) \times_{\mathbb{P}(\mathcal{K})} \mathbb{P}(S_2\mathcal{Q})$$

be the natural map. Let the sheaf homomorphism,

$$\pi^*(\mathcal{O}) \longrightarrow \mathcal{O}_{2\mathbf{D}}(\mathbf{Q}) \tag{3.5}$$

be defined by composition of the universal section $\mathcal{O} \longrightarrow \mathcal{O}(\mathbf{Q})$ with restriction to $2\mathbf{D}$. The requirement that it vanish along a fibre of π, say over (q, d, h), means that the equation of q lies in the square of the ideal of d in h. As in Altman-Kleiman [AK], the set of all such triples is the support of the scheme of zeros $\mathbf{Z}(s)$ of the section,

$$\mathcal{O} \xrightarrow{s} \pi_*(\mathcal{O}_{2\mathbf{D}}(\mathbf{Q})). \tag{3.6}$$

Put $\mathcal{E}' = \pi_*(\mathcal{O}_{2\mathbf{D}}(\mathbf{Q})) \otimes \mathcal{O}_{S_4\mathcal{H}}(-1)$. One checks that $\mathcal{E}'$ fits into the diagram with exact sequence,

$$\begin{array}{ccccc} & & \mathcal{O}_{S_4\mathcal{H}}(-1) & & \\ & & \downarrow & & \\ \mathcal{E} & \rightarrowtail & S_4\mathcal{H} & \twoheadrightarrow & \mathcal{E}' \end{array}$$

and $\mathbf{Z}(s) = \mathbb{P}(\mathcal{E})$.

For the formula stated for $\mathcal{E}$ we proceed first with the computation of $\mathcal{E}'$. We have,

$$\begin{aligned} \mathcal{E}' &= \pi_*\mathcal{O}_{\mathbf{D}}(\mathbf{Q})(-1) + \pi_*(\mathcal{N}^\vee_{\mathbf{D}/\mathbb{P}(\mathcal{H}^\vee)}(\mathbf{Q}))(-1) \\ &= S_4\mathcal{Q} - S_2\mathcal{Q}\mathcal{O}_{S_2\mathcal{Q}}(-1) + \\ &\quad \pi_*(\mathcal{R}\mathcal{O}^\vee_{\mathcal{Q}}(3)_{|\mathbf{D}})(-1) + \pi_*(\mathcal{O}^\vee_{\mathcal{Q}}(2)\mathcal{O}_{S_2\mathcal{Q}}(-1)_{|\mathbf{D}})(-1) \qquad (by\ 3.3) \\ &= S_4\mathcal{Q} + \mathcal{R}S_3\mathcal{Q} - \mathcal{R}\ \mathcal{Q}\ \mathcal{O}_{S_2\mathcal{Q}}(-1) + \mathcal{O}_{S_2\mathcal{Q}}(-2). \end{aligned}$$

Recalling the formula $S_m\mathcal{Q} = S_m\mathcal{H} - \mathcal{R}S_{m-1}\mathcal{H}$, and taking into account the isomorphism $\mathbb{P}(\mathcal{V}) \cong \mathbb{P}(S_2\mathcal{Q})$ under which $\mathcal{O}_{\mathcal{V}}(1)$ and $\mathcal{O}_{S_2\mathcal{Q}}(1)\mathcal{L}\ \mathcal{R}$ correspond, the formula for $\mathcal{E}$ follows. ■

4. MULTIPLICATION MAP.

The key ingredient in the sequel is a carefull analysis of singularities of the map of bundles introduced below.

4.1. PROPOSITION. Let $\mu : \mathcal{A} \otimes S_2\mathcal{F} \longrightarrow S_4\mathcal{F}$ be the natural map induced by multiplication. Then (1) the rank of μ is 19 outside **B**;
(2) **B** is equal to the scheme of zeros of $\Lambda^{17}\mu$.

Proof. Clearly $ker(\mu)$ contains the trivial relations $a \otimes b - b \otimes a$ whence rank μ is at most 19. Any orbit of $\mathbf{G} - \mathbf{B}$ contains in its closure the pencil (x_1^2, x_2^2). At this point one checks at once that the rank is 19. To prove the last assertion we notice first that equality certainly holds as sets. We'll finish by virtue of the following.

4.2 LEMMA. The scheme of zeros $\mathbf{Z} = Z(\Lambda^{17}\mu)$ is smooth.

Proof. It suffices to work it out in a neighborhood of the pencil $\pi_0 = (x_2^2, x_1x_2)$ (for it lies in the closed orbit **C**).

Let $\mathbf{X} \cong \mathbf{A}^{16}$ denote the standard coordinate neighborhood of π_0 in **G** with coordinate functions $a_{ij} (i = 1, 2; j = 1, ..., 8)$ so that a general point in $\mathbf{X} \subseteq \mathbf{G}$ may be thought of as the 2×8 matrix

$$\begin{matrix} x_2^2 & x_2x_1 & x_1^2 & x_1x_3 & x_1x_4 & x_2x_3 & x_2x_4 & x_3^2 & x_3x_4 & x_4^2 \end{matrix}$$
$$\begin{pmatrix} 1 & 0 & a_{11} & a_{12} & a_{13} & a_{14} & a_{15} & a_{16} & a_{17} & a_{18} \\ 0 & 1 & a_{21} & a_{22} & a_{23} & a_{24} & a_{25} & a_{26} & a_{27} & a_{28} \end{pmatrix}$$

where the row at the top indicates the ordered basis we've chosen for $S_2\mathcal{F}$. Let v_1, v_2 be the local basis of $\mathcal{A}_{|\mathbf{X}}$ defined by the rows of the above matrix, i.e.,

$$\begin{matrix} v_1 = & x_2^2 & & +a_{11}x_1^2 & +a_{12}x_1x_3 & +a_{13}x_1x_4 & +a_{14}x_2x_3 & +\dots \\ v_2 = & & x_2x_1 & +a_{21}x_1^3 & +a_{22}x_1x_3 & +a_{23}x_1x_4 & +a_{24}x_2x_3 & +\dots \end{matrix}$$

Let $\mathbf{G}_0 = \mathbf{A}^{11} \subset \mathbf{A}^{16}$ be the affine subspace defined by $a_{21} = a_{22} = a_{23} = a_{24} = a_{25} = 0$. A local matrix representation of $\mu_{|\mathbf{G}_0}$ with respect to the suitably ordered basis,

$$\left\{v_i \otimes x_1^a x_2^b x_3^c x_4^d | a + b + c + d = 2\right\} \quad \text{and} \quad \left\{x_1^a x_2^b x_3^c x_4^d | a + b + c + d = 4\right\}$$

of $\mathcal{A} \otimes S_2\mathcal{F}$ and $S_4\mathcal{F}$ may easily be written down (especially with the aid of REDUCE). We agree to put in the rows the 20 vectors of coordinates spanning the image of $\mu_{|\mathbf{G}_0}$. Reordering the basis of $S_4\mathcal{F}$ as necessary and performing elementary row operations, we achieve a 20×35-matrix representation displaying an identity block of size 16. The ideal generated by the entries of the complementary block is the same as the ideal of minors of order 17. Straightforward inspection reveals it is equal to

$$(a_{11}, a_{12}, a_{13}, a_{16}, a_{17}, a_{18}, a_{26}, a_{27}, a_{28}). \tag{4.2.1}$$

This implies that $\mathbf{Z} \cap \mathbf{G}_0 = \mathbf{A}^2$ as schemes, whence **Z** is smooth. ∎

4.2.2 REMARK. We sketch a slightly different proof for (4.2). To check smoothness of **Z**, it suffices to show its ideal is locally generated around π_0 by the right number $(= 9)$ of elements with independent differentials. For this, one computes the ideal of 17×17 minors mod. the square of the ideal of the coordinate functions (a_{ij}). This can be conveniently done by row operations mod. squares, bringing the matrix to triangular form.

4.3. PROPOSITION. Consider the blowingup diagram with center **B**,

$$\begin{matrix} \mathbf{P}(\mathcal{N}) = & \mathbf{B}' & \hookrightarrow & \mathbf{G}' \\ & \downarrow & & \downarrow \\ & \mathbf{B} & \hookrightarrow & \mathbf{G}. \end{matrix}$$

Set as above $\mathbf{C}' = \mathbb{P}(\mathcal{V}) \hookrightarrow \mathbf{B}'$. Put

$$\mathcal{P}_0 = (Image\ \mu_{|\mathbf{G}'}) : \mathcal{O}(-\mathbf{B}'), \qquad \mathcal{P} = S_4\mathcal{F}/\mathcal{P}_0.$$

Then we have that,
(1) $\mathcal{P}_0 (resp. \mathcal{P})$ is locally free of rank 19 (resp.16) off $\mathbf{C}'$;
(2) $\mathbf{C}'$ is equal to the subscheme defined by the Fitting ideal of minors of order 19 of a presentation of $\mathcal{P}$ starting with $S_4\mathcal{F} \longrightarrow \mathcal{P}$;
Proof. Let $\mathbf{C}''$ be the subscheme defined by the Fitting ideal $\mathcal{I}$ referred to in the statement. We'll show that $\mathbf{C}' = \mathbf{C}''$.
First observe $\mathbf{C}''$ lies over $\mathbf{B}$. Indeed, μ is of constant rank 19 off $\mathbf{B}$, hence its image is locally free over $\mathbf{G} - \mathbf{B}$ and $\mathcal{I} = \mathcal{O}_{\mathbf{G}'-\mathbf{B}'}$. Notation as in the proof of (4.2), notice the "slice" $\mathbf{G}_0$ meets $\mathbf{B}$.
Since $\mathbf{C}''$ (as well as $\mu, \mathbf{B}', \mathcal{P}$ and $\mathcal{P}_0$) is invariant under the natural $\mathbf{Gl}(4)$ action, it suffices to check $\mathbf{C}' = \mathbf{C}''$ (and the remaining assertions) over $\mathbf{G}_0$. Since $\mathbf{G}_0$ meets the blowup center properly, we have the fiber squares,

$$\begin{array}{ccccc} \mathbf{B}'_0 & \hookrightarrow & \mathbf{G}'_0 & \hookrightarrow & \mathbf{G}' \\ \downarrow & & \downarrow & & \downarrow \\ \mathbf{B}_0 = \mathbf{B}\bigcap \mathbf{G}_0 & \hookrightarrow & \mathbf{G}'_0 & \hookrightarrow & \mathbf{G} \end{array}$$

where $\mathbf{A}^2 \subset \mathbf{G}_0 = \mathbf{A}^{11}$ with equations (4.2.1) as in the proof of (4.2) and with $\mathbf{C}\bigcap\mathbf{G}_0 = \{\pi_0\}$. We have the imbedding, $\mathbf{G}'_0 \hookrightarrow \mathbf{G}_0 \times \mathbb{P}^8$. Write

$$b_{11}, b_{12}, b_{13}, b_{16}, b_{17}, b_{18}, b_{26}, b_{27}, b_{28}$$

(keeping in mind 4.2.1 for consistent choice of indices) for the homogeneous coordinates in $\mathbb{P}^8$. We may write the equations for $\mathbf{G}'_0$ in $\mathbf{G}_0 \times \mathbb{P}^8$, $b_{ij}a_{rs} = a_{ij}b_{rs}$. Let $\mathbf{G}''_0$ be the intersection of $\mathbf{G}'_0$ with an affine patch, say with $b_{26} = 1$, so that we have $\mathbf{G}''_0 = \mathbf{A}^{11}$ with coordinate functions

$$b_{11}, b_{12}, b_{13}, b_{16}, b_{17}, b_{18}, b_{27}, b_{28}, a_{14}, a_{15}, a_{26}, \quad \text{and} \quad a_{ij} = b_{ij}a_{26}.$$

Now the matrix of $\mu_{|\mathbf{G}''_0}$ with respect to suitable basis (cf. proof of (4.2)) presents a diagonal block of size 18 with an identity block I_{16}, entries (17,17) and (18,18) equal to a_{26} (the generator of the exceptional ideal) and rows 17 till 20 multiples of a_{26} . Dividing these rows by a_{26} yields a matrix

$$\mu' \tag{4.3.1}$$

wich gives us a local presentation of $\mathcal{P}$. The Fitting ideal of $\mathcal{P}$ defined by the minors of order 19 of μ' is generated by the entries on rows 19 and 20. We find it is generated by

$$b_{11}, b_{12}, b_{13}, b_{16}, b_{17}, b_{18}, a_{14}, a_{15}, a_{26} \tag{4.3.2}$$

Thus we see that $\mathbf{C}''$ is smooth and $\mathcal{P}$ is locally free of rank 16 (=35-19) off $\mathbf{C}''$. Moreover, $\mathbf{C}''$ lies over $\mathbf{C}$ as the last 3 generators above show.
Consider the curve κ in $\mathbf{G}$, (actually contained in $\mathbf{G}_0$), defined by

$$\kappa(t) = (x_2^2, x_2x_1 + t(ax_3^2 + bx_3x_4 + cx_4^2)),$$

where a, b, c denote constants not all zero. Set $v = d\kappa/dt$, a tangent vector to $\mathbf{G}$ at $\pi = (x_2^2, x_2x_1)$. One checks that v is not tangent to $\mathbf{B}$, hence defines a normal direction $v' \in \mathbf{B}' = \mathbb{P}(\mathcal{N})$. Moreover, notation as in (2.1), φ kills v, so that v' lies in the fibre of

$\mathbf{C}' = \mathbb{P}(\mathcal{V})$ over $\kappa(0)$. In fact, as we vary a, b, c the v' fill in the whole fibre. The curve κ lifts uniquely to a curve κ' in $\mathbf{G}'$. We have $\kappa'(0) = v'$. On the other hand, since formation of Fitting ideals commutes with base change, the ideal $\mathcal{J}$ of the pullback of $\mathbf{C}''$ under κ' is spanned by the $\kappa'^* b_{ij}$ and $\kappa'^* a_{rs}$ (from 4.3.2). The computation of μ' shows that $\mathcal{J} = (t)$, whence v' lies in $\mathbf{C}''$. We've shown that $\mathbf{C}'$ is contained in $\mathbf{C}''$. Counting dimension and recalling $\mathbf{C}''$ is smooth, we achieve the desired equality. ∎

4.4 REMARKS. (1) Notation as in the proof of 4.3, notice $\mathbf{B}'_0 = \mathbf{B}' \bigcap \mathbf{G}'_0$. For each $\mathbf{b}$ in $\mathbf{B}_0$, the fibres of $\mathbf{B}'_0$ and $\mathbf{B}'$ over $\mathbf{b}$ are equal.
(2) The intersection $\mathbf{C}' = \mathbf{C}' \bigcap \mathbf{G}'_0$ is transversal. Therefore the normal bundle $\mathcal{N}_{\mathbf{C}'_0/\mathbf{G}'_0}$ is equal to the restriction of $\mathcal{N}_{\mathbf{C}'/\mathbf{G}'}$ to $\mathbf{C}'_0$.

4.5 PROPOSITION. Consider the blowingup diagram with center $\mathbf{C}'$,

$$\begin{array}{ccc} \mathbf{C}^\wedge & \hookrightarrow & \mathbf{G}^\wedge \\ \downarrow & & \downarrow \\ \mathbf{C}' & \hookrightarrow & \mathbf{G}'. \end{array}$$

Notation as in (4.3), put

$$\mathcal{P}_0^\wedge = (image\ (\mathcal{P}_{0|\mathbf{G}^\wedge}^\wedge \longrightarrow S_4\mathcal{F})) : \mathcal{O}(-\mathbf{C}^\wedge), \qquad \mathcal{P}_0^\wedge = S_4\mathcal{F}/\mathcal{P}_0^\wedge.$$

Then $\mathcal{P}^\wedge$ and $\mathcal{P}_0^\wedge$ are locally free of ranks 16 and 19.

Proof. Notation as in the proof of (4.3), let $\mathbf{G}_0^\wedge$ be the blowup of $\mathbf{G}''_0$ along $\mathbf{C}''_0 = \mathbf{G}''_0 \bigcap \mathbf{C}'$. Thus $\mathbf{G}_0^\wedge$ imbeds in $\mathbf{G}''_0 \times \mathbb{P}^8$. Recalling (4.3.2), write

$$d_{11}, d_{12}, d_{13}, d_{16}, d_{17}, d_{18}, c_{14}, c_{15}, c_{26}.$$

for the homogeneous coordinates in $\mathbf{P}^8$. Let $\mathbf{G}_0^\sim$ be the intersection of $\mathbf{G}_0^\wedge$ with an affine patch, say defined by $d_{11} = 1$.We have $\mathbf{G}_0^\sim \cong \mathbf{A}^{11}$ with coordinate functions

$$d_{12}, d_{13}, d_{16}, d_{17}, d_{18}, c_{14}, c_{15}, c_{26}, b_{11}, b_{27}, b_{28},$$

and we may write, with indices running as in (4.3.2),

$$b_{ij} = b_{11} d_{ij}, \quad \text{and} \quad a_{ij} = b_{11} c_{ij}$$

Notation as in the proof of 4.3, the rows of the matrix (4.3.1) $\mu'_{|\mathbf{G}_0^\sim}$ span the image of $(\mathcal{P} \longrightarrow S_4\mathcal{F})_{\mathbf{G}_0^\sim}$; the last 2 rows are divisible by the local equation (b_{11}) of the exceptional ideal. Let

$$\mu^\wedge \tag{4.5.1}$$

denote the matrix obtained by dividing these rows by b_{11}. The rows of $\mu^\wedge$ span the split free submodule $\mathcal{P}^\wedge_{0\,|\mathbf{G}_0^\sim}$ of $S_4\mathcal{F}_{\,|\mathbf{G}_0^\sim}$ of rank 19. It follows that $\mathcal{P}^\wedge$ and $\mathcal{P}_0^\wedge$ are as asserted.

5. MAIN RESULT.

Let $\mathbf{H}$ denote the Hilbert scheme component of elliptic quartic curves. We show in this section that $\mathbf{H}$ is isomorphic to the double blowup of $\mathbf{G}$, first along $\mathbf{B}$ then along $\mathbf{C}'$.

For a pencil of quadrics $\pi \in \mathbf{G}$ let $\beta(\pi)$ denote the base locus. Form the incidence correspondence, $\mathbf{Q} = \{(\pi, P) \in \mathbf{G} \times \mathbb{P}^3 | P \in \beta(\pi)\}$.

Notice $\mathbf{Q}$ is the scheme of zeros of the map $\mathcal{A} \longrightarrow \mathcal{O}(2)$, composition of $\mathcal{A} \rightarrowtail S_4\mathcal{F}$ (1.3) with evaluation $S_2\mathcal{F} \twoheadrightarrow \mathcal{O}(2)$. Clearly $\mathbf{Q} \longrightarrow \mathbf{G}$ is flat off $\mathbf{B}$ with geometric fibres $\beta(\pi)$ a complete intersection. Thus $\mathbf{G} - \mathbf{B}$ imbeds in $\mathbf{H}$. By general principles, some modification of $\mathbf{G}$ dominates $\mathbf{H}$. We obtain one such explicitly.

The idea is to mimic the usual construction of Hilb as a closed subscheme of a grassmannian of linear systems of hypersurfaces of suitable degree (cf Mumford [M], Sernesi [S]).

Were we interested on the full Hilb associated to the Hilbert polyinomial $4n$, we'd need degree 6 or higher. Indeeed, the homogeneous ideal of a subscheme defined by the union of a plane quartic and 2 points requires a generator of degree 6 in general. For the good component $\mathbf{H}$ however, quartic forms suffice, as will become apparent. The first blowup $\mathbf{G}'$ obtains a partial flattenning. More precisely, notation as in (4.3), the subsheaf $\mathcal{P}$ of $S_4\mathcal{F}$ defines a rational map

$$\mathbf{G}' \cdots\longrightarrow \mathbf{G}(19, S_4\mathcal{F}), \tag{5.1}$$

which is regular off $\mathbf{C}'$ and is compatible with the map $\mathbf{G} \cdots\longrightarrow \mathbf{G}(19, S_4\mathcal{F})$, defined by (4.1(1)).

Consider the incidence correspondence, $\mathbf{Q}' \subset \mathbf{G}' \times \mathbf{P}^3$ defined by the zeros of the composite map (cf. 4.3),

$$\begin{array}{ccc} \mathcal{P}_0 \hookrightarrow S_4\mathcal{F} & & \\ & \searrow & \\ & & \mathcal{O}(4) \end{array}$$

Let π' be a point of $\mathbf{G}'$ off $\mathbf{C}'$ lying over $\pi \in \mathbf{G}$. The fibre of $\mathbf{Q}'$ over π' is the closed subscheme of $\mathbf{P}^3$ cut out by the linear system of 19 independent quartic forms defined by the fibre of $\mathcal{P}_0$ at π'. In order to compute it, we look at the matrix μ' (4.3.1). The coefficients of our 19 quartic forms are the entries of the first 19 rows of μ'. We find, for π in $\mathbf{B} - \mathbf{C}$, the ideal they span defines the same subscheme as the union of a cubic curve in the distinguished plane of π with the (incident!) distinguished line. For π in $\mathbf{C}$, the subscheme cut out by the system of quartics is, set theorethically, the union of a conic in the distinguished plane with the distinguished line; the latter is endowed with a non planar degree 2, genus -2 structure.♡

Since the fibres of $\mathbf{Q}'$ over $\mathbf{G}' - \mathbf{C}'$ have constant Hilbert polynomial ($4n$), we see $\mathbf{G}'$ yields the partial flattenning referred to above.

Notation as in 4.3, we have the following.

5.2 Theorem. Consider the blowingup diagram with center $\mathbf{C}'$,

$$\begin{array}{ccc} \mathbf{C}^\wedge & \hookrightarrow & \mathbf{G}^\wedge \\ \downarrow & & \downarrow \\ \mathbf{C}' & \hookrightarrow & \mathbf{G}'. \end{array}$$

Let $\mathbf{B}^\wedge$ denote the strict transform of $\mathbf{B}'$. Then we have that,

(1) $\mathbf{G}^\wedge$ is isomorphic to the component of the Hilbert scheme of complete intersections of quadric surfaces;

(2) $\mathbf{C}^\wedge$ consists of singular plane quartic curves with imbedded subscheme of length 2 supported at a doublet, with homogeneous ideal projectively equivalent to either

$$(x_2^2, x_2x_1, x_2x_3(x_3 - x_4), f_4(x_1, x_3, x_4)) \quad \text{or} \quad (x_2^2, x_2x_1, x_2x_3^2, f_4(x_1, x_3, x_4))$$

♡ We thank Ph. Ellia for explaining it to us while enjoying icecream along the hills of Trieste...

w—e f_4 stands for a quartic form in the fibre of $\mathbf{P}(\mathcal{E})$ over the corresponding doublet, i.e., contained in the square of the ideal of the corresponding doublet;

(3) a general element of $\mathbf{B}\hat{} \cap \mathbf{C}\hat{}$ represents a plane quartic curve which decomposes as a line and a cubic, with distinguished doublet with imbedded subscheme supported at their intersection;

(4) a general element of $\mathbf{B}' - \mathbf{B}'_{\mathbf{C}} - \mathbf{C}'$ corresponds to the union of a plane cubic and an incident line;

(5) a general element of $\mathbf{B}'_{|\mathbf{C}} - \mathbf{C}'$ consists of the union of a conic contained in the distinguished plane and a non planar degree 2, genus -2 structure on the distinguished line.

Precisely, let the distinguished plane and line be defined by $x_2 = 0$ and $x_1 = x_2 = 0$; set

$$g_1 := h_7x_1x_4 + h_6x_1x_3 + h_5x_1^2 + h_4x_3^2 + h_3x_3x_4 + h_2x_4^2,$$

$$g_2 := h_9x_3x_4 + h_8x_3^2 + x_4^2, \qquad g_3 := -x_1g_1 + x_2g_2,$$

where the h_i denote constants. Then the ideal is defined by (x_2x_1, x_2^2, g_3).

Proof. By (4.5), $\mathcal{P}_{|\mathbf{G}\hat{}}$ admits of a locally free quotient sheaf $\mathcal{P}\hat{}$ of rank 16. Denote by

$$\underline{\mu}\hat{} : \mathbf{G}\hat{} \longrightarrow \mathbf{G}(19, S_4\mathcal{F})$$

the induced map. It is clearly $\mathbf{Gl}(4)$-equivariant. Therefore the double locus (cf. Fulton-Laksov [FL], Laksov [L]) $\mathbf{D}$ of $\mu\hat{}$ is an invariant closed subset. We proceed to show $\mathbf{D}$ is empty. If non empty, it contains a closed orbit of $\mathbf{Gl}(4)$ in $G\hat{}$. Any such must lie over a closed orbit in $\mathbf{G}'$ and a closed orbit in $\mathbf{G}$. The sole closed orbit in $\mathbf{G}$ is $\mathbf{C}$.

Let $\pi\hat{}$ in $\mathbf{G}\hat{}$ lie over π in $\mathbf{G}$. We have the formula $\underline{\mu}\hat{}(\pi\hat{}) : S_4\mathcal{F} = \pi$, (the *lhs* meaning the space of quadratic forms that multiplied by all of $S_2\mathcal{F}$ land in the space of quartic forms $\underline{\mu}\hat{}(\pi\hat{})$). Indeed, the formula certainly holds for π off $\mathbf{B}$ (since π is generated by a regular sequence) whereas over $\mathbf{B}$ one is reduced to a direct checking say for $\pi_0 = (x_4x_3, x_4^2)$. In this case, let q be a quadratic form with indeterminate coefficients. We adjoin to the matrix $\mu\hat{}$ introduced at (4.5.1), 10 rows of coordinates of the quartic forms obtained as q times the elements of a basis of $S_4\mathcal{F}$. Imposing the condition that the rank remain equal to 19, one checks that the quadratic form must be in the span of π.

Thus, if $\pi\hat{}$ and $\eta\hat{}$ in $\mathbf{G}\hat{}$ have the same image under $\underline{\mu}\hat{}$ they lie on the same fiber over π.

On the other hand, we may compute $\mu\hat{}$ restricted to the fiber over π in $\mathbf{C}$ as just above. Notation as in the proofs of (4.3), (4.5), recall $\underline{\mu}(\pi\hat{})$ is the span of the rows of the matrix $\mu\hat{}$. One finds that the echelon form of $\mu\hat{}$ presents among its entries each of the 11 coordinate functions of $\mathbf{G}_0^{\sim}$ whence $\underline{\mu}\hat{}_{|\mathbf{G}_0^{\sim}}$ is injective. It follows that $\underline{\mu}\hat{}$ is injective on closed points.

Direct inspection reveals that the subscheme cut out by the 19 quartic forms defined by the rows of $\mu\hat{}$ fits the descriptions (2)-(5) in the statement and its Hilbert-Samuel polynomial is $4n$. Acting with suitable 1-parameter subgroups of $\mathbf{Gl}(4)$, one sees there are just 2 closed orbit in $\mathbf{H}$: one contains

$$\mathfrak{c}_1 = (x_2^2, x_2x_1, x_2x_3^2, x_1^4). \tag{5.1.1}$$

(Fourfold line in a plane plus a length 2, spatial imbedded point); the other,

$$\mathfrak{c}_2 = (x_2^2, x_2x_1, x_1^3). \tag{5.1.2}$$

Computing a resolution (MACAULAY!), we find both $\mathbf{c}_i$ are 4-regular. By semicontinuity of cohomology (Hartshorne [H], thm.12.8) it follows that any ideal sheaf corresponding to a point in **H** is 4-regular, whence **H** imbedds in $\mathbf{G}(19, S_4\mathcal{F})$ (cf. Mumford [M]).

This shows $\mathbf{G}^\wedge$ maps bijectively to **H** imbedded in $\mathbf{G}(19, S_4\mathcal{F})$.

It remains to control ramification. It suffices to calculate $\mu^\wedge$ modulo the square of the ideal of coordinate functions of an affine patch of $\mathbf{G}^\wedge$ with the point corresponding to $\mathbf{c}_i$ at the origin and realize its echelon form presents 16 independent entries as desired. ■

6. ENUMERATIVE STUDY

The Chow ring of **H** may now be worked out by general principles ([F] S6.7) Indeed, we have all information one needs for the blowup centers **B** and $\mathbf{C}'$. For the latter, notation as in §3, we have.

6.1 PROPOSITION. The normal bundle of $\mathbf{C}'$ in $\mathbf{G}'$ is isomorphic to $\mathcal{E}$.

Proof. Put $\mathbf{C}^\sim = \mathbb{P}(\mathcal{E})$. Assertion (2) of thm. 5.1 shows $\mathbf{C}^\sim$ is isomorphic to $\mathbf{C}^\wedge$ as projective bundles / $\mathbf{C}'$, thus proving the proposition up to a linebundle twist eventually detected to be trivial. ■

We may state the following.

6.2 PROPOSITION. *Pic* $\mathbf{H} \cong A^1\mathbf{G}^\wedge$ is freely generated by (the classes of) $\mathbf{B}^\wedge, \mathbf{C}^\wedge$ and $c_1\mathcal{A}$ (1.3). ■

Details for the results below will appear later.

6.3 PROPOSITION. $A_1\mathbf{H}$ is freely generated by the classes of the following 1-parameter families, defined by varying each of the 3 aspects of $\mathbf{c}_1$(5.1.1):

$\alpha_1 :=$ turn around the plane thru the support of $\mathbf{c}_1$;

$\alpha_2 :=$ turn around the support of $\mathbf{c}_1$ thru the imbedded point;

$\alpha_3 :=$ move the imbedded point inside the support of $\mathbf{c}_1$.

6.4.PROPOSITION. Fix a line l in $\mathbf{P}^3$. Let $\text{Ç} \subseteq \mathbf{G}$ be the divisor parametrizing the pencils the base loci of which meet l. Then we have,

(1) the class of Ç is $-2c_1\mathcal{A}$;

(2) Ç contains **B** with multiplicity 1;

(3) the strict transform Ç$'$ of Ç in $\mathbf{G}'$ contains $\mathbf{C}'$ with multiplicity 1;

(4) the strict transform Ç$^\wedge$ of Ç$'$ in $\mathbf{G}^\wedge$ parametrizes the set of (specializations of) elliptic quartics incident to l;

(5) $\int \text{Ç}^{\wedge 16} = \int(-2c_1\mathcal{A} - 2\mathbf{C}^\wedge - \mathbf{B}^\wedge)^{16} = 8 * 7 * 1009 * 112979.$

It seems that the case of canonical curves in $\mathbb{P}^3$ may be treated similarly. Similar results hold, with obvious changes, for pencils of higher dimensional quadrics.

acknowledgements

We wish to dedicate this work to S.L.Kleiman; his encouragement has spurred us in more ways than we could possibly express.Inspiration to detect the 2nd. blowup center came about during a visit (May'88) at the Oslo University Math. Institute where the 1st author had the oportunity to maintain with R. Piene many stimulating conversations.

REFERENCES

[AK] A.Altman & S.Kleiman, *Introduction to Grothendieck duality theory*, LNM #146, Springer-Verlag (1970).

[AV] D.Avritzer & I.Vainsencher, $Hilb^4 P^2$, to appear in Proc. Sitges Conference (1986), S.Xambó ed., LNM ?

[EPS] G. Ellingsrud, R. Piene & S.A. Strφmme, *On the variety of nets of quadrics defining twisted cubics*, in Space Curves (Rocca di Papa 1985), F.Ghione, C.Peskine, E.Sernesi eds., LNM #1266, Springer-Verlag (1987).

[FL] W. Fulton and D. Laksov, *Residual intersections and the double point formula*, in Real and complex singularities, Oslo 1976, P. Holm ed. Sijthoff & Noordhoff(1977).

[KSX] S. Kleiman, A. Strφmme and S. Xambó ,*Sketch of a verification of Schubert's number...of twisted cubics, in* Space Curves (Rocca di Papa 1985), F.Ghione, C.Peskine, E.Sernesi eds., LNM #1266, Springer-Verlag (1987).

[L] D. Laksov, *Residual intersections and Todd's formula for the double locus of a morphism*, Acta Math. 140,(1978) 75-92.

[M] D. Mumford, *Lectures on curves on an algebraic surface*, Annals of Math. Studies 59, Princeton (1966).

[H] R. Hartshorne , *Algebraic Geometry*, GTM 52, Springer-Verlag (1977).

[PS] R. Piene and M. Schlessinger, *On the Hilbert scheme compactification of the space of twisted cubics*, Am.J.Math. 107 (1985) 766-774.

[S] E.Sernesi, *Topics on families of projective schemes*, Kingston (1986)

[V] I. Vainsencher, *A note on the Hilbert scheme of twisted cubics*, Bol.Soc.Bras.Mat., 18, $n^{\underline{o}}1$ (1987), 81-89.

[V'] ———, *Classes características em geometria algébrica*, (book) XV Colóq. Brasileiro Mat., IMPA (1985).

[V"] ———,*Counting divisors with prescribed singularities*, Transaction AMS, 267 #2 (1981).

UNIVERSIDADE FEDERAL DE PERNAMBUCO
DEPARTAMENTO DE MATEMÁTICA
CIDADE UNIVERSITÁRIA
50738 RECIFE, PE, BRASIL.

UNIVERSIDADE FEDERAL DE MINAS GERAIS
DEPARTAMENTO DE MATEMÁTICA - ICEX
BELO HORIZONTE, MG, BRASIL

THREEFOLDS OF DEGREE 11 IN P^5

by

Mauro Beltrametti, Michael Schneider, Andrew J.Sommese

Contents

Introduction. This article is devoted to the classification of smooth projective threefolds in P^5.

In [BSS] we classified degree 9 and 10 threefolds in P^5; the lower degree varieties had already been classified (see [I1], [I2], [I3], [O1], [O2]). In that article, we used known constraints and new results from adjunction theory ([BBS], [S1], [S5], [SV]) to restrict the possible invariants. We then used liaison to construct examples with the possible invariants. Uniqueness of the examples satisfying the invariants was also shown.

In this paper we extend the methods of [BSS] to deal with the degree 11 case. The list we obtain of the possible invariants is again short, and we have examples for every possible set of invariants. We refer the reader to the start of § 4 where there is a table giving the degree 11 classification. For the reader's convenience we have given a one page appendix to this paper with a table giving the known classification of degree ≤ 10 threefolds in P^5.

Degree 11 is especially interesting because our calculations show that the number of possible sets of invariants begins to increase quite fast from degree 12 on. This is discussed in (4.6).

We would like to thank the DFG-Schwerpunktprogram "Komplexe Mannigfaltigkeiten" for making it possible for us to work together at the University of Bayreuth in the summer of 1988. The third author would like to thank the National Science Foundation (DMS 87-22330 and DMS 89-21702). The first and the third author would like to thank the University of Notre Dame for its support. We would like to thank Ms. Cinzia Matrì for the excellent typing.

§ 0. Some background material.

We work over the complex numbers C. By *variety* (*n-fold*) we mean an irreducible and reduced projective scheme V of dimension n. We denote its structure sheaf by $\mathcal{O}_V$. For any coherent sheaf $\mathcal{F}$ on V, $h^i(\mathcal{F})$ denotes the complex dimension of $H^i(V, \mathcal{F})$.

If V is normal, the *dualizing sheaf* K_V, sometimes denoted by ω_V, is defined to be

$j_*K_{Reg(V)}$ where $j : Reg(V) \to V$ is the inclusion of the smooth points of V and $K_{Reg(V)}$ is the canonical sheaf of holomorphic n-forms. Note that K_V is a line bundle if V is Gorenstein.

Let $\mathcal{L}$ be a line bundle on a normal variety V. $\mathcal{L}$ is said to be *numerically effective* (*nef*, for short) if $\mathcal{L} \cdot C > 0$ for all effective curves C on V, and in this case $\mathcal{L}$ is said to be *big* if $c_1(\mathcal{L})^n > 0$ where $c_1(\mathcal{L})$ is the first Chern class of $\mathcal{L}$. We shall denote by $|\mathcal{L}|$ the complete linear system associated to $\mathcal{L}$ and by $\Gamma(\mathcal{L})$ the space of the global sections.

(0.1) We fix some more notation.

$\sim$ (respectively $\approx$) the numerical (respectively linear) equivalence of line bundles;

$\chi(\mathcal{L}) = \Sigma(-1)^i h^i(\mathcal{L})$, the Euler characteristic of a line bundle $\mathcal{L}$;

$q(V) = h^1(\mathcal{O}_V)$, the *irregularity* and $p_g(V) = h^0(K_V)$, the *geometric genus*, for V smooth;

$e(V) = c_n(V)$, the topological Euler characteristic of V, for V smooth, where $c_n(V)$ is the nth Chern class of the tangent bundle and hence of V;

$\kappa(V)$, the *Kodaira dimension* of V.

Abuses. Line bundles and divisors are used with little (or no) distinction. Hence we shall freely switch from the multiplicative to the additive notation and vice versa. Sometimes symbol "$\cdot$" of intersection of cycles is understood.

(0.2) For a line bundle $\mathcal{L}$ on a normal variety V, the *sectional genus* $g(\mathcal{L})$ of $(V, \mathcal{L})$ is defined by

(0.2.1) $$2g(\mathcal{L}) - 2 = (K_V + (n-1)\mathcal{L}) \cdot \mathcal{L}^{n-1}.$$

(0.3) Assumptions. Throughout this paper it will be assumed that X^ is a smooth connected variety of dimension $n \geq 2$ and L^ is a very ample line bundle on X^. Further if $n \geq 3$, we denote by S^ a smooth surface (respectively by C^ a smooth curve) obtained as transverse intersection of n-2 (respectively n-1) general elements of |L^| .

(0.4) Reductions. ([S5], (0.5)). Let (X^, L^) be as in (0.3). We say that a pair (X, L) with X smooth is a *reduction* of (X^, L^) if L is ample and

(0.4.1) there exists a morphism $\pi : X\hat{} \to X$ expressing X^ as X with a finite set B blown up, L = $(\pi_* L\hat{}\,)^{**}$;

(0.4.2) $L\hat{} \approx \pi^*L - [\pi^{-1}(B)]$ (equivalently, $K_{X\hat{}} + (n-1)L\hat{} \approx \pi^*(K_X + (n-1)L)$).

(0.4.3) Remark. Note that the positive dimensional fibers of π are precisely the linear $P^{n-1} \subset X\hat{}$ with normal bundle $\mathcal{O}_{P^{n-1}}(1)$. Furthermore by sending each element of |L| to its proper transform, we obtain a one to one correspondence between the smooth divisors in |L| that contain B and the smooth elements of |L^|. ■

Recall also that if $K_{X\hat{}} + (n-1)L\hat{}$ is nef and big, then there exists a reduction π, (X, L) of (X^, L^) and $K_X + (n-1)L$ is ample [S5], (4.5). Note that in this case such a reduction, (X, L), is unique up to isomorphism. In this paper we will refer to this reduction (X, L) as *the reduction* of (X^, L^). Indeed, $K_{X\hat{}} + (n-1)L\hat{}$ is nef and big unless (X^, L^) is one of the following very special pairs (see [S2], [S5], [SV]). See also (0.8) below for the definition of special varieties.

a) $(X^\wedge, L^\wedge)$ is either $(P^n, \mathcal{O}_{P^n}(1))$, a scroll over a curve, or a quadric Q in P^{n+1} with $L^\wedge_Q = \mathcal{O}_{P^{n+1}}(1)_Q$;

b) $(X^\wedge, L^\wedge)$ is a Del Pezzo variety, i.e. $K_{X^\wedge} \approx (1-n)L^\wedge$;

c) $(X^\wedge, L^\wedge)$ is a quadric bundle over a smooth curve;

d) $(X^\wedge, L^\wedge)$ is a scroll over a surface.

Thus except for the explicit list above we can assume that the reduction (X, L) of $(X^\wedge, L^\wedge)$ exists.

Furthermore except for a second list of well understood pairs we make explicit in § 3 we can also assume that $K_X + (n-2)L$ is nef and big. Then from the Kawamata-Shokurov base point free theorem we know that $|m(K_X + (n-2)L)|$ defines a morphism for $m >> 0$, say $\varphi : X \to X'$, such that for m large enough φ has connected fibers and normal image. Therefore there is an ample line bundle $\mathcal{K}_{X'}$ on X' such that $\varphi^*\mathcal{K}_{X'} \approx K_{X'} + (n-2)L$. The pair $(X', \mathcal{K}_{X'})$ is known as the *second reduction* of $(X^\wedge, L^\wedge)$. The morphism φ is very well understood in [S4], [F], [FS] and [BFS]. Let us recall that X' has rational terminal singularities. If $n = 3$, X is 2-Gorenstein while, for $n \geq 4$, X is 2-factorial and it is Gorenstein in even dimension (see [BFS], (0.2.4)). Furthermore $\mathcal{K}_{X'} \approx K_{X'} + (n-2)L'$ where $L' = (\varphi_*L)^{**}$ ([BFS], (0.2.6)).

(0.5) The adjunction map. The following theorem is an easy consequence of [S1] and [V] (see also [SV], (0.1)).

(0.5.1) Theorem. *Let* $(X^\wedge, L^\wedge)$ *be as in* (0.3). *Then* $K_{X^\wedge} + (n-1)L^\wedge$ *is spanned by global sections unless either:*

i) $(X^\wedge, L^\wedge) \cong (P^n, \mathcal{O}_{P^n}(1))$ *or* $(P^2, \mathcal{O}_{P^2}(2))$;

ii) $(X^\wedge, L^\wedge) \cong (Q, \mathcal{O}_Q(1))$ *where* Q *is a smooth quadric in* P^{n+1};

iii) $X^\wedge$ *is a* P^{n-1} *bundle over a smooth curve and the restriction of* L *to a fiber is* $\mathcal{O}_{P^{n-1}}(1)$. ■

Now suppose $K_{X^\wedge} + (n-1)L^\wedge$ to be spanned. Then we shall call the morphism $\phi : X^\wedge \to P^m$ determined by $K_{X^\wedge} + (n-1)L^\wedge$ the adjunction map. We shall write $\phi = s \circ r$ for the Remmert-Stein factorization of ϕ, so $r : X^\wedge \to Y$ is a morphism with connected fibers onto a normal variety $Y = r(X^\wedge)$ and s is a finite map. Note that if $\dim \phi(X^\wedge) = n$, then $r : X^\wedge \to r(X^\wedge)$, $L = (r_*L^\wedge)^{**}$ is the reduction of $(X^\wedge, L^\wedge)$ (see [SV], (0.3)).

Note also that, if $n \geq 3$, for a smooth $A \in |L^\wedge|$, the restriction of r to A is the adjunction map of the pair $(A, L^\wedge_A)$ given by $\Gamma(K_A + (n-2)L^\wedge_A)$. To see this note that from the Kodaira vanishing theorem it follows that $h^1(K_{X^\wedge} + (n-2)L^\wedge) = 0$, and therefore that the restriction

$$\Gamma(K_{X^\wedge} + (n-1)L^\wedge) \to \Gamma(K_A + (n-2)L^\wedge_A)$$

is onto.

In the following particular situation we are dealing with through the paper, ϕ is in fact a morphism.

(0.5.2) Proposition. *Let* $(X^\wedge, L^\wedge)$, $S^\wedge$ *be as in* (0.3) *with* $n \geq 3$. *Assume that* $\Gamma(L^\wedge)$ *embeds*

$X^\wedge$ *in* P^{n+2} *and let* $d^\wedge = L^{\wedge n} \geq 4$. *Then* $K_{X^\wedge}+(n-1)L^\wedge$ *is spanned by its global sections.*

Proof. Assume $K_{X^\wedge}+(n-1)L^\wedge$ not to be spanned. By (0.5.1) it thus follows that $X^\wedge$ is a P^{n-1} bundle $p : X^\wedge \to B$ over a smooth curve B and $L^\wedge_F \cong \mathcal{O}_{P^1}(1)$ for any fiber F. Now, a standard consequence of the Barth Lefschetz theorem implies that $q(X^\wedge) = 0$, therefore $B \cong P^1$ and hence $g(L^\wedge) = g(B) = 0$. Furthermore the restriction $p : S^\wedge \to B$ is a P^1 bundle over P^1 so that $K_{S^\wedge} \cdot K_{S^\wedge} = 8$ and $\chi(\mathcal{O}_{S^\wedge}) = 1$. Thus by Lemma (0.6) below one has $d^{\wedge 2} - 5d^\wedge + 6 = 0$, whence the contradiction $d = 2, 3$. Q.E.D.

We shall use over and over the following relation

(0.6) Lemma ([Ha], p. 434). *Let* $(X^\wedge, L^\wedge)$, $S^\wedge$ *be as in* (0.3). *Assume that* $|L^\wedge|$ *embeds* $X^\wedge$ *in a projective space* P^N *with* $N = n+2$ *and let* $d^\wedge = L^{\wedge n}$. *Then*

$$d^{\wedge 2} - 5d^\wedge - 10(g(L^\wedge) - 1) + 12\,\chi(\mathcal{O}_{S^\wedge}) = 2K_{S^\wedge} \cdot K_{S^\wedge}.$$

(0.7) Castelnuovo's bound. Let $(X^\wedge, L^\wedge)$, $S^\wedge$, $C^\wedge$ be as in (0.3). Assume that $|L^\wedge|$ embeds $X^\wedge$ in a projective space P^N, $N \geq 4$, and let $d^\wedge = L^{\wedge n}$. Then $g(C^\wedge) = g(L^\wedge)$ and Castelnuovo's Lemma (see e.g. [BSS], (0.11)) says that

(0.7.1) $$g(C^\wedge) \leq [(d^\wedge-2)/(N-n)]\,(d^\wedge - N+n-1 - ([(d^\wedge-2)/(N-n)] - 1)(N-n)/2)$$

where $N = h^0(L^\wedge) - 1$ and $[x]$ means the greatest integer $\leq x$.

Note that, for $n = 3$, by writing $[(d^\wedge-2)/(N-3)] = (d^\wedge-2-\varepsilon)/(N-3)$, $0 \leq \varepsilon \leq N-4$, we infer that

$$d^\wedge \geq (N-1)/2 + (2(N-3)\,g(L^\wedge) + ((N-5)/2) - \varepsilon)^2)^{1/2}, \quad 0 \leq \varepsilon \leq N-4$$

which leads to

(0.7.2) $$d^\wedge \geq \begin{cases} (N-1)/2 + (2(N-3)\,g(L^\wedge) + 1/4)^{1/2} & \text{if } N-5 \text{ is odd,} \\ (N-1)/2 + (2(N-3)\,g(L^\wedge))^{1/2} & \text{if } N-5 \text{ is even.} \end{cases}$$

Note also that if $g(L^\wedge)$ does not reach the maximum with respect to (0.7.1) then the stronger bound

(0.7.3) $$g(L^\wedge) \leq d^\wedge\,(d^\wedge-3)/6 + 1$$

holds true (see [GP]).

(0.8) Some special varieties. We say that a polarized pair (V, L), L an ample line bundle on a n-dimensional normal Gorenstein variety V, is a *scroll* over a variety Y (respectively *quadric bundle*; respectively *Del Pezzo bundle*) if there exists a surjective morphism with connected fibres $p : V \to Y$ such that $K_V+(n - \dim Y+1)L \approx p^*\mathcal{L}$ (respectively $K_V+(n - \dim Y)L \approx p^*\mathcal{L}$; respectively $K_V+(n - \dim Y-1)L \approx p^*\mathcal{L}$) for some ample line bundle $\mathcal{L}$ on Y.

Note that if (V, L) is a scroll over Y with $\dim Y \leq 2$, then $(F, L_F) \cong (P^k, \mathcal{O}(1))$, $k = n - \dim Y$, for every fibre F of p and Y is smooth (see [S5], § 3).

If $-K_V$ is ample we say that X is a *Fano variety*. The largest positive integer r such that $-K_V \approx rH$ for some ample line bundle H on V, is the *index* of V. The integer $\dim V+1-r$ is called the *coindex* of V. A polarized pair (V, L) is said to be a *Del Pezzo variety* if $K_V \approx (1-n)L$ and a *Mukai variety* if $K_V \approx (2-n)L$ (see [Mu]).

(0.9) On smooth 2-codimensional subvarieties of $\mathbf{P}^N$. We recall here some well known general facts we need. Besides [PS] we also refer to the nice survey paper [O3].

First, recall that a smooth 2-codimensional variety $V \subset \mathbf{P}^N$ is projectively Cohen-Macaulay (sometimes also called arithmetically Cohen-Macaulay) if and only if $H^i(\mathbf{P}^N, \mathfrak{I}_V(\ell)) = (0)$, $i = 1, \ldots, N-2$, $\ell \in \mathbf{Z}$, where $\mathfrak{I}_V$ is the ideal sheaf defining V in $\mathbf{P}^N$. This is equivalent to the existence of a resolution (see [PS]):

(0.9.1) $$0 \to \oplus_{i=1}^{r} \mathcal{O}_{\mathbf{P}^N}(-a_i) \to \oplus_{j=1}^{r+1} \mathcal{O}_{\mathbf{P}^N}(-b_j) \to \mathfrak{I}_V \to 0$$

where $r \geq 1$, $a_i > b_j$ for any pair i, j and $\Sigma_i a_i = \Sigma_j b_j$.

(0.9.2) Lemma. *Let $V \subset \mathbf{P}^N$ be a 2-codimensional smooth variety and let $Y = V \cap H$ be the general hyperplane section. Assume Y projectively Cohen-Macaulay with minimal resolution*

$$0 \to \oplus_{i=1}^{r} \mathcal{O}_{\mathbf{P}^{N-1}}(-a_i) \to \oplus_{j=1}^{r+1} \mathcal{O}_{\mathbf{P}^{N-1}}(-b_j) \to \mathfrak{I}_Y \to 0.$$

Then V is projectively Cohen-Macaulay and $\mathfrak{I}_V$ has the same type of resolution as $\mathfrak{I}_Y$.

Proof. See [BSS], (0.17.2). Q.E.D.

Consider two subvarieties V, V' of $\mathbf{P}^N$ locally Cohen-Macaulay and of codimension 2. If V and V' have no common irreducible components, they are said to be *linked geometrically* if there exist hypersurfaces F, G of degrees a, b in $\mathbf{P}^N$ such that

$$V \cup V' = F \cap G.$$

Clearly the degrees d, d' of V, V' are related by

$$d + d' = ab.$$

Liaison can be used as a technique to produce new codimension 2 subvarieties starting from given ones. The following statement, which is contained in a more general result of Peskine and Szpiro [PS], says under which conditions the residual intersection V' will be nonsingular, whenever $N \leq 5$.

(0.9.3) Theorem (Peskine-Szpiro). *Consider a local complete intersection* V *of codimension* 2 *in* $\mathbf{P}^N$, $N \leq 5$. *Let* e *denote the smallest number such that the twisted ideal sheaf* $\mathfrak{I}_V(e)$ *is globally generated. For any given* $a_i \geq e$, *there exist forms* $f_i \in H^0(\mathfrak{I}_V(a_i))$, $i = 1, 2$, *such that the hypersurfaces* $F_i = \{f_i = 0\}$ *intersect properly:* $V \cup V' = F_1 \cap F_2$.
Furthermore it is possible to choose f_1 *and* f_2 *with the following properties:*

i) V' *is a local complete intersection;*

ii) V *and* V' *have no common components;*

iii) V' *is nonsingular outside a set of positive codimension in* Sing(V). ■

(0.9.4) Genus formula for 3-folds in $\mathbf{P}^5$. See [BSS], (0.17.4). Let X^ be a smooth threefold in $\mathbf{P}^5$, $L\hat{} = \mathcal{O}_{X\hat{}}(1)$, $S\hat{} \in |L\hat{}|$ and C^ as in (0.3). Assume that the ideal sheaf $\mathfrak{I}_{X\hat{}}$ of X^ in $\mathbf{P}^5$ has a resolution of type (0.9.1). Then the genus g(C^) of C^ is given by

$$g(C\hat{}) = \Sigma_{i=1}^{r} h^0(\mathcal{O}_{\mathbf{P}^3}(a_i-4)) - \Sigma_{j=1}^{r+1} h^0(\mathcal{O}_{\mathbf{P}^3}(b_j-4)).$$

(0.10) Pluridegrees. Let (X^, L^) be a smooth threefold embedded by $\Gamma(L\hat{})$ in $\mathbf{P}^5$ and of degree $d^\wedge = L^{\wedge 3}$. Following [BSS] let us denote $\mathcal{K}_{X^\wedge} = K_{X^\wedge} + L^\wedge$ and define

$$d_j^\wedge = \mathcal{K}_{X^\wedge}{}^j \cdot L^{\wedge 3-j} \quad , \quad j = 0, 1, 2, 3, d_0^\wedge = d^\wedge.$$

Now, assume that $(X^\wedge, L^\wedge)$ has a reduction (X, L). Similarly denote $\mathcal{K}_X = K_X + L$ and define

$$d_j = \mathcal{K}_X{}^j \cdot L^{3-j} \quad , \quad j = 0, 1, 2, 3, d_0 = d.$$

If γ denotes the numbers of points blown up under $r : X^\wedge \to X$, the numbers $d_j^\wedge$, d_j are related by

$$d^\wedge = d - \gamma \; ; \; d_1^\wedge = d_1 + \gamma \; ; \; d_2^\wedge = d_2 - \gamma \; ; \; d_3^\wedge = d_3 + \gamma.$$

For a smooth $S^\wedge \in |L^\wedge|$, let S be the corresponding smooth element of $|L|$. Then note that

$$d = L \cdot L \cdot L = L_S \cdot L_S \; ; \; d_1 = L_S \cdot K_S \; ; \; d_2 = K_S \cdot K_S$$

and the genus formula (0.2) becomes

$$d_1 + d = 2g(L) - 2.$$

In the particular case when $\mathcal{K}_X$ is nef and big some further relations hold true. First, the numbers d_j's are positive. Moreover by the generalized Hodge index theorem (see [BBS], (0.15)) one has

$$d_1^2 \geq dd_2 \; ; \; d_2^2 \geq d_1 d_3$$

and parity Lemma (1.4) of [BBS] says that

$$d = d_1 \bmod(2) \; ; \; d_2 = d_3 \bmod(2).$$

As a consequence of the Riemann-Roch theorem we get the following bound for $\chi(O_X)$.

(0.10.1) Lemma ([BSS], (3.3)).*With the notation as above, let $\mathcal{K}_X$ be nef and big and let $\lambda = h^0(K_X + \mathcal{K}_X)$. Then*

$$6\chi(O_X) + 2\lambda = 2\chi(O_S) + d_3 - d_2 .$$

Furthermore $\lambda = 0$ whenever $2d_3 - d_2 < 0$. ∎

For further numerical conditions the invariants d_j, $d_j^\wedge$, $j = 0, 1, 2, 3$ satisfy we refer to [BSS].

If $\mathcal{K}_X$ is nef and big we can also consider the 2nd reduction $(X', \mathcal{K}_{X'})$ of $(X^\wedge, L^\wedge)$ as in (0.4). Let $\varphi : X \to X'$ be the 2nd reduction map, $L' = (\varphi_* L)^{**}$. Then wen can define $d_j' = \mathcal{K}_{X'}{}^j \cdot L'^{3-j}$. Since $\varphi^* \mathcal{K}_{X'} \approx \mathcal{K}_X$, one has $d_j = d_j'$ for $j \geq 2$.

(0.11) Congruences for smooth threefolds in P^5. Simply as a consequence of the Riemann-Roch theorem we find very strong numerical conditions for a smooth $X^\wedge$ in P^5. Indeed the Riemann-Roch formula for a coherent rank r sheaf $\mathcal{F}$ on P^5 is (see [O1], p. 193)

$$\chi(P^5, \mathcal{F}) = r + 1/120 \, (c_1^5 - 5c_1^3 c_2 + 5c_1^2 c_3 + 5c_1 c_2^2 - 5c_1 c_4 - 5c_2 c_3 + 5c_5) +$$

$$+ 15/120 \, (c_1^4 - c_1^2 c_2 + 4c_1 c_3 + 2c_2^2 - 4c_4) +$$

$$+ 85/120 \, (c_1^3 - 3c_1 c_2 + 3c_3) + 225/120 \, (c_1^2 - 2c_2) + 274/120 \, c_1,$$

where $c_i = c_i(\mathcal{F})$ are the Chern classes of $\mathcal{F}$, $i = 1,..., 5$. Now, let $L^\wedge = O_{X^\wedge}(1)$ and $d^\wedge =$

deg(X^). The sheaf $\mathfrak{I}_{X^\wedge}(5)$ has Chern classes (see [O1], p. 192)

$$c_1 = 5 \;;\; c_2 = d^\wedge \;;\; c_3 = (K_{X^\wedge} + L^\wedge)\cdot L^{\wedge 2};$$
$$c_4 = (K_{X^\wedge} + L^\wedge)^2 \cdot L^\wedge \;;\; c_5 = (K_{X^\wedge} + L^\wedge)^3.$$

A purely mechanical check shows that the formula above leads to the following congruence, in terms of the invariants $d_j^\wedge$, j = 0, 1, 2, 3,

$$5(11d^{\wedge 2} - d_1^\wedge d^\wedge - 770d^\wedge + 136d_1^\wedge - 17d_2^\wedge + d_3^\wedge) \equiv 0\ (120).$$

It thus follows that -

$$11d^{\wedge 2} - d_1^\wedge d^\wedge - 2d^\wedge - 8d_1^\wedge + 7d_2^\wedge + d_3^\wedge \equiv 0\ (24).$$

Note that the congruence above can be expressed in terms of the d_j's as

$$11d^{\wedge 2} - \gamma d^\wedge - d^\wedge d_1 - 2d^\wedge - 8d_1 + 7d_2 + d_3 + 10\gamma \equiv 0\ (24).$$

In particular in the case d^ = 11 we mainly deal with through the paper, the congruence above becomes

(0.11.1) $$d_3 + 7d_2 + 5d_1 - \gamma + 13 \equiv 0\ (24).$$

Note. Throughout the paper we'll mainly deal with smooth threefolds (X^, L^) embedded by Γ(L^) in $\mathbf{P}^5$. We will use without mentioning that X^ has irregularity q(X^) = 0 and hence q(S^) = 0 for a smooth S^ ∈ |L^|. Note that if (X, L) is a reduction of (X^, L^) and S is a smooth element of |L| one has also q(X) = q(S) = 0. ■

For any further background material we refer to [BSS].

§ 1. The case dimϕ(X^) ≤ 2.

From now on, we deal with a polarized pair (X^, L^) where X^ is a smooth 3-fold embedded in $\mathbf{P}^5$ and $L^\wedge = \mathcal{O}_{X^\wedge}(1)$ is the line bundle associated to the hyperplane section. We assume that (X^, L^) has degree $d^\wedge = L^{\wedge 3} \geq 9$. Let $\phi : X^\wedge \to \mathbf{P}^m$ be the adjunction map for (X^, L^). From (0.5.2) we know that ϕ is a morphism. Let ϕ = s∘r be the Remmert-Stein factorization of ϕ. In this section we consider the case when dimϕ(X^) ≤ 2.

(1.1) Theorem. *Let* (X^, L^)*be a smooth threefold embedded by* Γ(L^)*in* $\mathbf{P}^5$ *and of degree* d^ ≥ 9. *Then* dimϕ(X^) ≥ 2. *If* dimϕ(X^) = 2, *then* r : X^ → Y *is a* $\mathbf{P}^1$ *bundle over a smooth surface* Y *and the numerical invariants of all possible cases are as in the following table*

d^	d'	g(L^)	#	number of possible cases according to the values of d_1', d_2', $(K_Y+L')^2$ and $\chi(\mathcal{O}_Y)$
9	14	8	5	1
12	20	17	8	1
15	23	29	8	8
21	42	64	21	30
23	31	77	8	83
24	36	85	12	84

Here $\# = e(S^\wedge) - e(Y)$ *denotes the number of positive dimensional fibres of the restriction* $p = r_{S^\wedge} : S^\wedge \to Y$, $d' = L' \cdot L'$, $d_1' = K_Y \cdot L'$, $d_2' = K_Y \cdot K_Y$ *where* $L' = (p_* L^\wedge_{S^\wedge})^{**}$.

Proof. Let $\dim\phi(X^\wedge) = 0$. Then $K_{X^\wedge} \approx -2L^\wedge$ (see also [SV], (0.2)) so that $g(L^\wedge) = 1$. Therefore $-K_{S^\wedge}$ is ample and hence $S^\wedge$ is a Del Pezzo surface so that $\chi(\mathcal{O}_{S^\wedge}) = 1$. Hence Lemma (0.6) leads to $d^{\wedge 2} - 5d^\wedge + 12 = 2 K_{S^\wedge} \cdot K_{S^\wedge} = 2d^\wedge$, which contradicts $d^\wedge \geq 9$.

Let $\dim\phi(X^\wedge) = 1$. Then $K_{X^\wedge} + 2L^\wedge \approx r^*\mathcal{L}$, $\mathcal{L}$ a line bundle on smooth curve Y, that is $(X^\wedge, L^\wedge)$ is a quadric bundle over Y. Look at the restriction p of r to a general element $S^\wedge \in |L^\wedge|$. From [S5], (0.3.2) we know that $p: S^\wedge \to Y$ is onto; furthermore $K_{S^\wedge} + L^\wedge_{S^\wedge}$ is a conic bundle over Y via p. Then $(K_{S^\wedge} + L^\wedge_{S^\wedge})^2 = 0$ which becomes

(1.1.1) $$K_{S^\wedge} \cdot K_{S^\wedge} + 4(g(L^\wedge) - 1) = d^\wedge$$

by using the genus formula. Moreover $p_g(S^\wedge) = 0$ and $\chi(\mathcal{O}_{S^\wedge}) = 1$, so that (1.1.1) and Lemma (0.6) lead to

(1.1.2) $$g(L^\wedge) = (d^{\wedge 2} - 7d^\wedge + 12)/2 + 1.$$

Now a straightforward check shows that

$$(d^{\wedge 2} - 7d^\wedge + 12)/2 + 1 > d^\wedge(d^\wedge - 3)/6 + 1$$

unless $3 \leq d^\wedge \leq 6$, which is not the case. Recalling (0.7) it thus follows that $g(L^\wedge)$ reaches the maximum with respect to the Castelnuovo's bound. Let $d^\wedge$ be even; then $g(L^\wedge) = d^{\wedge 2}/4 - d^\wedge + 1$, so (1.1.2) yields $d^\wedge = 6$ or 4. Let $d^\wedge$ be odd; then $g(L^\wedge) = d^{\wedge 2}/4 - d^\wedge + 3/4$ and (1.1.2) gives $d^\wedge = 5$. Since $d^\wedge \geq 9$, we conclude that $\dim\phi(X^\wedge) \geq 2$. If $\dim\phi(X^\wedge) = 2$, $p: S^\wedge \to Y$ is a P^1 bundle over a smooth surface Y (see (0.8)) so [BS], § 4 applies to give the result.

Q.E.D.

§ 2. The case $\dim\phi(X^\wedge) = 3$ and of log-general type.

Let $X^\wedge$ be a smooth threefold in P^5, $L^\wedge = \mathcal{O}_{P^5}(1)|_{X^\wedge}$ and let $\phi : X^\wedge \to P^m$ be the adjunction map for $(X^\wedge, L^\wedge)$ as in § 1. Let $\phi = s \circ r$ be the Remmert-Stein factorization of ϕ. From now on, we shall consider the remaining case when $\dim\phi(X^\wedge) = 3$. Then $r : X^\wedge \to X$ is the blowing up of a smooth threefold X at a finite number, say γ, of distinct points. In fact, since $K_{X^\wedge} + 2L^\wedge$ is nef and big, (X, L) is the reduction of $(X^\wedge, L^\wedge)$ where $L = (r_* L^\wedge)^{**}$ (see (0.4)). Let $\mathcal{K}_X = K_X + L$. In this section we shall assume that $\mathcal{K}_X$ is *nef* and *big*, which means that $(X^\wedge, L^\wedge)$ and (X, L) are of *log-general type*. For a smooth $S^\wedge \in |L^\wedge|$, let S be the corresponding smooth element of $|L|$. Then the restriction $r : S^\wedge \to S$ is the map onto its minimal model and S is a surface of general type (see [S5], (4.5)).

Let d_j's be the pluridegrees of (X, L) introduced in (0.10). First, we need the following fact.

(2.1) Proposition. *With the notation as above, let* $(X^\wedge, L^\wedge)$ *be a smooth threefold of log-general type embedded by* $\Gamma(L^\wedge)$ *in* P^5 *and of degree* $d^\wedge \geq 11$, *and let* (X, L) *be the reduction of* $(X^\wedge, L^\wedge)$. *Then* $h^0(K_X + 3\mathcal{K}_X) > 0$ *unless* $d^\wedge = 23$. *If* $h^0(K_X + 3\mathcal{K}_X) = 0$ *then* $\Gamma(\mathcal{K}_X)$ *gives a birational* *morphism* $\psi : X \to P^3$ *and the numerical invariants of* (X,L) *are* $d_3 = 1, d_2 = 5, d_1 = 23, d = L^3 = 101$ *and* $\chi(\mathcal{O}_S) = 5$, *S a smooth element of* $|L|$.

Proof. Assume $h^0(K_X + 3\mathcal{K}_X) = 0$. Then by [BBS], (2.3) we know that $d_3 = 1, d_2 = 5$, $3 \le d_1 \le 25, 0 < d < 125$. By using the above relations together with Lemma (0.6), the congruence (0.11), the Hodge inequality $d \le d_1^2/d_2 = d_1^2/5$ (see (0.10)) and the bound $0 \le \gamma = d - d\hat{} \le 126$ a purely mechanical procedure gives the result. We have carried these computations out using a simple Pascal program. Q.E.D.

We can prove now the main result of this section, which concerns with the degree $d\hat{} = 11$ case.

(2.2) Theorem. *Let* $(X\hat{}, L\hat{})$ *be a smooth threefold of log-general type embedded by* $\Gamma(L\hat{})$ *in* P^5 *and of degree* $d\hat{} = 11$. *Let* (X, L) *be the reduction of* $(X\hat{}, L\hat{})$. *Then the numerical invariants of all possible cases are listed as in the table below*

Case	g(L)	d	d_1	d_2	d_3	γ	$\chi(\mathcal{O}_S)$
i)	13	11	13	9	3	0	6
ii)				15	9		7
iii)	14	11	15	16	16	0	8
iv)	15	14	14	14	14	3	8
v)		11	17	23	29	0	10
vi)	20	11	27	64	148	0	21

Proof. First note that, by (0.7), either g(L) = 20 or g(L) ≤ 15.

Let g(L) = 20. Then $(X\hat{}, L\hat{})$ is a Castelnuovo's variety and hence from [H] we know that $X\hat{}$ is linked to a P^3 in the complete intersection of a quadric and a sextic hypersurface. Furthermore, from the resolution

$$0 \to \mathcal{O}_{P^5}(-2) \to \mathcal{O}_{P^5}(-1)^{\oplus 2} \to \mathfrak{I}_{P^3} \to 0$$

we get the following locally free resolution for the ideal sheaf $\mathfrak{I}_{X\hat{}}$ defining $X\hat{}$ in P^5:

(2.2.1) $$0 \to \mathcal{O}_{P^5}(-7)^{\oplus 2} \to \mathcal{O}_{P^5}(-6)^{\oplus 2} \oplus \mathcal{O}_{P^5}(-2) \to \mathfrak{I}_{X\hat{}} \to 0.$$

From (2.2.1) we easily find $\chi(\mathfrak{I}_{X\hat{}}) = 10$ and hence $\chi(\mathcal{O}_{X\hat{}}) = 1 - \chi(\mathfrak{I}_{X\hat{}}) = -9$. To compute $\chi(\mathcal{O}_S)$ look at the resolution

(2.2.2) $$0 \to \mathcal{O}_{P^4}(-7)^{\oplus 2} \to \mathcal{O}_{P^4}(-6)^{\oplus 2} \oplus \mathcal{O}_{P^4}(-2) \to \mathfrak{I}_{S\hat{}} \to 0$$

which gives $\chi(\mathcal{O}_S) = \chi(\mathcal{O}_{S\hat{}}) = 21$. Now dualize (2.2.2):

$$0 \to \mathcal{O}_{P^4} \to \mathcal{O}_{P^4}(6)^{\oplus 2} \oplus \mathcal{O}_{P^4}(2) \to \mathcal{O}_{P^4}(7)^{\oplus 2} \to \omega_{S\hat{}}(5) \to 0.$$

Thus $\omega_{S\hat{}}(-2)$ is spanned, so in particular $K_{S\hat{}}$ is nef. Therefore $\gamma = 0$ and $d = d\hat{} = 11$. Moreover $d_1 = 2g(L) - 2 - d = 27$, while (0.6) yields $d_2 = 64$. By dualizing (2.2.1) one has

$$0 \to \mathcal{O}_{P^5} \to \mathcal{O}_{P^5}(6)^{\oplus 2} \oplus \mathcal{O}_{P^5}(2) \to \mathcal{O}_{P^5}(7)^{\oplus 2} \to \omega_X(6) \to 0.$$

It thus follows that $\mathcal{K}_X - 2L \cong \omega_X(-1)$ is spanned by 2 sections, so that it is nef and big. Then $(\mathcal{K}_X - 2L)^3 = 0$ which gives $d_3 - 6d_2 + 12d_1 - 8d = 0$, whence $d_3 = 148$. So we find case vi).

Let $g(L) \le 15$. Note that $h^0(K_X + 3\mathcal{K}_X) \ne 0$ by Proposition (2.1). Therefore the same program used in [BBS], § 3, to find out all possible invariants for threefolds $(X^\wedge, L^\wedge)$ in P^5 of log-general type with $g(L) \le 14$ and such that $h^0(K_X + 3\mathcal{K}_X) > 0$, can be used again, running now for $g(L) \le 15$. This program includes all the numerical conditions summarized in Table 1 of [BBS], § 2 together with Castelnuovo's bound (0.7), but it does not include congruences (0.11). By looking over the lists the program gives, it is a purely mechanical check to show that such congruences reduce the list of all possible numerical invariants, for $d^\wedge = 11$ and $g(L) \le 15$, to the following table (compare with [BSS], (3.7)).

$g(L)$	d	d_1	d_2	d_3	γ	$\chi(\mathcal{O}_S)$
13	11	13	9	3	0	6
	11	13	15	9	0	7
14	11	15	16	16	0	8
	13	13	6	2	2	6
15	14	14	14	14	3	8
	11	17	23	29	0	10

Note that $(g(L), d) = (14, 13)$ does not occur. Indeed in this case $2d_3 - d_2 < 0$, so Lemma (0.10.1) applies to give the numerical contradiction $6\chi(\mathcal{O}_S) = 8$. The remaining cases lead to i), ii), iii), iv) and v), so we are done. We carried these computations out using a revised version of the program of [BBS], which also contains congruences (0.11) and Lemma (0.10.1). We don't include here this program which is, however, available on request. Q.E.D.

§ 3. The case $\dim\phi(X^\wedge) = 3$, not of log-general type.

Let $X^\wedge$ be a smooth threefold in P^5, $L^\wedge = \mathcal{O}_{P^5}(1)|_{X^\wedge}$. As in § 2, assume that the adjunction map $\phi : X^\wedge \to P^m$ has a 3-dimensional image. Hence in particular $K_{X^\wedge} + 2L^\wedge$ is nef and big and let (X, L) be the reduction of $(X^\wedge, L^\wedge)$. As usual, we shall denote by $S^\wedge$ a general smooth member of $|L^\wedge|$ and by S the corresponding smooth element of $|L|$ passing through the γ points blown up under $r : X^\wedge \to X$. Recall that $\gamma = d - d^\wedge$ where $d^\wedge = L^{\wedge 3}$, $d = L^3$.

In this section we shall consider the case when $\mathcal{K}_X = K_X + L$ is not nef and big. Precisely, we shall prove that there are no threefolds $(X^\wedge, L^\wedge)$ in P^5 of degree $d^\wedge = 11$ admitting a reduction (X, L) with $\mathcal{K}_X$ not nef and big up to the only (possible) exception of a conic bundle of sectional genus $g(L) = 14$.

If $\mathcal{K}_X$ is not nef and big we know from [S5] that either:

i) $(X, L) \cong (P^3, \mathcal{O}_{P^3}(3))$;

ii) $(X, L) \cong (Q, \mathcal{O}_Q(2))$, Q hyperquadric in P^5;

iii) there is a holomorphic surjection $\phi : X \to B$, onto a smooth curve B, whose general fiber is $(P^2, \mathcal{O}_{P^2}(2))$ and $K_X^2 \otimes L^3 \approx \phi^*\mathcal{L}$ for some ample line bundle $\mathcal{L}$ on B;

iv) (X, L) is a Mukai variety, i.e. $K_X \approx (2-n)L$;

v) (X, L) is a Del Pezzo fibration over a smooth curve;

vi) (X, L) is a conic bundle over a surface (in the sense of (0.8)).

     First of all, let us recall the following results on the structure of (X, L) we need. We refer to [BS], § 1 for the proof. Note that such results are quite general and do not require the assumption that $X^\wedge$ is embedded in $\mathbf{P}^5$.

(3.1) Proposition ([BS], (1.2)). *Let $X^\wedge$ be a smooth threefold, $L^\wedge$ a very ample line bundle on $X^\wedge$ such that $\Gamma(L^\wedge)$ embeds $X^\wedge$ in $\mathbf{P}^N$. Assume $K_{X^\wedge} + 2L^\wedge$ nef and big and let (X, L) be the reduction of $(X^\wedge, L^\wedge)$. Assume that there is a holomorphic surjection $\phi : X \to B$, onto a smooth curve B, as in iii) above. Then we have*

(3.1.1) $(F, L_F) \cong (\mathbf{P}^2, O_{\mathbf{P}^2}(2))$ *for any fibre F of ϕ;*

(3.1.2) $N \geq 6$ *if $d^\wedge = L^{\wedge 3} \neq 5$.*

(3.2) Proposition ([BS], (1.1)). *Let $X^\wedge$ be a smooth threefold, $L^\wedge$ a very ample line bundle on $X^\wedge$ such that $\Gamma(L^\wedge)$ embeds $X^\wedge$ in $\mathbf{P}^N$. Assume $K_{X^\wedge} + 2L^\wedge$ is nef and big and let (X, L) be the reduction of $(X^\wedge, L^\wedge)$. Assume that (X, L) is a Del Pezzo fibration $\phi : X \to B$ over a smooth curve B.*

(3.2.1) *Let F be the general fiber of ϕ and let $\deg F = K_F \cdot K_F = f$. Then there are no fibers containing more than $f-3$ distinct points blown up under $r : X^\wedge \to X$ (hence in particular $X^\wedge \cong X$ if $f = 3$).*

(3.2.2) *One has $N \geq 6$ if $f < 5$ and $d^\wedge = L^{\wedge 3} \geq 9$.* ∎

     We proceed by analyzing all the cases when $\mathcal{K}_X$ is not nef and big. The following general result holds true for any $d^\wedge \geq 11$.

(3.3) Proposition. *Let $(X^\wedge, L^\wedge)$ be a smooth threefold embedded by $\Gamma(L^\wedge)$ in $\mathbf{P}^5$, of degree $d^\wedge \geq 11$ and with $K_{X^\wedge} + 2L^\wedge$ nef and big. Let (X, L) be the reduction of $(X^\wedge, L^\wedge)$. Then (X, L) cannot be as in one of the cases i), ii), iii) or iv) above.*

Proof. Let (X, L) be as in case i). Then $d = 27$ and S is a cubic surface in $\mathbf{P}^3$. Therefore $K_S \approx O_S(-1)$, so that $p_g(S) = 0$, $\chi(O_S) = 1$. Hence $\gamma = 27 - d^\wedge$, $d_2^\wedge = d_2 - \gamma = d^\wedge - 24$ and Lemma

(0.6) leads to $d^{\wedge 2} - 7d^\wedge - 30 = 0$, which is a contradiction.

     If (X, L) is as in case ii), we have $d = 16$ and exactly the same argument as in [BSS], (4.3) yields $\gamma \geq 8$. Therefore we get the contradiction $d = d^\wedge + \gamma \geq 19$. The case when (X, L) is as in iii) is ruled out by (3.1.2).

     Let (X, L) be as in case iv). We have $K_X \approx -L$ and by the genus formula, $2g(L) - 2 = d$. Furthermore $K_S \approx O_S$ so that $d_2 = K_S \cdot K_S = 0$, $d_2^\wedge = K_{S^\wedge} \cdot K_{S^\wedge} = -\gamma$, $p_g(S) = 1$, $\chi(O_S) = 2$.

Thus Lemma (0.6) gives

(3.3.1) $\qquad\qquad\qquad d^{\wedge 2} - 10d^\wedge + 24 - 3\gamma = 0$.

Since $\mathcal{K}_X \approx O_X$, we also have $d_3 = d_1 = 0$, so the congruence (0.11) becomes

(3.3.2) $$11\hat{d}^2 - \gamma\hat{d} - 2\hat{d} + 10\gamma \equiv 0\ (24).$$

By combining (3.3.1) and (3.3.2) we find

$$\hat{d}^3 - 53\hat{d}^2 + 130\hat{d} - 240 \equiv 0\ (72).$$

Therefore

$$\hat{d}^3 + 19\hat{d}^2 - 14\hat{d} \equiv 0\ (24)$$

and also

(3.3.3) $$\hat{d}^3 - 5\hat{d}^2 + 10\hat{d} \equiv 0\ (24).$$

Moreover, from [BSS] (see the proof of (4.3), case iii)) we know that

(3.3.4) $$-e(X) = 120 + d^2 - 2\gamma d + 2\gamma + 15\gamma - 20d.$$

Note that $\hat{d} \leq 17$. Indeed from (3.3.1) we have

(3.3.5) $$\hat{d} = 5 + (1 + 3\gamma)^{1/2}$$

so that, if $\hat{d} \geq 18$, we would have $\gamma \geq 56$ and hence $d = \hat{d} + \gamma \geq 74$. This contradicts the classification of smooth Fano threefolds worked out in [IS] and [MM]. Now a straightforward check, simply using (3.3.3), (3.3.4) and (3.3.5), shows that, for $11 \leq \hat{d} \leq 17$, the only possible cases are listed as in the table below

$\hat{d}$	γ	d	e(X)
12	16	28	56
15	33	48	120
16	40	56	144

Write e(X) in terms of the Betti numbers of X, as

$$e(X) = 2 + 2b_2 - b_3.$$

Here $b_2 = h^{1,1}$, $b_3 = h^{2,1} + h^{1,2}$ where $h^{p,q} = h^q(\Omega^p_X)$ are the Hodge numbers, Ω^p_X denoting the sheaf of regular p-forms on X.

From [IS] we know that there are no Fano 3-folds with $b_2 = 1$ and of degree $d \geq 28$. Thus we have $b_2 \geq 2$. Hence the results of [MM] apply to give the result. Indeed, by looking over the lists given in [MM], we see that $b_2 \leq 5$ and either $b_3 = 0$ or 2 for d = 28, 48 or 56. This contradicts the values of e(X) in the table above. Q.E.D.

From now on, we assume that $\hat{d} = 11$. The two remaining cases when (X, L) is either a Del Pezzo fibration or a conic bundle will be treated separately.

(3.4) Proposition. *Let* $(\hat{X}, \hat{L})$ *be a smooth threefold embedded by* $\Gamma(\hat{L})$ *in* $\mathbf{P}^5$, *of degree* $\hat{d} = 11$ *and with* $K_{\hat{X}} + 2\hat{L}$ *nef and big. Let* (X, L) *be the reduction of* $(\hat{X}, \hat{L})$. *Then* (X, L) *cannot be a Del Pezzo fibration* $\phi : X \to B$ *over a smooth curve* B.

Proof. Assume otherwise. Clearly $B \cong \mathbf{P}^1$ since $q(X) = 0$. From [S6] we know that

(3.4.1) $$K_X + L \approx \phi^*\mathcal{O}_B(p_g(S)-1).$$

Note also that in view of Proposition (3.3) we can assume $p_g(S) \geq 2$. Otherwise, since $K_X + L$ is nef (see [S5]), $p_g(S) = 1$ and hence $K_X \approx -L$, that is (X, L) would be a Mukai variety as in case iv). Let F be a general fiber of ϕ and let $f = K_F \cdot K_F$ be the degree of F. Then $f = L_F \cdot L_F$ since $K_F \sim -L_F$ and from (3.2.2) we can assume $f \geq 5$. By (3.4.1) and the genus formula we find

(3.4.2) $$2g(L)-2=(p_g(S)-1)f+\gamma+d^\wedge.$$
Note also that $K_S \cdot K_S = 0$ since K_S is concentrated on the fibers of ϕ. Therefore $K_{S^\wedge} \cdot K_{S^\wedge} = -\gamma$. Thus by combining Lemma (0.6) with (3.4.2) we find, for $d^\wedge = 11$,

(3.4.3) $$12p_g(S)+23=5f(p_g(S)-1)+3\gamma.$$
Furthermore, since $d_3 = d_2 = 0$, the congruence (0.11) becomes

(3.4.4) $$5d_1-\gamma-11 \equiv 0\ (24).$$
Let $f = 5$. Then (3.4.3) gives $3\gamma = 48 - 13p_g(S)$, whence $p_g(S) \le 3$. The case $p_g(S) = 2$ is clearly not possible. If $p_g(S) = 3$ one has $\gamma = 3$ and hence by (3.4.2), $g(L) = 13$, $d_1 = 10$, which contradicts (3.4.4).

Let $f \ge 6$. Then $p_g(S) = 2$ since $5f \le (12p_g(S)+23)/(p_g(S)-1)$ by (3.4.3). Hence (3.4.3) becomes $3\gamma = 47 - 5f$ which implies that $f = 7$, $\gamma = 4$. Therefore $g(L) = 12$, $d_1 = 7$. This contradicts again (3.4.4) and we are done. Q.E.D.

Now we consider the case when the reduction (X, L) of $(X^\wedge, L^\wedge)$ is a conic bundle $p: X \to Y$ over a surface Y, that is $K_X + L \approx p^*\mathcal{L}$ for some ample line bundle $\mathcal{L}$ on Y. Note that by the main theorem of [BS], the only divisorial fibres of p are isomorphic to either $F_0 = P^1 \times P^1$ or $F_0 \cup F_1$ where F_1 is $P(\mathcal{O}_{P^1} \oplus \mathcal{O}_{P^1}(1))$ and Y has at worst Gorenstein rational singular points y of type A_1, such that $p^{-1}(y)$ is a 1-dimensional, non-reduced fiber.

First of all note that, for a smooth $S \in |L|$, one as $K_S \approx p^*\mathcal{L}$ so K_S is nef and big and S is a minimal surface of *general type*. This gives a bound for the number γ of points blown up under $r: X^\wedge \to X$. Indeed the genus formula $K_S \cdot K_S = 2g(L)-2-\gamma-d^\wedge$ leads to

(3.5) $$\gamma \le 2g(L)-2-d^\wedge.$$
Furthermore, for a general fiber of p, one has $L \cdot f = -\deg K_f = 2$. Therefore the restriction $p: S \to Y$ is a generically finite covering of degree 2. Thus $K_S \cdot K_S = 2(p^*\mathcal{L})^2$ is even.

The following is a consequence of the Hodge index theorem and Lemma (0.6).

(3.6) Lemma. *With the assumptions and notation as above, we have* $g(L) \ge 11$ *for* $d^\wedge = 11$.

Proof. First, assume $g(L) \le 9$. Lemma (0.6) gives

(3.6.1) $$38-5g(L)+6\chi(\mathcal{O}_S) = K_S \cdot K_S - \gamma$$
while (4.5) yields $\gamma \le 5$.

Let $5g(L)-38$ be even. Therefore γ is even and $g(L) \le 8$. Hence $K_S \cdot K_S = 2g(L)-13-\gamma \le 3$. Then, since $K_S \cdot K_S$ is even, for each possible value $\gamma = 0, 2, 4$ we get the contradiction
$$22 \le K_S \cdot K_S\,(11+\gamma) \le (K_S \cdot L_S)^2 \le 9.$$
Let $5g(L)-38$ be odd. The only possible cases are $\gamma = 1, 3, 5$ and we have $K_S \cdot L_S = 2g(L)-13-\gamma \le 5$. The inequalities
$$2(11+\gamma) \le K_S \cdot K_S\,(11+\gamma) \le (K_S \cdot L_S)^2 \le 25$$
rule out $\gamma = 3, 5$ while for $\gamma = 1$ we find $K_S \cdot K_S = 2$, $d = 12$ and $g(L) = 9$. This contradicts (3.6.1).

If $g(L) = 10$, Lemma (0.6) yields

(3.6.2) $$\gamma+6\chi(\mathcal{O}_S) = K_S \cdot K_S + 12.$$
Hence γ is even and $\gamma \le 6$. By the Hodge index theorem one has

$$2(11+\gamma) \le K_S \cdot K_S (11+\gamma) \le (K_S \cdot L_S)^2 = (7-\gamma)^2,$$

therefore $\gamma = 0$ and $K_S \cdot K_S = 2$ or 4. Both of the cases again contradict (3.6.1), so we are done.

Q.E.D.

(3.7) Proposition. *Let* $(X^\wedge, L^\wedge)$ *be a smooth threefold embedded by* $\Gamma(L^\wedge)$ *in* $\mathbf{P}^5$, *of degree* $d^\wedge \ge 11$ *and with* $K_{X^\wedge} + 2L^\wedge$ *nef and big. Let* (X, L) *be the reduction of* $(X^\wedge, L^\wedge)$ *and suppose that* (X, L) *is a conic bundle over a normal surface* Y. *Then* g(L) = 14, d = 15, $\chi(\mathcal{O}_S) = 6$ *and* $K_S \cdot K_S = 8$.

Proof. We use over and over (3.5), the congruence (0.11) with $d_3 = 0$, Lemma (0.6) as well as the Hodge index theorem which imply $2d \le dK_S \cdot K_S \le (K_S \cdot L_S)^2$. By (3.6) and Castelnuovo's bounds (0.7) either g(L) = 20 or $11 \le g(L) \le 15$.

Let g(L) = 20. Then $(X^\wedge, L^\wedge)$ is a Castelnuovo's variety and $X^\wedge$ is linked to a $\mathbf{P}^3$ in the complete intersection of a quadric and a sextic hypersurfaces (see [H]). By looking at the resolution of the ideal sheaf $\mathfrak{I}_{X^\wedge}$ defining $X^\wedge$ in $\mathbf{P}^5$, the same argument as in the proof of (2.2) shows that $h^0(K_{X^\wedge}) > 0$, which contradicts $\kappa(X^\wedge) = \kappa(X) = -\infty$.

Let g(L) = 15. Lemma (0.6) yields

(3.7.1) $$\gamma + 6\chi(\mathcal{O}_S) = K_S \cdot K_S + 37$$

so that $\gamma = 2\tau + 1$ is odd. Since $d_1 = 17 - \gamma$ and $d_2 = 2x$, $x \ge 1$, is even the congruence (0.11) becomes $7x - 6\tau + 46 \equiv 0$ (12). Therefore x = 2y is even and

$$7y - 3\tau + 23 \equiv 0 \ (6).$$

Furthermore $K_S \cdot K_S = 4y \ge 4$, so that from

$$4(11-\gamma) \le K_S \cdot K_S(11+\gamma) \le (17-\gamma)^2$$

we find the possible values $\gamma = 1, 3, 5$ or 7.

Let $\gamma = 1$. Then $K_S \cdot K_S = 4y \le 20$, $\tau = 0$ and $7y + 23 \equiv 0$ (6). Hence y = 2z + 1 is odd and $K_S \cdot K_S = 8z + 4$ with $0 \le z \le 2$. If z = 1 or 2 we contradict $7z + 15 \equiv 0$ (3); if z = 0, $K_S \cdot K_S = 4$ which contradicts (3.7.1). Let $\gamma = 3$. Then $K_S \cdot K_S \le 14$, $\tau = 1$ and $7y + 20 \equiv 0$ (6) where $1 \le y \le 3$, a contradiction. Let $\gamma = 5$. Then $K_S \cdot K_S \le 8$, $\tau = 2$ and $7y + 17 \equiv 0$ (6) with $1 \le y \le 2$. Thus y = 1, $K_S \cdot K_S = 4$ and (3.7.1) gives $\chi(\mathcal{O}_S) = 6$. To rule out this case use the inequality $K_S \cdot K_S \ge 2p_g(S) - 4$ (see [B], p. 159). Let $\gamma = 7$. Then $K_S \cdot K_S = 4$, y = 1, $\tau = 3$ which contradicts again the congruence.

Let g(L) = 14. Lemma (0.6) gives

(3.7.2) $$\gamma + 6\chi(\mathcal{O}_S) = K_S \cdot K_S + 32$$

so that $\gamma = 2\tau$ is even. Since $d_1 = 15 - \gamma$ and $d_2 = 2x$, $x \ge 1$, is even, the congruence (0.11) becomes

$$7x - 6\tau + 44 \equiv 0 \ (12).$$

Therefore x = 2y is even and

$$7y - 6\tau + 22 \equiv 0 \ (6).$$

Furthermore $K_S \cdot K_S = 4y \ge 4$ so that from

$$4(11+\gamma) \le K_S \cdot K_S(11+\gamma) \le (15-\gamma)^2$$

we find the possible values $\gamma = 0, 2, 4$ or 6.

Let $\gamma = 0$. Then $K_S \cdot K_S \le 20$, $\tau = 0$ and $7y + 22 \equiv 0$ (6). Hence y = 2z is even and $K_S \cdot K_S$

$= 8z$ with $1 \le z \le 2$. If $z = 1$, $K_S \cdot K_S = 8$ which contradicts (3.7.2) while $z = 2$ is ruled out by $7z + 11 \equiv 0\ (3)$. Let $\gamma = 2$. Then $K_S \cdot K_S \le 12$, $\tau = 1$ and $7y + 19 \equiv 0\ (6)$ where $1 \le y \le 3$, which is not possible. Let $\gamma = 4$. Then $K_S \cdot K_S \le 8$, $\tau = 2$ and $7y + 16 \equiv 0\ (6)$ where $1 \le y \le 2$. Thus $y = 2$, which leads to the case $K_S \cdot K_S = 8$, $\chi(\mathcal{O}_S) = 6$, $d = 15$. Let $\gamma = 6$. Then $K_S \cdot K_S = 4$, $y = 1$. This contradicts again the congruence.

Now a case by case analysis which runs parallel to the above and which we omit for shortness rules out the remaining cases $g(L) = 13, 12, 11$. Q.E.D.

§ 4. Summary of results and examples.

In this section we summarize all the results we have proved and we produce a number of examples. In view of (1.1), (2.1), (2.2) and the results of § 3 the only possible invariants for *(non degenerate) threefolds* $(X\char`^, L\char`^)$ *in* $\mathbf{P}^5$, *of degree* $d\char`^ = 11$, are listed in the table below. The notation are as in the previous sections. Here $X \overset{(a,b)}{\sim} V$ means that X is linked to V inside the complete intersection of two hypersurfaces of degree a and b.

Case	g(L)	d	d_1	d_2	d_3	γ	$\chi(\mathcal{O}_S)$	$\chi(\mathcal{O}_X)$	κ(X)	liaison class (resolution of $\mathfrak{I}_X$)	Example	Comments (all but case 7) are of log-general type with $h^0(K_X+3\mathcal{K}_X) \neq 0$)
1)	13	11	13	9	3	0	6	1	−∞	$\mathcal{O}^{\oplus 5} \to \Omega(2) \oplus \mathcal{O}(1) \to \mathfrak{I}_X(5)$	(4.3.4)	$\mathcal{K}_X$ is spanned and defines a birational morphism $X \to V$, V cubic 3-fold in P^4
2)				15	9		7				does not exist	
3)	14	11	15	16	16	0	8	0 1 or 2	0 −∞	$X \overset{(4,4)}{\sim}$ quintic 3-fold in P^5	(4.3.3)	
4)	15	14	14	14	14	3	8	0			does not exist	
5)		11	17	23	29	0	10	$-2 \le$ and ≤ 3	?	$X \overset{(3,4)}{\sim} P^3$ $\chi(\mathcal{O}_X) = -1$, $\kappa(X) = 1, 2$	(4.3.2)	
6)	20	11	27	64	148	0	21	−9	3	$X \overset{(2,6)}{\sim} P^3$	(4.3.1)	$\mathcal{K}_X$-2L is spanned by 2 sections and gives a morphism $X \to P^1$ whose general fibers are quintic surfaces in P^3
7)	14	15	11	8	0	4	6	1 or 2	−∞		does not exist	

We need some further argument to complete the proof of the table above. First, let us state the following rather general results.

(4.1) Lemma. *Let* $(X\char`^, L\char`^)$ *be a smooth threefold,* $L\char`^$ *a very ample line bundle on* $X\char`^$ *and let*

(X, L) *be the reduction of* $(X^\wedge, L^\wedge)$. *Assume that* $\mathcal{K}_X$ *is nef and big and* $d_2 = d_3$. *Then* $h^0(mK_X) \le 1$ *for all* $m > 0$.

Proof. Since $\mathcal{K}_X$ is nef and big, the Kawamata-Shokurov base point free theorem says that $t\mathcal{K}_X$ is spanned by $\Gamma(t\mathcal{K}_X)$ for all $t >> 0$. Let $\mathcal{D}$ be a smooth element in $|t\mathcal{K}_X|$ and let $\mathcal{K}_\mathcal{D}$ be the restriction of $\mathcal{K}_X$ to $\mathcal{D}$. Assume $h^0(mK_X) \ge 2$ for some $m > 0$ and let D be an effective divisor in $|mK_X|$. Then $\mathcal{D}$ and D meet along an effective curve C such that C is not numerically trivial and $C \cdot C \ge 0$ since $\mathcal{D}$ moves and it is big. The assumption $d_2 = d_3$ reads $\mathcal{K}_X{}^2 \cdot K_X = 0$ and hence $C \cdot \mathcal{K}_\mathcal{D} = 0$. Therefore either $C \sim 0$ or $C \cdot C < 0$ by the Hodge index theorem, which is a contradiction. Q.E.D.

(4.2) Lemma. *Let* $(X^\wedge, L^\wedge)$ *be a smooth threefold embedded by* $\Gamma(L^\wedge)$ *in* $\mathbf{P}^5$. *Assume that* $\kappa(X^\wedge) \ge 0$ *and let* (X, L) *be the reduction of* $(X^\wedge, L^\wedge)$. *Let S be a smooth element of* $|L|$. *Then if* $\mathcal{K}_X$ *is nef and big,*

$$d_3 \ge 3(\chi(\mathcal{O}_S) - \chi(\mathcal{O}_X)) - 10.$$

Furthermore the inequality is strict if $\kappa(X^\wedge) > 0$.

Proof. Since $\kappa(X) \ge 0$ we know from [M], (8.1) that $\chi(\mathcal{O}_X) \le 0$ and hence $h^0(K_X) > 0$. Since $\mathcal{K}_X$ is nef and big, it thus follows that $\Gamma(\mathcal{K}_X)$ defines a birational map $\psi : X \to \mathbf{P}^N$.

Consider the resolution, $\sigma : V \to X$ of the fundamental locus of $\psi : X \to \mathbf{P}^N$. Then $\sigma^* K_X \approx M + \Delta$ where M is spanned by sections of $\mathcal{K}_X$ and Δ is an effective divisor. Compute

(4.2.1) $$\mathcal{K}_X{}^3 = (\sigma^*\mathcal{K}_X)^2 \cdot (M+\Delta) \ge (\sigma^*\mathcal{K}_X)^2 \cdot M \ge \sigma^*\mathcal{K}_X \cdot M^2 \ge M^3$$

Furthermore by [LS], (0.6) one has

(4.2.2) $$M^3 \ge 3(N-3) + 2,$$

with strict inequality if $\kappa(X) > 0$. From the exact standard sequence

(4.2.3) $$0 \to K_X \to K_X \otimes L \to K_S \to 0$$

we get $h^0(K_X + L) = \chi(\mathcal{O}_S) - \chi(\mathcal{O}_X)$ and hence (4.2.2) becomes

(4.2.4) $$M^3 \ge 3(\chi(\mathcal{O}_S) - \chi(\mathcal{O}_X) - 4) + 2.$$

By combining (4.2.1) and (4.2.4) we get the result. Q.E.D.

(4.3) Existence results. The following existing classes are arranged in increasing "difficulty" and in decreasing sectional genus. Let $\mathfrak{I}_{X^\wedge}$ be the ideal sheaf defining $X^\wedge$ in $\mathbf{P}^5$. We refer to (0.9) for properties of 3-folds in $\mathbf{P}^5$.

(4.3.1) *Case* 6) *of the table*. Take the resolution

$$0 \to \mathcal{O}_{\mathbf{P}^5}(-2) \to \mathcal{O}_{\mathbf{P}^5}(-1)^{\oplus 2} \to \mathfrak{I}_{\mathbf{P}^3} \to 0$$

and take a smooth 3-fold $X^\wedge$ in $\mathbf{P}^5$ which is linked to $\mathbf{P}^3$ inside the intersection of a quadric and a sextic hypersurface, $X^\wedge \overset{(2,6)}{\sim} \mathbf{P}^3$. This is possible by [PS]. The locally free resolution of $\mathfrak{I}_{X^\wedge}$ is obtained in the usual way (cfr. [O3]).

(1) $$0 \to \mathcal{O}_{\mathbf{P}^5}(-7)^{\oplus 2} \to \mathcal{O}_{\mathbf{P}^5}(-6)^{\oplus 2} \oplus \mathcal{O}_{\mathbf{P}^5}(-2) \to \mathfrak{I}_{X^\wedge} \to 0.$$

The genus formula (0.9.4) gives $g(L^\wedge) = 20$ that is $X^\wedge$ a Castelnuovo's variety (see also [H]). By the list or direct computation we get the other values of the invariants in the table. For this recall that $d_j^\wedge = \mathcal{K}_{X^\wedge}{}^j \cdot L^{\wedge 3-j} = c_{2+j}(\mathfrak{I}_{X^\wedge}(5))$ for $j = 0, 1, 2, 3$. In order to compute the Kodaira dimension we calculate from (1) the resolution

(2) $$0 \to \mathcal{O}_{P^5}(-6) \to \mathcal{O}_{P^5}{}^{\oplus 2} \oplus \mathcal{O}_{P^5}(-4) \to \mathcal{O}_{P^5}(1)^{\oplus 2} \to \omega_{X^\wedge} \to 0.$$

This shows that $\omega_{X^\wedge}$ is a quotient of $\mathcal{O}_{P^5}(1)^{\oplus 2}$, and therefore it is ample. Hence in particular $\gamma = 0$ and $\kappa(X) = 3$.

From (2) it follows that $\mathcal{K}_X - 2L$ is spanned by 2 sections, so that it defines a morphism $\psi : X \to P^1$ (see the proof of (2.2)). Let F be a general fiber of ψ and let L_F be the restriction of L to F. Then

$$L_F \cdot L_F = (\mathcal{K}_X - 2L) \cdot L \cdot L = d_1 - 2d = 5,$$

therefore F is a quintic surface. A standard computation, by using the spectral sequences associated to (2) and

(2') $$0 \to \mathcal{O}_{P^5}(-7) \to \mathcal{O}_{P^5}(-1)^{\oplus 2} \oplus \mathcal{O}_{P^5}(-5) \to \mathcal{O}_{P^5}{}^{\oplus 2} \to \omega_{X^\wedge}(-1) \cong \mathcal{O}_X(F) \to 0,$$

shows that $h^1(\omega_X) = h^2(\mathcal{O}_X) = 0$, $h^0(F) = 2$, $h^3(F) = 6$. Hence in particular $h^3(\mathcal{O}_X) = \chi(\mathcal{O}_X) + 1 = 10$. Note also that $\mathcal{O}_F(F) \cong \mathcal{O}_F$ since the image of X under ψ is P^1. Thus the exact sequence

$$0 \to \mathcal{O}_X \to F \to \mathcal{O}_F \to 0$$

gives $h^2(\mathcal{O}_F) = h^0(K_F) = 4$. Since $K_X - L \approx F$ we have $L_F \approx K_F$. Therefore $h^0(L_F) = 4$ and F is a quintic surface in P^3.

(4.3.2) *Case* 5) *of the table*. From the exact sequence (4.2.3) we find

$$2h^0(K_X + L) - \chi(\mathcal{O}_S) = \chi(\mathcal{O}_S) - 2\chi(\mathcal{O}_X).$$

Furthermore by [S6], (1.2) one has

$$32(2h^0(K_X + L) - \chi(\mathcal{O}_S)) \geq d_3 + 8d_1/3.$$

Therefore, by combining the two relations above, we get

$$\chi(\mathcal{O}_S) - 2\chi(\mathcal{O}_X) \geq d_3 / 32 + d_1/12.$$

which leads to $2\chi(\mathcal{O}_X) \leq 7$, that is $\chi(\mathcal{O}_X) \leq 3$. Note that $\chi(\mathcal{O}_X) \geq 1$ if $\kappa(X) = -\infty$. If $\kappa(X) \geq 0$ the lower bound $\chi(\mathcal{O}_X) \geq -2$ follows from Lemma (4.2).

Here one example is obtained by linking to P^3 inside a complete intersection of a cubic and a quartic hypersurface, $X^\wedge \overset{(3,4)}{\sim} P^3$. This gives a resolution

(3) $$0 \to \mathcal{O}_{P^5}(-6) \to \mathcal{O}_{P^5}(-1) \oplus \mathcal{O}_{P^5}(-2) \oplus \mathcal{O}_{P^5}(-3) \to \mathcal{O}_{P^5}{}^{\oplus 2} \to \omega_{X^\wedge} \to 0.$$

From (2) we compute $g(L^\wedge) = 15$, again using (0.9.1), and all the pluridegrees. Also we obtain $\chi(\mathcal{O}_{S^\wedge}) = 10$ and $\chi(\mathcal{O}_{X^\wedge}) = -1$. From (3) we get that $K_{X^\wedge}$ is spanned, hence $X^\wedge \cong X$ and $h^0(K_X) = 2$. Therefore $\kappa(X) \geq 1$ and from $K_X^3 = d_3 + 3d_1 - 3d_2 - d = 0$ we infer $\kappa(X) \leq 2$.

(4.3.3) *Case* 3) *of the table*. From Lemma (0.10.1) we get $\chi(\mathcal{O}_X) \leq 2$. Since $d_2 = d_3 = 16$, Lemma (4.1) applies to give $\chi(\mathcal{O}_X) \geq 0$ and $\kappa(X) \leq 0$. From [M], (8.1) it thus follows that $\chi(\mathcal{O}_X) = 0$ if $\kappa(X) = 0$.

To find an example, note that there is a class of 3-folds V of degree 5 in P^5 having a resolution (cf. [O1])

$$0 \to \mathcal{O}_{P^5}(-4)^{\oplus 2} \to \mathcal{O}_{P^5}(-3)^{\oplus 2} \oplus \mathcal{O}_{P^5}(-2) \to \mathfrak{I}_V \to 0.$$

Take $X^\wedge$ linked to such a V inside a complete intersection of two quartic hypersurfaces, $X^\wedge \overset{(4,4)}{\sim} V$. We get a resolution

(4) $$0 \to \mathcal{O}_{P^5}(-6) \oplus \mathcal{O}_{P^5}(-5)^{\oplus 2} \to \mathcal{O}_{P^5}(-4)^{\oplus 4} \to \mathfrak{I}_{X^\wedge} \to 0$$

leading to a resolution

(5) $$0 \to \mathcal{O}_{P^5}(-6) \to \mathcal{O}_{P^5}(-2)^{\oplus 4} \to \mathcal{O}_{P^5}(-1)^{\oplus 2} \oplus \mathcal{O}_{P^5} \to \omega_{X^\wedge} \to 0.$$

From (5) we have $h^0(K_{X^\wedge}) = 1$ and therefore $\kappa(X^\wedge) \geq 0$. From (0.9.1) we find $g(L^\wedge) = 14$ and hence the invariants are as in the table (or compute by using (4)). In particular $\gamma = 0$, so that $X^\wedge \cong X$. For $\chi(\mathcal{O}_X)$ we obtain $\chi(\mathcal{O}_X) = 0$. We compute $\mathcal{K}_X{}^2 \cdot K_X = d_3 - d_2 = 0$. Since $\mathcal{K}_X{}^3 = d_3 > 0$ we conclude $\kappa(X) = 0$ (use [C], Prop. 3.1).

(4.3.4) *Case* 1) *of the table*. Lemma (0.10.1) gives $\chi(\mathcal{O}_X) = 1$. Furthermore $\kappa(X) = -\infty$ since $d_3 < d_2$ (see [S3], § 1). Here we simply refer to Chang's paper [C] (see in particular (3.8)), also for the computation of further numerical invariants. There is a resolution

$$0 \to \mathcal{O}_{P^5}{}^{\oplus 5} \to \Omega_{P^5}(2) \oplus \mathcal{O}_{P^5}(1) \to \mathfrak{I}_X(5) \to 0$$

which leads to the resolution

$$0 \to \mathcal{O}_{P^5} \to \Omega_{P^5}{}^{\vee}(3) \oplus \mathcal{O}_{P^5}(4) \to \mathcal{O}_{P^5}(5)^{\oplus 5} \to \omega_X(6) \to 0.$$

In thus follows that $K_X + L$ is spanned by 5 sections. Therefore, since $d_3 = 3$, $\Gamma(\mathcal{K}_X\ )$ defines a birational morphism $X \to V \subset P^4$, V cubic threefold in P^4. We should recall that X is unirational but not rational.

(4.4) Non-existence results. It remains to prove that the cases 2), 4), and 7) of the table do not occur.

(4.4.1) Theorem. *Let* $X^\wedge$ *be a smooth* n*-fold embedded by* $\Gamma(L^\wedge)$ *in* P^N *with* $2n > N$. *Let* (X, L) *and* (X', L') *be the first and second reduction respectively.*

i) *If* $d^\wedge \neq 2^n - 1$ *then* $X^\wedge \cong X$;

ii) *if further* $d^\wedge \neq 3^n - 1$, *then* X' *is Gorenstein.*

Proof. i) Assume otherwise. Then there exists a linear $P^{n-1} := E$ in $X^\wedge$ with $L^\wedge - E$ spanned by global sections. The morphism associated to $|L^\wedge - E|$ is the projection, say π, of $X^\wedge$ from E onto a P^{N-n}. Since the image is P^{N-n} and $N < 2n$ we get $(L^\wedge - E)^n = 0$ and hence $L^{\wedge n} = 2^n - 1$.

ii) We have $X^\wedge \cong X$ by i). If X' is not Gorenstein, then there exists a $P^{n-1} := E$ in $X^\wedge$ with normal bundle $\mathcal{N}_E^{X^\wedge} \cong \mathcal{O}_E(-2)$ (see [FS]). By looking again at the projection of $X^\wedge$ from E onto P^{N-n} the same argument as above gives the result. Q.E.D.

The assertion i) of the Theorem rules out cases 4) and 7) of the table.

The following argument, suggested to us by K. Ranestad, rules out the remaining case 2) of the table.

(4.4.2) Lemma (K. Ranestad). *The case* 2) *of the table does not occur.*

Proof. Let X ($\cong X^\wedge$), L, S, $C = L \cdot L$ as in case 2) of the table. Consider the short exact sequence

$$0 \to K_S \otimes L_S^{-1} \to K_S \to K_{S|C} \to 0.$$

Since $h^1(K_S) = 0$, $h^2(K_S) = 1$, $h^2(K_S - L_S) = h^0(L_S) = 5$ and $h^2(K_{S|C}) = 0$ we infer that $h^1(K_{S|C}) = 4$. Then the Riemann-Roch theorem on the curve C gives

$$h^0(K_{S|C}) = K_S \cdot L_S - g(L) + 5 = 5.$$

Since $h^0(K_S) = 6$ it follows that $h^0(K_S - L_S) \geq 1$. Let D be an effective curve in $|K_S - L_S|$. We have $D \cdot D = d_2 - 2d_1 + d = 0$; $K_S \cdot D = d_2 - d_1 = 2$; $L_S \cdot D = d_1 - d = 2$. Hence in particular D is a conic and if D is either reduced or is a double line we contradict the genus formula, so we are done. Q.E.D.

(4.5). Let us recall the following bound for the sectional genus, even if it is not the best possible. This is an easy consequence of the results of this paper. Note that no similar bound is known for surfaces in $\mathbf{P}^4$. Indeed there is a surface in $\mathbf{P}^4$ of degree 8 and sectional genus 5 which does not satisfy such a bound.

(4.5.1) Theorem. Let $X^\wedge$ *be a smooth threefold embedded by* $\Gamma(L^\wedge)$ *in* $\mathbf{P}^5$ *and of degree* $d^\wedge \geq 8$. *Then* $g(L^\wedge) > (d^\wedge + 2)/2$.

Proof. If $d^\wedge \leq 10$ see the enclosed page appendix. So $d^\wedge \geq 11$ and let ϕ be the adjunction map (see (0.5.2)). If $\dim \phi(X^\wedge) \leq 2$ use Theorem (1.1). Thus we can assume that $\dim \phi(X^\wedge) = 3$. The genus formula gives

$$2g(L^\wedge) - 2 = d + d_1 \geq d^\wedge + d_1.$$

Then the result follows as soon as $d_1 > 0$. If (X,L) is of log-general type this is clear. Otherwise we know from (3.3) that (X,L) is either a Del Pezzo fibration $p : X \to B$ over a smooth curve B or a conic bundle $p : X \to Y$ over a surface Y. In both cases we have $K_X + L \approx p^*\mathcal{L}$ for some ample line bundle $\mathcal{L}$. Hence $d_1 = (K_X + L) \cdot L^2 > 0$ and we are done. Q.E.D.

Let us point out the following concluding fact.

(4.6) Remark. The degree $d^\wedge = 11$ seems to be the last doable case. To make this clear we note that the same Pascal program used in § 2 gives a very long list of 32 possible cases for 3-*folds* $(X^\wedge, L^\wedge)$ *in* $\mathbf{P}^5$ *of log-general type, of degree* $d^\wedge = 12$ *and sectional genus* $g(L^\wedge) \leq 19$ (i.e. not of maximal genus with respect to the Castelnuovo's bound). We summarize this list in the following table.

g(L)	numbers of possible cases according to the values of $d = 12 + \gamma$, d_1, d_2, d_3, $\chi(\mathcal{O}_X)$.
15	2
16	6
17	3
18	9
19	12

APPENDIX

It should be worth to report the list of all *smooth threefolds in* $\mathbf{P}^5$ *of degree* $d \le 10$ (up to complete intersections) which comes out from [I1], [I2], [I3], [O1], [O2] for $d \le 8$ and [BSS] for $d = 9, 10$.

d	Smooth threefolds (X, L) embedded by $\Gamma(L)$ in $\mathbf{P}^5$
3	$X \cong \mathbf{P}(\mathcal{O}_{\mathbf{P}^1}(1)^{\oplus 3})$, the Segre embedding of $\mathbf{P}^1 \times \mathbf{P}^2$; L the tautological bundle, $g(L) = 0$.
5	X quadric fibration over $\mathbf{P}^1$ (any fibre F is embedded by L_F as an irreducible quadric) with 8 singular fibres and Betti numbers $b_1 = 0$, $b_2 = 2$, $b_3 = 6$, $g(L) = 2$.
6	$X = \mathbf{P}(\mathcal{E})$, $\mathcal{E}$ rank-2 locally free sheaf on $\mathbf{P}^2$ given by $0 \to \mathcal{O}_{\mathbf{P}^2} \to \mathcal{E} \to \mathfrak{I}_Z(4) \to 0$, $Z = \{x_1, \ldots, x_{10}\}$; L the tautological line bundle, $g(L) = 3$.
	$X = \mathbf{P}(\mathcal{E})$, $\mathcal{E}$ rank-2 locally free sheaf on $\mathbf{P}^2(x_1, \ldots, x_6)$, the blowing up of $\mathbf{P}^2$ at 6 points ; L the tautological line bundle, $g(L) = 4$.
7	$X = S_{(2,2,2)}(x_0)$, the blowing up at a point x_0 of a complete intersection S of type (2,2,2) in $\mathbf{P}^6$; L the proper transform of $\mathcal{O}_S(1)$, $g(L) = 5$.
	$\Gamma(K_X + L)$ gives a morphism $X \to \mathbf{P}^1$ with $\mathbf{P}^2(x_1, \ldots, x_6)$ as a general fibre, $g(L) = 6$.
8	$\Gamma(K_X + L)$ gives a morphism $X \to \mathbf{P}^1$ whose general fibres are complete intersections of type (2,2), $g(L) = 7$.
	$X \overset{(2,5)}{\sim} \mathbf{P}^3$, $K_X + L$ nef and big (i.e. (X,L) of log-general type), $g(L) = 12$.
9	$X \overset{(3,4)}{\sim}$ cubic scroll, (X, L) conic bundle over $\mathbf{P}^2$, $g(L) = 9$.
	(X, L) $\mathbf{P}^1$ bundle over a minimal K3 surface, $K_S^2 = -5$ for a smooth $S \in \lvert L \rvert$, $g(L) = 8$.
10	$X \overset{(4,4)}{\sim}$ Bordiga 3-fold, (X, L) of log-general type; $\Gamma(K_X + L)$ gives a birational morphism $X \to \mathbf{P}^3$, $g(L) = 11$.
	$X \overset{(3,4)}{\sim}$ quadric, (X, L) of log-general type, $g(L) = 12$.

REFERENCES

[B] A.Beauville, *Surfaces Algébriques complexes*, Astérisque **54**, (1978).

[BBS] M.Beltrametti, A.Biancofiore, A.J.Sommese, *Projective* N-*folds of log-general type*, I, Transaction of the A.M.S., vol. **314**, n. 2, (1989), 825-849.

[BS] M.Beltrametti, A.J.Sommese, *New properties of special varieties arising from the adjunction theory*, J. Math. Soc. Japan, 43 (1991), 381–412.

[BFS] M.Beltrametti, M.L.Fania, A.J.Sommese, *Projective classification of varieties via adjunction theory*, Math. Ann., **290** (1991), 31–62.

[BSS] M.Beltrametti, M.Schneider, A.J.Sommese, *Threefolds of degree* 9 *and* 10 *in* $\mathbf{P}^5$, Math. Ann., **288** (1990), 413–444.

[C] M.C.Chang, *Classification of Buchsbaum subvarieties of codimension* 2 *in projective space*, Jour für die reine und angew. Math. **401** (1989), 101–112.

[F] M.L.Fania, *Configurations of* −2 *Rational Curves on Sectional Surfaces of* n-*Folds*, Math. Ann., **275** (1986), 317-325.

[FS] M.L.Fania, A.J.Sommese, *On the projective classification of smooth* n-*folds with* n *even*, Arkiv för Mathematik, **27** (1989), 245–256.

[GP] L.Gruson, C.Peskine, *Genre des courbes de l'espace projectif*, Algebraic Geometry, Proceedings Tromsö, Norway 1977, Lecture Notes in Math. **687**, Springer-Verlag (1978).

[H] J.Harris, *A bound on the geometric genus of projective varieties*, Annali Scuola Normale Superiore, Pisa, **8**, (1981), 35-68.

[Ha] R.Hartshorne, *Algebraic Geometry*, G.T.M. **52**, Springer-Verlag (1977).

[I1] P.Ionescu, *Embedded projective varieties of small invariants*, Proceedings of the Week of Algebraic Geometry, Bucharest 1982, Lecture Notes in Math., **1056**, Springer-Verlag (1984), 142-187.

[I2] P.Ionescu, *Embedded projective varieties of small invariants*, II, Rev. Roumaine Math. Pures Appl. **31**, (1986), 539-544.

[I3] P.Ionescu, *Embedded projective varieties of small invariants*, III, Algebraic Geometry, Proceedings L'Aquila 1988, Lecture Notes in Math., **1417**, Springer-Verlag (1990) 138-154.

[IS] V.A.Iskovskih, V.V.Shokurov, *Biregular theory of Fano* 3-*folds*, Proceedings Algebraic Geometry, Copenhagen, 1978, Lecture Notes in Math., **732**, Springer-Verlag (1984), 171-182.

[LS] E.L.Livorni, A.J.Sommese, *Threefolds of Non Negative Kodaira Dimension with Sectional Genus Less than or Equal to* 15, Annali Scuola Normale Superiore, Pisa, Serie IV, **13**, n. 4 (1986), 537-558.

[M] Y.Miyaoka, *The Chern Classes and Kodaira Dimension of a Minimal Variety*, Advanced Studies in Pure Math., **10** (1987), Algebraic Geometry, Sendai 1985, 449-476.

[MM] S.Mori, S.Mukai, *Classification of Fano* 3-*folds with* $B_2 \geq 2$, manuscripta math. **36**, (1981), 147-162.

[Mu] S.Mukai, *New classification of Fano threefolds and Fano manifolds of coindex* 3, preprint (1988).

[O1] C.Okonek, 3-*Mannigfaltigkeiten im* $\mathbf{P}^5$ *und ihre zugehörigen stabilen Garben*, manuscripta math. **38**, (1982), 175-199.

[O2] C.Okonek, *Uber 2-codimensionale Untermannigfaltigkeiten vom Grad* 7 *in* $\mathbf{P}^4$ *und* $\mathbf{P}^5$, Math. Z. **187**, 209-219 (1984).

[O3] C.Okonek, *On codimension -2 submanifolds in* $\mathbf{P}^4$ *and* $\mathbf{P}^5$, Mathematica Göttingensis, Schriftenreihe des Sonderforschungsbereich, Geometrie und Analysis, Heft 50 (1986).

[PS] C.Peskine, L.Szpiro, *Liaison des variéties algébriques*, I, Invent. Math. 26, (1974), 271-302.

[S1] A.J.Sommese, *Hyperplane sections of projective surfaces*, I, *The adjunction mapping*, Duke math. J., 46 (1979), 377-401.

[S2] A.J.Sommese, *Hyperplane sections*, Proceedings of the Algebraic Geometry Conference, University of Illinois at Chicago Circle, 1980, Lecture Notes in Math., 862, Springer-Verlag (1981), 232-271.

[S3] A.J.Sommese, *On the minimality of hyperplane sections of projective threefolds*, Jour. für die reine und angew. Math. 329, (1981), 16-41.

[S4] A.J.Sommese, *Configurations of −2 rational curves on hyperplane sections of projective threefolds*, Classification of Algebraic and Analytic Manifolds (ed. K.Ueno), Progress in Mathematics 39, (1983), 465-497, Birkhäuser.

[S5] A.J.Sommese, *On the Adjunction Theoretic Structure of Projective Varieties*, Complex Analysis and Algebraic Geometry, Proceedings Göttingen 1985, Lecture Notes in Math., 1194, (1986), 175-213.

[S6] A.J.Sommese, *On the Nonemptiness of the Adjoint Linear System of a Hyperplane Section of a Threefold*, Jour. für die reine und angew. Math. 392, (1989), 211–220; Erratum, Jour. für die reine und angew. Math. 411, (1990), 122–123.

[SV] A.J.Sommese, A.Van de Ven, *On the adjunction mapping*, Math. Ann., 278, (1987), 593-603.

[V] A.Van de Ven, *On the 2-connectedness of very ample divisors on a surface*, Duke Math. J., 46, (1979), 403-407.

Mauro Beltrametti
Dipartimento di Matematica
Via L.B.Alberti 4, I-16132 GENOVA (ITALY)

Michael Schneider
Lehrstuhl VIII für Mathematik
Universität Bayreuth, D-8580 BAYREUTH (GERMANY)

Andrew J.Sommese
Department of Mathematics
University of Notre Dame, NOTRE DAME, Indiana 46556 (U.S.A.)
SOMMESE@IRISHMVS.bitnet

Complete Extensions and their Map to Moduli Space

Aaron Bertram
Harvard University
Cambridge, MA 02138

1 Introduction

Let C be a smooth, irreducible, projective curve of genus g defined over an algebraically closed field. Let ω_C be the canonical line bundle of one forms, and L a line bundle of degree $d > 2g-2$. Recall that a vector bundle E of rank 2 on C is stable (resp. semi-stable) if $\deg(M) > 1/2(\deg(E))$ (resp. $\deg(M) \geq 1/2(\deg(E))$) for all quotient line bundles M of E. (See [S] for more details and the construction of the moduli space of semi-stable vector bundles on C.)

We let $\mathcal{SU}(2, \omega_C \otimes L)$ be the moduli space of semi-stable rank 2 vector bundles with determinant isomorphic to $\omega_C \otimes L$. If we let $\mathbf{P}_L := \mathbf{P}(\mathrm{Ext}^1(L, \omega_C)^*)$ parametrize the space of non-split short exact sequences

$$\varepsilon : 0 \to \omega_C \to E_\varepsilon \to L \to 0$$

(modulo isomorphism and scalars), then each extension class $\varepsilon \in \mathbf{P}_L$ determines a well-defined rank 2 vector bundle E_ε, and we let $U_L \subset \mathbf{P}_L$ consist of those extension classes ε for which E_ε is semi-stable.

One can alternatively think of U_L in the following sense. There is a canonical rank 2 vector bundle $\mathcal{E}^0_{\mathbf{P}_L}$ on $C \times \mathbf{P}_L$, whose restriction, $\mathcal{E}^0_{U_L}$, to $C \times U_L$ determines a morphism

$$\phi : U_L \to \mathcal{SU}(2, \omega_C \otimes L)$$

For each $E \in \mathcal{SU}(2, \omega_C \otimes L)$, the set-theoretic fibre $\phi^{-1}(E)$ is just the set of injective bundle maps $\omega_C \hookrightarrow E$. (A map $\omega_C \to E$ is an injective bundle map if it is injective on fibres.) Thus, set-theoretically,

$$U_L = \{(E, f) | E \in \mathcal{SU}(2, \omega_C \otimes L), f : \omega_C \hookrightarrow E(\text{mod scalars})\}$$

The object of this paper is to describe a particular compactification of U_L to a projective variety $\widetilde{P}_L$, and to exhibit a vector bundle $\mathcal{E}^{k(L)}_{\widetilde{P}_L}$ on $C \times \widetilde{P}_L$ that extends $\mathcal{E}^0_{U_L}$ in such a way that $\mathcal{E}^{k(L)}_{\widetilde{P}_L}$ determines a morphism

$$\phi_{k(L)} : \widetilde{P}_L \to \mathcal{SU}(2, \omega_C \otimes L)$$

In particular, a point $\varepsilon \in \widetilde{P}_L$ determines an injective map *of sheaves* $\omega_C \to E_\varepsilon$, and the morphism $\phi_{k(L)}$ will send ε to E_ε. It should be emphasized that unlike the description of U_L, however, the pair $(E, f : \omega_C \to E)$ will not be enough in general to determine a unique point of $\widetilde{P}_L$.

In §2 , we construct $\widetilde{P}_L$ by blowing up $\mathbf{P}_L$ along a series of naturally occurring smooth centers. Since the construction of $\widetilde{P}_L$ closely resembles the classical theory of complete quadrics, we include a description of that theory. Theorem 2.4, the "complete extension" theorem, lists the properties of $\widetilde{P}_L$ that make it analogous to the space of complete quadrics. We prove theorem 2.4 in the first few special cases, and outline the proof in general. (The interested reader may refer to [B1] or [B2] for the full details.)

In §3, we define the canonical vector bundle $\mathcal{E}^0_{\mathbf{P}_L}$ on $C \times \mathbf{P}_L$ and create $\mathcal{E}^{k(L)}_{\widetilde{P}_L}$ by modifying $\mathcal{E}^0_{\mathbf{P}_L}$ along the exceptional divisors in the blow-up. Theorem 3.1 lists the properties of $\mathcal{E}^{k(L)}_{\widetilde{P}_L}$, which in particular imply the existence of the map to moduli space. We start the section with a key lemma, which is really a local version of the theorem, then we prove the theorem in the first few special cases. Again, because the notation becomes unwieldy and because there are really no new ideas involved, we only outline the proof in general. (See [B2] for all the details.)

I'd like to thank Mark Green, David Gieseker, and especially Robert Lazarsfeld for their helpful comments while I worked on my PhD thesis, which led to the results presented here. I'd also like to thank the Sloan foundation for its financial support at that time.

2 Construction of $\widetilde{P}_L$ and analogy with complete quadrics

We remark first of all that

$$\mathbf{P}_L = \mathbf{P}(\mathrm{Ext}^1(L, \omega_C)^*) = \mathbf{P}(H^1(C, L^* \otimes \omega_C)^*) = \mathbf{P}(H^0(C, L))$$

Thus, in particular, there is the usual "linear series" map : $\phi_L : C \to \mathbf{P}_L = \mathbf{P}(H^0(C, L))$ and we can describe U_L in terms of $\phi_L(C)$. From now on, let $k(L) = [1/2(d - 2g + 1)]$ (where "[ ]" denotes "integer part of"). Then we have the following:

Lemma 2.1: $\varepsilon \in \mathbf{P}_L - U_L$ if and only if $\exists D \subset C$, D an effective divisor of degree at most $k(L)$, so that $\varepsilon \in \overline{D}$ (where $\overline{D}$ denotes the linear span of D).

Proof: From the description in §1 of a semi-stable vector bundle, $\varepsilon \in \mathbf{P}_L - U_L$ if and only if there is a quotient line bundle M of E_ε with $\deg(M) \leq 2g - 2 + k(L)$. It follows immediately that the induced map f below:

$$\begin{array}{ccccccccc} \varepsilon: & 0 & \to & \omega_C & \to & E_\varepsilon & \to & L & \to 0 \\ & & & & \underset{f}{\searrow} & \downarrow & & & \\ & & & & & M & & & \end{array}$$

is nonzero, hence $M \cong \omega_C(D)$, D effective, $\deg(D) \leq k(L)$. But this holds if and only if $\varepsilon \in \ker(\psi_D)$, where $\psi_D : \mathrm{Ext}^1(L, \omega_C) \to \mathrm{Ext}^1(L, \omega_C(D))$ or dually, $\varepsilon \in \ker(H^0(C, L)^* \to H^0(C, L(-D))^*)$, which is precisely to say that $\varepsilon \in \overline{D}$.

We immediately get:

Corollary 2.2: If $k(L) = 0$, then $U_L = \mathbf{P}_L$. If $k(L) = 1$, then $U_L = \mathbf{P}_L - \phi_L(C)$.

Remark: If $k(L) \geq 1$, then by Riemann-Roch, ϕ_L is an embedding, and we will not distinguish between C and its image $\phi_L(C)$.

Let $C_k := \mathrm{Sym}^k(C)$, the kth symmetric product of C. Let $\mathcal{D}_k = C \times C_{k-1} \subset C \times C_k$ be the "universal" divisor, embedded via $(p, D) \mapsto (p, D + p)$, and let $\pi_C : C \times C_k \to C$, $\pi_{C_k} : C \times C_k \to C_k$ be the projections. Then following [Sc], we make the following:

Definition: $B^k(L) := \mathbf{P}(\pi_{C_{k+1}}{}_*(\pi_C^* L \otimes \mathcal{O}_{\mathcal{D}_{k+1}}))$ is the secant bundle over C_{k+1}, and $\beta_k : B^k(L) \to \mathbf{P}_L$ is the rational map determined by the push-down to C_{k+1} of the restriction map: $\pi_C^* L \to \pi_C^* L \otimes \mathcal{O}_{\mathcal{D}_{k+1}}$.

As soon as $\deg(L) \geq 2g - 2 + k$, the pushed down map is surjective, and β_k is a morphism. In that case, we define $\mathrm{Sec}^k(C) := \beta_k(B^k(L))$, the secant variety spanned by $\mathbf{P}^k$'s through $k+1$ points of C. As one can readily check, $B^0(L) = C$, $\beta_0 = \phi_L$, and $\mathrm{Sec}^1(C)$ is the usual variety of secant lines.

In terms of these definitions, we can rephrase lemma 2.1:

Lemma 2.1 (rephrased): $\mathbf{P}_L - U_L = \mathrm{Sec}^{k(L)-1}(C)$.

In particular, of course, U_L is open.

We break now to discuss the classical theory of complete quadrics.

Let V be a vector space over k of dimension n, and $\mathbf{P}_{V,2} := \mathbf{P}(\mathrm{Sym}^2(V))$ be the projective space of symmetric bilinear forms on V. Then the space $\mathbf{P}_{V,2}$ is naturally stratified as $W_1 \subset W_2 \subset \ldots \subset W_n = \mathbf{P}_{V,2}$, where

$W_i = \{$ bilinear forms of rank $\leq i\}$.

W_1 is just the image of the Veronese embedding $\mathbf{P}(V) \hookrightarrow \mathbf{P}_{V,2}$. More generally, if $w_1, \ldots, w_i \in W_1$ are distinct, then $w \in < w_1, \ldots, w_i >$, (the linear span of $w_1, \ldots, w_i$) implies that $w \in W_i$. (This is just simple linear algebra. A sum of i matrices of rank one has rank at most i). Furthermore, $W_i = \bigcup < w_1, \ldots, w_i >$ where the union is taken over all distinct collections of w_j's. Thus, if we define $\mathrm{Sec}^k(\mathbf{P}(V)) = \overline{\bigcup < w_1, \ldots, w_k >}$, then $\mathrm{Sec}^k(\mathbf{P}(V)) = W_k$ since W_k is already closed, and we have:

Lemma 2.2 (quadric version): Let $U_{V,2} \subset \mathbf{P}_{V,2}$ be the set of non-degenerate bilinear forms. Then $\mathbf{P}_{V,2} - U_{V,2} = \mathrm{Sec}^{n-1}(\mathbf{P}(V))$.

The theory of complete quadrics gives a compactification of $U_{V,2}$ to $\tilde{P}_{V,2}$, the space of complete quadrics, and a way of constructing $\tilde{P}_{V,2}$ from $\mathbf{P}_{V,2}$. Namely:

Definition: Let $bl_1(\mathbf{P}_{V,2})$ denote the blow-up of $\mathbf{P}_{V,2}$ along W_1. For $k < n$, let $bl_k(\mathbf{P}_{V,2})$ denote the blow-up of $bl_{k-1}(\mathbf{P}_{V,2})$ along the proper transform of W_k.

Theorem 2.3 (complete quadrics): Let $\tilde{P}_{V,2} := bl_{n-1}(\mathbf{P}_{V,2})$. Then

(1) $\tilde{P}_{V,2}$ is smooth, and under the natural inclusion $U_{V,2} \hookrightarrow \tilde{P}_{V,2}$, $\tilde{P}_{V,2} - U_{V,2}$ is a divisor consisting of $n-1$ smooth components $D_1, \ldots, D_{n-1}$ with normal crossings.

(2) Set-theoretically, $\tilde{P}_{V,2}$ consists of the set of sequences $\{Q_1, \ldots, Q_k\}$ with the property that:

(a) $Q_1 \in \mathbf{P}_{V,2}$

(b) $Q_j \in \mathbf{P}(\mathrm{Sym}^2(\ker(Q_{j-1})))$ for $j > 1$, and

(c) Q_k is nondegenerate.

Furthermore, $\{Q_1, \ldots, Q_k\} \in D_i$ if and only if $\dim(\ker(Q_j)) = n - i$ for some j.

For proofs of these, see e.g. [V].

The discussion of complete quadrics was meant to motivate the following definition and theorem:

Definition: $bl_1(\mathbf{P}_L)$ is the blow-up of $\mathbf{P}_L$ along C, and if $k \leq k(L)$, then $bl_k(\mathbf{P}_L)$ is the blow-up of $bl_{k-1}(\mathbf{P}_L)$ along the proper transform of $\mathrm{Sec}^{k-1}(C)$.

Theorem 2.4 (complete extensions): Let $\widetilde{P}_L = bl_{k(L)}(\mathbf{P}_L)$. Then
(1) $\widetilde{P}_L$ is smooth, and under the natural inclusion $U_L \hookrightarrow \widetilde{P}_L$, $\widetilde{P}_L - U_L$ is a divisor consisting of $k(L)$ smooth components $D_1, ..., D_{k(L)}$ with normal crossings.
(2) Set-theoretically, $\widetilde{P}_L$ consists of the set of sequences $\{\epsilon_1, ..., \epsilon_k\}$ of extension classes with the following properties:

(a) $\epsilon_1 \in \mathbf{P}_L$

(b) If $\epsilon_{j-1} \in \mathbf{P}_{L(-2A)}$ and $E_{\epsilon_{j-1}}$ is not semi-stable, let $A_j \subset C$ be the divisor of minimal degree such that $\epsilon_{j-1} \in \overline{A_j}$ (recall lemma 2.1). Then $\epsilon_j \in \mathbf{P}_{L(-2A-2A_j)}$.

(c) E_{ϵ_k} is semi-stable.

Furthermore, $\{\epsilon_1, ..., \epsilon_k\} \in D_i$ if and only if $\sum_{r=1}^{j} \deg(A_r) = i$ for some j.

Let's look at the first few cases of the theorem:

k(L) = 0: (i.e. $\deg(L) = 2g - 1$ or $2g$).
Then $U_L = \mathbf{P}_L$, and there is nothing to prove.

k(L) = 1: (i.e. $\deg(L) = 2g + 1$ or $2g + 2$).
Then $U_L = \mathbf{P}_L - \phi_L(C)$. Since ϕ_L is an embedding, the exceptional divisor $D_1 \subset bl_1(\mathbf{P}_L)$ is smooth, and we have (1).

Since $D_1 = \mathbf{P}(N^*_{C/\mathbf{P}_L})$, ($N^*_{C/\mathbf{P}_L}$ is the conormal bundle of C in $\mathbf{P}_L$), (2) follows from the fact that over each point $p \in C$, the fibre of $\mathbf{P}(N^*_{C/\mathbf{P}_L})$ is isomorphic to $\mathbf{P}(H^0(C, L(-2p))$.

k(L) = 2: (i.e. $\deg(L) = 2g + 3$ or $2g + 4$).
$U_L = \mathbf{P}_L - \mathrm{Sec}^1(C)$. Consider the map $\beta_1 : B^1(L) \to \mathbf{P}_L$. (1) follows from the two facts:
(a) $\beta_1^{-1}(\phi_L(C)) \subset B^1(L)$ is a smooth divisor, isomorphic to $C \times C$.
(b) When we lift (which is possible by (a)):

$$\begin{array}{ccc} & & bl_1(\mathbf{P}_L) \\ & \tilde{\beta}_1 \nearrow & \downarrow \\ B^1(L) & \rightarrow & \mathbf{P}_L \end{array}$$

the induced map $\widetilde{\beta}_1$ is an embedding.

Let's grant (a) and (b) for the moment. Then $\widetilde{P}_L = bl_2(\mathbf{P}_L)$ is smooth, since it is the blow-up of $\mathbf{P}_L$ along two smooth centers. Furthermore, $\widetilde{P}_L - U_L$ consists of the two divisors:
$D_2 =$ the exceptional divisor in the blow-up of $bl_1(\mathbf{P}_L)$ along $B^1(L)$,

D_1 = the blow-up of the exceptional divisor $D \subset bl_1(\mathbf{P}_L)$ along $\tilde{\beta}_1(B^1(L)) \cap D \cong C \times C$. Finally,

$$\begin{aligned} D_1 \cap D_2 &= \text{the exceptional divisor in } D_1 \\ &= \text{the preimage in } D_2 \text{ of } C \times C \subset B^1(L). \end{aligned}$$

Proof of (a): The map

$$\begin{aligned} \alpha: \ C \times C &\rightarrow B^1(L) \\ (p,q) &\mapsto \mathbf{P}(H^0(C, L_p)) \in \mathbf{P}(H^0(C, L_{p+q})) \end{aligned}$$

is an embedding. Furthermore, one readily checks that $\alpha(C \times C) = \beta_1^{-1}(\phi_L(C))$ as sets. To conclude that this is an equality of schemes, we have to show that

$$d\beta_1 : \beta_1^* N^*_{C/\mathbf{P}_L} \rightarrow N^*_{C \times C/B^1(L)}$$

is surjective. But at a point $(p,q) \in C \times C$, there are isomorphisms $\beta_1^* N^*_{C/\mathbf{P}_L}(p,q) \cong H^0(C, L(-2p))$, $N^*_{C \times C/B^1(L)}(p,q) \cong H^0(C, L(-2p) \otimes \mathcal{O}_q)$, and $d\beta_1$ is the restriction mapping, so since $k(L(-2p)) \geq 1$, this is surjective.

Proof of (b): Away from $\alpha(C \times C)$, the map $\beta_1 : B^1(L) \rightarrow \mathbf{P}_L$ is an embedding. For each fixed $p \in C$, $\tilde{\beta}_1$ embeds $p \times C$ in the fibre, $\mathbf{P}_{L(-2p)}$, of D over p. So $\tilde{\beta}_1$ is set-theoretically injective. From the commuting diagram:

$$\begin{array}{ccccccccc} 0 & \rightarrow & T_{C \times C}(p,q) & \rightarrow & T_{B^1(L)}(p,q) & \rightarrow & N_{C \times C/B^1(L)}(p,q) & \rightarrow & 0 \\ & & \downarrow & & \downarrow \tilde{\beta}_{1*} & & \downarrow \wr & & \\ 0 & \rightarrow & T_D(p,q) & \rightarrow & T(bl_1(\mathbf{P}_L))(p,q) & \rightarrow & N_{D/bl_1(\mathbf{P}_L)}(p,q) & \rightarrow & 0 \end{array}$$

and the fact that γ is injective (γ is just the tangent map $\phi_{L(-2p)\,*}$ at $(p,q) \in p \times C$), it follows that $\tilde{\beta}_{1\,*}$ is injective on fibres, so $\tilde{\beta}_1$ is an embedding.

Finally, (2) follows from the fact that over a point $x \in B^1(L)$, (spanned by $A \in C_2$), the fibre of D_2 is isomorphic to $\mathbf{P}_{L(-2A)}$, while over a point $p \in C$, the fibre of D_1 is isomorphic to $\tilde{P}_{L(-2p)}$.

To prove the theorem in general, we first note that there are k "α maps" to each $B^k(L)$:

$$\begin{array}{rccc} \alpha_{0,k}: & C \times C_k & \rightarrow & B^k(L) \\ \alpha_{1,k-1}: & B^1(L) \times C_{k-1} & \rightarrow & B^k(L) \\ & & \vdots & \\ \alpha_{k-1,1}: & B^{k-1}(L) \times C & \rightarrow & B^k(L) \end{array}$$

Exactly as we defined $\widetilde{P}_L$, we define $\widetilde{B}^k(L)$ to be the blow-up of $B^k(L)$ along the images of the α-maps. The first step of the proof is to show that $\widetilde{B}^k(L)$ is smooth, and the exceptional divisor has smooth components with normal crossings.

Fibre of $B^2(L)$ over $p+q+r \in C_3$

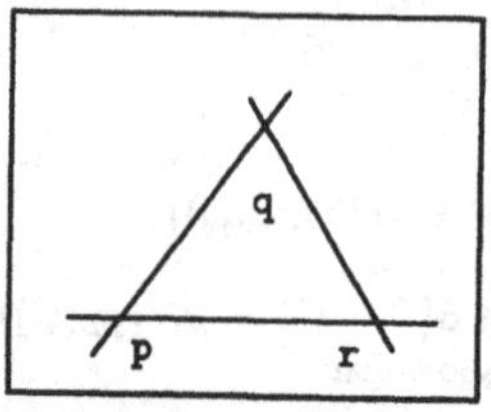

$p \cup q \cup r =$ image of $\alpha_{0,2}$

$\overline{pq} \cup \overline{qr} \cup \overline{pr} =$ image of $\alpha_{1,1}$

Next, we turn to the maps $\widetilde{B}^k(L) \to \mathbf{P}_L$. By induction and a series of applications of generalized versions of (a) and (b) from the case $k(L) = 2$, we find that $\beta_k : B^k(L) \to \mathbf{P}_L$ lifts to an embedding $\widetilde{\beta}_k : \widetilde{B}^k(L) \to bl_{k-1}(\mathbf{P}_L)$ for all $k < k(L)$. Finally, an analysis of the normal bundles to the embeddings yields the theorem.

Let $\sigma : \widetilde{P}_L \to \mathbf{P}_L$ be the "multiple" blow-down map. We end this section by describing the fibres of σ.

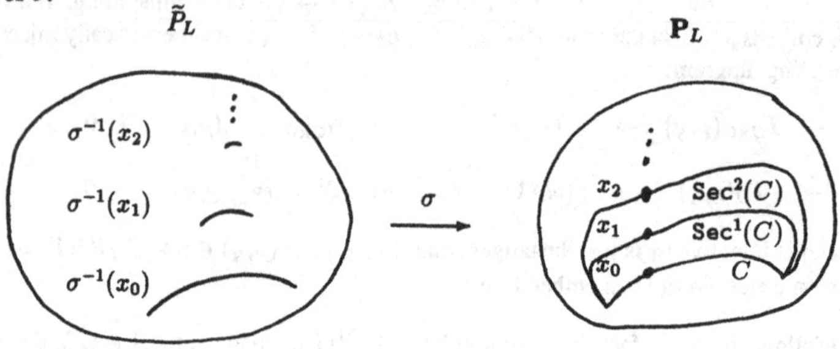

Let $x_l \in \mathrm{Sec}^l(C) - \mathrm{Sec}^{l-1}(C)$ for $l < k(L)$, D_x the spanning divisor. By theorem 2.4(2), $\sigma^{-1}(x_l)$ may be described set-theoretically as the set of sequences $\{\varepsilon_2, \ldots, \varepsilon_k\}$ with $\varepsilon_2 \in \mathbf{P}_{L(-2D_x)}$ and the properties described in the theorem. But this also gives $\widetilde{P}_{L(-2D_x)}$, and indeed, from the proof it falls out that $\sigma^{-1}(x_l) \cong \widetilde{P}_{L(-2D_x)}$. Thus $\widetilde{P}_L$ is recursive in the sense that the fibres of σ each correspond to a $\widetilde{P}_{L(-2A)}$ for some effective divisor $A \subset C$.

3 The map to moduli space

Recall that $\mathbf{P}_L = \mathbf{P}(\mathrm{Ext}^1(L, \omega_C)^*)$. As a result, there is a canonical "Euler" map $\mathcal{O}_{\mathbf{P}_L}(-1) \to \mathrm{Ext}^1(L, \omega_C) \otimes \mathcal{O}_{\mathbf{P}_L}$, hence a canonical section $s \in H^0(\mathbf{P}_L, \mathrm{Ext}^1(L, \omega_C) \otimes \mathcal{O}_{\mathbf{P}_L}(1))$.

Let $\pi_C : C \times \mathbf{P}_L \to C$ and $\pi_{\mathbf{P}_L} : C \times \mathbf{P}_L \to \mathbf{P}_L$ be the projections. Then since

$$\begin{aligned} \mathrm{Ext}^1(\pi_C^* L, \pi_C^*\omega_C \otimes \pi_{\mathbf{P}_L}^*\mathcal{O}(1)) &= H^1(C \times \mathbf{P}_L, \pi_C^*(L^* \otimes \omega_C) \otimes \pi_{\mathbf{P}_L}^*\mathcal{O}(1)) \\ &= H^1(C, L^* \otimes \omega_C) \otimes H^0(\mathbf{P}_L, \mathcal{O}_{\mathbf{P}_L}(1)) \\ &= H^0(\mathbf{P}_L, \mathrm{Ext}^1(L, \omega_C) \otimes \mathcal{O}_{\mathbf{P}_L}(1)) \end{aligned}$$

we get a canonical extension class on $C \times \mathbf{P}_L$:

$$(*):\ 0 \to \pi_C^*\omega_C \otimes \pi_{\mathbf{P}_L}^*\mathcal{O}(1) \to \mathcal{E}^0_{\mathbf{P}_L} \to \pi_C^* L \to 0$$

Furthermore, unraveling the definitions, it follows that on $C \times \{\epsilon\}$, the extension class $\epsilon \in \mathbf{P}_L$ gives the restriction of $(*)$:

$$0 \to \omega_C \to \mathcal{E}^0_{\mathbf{P}_L}|_{C\times\{\epsilon\}} \to L \to 0$$

so in particular, $\mathcal{E}^0_{\mathbf{P}_L}|_{C\times\{\epsilon\}}$ is semi-stable for $\epsilon \in U_L$.

As at the end of last section, let $\sigma : \widetilde{P}_L \to \mathbf{P}_L$ denote the blow-down. In this section, we outline the proof of:

Theorem 3.1 : There is a rank 2 vector bundle $\mathcal{E}^{k(L)}_{\widetilde{P}_L}$ on $C \times \widetilde{P}_L$ that satisfies

$$\mathcal{E}^{k(L)}_{\widetilde{P}_L}|_{C\times U_L} \cong \mathcal{E}^0_{\mathbf{P}_L}|_{C\times U_L}$$

and is recursive in the sense that

$$\mathcal{E}^{k(L)}_{\widetilde{P}_L}|_{C\times\sigma^{-1}(x)} \cong \mathcal{E}^{k(L(-2D_x))}_{\widetilde{P}_{L(-2D_x)}} \otimes \pi_C^*\mathcal{O}(D_x)$$

for all $x \in \mathrm{Sec}^l(C) - \mathrm{Sec}^{l-1}(C), l < k(L), D_x$ the divisor of degree $l+1$ spanning x.

As an immediate corollary, we get the result mentioned in the introduction:

Corollary 3.2 : $\mathcal{E}^{k(L)}_{\widetilde{P}_L}$ determines a morphism $\phi_{k(L)} : \widetilde{P}_L \to \mathcal{SU}(2, \omega_C \otimes L)$

Proof of the corollary : Induction on $k(L)$.
We need to show that for all $y \in \widetilde{P}_L$, $\mathcal{E}^{k(L)}_{\widetilde{P}_L}|_{C\times\{y\}}$ is semi-stable. If $k(L) = 0$, then $\widetilde{P}_L = \mathbf{P}_L = U_L$, $\mathcal{E}^{k(L)}_{\widetilde{P}_L} = \mathcal{E}^0_{\mathbf{P}_L}$, and the $\mathcal{E}^0_{\mathbf{P}_L}|_{C\times\{y\}}$ are semi-stable.

If $k(L) > 0$ and $y \in U_L$, then again $\mathcal{E}^{k(L)}_{\widetilde{P}_L}|_{C\times\{y\}} \cong \mathcal{E}^0_{\mathbf{P}_L}|_{C\times\{y\}}$ is semi-stable. If $y \notin U_L$, then $y \in \sigma^{-1}(x)$ for some $x \in \mathrm{Sec}^l(C) - \mathrm{Sec}^{l-1}(C)$, $l < k(L)$, by lemma 2.1. By theorem 3.1, then, $\mathcal{E}^{k(L)}_{\widetilde{P}_L}|_{C\times\{y\}} \cong \mathcal{E}^{k(L(-2D_x))}_{\widetilde{P}_{L(-2D_x)}} \otimes \pi_C^*\mathcal{O}(D_x)|_{C\times\{y\}}$, which is semi-stable by induction.

Let X be a variety, $\mathcal{E}$ a vector bundle on X and M a line bundle on a Cartier divisor $A \subset X$, which we regard as a coherent sheaf on X. If there is a surjective map $\mathcal{E} \to M$ of sheaves on X, then the kernel, which we call $\mathcal{E}^1$, is also a vector bundle, of the same rank as $\mathcal{E}$.

$\mathcal{E}^{k(L)}_{\widetilde{P}_L}$ is constructed by modifying the canonical rank 2 vector bundle $\mathcal{E}^0_{\widetilde{P}_L} := \sigma^*\mathcal{E}^0_{\mathbf{P}_L}$ in this way. Specifically, there are line bundles M^k_L on the divisors $C \times D_k$, and $\mathcal{E}^k_{\widetilde{P}_L}$ is defined inductively as the kernel of a natural sheaf map $\mathcal{E}^{k-1}_{\widetilde{P}_L} \to M^k_L$. Finally, the bundle $\mathcal{E}^{k(L)}_{\widetilde{P}_L}$ constructed in this way satisfies the conditions of the theorem. We start with the key local lemma in the proof of all this.

Suppose $x \in \mathrm{Sec}^l(C) - \mathrm{Sec}^{l-1}(C), l < k(L)$, and D_x is the spanning divisor, so $\overline{2D_x} \subset \mathbf{P}_L$ is the (projective) tangent plane to $\mathrm{Sec}^l(C)$ at x.

Then there is a surjective lift f of bundles on $C \times \{x\}$:

$$\begin{array}{ccccccccc} 0 & \to & \omega_C & \to & E_x & \to & L & \to & 0 \\ & & \downarrow & \swarrow_f & & & & & \\ & & \omega_C(D_x) & & & & & & \end{array}$$

Let $H \subset \mathbf{P}_L$ be a plane of codimension $2l+1$, meeting $\mathrm{Sec}^l(C)$ transversally at x. Let $\varepsilon : \widetilde{H} \to H$ be the blow up at x, and let $A_H \subset \widetilde{H}$ denote the exceptional divisor. Let $\mathcal{E}_H := \mathcal{E}^0_{\mathbf{P}_L}|_{C\times H}$ and $\mathcal{E}_{\widetilde{H}} := \varepsilon^*\mathcal{E}_H$.

Then from the earlier remarks, we can construct a new vector bundle $\mathcal{E}^1_{\widetilde{H}}$ on $C \times \widetilde{H}$ via:

$$\begin{array}{ccccccc} 0 & \to & \mathcal{E}^1_{\widetilde{H}} & \to & \mathcal{E}_{\widetilde{H}} & \to & \varepsilon^*\omega_{C\times\{x\}}(D_x) & \to & 0 \\ & & & & \downarrow & \nearrow_{\varepsilon^* f} & & & \\ & & & & \varepsilon^* E_x & & & & \end{array}$$

Lemma 3.3: The vector bundle $\mathcal{E}^1_{\widetilde{H}}|_{C\times A_H}$ is canonical.

That is, under the natural identification $A_H \cong \mathbf{P}_{L(-2D_x)}$, the vector bundle $\mathcal{E}^1_{\widetilde{H}}|_{C\times A_H}$ is canonically isomorphic to $\mathcal{E}^0_{\mathbf{P}_{L(-2D_x)}} \otimes \pi^*_C\mathcal{O}(D_x)$

Proof: We construct a new vector bundle $\mathcal{F}_{\widetilde{H}}$ on $C \times \widetilde{H}$ by pushing forward the extension which gives $\mathcal{E}_{\widetilde{H}}$:

$$\begin{array}{ccccccccc} 0 & \to & \pi^*_C\omega_C \otimes \pi^*_{\widetilde{H}}\mathcal{O}(1) & \to & \mathcal{E}_{\widetilde{H}} & \to & \pi^*_C L & \to & 0 \\ & & \downarrow & & \downarrow & & \| & & \\ 0 & \to & \pi^*_C\omega_C(D_x) \otimes \pi^*_{\widetilde{H}}\mathcal{O}(1) & \to & \mathcal{F}_{\widetilde{H}} & \to & \pi^*_C L & \to & 0 \end{array}$$

The extension giving $\mathcal{F}_{\widetilde{H}}$ splits when restricted to $C \times A_H$, and we form $\mathcal{F}^1_{\widetilde{H}}$ as the kernel:

$$0 \to \mathcal{F}^1_{\widetilde{H}} \to \mathcal{F}_{\widetilde{H}} \to \varepsilon^*\omega_{C\times\{x\}}(D_x) \to 0$$

But $\mathcal{F}^1_{\widetilde{H}}$ can also be thought of as a pull-back of extensions, dividing by the equation for $C \times A_H$:

$$\begin{array}{cccccccccc} (**): & 0 & \to & \pi^*_C\omega_C(D_x) \otimes \pi^*_{\widetilde{H}}(\mathcal{O}(1) \otimes \mathcal{O}(-A_H)) & \to & \mathcal{F}^1_{\widetilde{H}} & \to & \pi^*_C L & \to & 0 \\ & & & \downarrow & & \downarrow & & \| & & \\ & 0 & \to & \pi^*_C\omega_C(D_x) \otimes \pi^*_{\widetilde{H}}\mathcal{O}(1) & \to & \mathcal{F}_{\widetilde{H}} & \to & \pi^*_C L & \to & 0 \end{array}$$

$\mathcal{E}_{\tilde{H}}, \mathcal{E}^1_{\tilde{H}}, \mathcal{F}_{\tilde{H}}$ and $\mathcal{F}^1_{\tilde{H}}$ all fit into the following commuting diagram:

$$\begin{array}{ccccccccc} & & 0 & & 0 & & & & \\ & & \downarrow & & \downarrow & & & & \\ 0 & \rightarrow & \mathcal{E}^1_{\tilde{H}} & \rightarrow & \mathcal{E}_{\tilde{H}} & \rightarrow & \epsilon^*\omega_C(D_x) & \rightarrow & 0 \\ & & \downarrow & & \downarrow & & \| & & \\ 0 & \rightarrow & \mathcal{F}^1_{\tilde{H}} & \rightarrow & \mathcal{F}_{\tilde{H}} & \rightarrow & \epsilon^*\omega_C(D_x) & \rightarrow & 0 \\ & & \downarrow & & \downarrow & & & & \\ & & \pi^*_{\tilde{H}}\mathcal{O}(1)|_{D_x \otimes \tilde{H}} & = & \pi^*_{\tilde{H}}\mathcal{O}(1)|_{D_x \otimes \tilde{H}} & & & & \\ & & \downarrow & & \downarrow & & & & \\ & & 0 & & 0 & & & & \end{array}$$

from which it follows that $\mathcal{E}^1_{\tilde{H}}|_{C \times A_H}$ fits into the pull-back of extensions:

$$\begin{array}{lccccccc} & 0 & \rightarrow & \pi^*_C\omega_C(D_x) \otimes \pi^*_{A_H}\mathcal{O}(1) & \rightarrow & \mathcal{E}^1_{\tilde{H}}|_{C \times A_H} & \rightarrow & \pi^*_C L(-D_x) & \rightarrow & 0 \\ & & & \| & & \downarrow & & \downarrow & & \\ (**): & 0 & \rightarrow & \pi^*_C\omega_C(D_x) \otimes \pi^*_{A_H}\mathcal{O}(1) & \rightarrow & \mathcal{F}^1_{\tilde{H}}|_{C \times A_H} & \rightarrow & \pi^*_C L & \rightarrow & 0 \end{array}$$

hence is canonical, and the lemma is proved.

Let's go through the first few cases of the theorem in detail:

k(L) = 0: $\tilde{P}_L = \mathbf{P}_L, \mathcal{E}^0_{\mathbf{P}_L} = \mathcal{E}^{k(L)}_{\tilde{P}_L}$, and there is nothing to prove.

k(L) = 1: $\tilde{P}_L = bl_1(\mathbf{P}_L)$, and the exceptional divisor $D_1 \subset \tilde{P}_L$ maps to $C \subset \mathbf{P}_L$. Let $\pi_C : C \times \phi_L(C) \rightarrow C, \pi_{\phi_L(C)} : C \times \phi_L(C) \rightarrow \phi_L(C)$ be the projections, and $\Delta \subset C \times \phi_L(C)$ be the diagonal. Consider the line bundle $M := \pi^*_C\omega_C \otimes \pi^*_{\phi_L(C)}\mathcal{O}(1)$, and let M^1_L be the lift of $M(\Delta)$ to $C \times D_1$.

Claim 1: There is a surjective lift f_1:

$$\begin{array}{lcccccccc} (*): & 0 & \rightarrow & M & \rightarrow & \mathcal{E}^0_{\mathbf{P}_L}|_{C \times \phi_L(C)} & \rightarrow & \pi^*_C L & \rightarrow & 0 \\ & & & \downarrow & \swarrow_{f_1} & & & & & \\ & & & M(\Delta) & & & & & & \end{array}$$

Remark: At each point $p \in \phi_L(C)$, this just says that

$$\begin{array}{ccccccccc} 0 & \rightarrow & \omega_C & \rightarrow & \mathcal{E}^0_{\mathbf{P}_L}|_{C \times p} & \rightarrow & L & \rightarrow & 0 \\ & & \downarrow & \swarrow & & & & & \\ & & \omega_C(p) & & & & & & \end{array}$$

lifts, a fact that we have already seen in the remarks leading up to lemma 3.3. This claim is meant to globalize those remarks.

Proof: The existence of f_1 is equivalent to the extension class $(*)$ being in the kernel of the map:

$$H^1(C \times \phi_L(C), \pi_C^* L^* \otimes M) \to H^1(C \times \phi_L(C), \pi_C^* L^* \otimes M(\Delta))$$

or, equivalently, in the image of

$$\delta : H^0(\Delta, \mathcal{O}_\Delta) \to H^1(C \times \phi_L(C), \pi_C^* L^* \otimes M)$$

But this image is precisely the "canonical" extension.

The fact that f_1 is surjective follows immediately from the "pointwise" case. Namely, $f_1|_{C\times p}$ is surjective for each point $p \in \phi_L(C)$ because otherwise $\mathcal{E}^0_{\mathbf{P}_L}|_{C\times p}$ would split, which is impossible.

Now that we have the claim, we construct $\mathcal{E}^1_{\tilde{P}_L}$ via:

$$\begin{array}{ccccccccc} (**): & 0 & \to & \mathcal{E}^1_{\tilde{P}_L} & \to & \mathcal{E}^0_{\tilde{P}_L} & \to & M^1_L & \to & 0 \\ & & & & & \downarrow & \nearrow_{\sigma_1^* f_1} & & & \\ & & & & & \mathcal{E}^0_{\tilde{P}_L}|_{C\times D_1} & & & & \end{array}$$

To complete the theorem in this case, we need to show:

Claim 2: $\mathcal{E}^1_{\tilde{P}_L}|_{C\times e^{-1}(p)} \cong \mathcal{E}_{\mathbf{P}^0_{L(-2p)}} \otimes \pi_C^* \mathcal{O}(p)$ for all $p \in \phi_L(C)$.

Proof: Pick a hyperplane H transverse to C at p, and apply lemma 3.3. The restriction of $(**)$ to $C \times \widetilde{H}$ remains exact, so $\mathcal{E}^1_{\tilde{P}_L}|_{C\times e^{-1}(p)} \cong \mathcal{E}^1_{\tilde{H}}|_{C\times A_H} \cong \mathcal{E}^0_{\mathbf{P}_{L(-2p)}} \otimes \pi_C^* \mathcal{O}(p)$ by the lemma.

k(L) = 2: $\tilde{P}_L = bl_2(\mathbf{P}_L)$. The line bundle M^1_L on $C \times D_1$ is defined exactly as in the previous case. We need to put the line bundle M^2_L on $C \times D_2$ and get a surjective map $\mathcal{E}^1_{\tilde{P}_L} \to M^2_L$.

Step 1: Let $\pi_C : C \times B^1(L) \to C$ and $\pi_{B^1(L)}$ be the projections, and in addition, let π_{C_2} : $C \times B^1(L) \to C \times C_2$ be the natural map. Let $M = \pi_C^* \omega_C \otimes \pi_{B^1(L)}^* \mathcal{O}(1)$. Then there is a lift f_2:

$$\begin{array}{ccccccc} 0 \to & M & \to & \mathcal{E}^0_{\mathbf{P}_L}|_{C\times B^1(L)} & \to & \pi_C^* L & \to 0 \\ & \downarrow & \swarrow_{f_2} & & & & \\ & M \otimes \pi_{C_2}^* \mathcal{O}(\mathcal{D}_2) & & & & & \end{array}$$

This is proved exactly as in the previous case. However, this time f_2 is not surjective.

Step 2: Let $\sigma_1 : bl_1(\mathbf{P}_L) \to \mathbf{P}_L$ be the blow-up. As in the case $k(L) = 1$, we get a map $\mathcal{E}^1_{bl_1(\mathbf{P}_L)} \to \sigma_1^* \mathcal{E}^0_{\mathbf{P}_L}$. Then f_2 lifts to a surjective map g_2:

$$\begin{array}{ccc} \mathcal{E}^1_{bl_1(\mathbf{P}_L)} & \stackrel{g_2}{\twoheadrightarrow} & M(-C \times C \times C) \otimes \pi_{C_2}^* \mathcal{O}(\mathcal{D}_2) \\ \downarrow & & \downarrow \\ \sigma_1^* \mathcal{E}^0_{\mathbf{P}_L} & \stackrel{f_2}{\to} & M \otimes \pi_{C_2}^* \mathcal{O}(\mathcal{D}_2) \end{array}$$

Proof: To verify that f_2 lifts, we only have to show that the composed map $h_2 : \mathcal{E}^1_{bl_1(\mathbf{P}_L)} \to M \otimes \pi_{C_2}^* \mathcal{O}(\mathcal{D}_2)$ is zero, when restricted to $C \times C \times C$. But $\pi_{C_2}^{-1}(\mathcal{D}_1) \cap C \times C \times C = \Delta_{1,2} \cup \Delta_{1,3}$, where $\Delta_{i,j}$ are the diagonals, and h_2 restricted to $C \times C \times C$ factors through a map to $M(\Delta_{1,2})$ which is zero by the construction of $\mathcal{E}^1_{\widetilde{P}_L}$.

g_2 is surjective for $x \notin C \times C \times C$ by the "pointwise" remarks preceding lemma 3.3. When restricted to $C \times p \times C$, g_2 becomes f_1 for the bundle $L(-2p)$, so is surjective by induction.

Step 3: Let M^2_L be the lift of $M(-C \times C \times C) \otimes \pi^*_{\widetilde{C}_2}\mathcal{O}(\mathcal{D}_2)$ to $C \times D_2$. We form $\mathcal{E}^2_{\widetilde{P}_L}$ as the kernel of the map to M^2_L. Then $\mathcal{E}^2_{\widetilde{P}_L}$ satisfies the conditions of the theorem.

Proof: We have only to prove the recursiveness of $\mathcal{E}^2_{\widetilde{P}_L}$. If $x \in \mathrm{Sec}^1(C) - C$, then apply lemma 3.3 to a transverse plane. If $x \in C$, then first remark that by the case $k(L) = 1$, the restriction of $\mathcal{E}^1_{bl_1(\mathbf{P}_L)}$ to the preimage of x in $bl_1(\mathbf{P}_L)$ is isomorphic to $\mathcal{E}^0_{\mathbf{P}_{L(-2p)}} \otimes \pi^*_C\mathcal{O}(p)$. Then note, as in step 2, that when restricted to this fibre, g_2 is just f_1 for the line bundle $L(-2p)$, so again by induction, we get the result.

The proof in general follows exactly these three steps. By induction, we only need to understand the "last" line bundle $M^{k(L)}_L$ on $C \times D_{k(L)}$. There is a natural bundle on $C \times \widetilde{B}^{k(L)}(L)$ involving ω_C, $\mathcal{O}_{B^{k(L)}}(1)$, $\mathcal{D}_{k(L)}$ and the exceptional divisor. We lift this bundle to $C \times D_{k(L)}$, and exhibit a surjective map $\mathcal{E}^{k(L)-1}_{\widetilde{P}_L} \to M^{k(L)}_L$. Finally, we apply the lemma and induction to show that $\mathcal{E}^{k(L)}_{\widetilde{P}_L}$ constructed in this way has all the desired properties.

References

[B1] Bertram, A., A compctification of the complement of a secant variety, PhD Thesis, UCLA.

[B2] Bertram, A., Moduli of rank 2 vector bundles, theta divisors, and the geometry of curves in projective space, to appear in Journal of Diff. Geo.

[S] Seshadri, C.S., Fibres vectoriels sur les courbes algebriques. Asterisque 96, 1982.

[Sc] Schwarzenberger, R.L.E. The secant bundle of a projective variety, Proc. of the London Math. Soc., Vol. XIV. 1964, 369-384.

[V] Vainsencher, I., Schubert calculus for complete quadrics, in *Enumerative Geometry and Classical Algebraic Geometry*, (P.LeBarz,Y. Hervier, eds.), Birkhauser, Boston, 1982, 199-236.

ON THE BETTI NUMBERS OF THE MODULI SPACE OF STABLE BUNDLES OF RANK TWO ON A CURVE

EMILI BIFET, FRANCO GHIONE, AND MAURIZIO LETIZIA

1. Introduction

The aim of this paper is to begin exploring a new algebro-geometric approach to the study of the geometry of the moduli space of stable bundles on a curve X over a field k. This approach establishes a bridge between the arithmetic approach of Harder-Narasimhan [13] and the gauge group approach of Atiyah-Bott [2]. In particular, it may help explain some of the mysterious analogies observed by Atiyah and Bott.

The basic idea is to consider the scheme of quotients $\mathrm{Quot}^d_{\mathcal{O}_X^r/X/k}$, introduced by A. Grothendieck [10], as the variety that parameterizes effective divisors of rank r and degree d in the sense of A. Weil [26]. The dual of the kernel of the universal quotient gives rise to a family of vector bundles of rank r and degree d. It is useful to collect these varieties into the system of effective divisors

$$\boldsymbol{Q}(r,d): \quad \mathrm{Quot}^d_{\mathcal{O}_X^r/X/k} \to \mathrm{Quot}^{d+r}_{\mathcal{O}_X^r/X/k} \to \mathrm{Quot}^{d+2\cdot r}_{\mathcal{O}_X^r/X/k} \to \cdots$$

(maps given by tensoring the submodules with $\mathcal{O}_X(-P)$, P a fixed point in X) and to stratify it according to Harder-Narasimhan type

$$\boldsymbol{Q}(r,d) = \boldsymbol{Q}^{ss}(r,d) \cup \bigcup_{\mu \neq ss} \boldsymbol{S}_\mu(r,d) \qquad (\, ss = \text{ semistable . })$$

Suppose that r and d are coprime, so that being semistable is equivalent to being stable. Taking the dual of the kernel of a quotient, defines an Abel-Jacobi map

$$\vartheta : \boldsymbol{Q}^{ss}(r,d) \to \boldsymbol{N}(r,d)$$

where $\boldsymbol{N}(r,d)$ is the system of moduli spaces of stable vector bundles

$$N(r,d) \to N(r,d+r) \to N(r,d+2\cdot r) \to \cdots$$

(maps given by tensoring with $\mathcal{O}_X(P)$.) Taking cohomology of these systems

$$H^*(-) = \varprojlim_n H^*(-_n)$$

(ℓ-adic if you wish,) one obtains

$$H^*(Q^{ss}(r,d)) = H^*(N(r,d))[x]$$

(polynomial algebra in one variable x of degree 2) and, in particular, the identity of Poincaré series

$$P(N(r,d);t) = (1-t^2)\cdot P(\boldsymbol{Q}^{ss}(r,d);t).$$

This is due to the existence of a locally free module $\boldsymbol{E}$ over the system $\boldsymbol{N}(r,d)$, and a map

$$\begin{array}{ccc} \boldsymbol{Q}^{ss}(r,d) & \xrightarrow{j} & \mathbb{P}(\boldsymbol{E}) \\ & \vartheta \searrow \quad \swarrow & \\ & \boldsymbol{N}(r,d) & \end{array}$$

which induces isomorphisms in cohomology.

The research of the first author at MSRI was supported in part by NSF grant DMS-8505550

Similarly we have, in complete analogy to [2],

$$P(\boldsymbol{Q}^{ss}(r,d);t) = P(\boldsymbol{Q}(r,d);t) - \sum_{\mu \neq ss} t^{2 \cdot d_\mu} \cdot P(\boldsymbol{S}_\mu(r,d);t)$$

and, if μ has $P_\mu = \{(0,0),(r_1,d_1),(r_1+r_2,d_1+d_2),\ldots,(r,d)\}$ as its Shatz polygon, then

$$P(\boldsymbol{S}_\mu(r,d);t) = \prod_i P(\boldsymbol{Q}^{ss}(r_i,d_i);t).$$

Therefore, the Betti numbers of $N(r,d)$ are found, via an inductive formula, as soon as we know those of $\boldsymbol{Q}(r,d)$ for general r and d. These can be easily deduced, for instance, from [17, 18, 5] and the natural action of the torus $\mathbb{G}^r_{m,k}$ on $\boldsymbol{Q}$ [3]. It turns out that they coincide with those of the classifying space of the gauge group $\mathcal{G}$ for a principal U(r)-bundle, and therefore we obtain the same formulæ as in [19, 11, 13, 6, 2, 25, 14, 4].

We have just mentioned that the Betti numbers of $\boldsymbol{Q}(r,d)$ coincide with those of the classifying space of the gauge group $\mathcal{G}$ for a principal U(r)-bundle. More generally, the Betti numbers of the strata $\boldsymbol{S}_\mu \subset \boldsymbol{Q}$ coincide with those of the Borel construction $(\mathcal{C}_\mu)_\mathcal{G}$ of the corresponding infinite dimensional strata $\mathcal{C}_\mu$ considered in [2]; one could say that effective divisors provide us, once stratified, with algebraic models of these Borel constructions. Thus equivariant cohomology is replaced by ordinary cohomology, the infinite dimensional strata by systems of algebraic varieties, and the theory can be developed algebraically and in any characteristic.

In this paper we shall consider the case where $r = 2$, $d = 1$ and $k = \mathbb{C}$. In this case, one has

$$P(\boldsymbol{Q}(2,1);t) = \frac{\{(1+t)\,(1+t^3)\}^{2g}}{(1-t^2)^2\,(1-t^4)},$$

and, since

$$P(\boldsymbol{Q}(1,d);t) = \frac{(1+t)^{2g}}{(1-t^2)},$$

we also have

$$P(\boldsymbol{Q}^{ss};t) = \frac{\{(1+t)\,(1+t^3)\}^{2g}}{(1-t^2)^2\,(1-t^4)} - \left(\frac{(1+t)^{2g}}{(1-t^2)}\right)^2 \cdot \sum_{m \geq 0} t^{2(2m+g)}.$$

Therefore

$$P(N(2,1);t) = (1+t)^{2g} \frac{(1+t^3)^{2g} - t^{2g}(1+t)^{2g}}{(1-t^2)\,(1-t^4)}.$$

We shall return to the general case in a subsequent paper. There, we shall also address some of the arithmetic aspects of this approach. Let us just mention here that it is possible to give, along these lines, a simple proof of the Siegel formula [9] and that there are interesting relations with the work of Harder [12], Laumon [15], ...

2. Rank one

The aim of this section is to provide motivation for the general methods and techniques used in this paper.

If X is a smooth projective algebraic curve, there exists an algebraic variety Pic X, called the Picard variety of X, parameterizing isomorphism classes of line bundles over X. The connected components $\mathrm{Pic}^d X$, $d \in \mathbf{Z}$, of this variety correspond to bundles of degree d. Moreover, there are bundles, named after Poincaré, $\mathcal{L}$ over $X \times \mathrm{Pic}^d X$ such that for every $[L] \in \mathrm{Pic}^d X$, the restriction of $\mathcal{L}$ to $X \times \{[L]\}$ is isomorphic to L. These bundles are unique up to tensoring with the inverse image under the natural projection

$$p : X \times \mathrm{Pic}^d X \longrightarrow \mathrm{Pic}^d X$$

of line bundles over $\mathrm{Pic}^d X$. In particular, once a point $P \in X$ has been chosen, there is a unique Poincaré bundle with the property that its restriction to $\{P\} \times \mathrm{Pic}^d X$ is trivial.

There is a well known relation between line bundles and effective divisors on X. Given any effective divisor D of degree d (an element of the symmetric product $X^{(d)}$) we consider the locally free submodule $\mathcal{O}_X(-D) \subset \mathcal{O}_X$ of functions vanishing on D with the right order. It is customary to associate with D the dual $\mathcal{O}_X(D)$ of $\mathcal{O}_X(-D)$. Thus, every $X^{(d)}$ gives rise to a family of line bundles of degree d over X. In particular, we have Abel-Jacobi maps

(2.1)
$$\begin{array}{rcl} \vartheta_d : X^{(d)} & \longrightarrow & \mathrm{Pic}^d X \\ D & \longmapsto & [\mathcal{O}_X(D)] \end{array}$$

Moreover, we can define vector bundles

$$E_d = p_* \mathcal{H}om(\mathcal{O}_{X \times \mathrm{Pic}^d X}, \mathcal{L})$$

over $\mathrm{Pic}^d X$, and morphisms

(2.2)
$$j_d : X^{(d)} \longrightarrow \mathbb{P}(E_d)$$

making the diagram

(2.3)
$$\begin{array}{ccccc} X^{(d)} & & \xrightarrow{j_d} & & \mathbb{P}(E_d) \\ & \vartheta_d \searrow & & \swarrow \pi_d & \\ & & \mathrm{Pic}^d X & & \end{array}$$

commutative for every d. These morphisms may be defined by sending D to the class of the dual $i^\vee : \mathcal{O}_X \to \mathcal{O}_X(D)$ of the inclusion $i : \mathcal{O}_X(-D) \to \mathcal{O}_X$. Note that we are identifying the fibre of E_d over the class $[\mathcal{O}_X(D)]$ with $\mathrm{Hom}_{\mathcal{O}_X}(\mathcal{O}_X, \mathcal{O}_X(D)) = H^0(X, \mathcal{O}_X(D))$.

If $d \geq 2g-1$, then the j_d's are isomorphisms (see [1] for example) and we obtain, as a corollary, the cohomology of $X^{(d)}$ in terms of that of $\mathrm{Pic}^d X$:

(2.4)
$$H^*(X^{(d)}) = H^*(\mathrm{Pic}^d X)[\xi_d]/(R_d)$$

with

$$R_d = \xi_d^r - \pi_d^*(c_1) \cdot \xi_d^{r-1} + \ldots + (-1)^r \pi_d^*(c_r)$$

where r is the rank of E_d, and the c_i's are its Chern classes. Recall that the varieties $\mathrm{Pic}^d X$, $d \in \mathbb{Z}$, are isomorphic to each other, and therefore have the same cohomology. A way of finding this cohomology suggests itself: let d go to infinity in (2.4). Since $\mathrm{Pic}^d X$ has finite dimension and the rank r of E_d goes to infinity with d, the relations on the right hand side disappear and one is left in the limit with the polynomial ring $H^*(\mathrm{Pic}^0 X)[x]$.

One way of making this passage to the limit precise is to consider for every i the limit as $d \to \infty$ of the sequence of Betti numbers $\{b^i(X^{(d)})\}_{d \in \mathbb{N}}$. If one takes this point of view, it is clear from (2.4) that $b^i = \lim_{d \to \infty} b^i(X^{(d)})$ exists, and it coincides with the ith Betti number of $H^*(\mathrm{Pic}^0 X)[x]$. On the other hand, the values of the $b^i(X^{(d)})$ are well known (see [17, 18] or (4.2) below) and one has

(2.5)
$$\sum_{i \in \mathbb{N}} b^i \cdot t^i = \frac{(1+t)^{2g}}{(1-t^2)}$$

(see also [7, 8, 16] .) We obtain, in the limit, the identity

(2.6)
$$\frac{(1+t)^{2g}}{(1-t^2)} = \frac{1}{(1-t^2)} \cdot P(\mathrm{Pic}^0 X; t)$$

and, therefore

$$P(\mathrm{Pic}^0 X; t) = (1+t)^{2g}$$

in full accord with $\mathrm{Pic}^0 X$ being an Abelian variety of dimension g. Recall that the cohomology ring of such a variety is isomorphic to the exterior algebra of a free $\mathbb{Z}$-module of rank $2g$.

Another way of doing this is to fix a point $P \in X$, and consider the system of varieties (an infinite dimensional algebraic variety in the sense of Shafarevich [22, 23])

$$Q: \quad X^{(0)} \to X^{(1)} \to X^{(2)} \to \cdots$$

where each map is given by sending an effective divisor D to $D+P$. By tensoring at every stage with $\mathcal{O}_X(P)$, we also have the system of Picard varieties

$$N: \quad \mathrm{Pic}^0 X \to \mathrm{Pic}^1 X \to \mathrm{Pic}^2 X \to \cdots$$

and it is clear that the Abel-Jacobi maps (2.1) commute with the structure maps and give a morphism of systems $\vartheta : Q \to N$. Similarly, we can arrange things so that we have a system of projectivized bundles

$$\mathbb{P}(E): \quad \mathbb{P}(E_0) \to \mathbb{P}(E_1) \to \mathbb{P}(E_2) \to \cdots$$

and a projection map $\pi : \mathbb{P}(E) \to N$. The morphisms (2.2) combine to give a morphism $j : Q \to \mathbb{P}(E)$ and a commutative diagram analogous to (2.3). Applying the cohomology functor to any of these systems, Q for example, we obtain a system of rings

$$H^*(X^{(0)}) \leftarrow H^*(X^{(1)}) \leftarrow H^*(X^{(2)}) \leftarrow \cdots \tag{2.7}$$

and we take its limit to be, by definition, the cohomology ring of the system. Recall [17] that the systems of rings (2.7) has the property that for every $i \in \mathbb{N}$, all but a finite number of the morphisms

$$H^i(X^{(0)}) \leftarrow H^i(X^{(1)}) \leftarrow H^i(X^{(2)}) \leftarrow \cdots$$

are isomorphisms. If we denote by $H^*(Q)$ the cohomology ring of the system Q, then we have

$$H^*(Q) \simeq H^*(\mathrm{Pic}^0 X)[x],$$

and therefore, as a consequence, identity (2.6). Note that although $j : Q \to \mathbb{P}(E)$ is not an isomorphism of systems, it induces an isomorphism of cohomology rings. Moreover, the morphisms of graded rings (recall that x and ξ_i have degree 2)

$$\begin{array}{ccccc}
H^*(N)[x] & = & H^*(N)[x] & = & \cdots \\
\downarrow{\scriptstyle \psi_0} & & \downarrow{\scriptstyle \psi_1} & & \\
H^*(\mathrm{Pic}^0 X)[x] & \longleftarrow & H^*(\mathrm{Pic}^1 X)[x] & \longleftarrow & \cdots \\
\downarrow{\scriptstyle \phi_0} & & \downarrow{\scriptstyle \phi_1} & & \\
H^*(\mathrm{Pic}^0 X)[\xi_0]/(R_0) & \longleftarrow & H^*(\mathrm{Pic}^1 X)[\xi_1]/(R_1) & \longleftarrow & \cdots
\end{array}$$

have the property that for every $i \in \mathbb{N}$, the components of degree i, namely ϕ^i_n and $\phi^i_n \circ \psi^i_n$, are isomorphisms for all but a finite number of values of n. Since $H^*(N)$ is clearly isomorphic to $H^*(\mathrm{Pic}^0 X)$, it follows immediately that $H^*(\mathbb{P}(E)) \simeq H^*(\mathrm{Pic}^0 X)[x]$.

In this paper we shall follow, for simplicity, the approach via limits of Betti numbers, but it is not difficult to switch from one approach to the other.

Note that (2.5) is precisely the Poincaré series $P(B\mathcal{G};t)$ of the classifying space $B\mathcal{G}$ of the gauge group $\mathcal{G}$ for a principal $U(1)$-bundle. In [2] identity (2.6) is deduced from the fibration

$$BU(1) \to B\mathcal{G} \to B\overline{\mathcal{G}}$$

where $\overline{\mathcal{G}}$ is the quotient of $\mathcal{G}$ by its constant central $U(1)$-subgroup.

3. Some notation

We shall fix some notation that should help organize the material in the sections that follow.

In this paper a *system of algebraic varieties* is a family of varieties $\boldsymbol{X} = \{X_n\}_{n\in\mathbf{N}}$ indexed by the natural numbers. Similarly for topological spaces, graded rings, abelian groups, ...

A system of algebraic varieties $\boldsymbol{X} = \{X_n\}_{n\in\mathbf{N}}$ is said to be *smooth* whenever X_n is smooth for all but a finite number of values of n.

The *cohomology* of a system $\boldsymbol{X}$ is by definition the system of graded rings

$$H^*(\boldsymbol{X}) = \{H^*(X_n)\}_{n\in\mathbf{N}}$$

We say that the ith Betti number of the system $\boldsymbol{X} = \{X_n\}_{n\in\mathbf{N}}$ *stabilizes* whenever the limit, as $n \to \infty$, of the sequence of Betti numbers $b^i(X_n)$ exists i.e. $b^i(X_n)$ becomes constant for $n \gg 0$. The value of this constant will be by definition the ith *Betti number* $b^i(\boldsymbol{X})$ of the system $\boldsymbol{X}$. Moreover, if all the Betti numbers of $\boldsymbol{X}$ stabilize, we define its *Poincaré series* by

$$P(\boldsymbol{X};t) = \sum_{i\geq 0} b^i(\boldsymbol{X})\cdot t^i.$$

Note that, under these hypothesis, if we write

$$E(u,t) = \sum b^i(X_n)\cdot u^n t^i$$

and let

$$\begin{aligned} f(u,t) &= (1-u)\cdot E(u,t) \\ &= \sum\{b^i(X_n) - b^i(X_{n-1})\}\cdot u^n t^i \end{aligned}$$

(where $b^i(X_{-1}) = 0$,) then we have

$$\begin{aligned} P(\boldsymbol{X};t) &= f(1,t) \\ &= -\operatorname*{Res}_{u=1} E(u,t). \end{aligned}$$

Let $\boldsymbol{A} = \{A_n^*\}_{n\in\mathbf{N}}$ and $\boldsymbol{B} = \{B_n^*\}_{n\in\mathbf{N}}$ be two systems of graded rings. Let $\varphi = \{\varphi_n\}_{n\in\mathbf{N}}$ be a sequence of morphisms of graded rings

$$\varphi_n : A_n^* \longrightarrow B_n^*.$$

We say that φ is an *asymptotic isomorphism* if for every $i \in \mathbf{N}$, the morphisms

$$\varphi_n^i : A_n^i \longrightarrow B_n^i$$

are isomorphisms for all but a finite number of values of n.

Let $\boldsymbol{X} = \{X_n\}_{n\in\mathbf{N}}$ and $\boldsymbol{Y} = \{Y_n\}_{n\in\mathbf{N}}$ be two systems of algebraic varieties. A sequence $f = \{f_n\}_{n\in\mathbf{N}}$ of morphisms

$$f_n : X_n \longrightarrow Y_n$$

is said to be a *quasi-isomorphism* whenever:

(1) $\boldsymbol{Y}$ is smooth;
(2) f_n is an open immersion for $n \gg 0$;
(3) the codimension of $Y_n - f_n(X_n)$ in Y_n goes to infinity as $n \to \infty$.

One pleasant property of these morphisms is that they induce asymptotic isomorphisms

$$f_n^* : H^*(Y_n) \longrightarrow H^*(X_n), \qquad n \in \mathbf{N}$$

of cohomology systems. In particular, if the Betti numbers of either system stabilize, then those of the other also do, and the two Poincaré series coincide.

4. THE SYSTEM OF EFFECTIVE DIVISORS OF HIGHER RANK

We introduce a notion of effective divisor of higher rank on a curve and compute the Poincaré series of the corresponding system.

Let X be a smooth projective curve of genus g over the complex numbers $\mathbb{C}$. Effective divisors of degree d on X can be thought of as elements of the symmetric product $X^{(d)}$. This in turn can be identified with the Hilbert scheme $\mathrm{Hilb}^d_{X/\mathbb{C}}$ of closed subschemes of X with constant Hilbert polynomial d. Recall that Grothendieck [10] defined $\mathrm{Hilb}_{X/\mathbb{C}}$ as the case $\mathcal{F} = \mathcal{O}_X$ of the more general scheme $\mathrm{Quot}_{\mathcal{F}/X/\mathbb{C}}$ of quotients of a coherent sheaf $\mathcal{F}$. Because of this, and other reasons, it seems natural to consider $\mathrm{Quot}^d_{\mathcal{O}^r_X/X/\mathbb{C}}$ as the variety that parameterizes effective divisors of rank r and degree d on X. In particular

Definition 4.1. The system of effective divisors of rank two is the system of smooth algebraic varieties

$$\boldsymbol{Q} = \{Q_n\}_{n\in\mathbb{N}}$$

where $Q_n = \mathrm{Quot}^n_{\mathcal{O}^2_X/X/\mathbb{C}}$.

If $[k,d]$ denotes the Hilbert polynomial $P(t) = k\cdot t + \{d + k(1-g)\}$, then the scheme $\mathrm{Quot}^{[k,d]}_{\mathcal{F}/X/\mathbb{C}}$ consists of quotients of $\mathcal{F}$ having rank k and degree d (or, equivalently, submodules of $\mathcal{F}$ having rank $\mathcal{F} - k$ as its rank and $\deg\mathcal{F} - d$ as its degree.) Note that, with this notation, we have $Q_d = \mathrm{Quot}^{[0,d]}_{\mathcal{O}^2_X/X/\mathbb{C}}$.

The torus $\mathbb{G}^2_m$ acts naturally on each Q_n and the fixed points under this action are the union of the products $X^{(n-i)} \times X^{(i)}$, $0 \le i \le n$ (cf. [3].) Here we consider $X^{(n-i)} \times X^{(i)}$ as embedded in Q_n by sending any pair of effective divisors (D_1, D_2) to the quotient of $\mathcal{O}^2_X$ determined by its submodule $\mathcal{O}_X(-D_1) \oplus \mathcal{O}_X(-D_2)$. It follows from this (cf. [3],) that the cohomology of Q_n is free of torsion and its Poincaré polynomial is given by

$$P(Q_n;t) = \sum_{0\le i\le n} t^{2i}\, P(X^{(n-i)};t)\cdot P(X^{(i)};t)\,. \tag{4.1}$$

Recall from [17, 18] that we have

$$\sum_{i\ge 0} P(X^{(i)};t)\cdot u^i = \frac{(1+ut)^{2g}}{(1-u)\,(1-ut^2)}\,. \tag{4.2}$$

¿From this and (4.1) we obtain that the formal power series in two variables

$$E(u,t) = \sum_{n\ge 0} P(Q_n;t)\cdot u^n$$

is, in fact, the rational function

$$E(u,t) = \frac{(1+ut)^{2g}\,(1+ut^3)^{2g}}{(1-u)\,(1-ut^2)^2\,(1-ut^4)}.$$

An immediate consequence is the following

Proposition 4.1. *The Betti numbers of the system* $\boldsymbol{Q} = \{Q_n\}_{n\in\mathbb{N}}$ *stabilize and its Poincaré series is given by*

$$\begin{aligned} P(\boldsymbol{Q};t) &= -\operatorname*{Res}_{u=1} E(u,t) \\ &= \frac{\{(1+t)\,(1+t^3)\}^{2g}}{(1-t^2)^2\,(1-t^4)}. \end{aligned}$$

Remark. Note that $P(\boldsymbol{Q};t)$ coincides with the Poincaré series $P(B\mathcal{G};t)$ of the classifying space $B\mathcal{G}$ of the gauge group $\mathcal{G}$ for a principal $U(2)$-bundle.

5. The Abel-Jacobi map

We define a generalization of the classical Abel-Jacobi map; then using it we relate the cohomology of the moduli space of stable bundles to that of the stable stratum $\boldsymbol{Q}_1^s$ in the system $\boldsymbol{Q}_1 = \{Q_{2n+1}\}_{n\in\mathbb{N}}$.

Recall that a vector bundle E of rank two is stable (resp. semistable) if for every line subbundle $L \subset E$ one has

$$\deg L < \frac{\deg E}{2} \qquad \text{(resp. } \deg L \leq \frac{\deg E}{2}\text{ .)}$$

There exists a coarse moduli space (see [21, 20] for example,) which we shall denote N_d, parameterizing isomorphism classes of stable vector bundles of rank 2 and degree d over the curve X. This is a smooth quasi-projective variety of dimension $1 + 2^2(g-1) = 4g-3$. When d is odd, the notion of stable and semistable bundles coincide and N_d is projective. Moreover, in this case, N_d is a fine moduli space. In particular, there are Poincaré bundles $\mathcal{P}$ over $X \times N_d$ such that for every $[E]$ in N_d the restriction $\mathcal{P}_{[E]}$ of $\mathcal{P}$ to $X \times \{[E]\}$ is isomorphic to E.

Let Q_d^s be the open subset of Q_d corresponding to the stable bundles in the family parameterized by Q_d. The restriction of this universal family to Q_d^s induces a natural morphism

$$\vartheta_d : Q_d^s \longrightarrow N_d$$

which generalizes the classical Abel-Jacobi map (2.1). Note that ϑ_d sends any stable quotient $M = \mathcal{O}_X^2/K$ to the isomorphism class of the dual vector bundle $K^\vee$. This is a vector bundle because K is torsion-free and X is smooth of dimension one.

Define, by analogy with the case of rank 1, vector bundles over N_d by

$$E_d = p_*\mathcal{H}om\,(\mathcal{O}_{X\times N_d}^2, \mathcal{P})$$

where $\mathcal{P}$ is a Poincaré bundle over N_d and $p : X \times N_d \to N_d$ is the canonical projection. Let $\mathbb{P}(E_d)$ be the projectivisation of E_d. There are morphisms

$$j_d : Q_d^s \to \mathbb{P}(E_d)$$

such that the diagram

$$\begin{array}{ccc} Q_d^s & \xrightarrow{j_d} & \mathbb{P}(E_d) \\ & \vartheta_d \searrow \quad \swarrow \pi_d & \\ & N_d & \end{array}$$

is commutative for every d. These are defined by sending $M = \mathcal{O}_X^2/K$ in Q_d^s to the class of

$$\mathcal{O}_X^2 \xrightarrow{i^\vee} K^\vee \xrightarrow{\gamma} \mathcal{P}_{\vartheta_d(K)}$$

where i is the inclusion $K \subset \mathcal{O}_X^2$ and γ is any isomorphism $K^\vee \to \mathcal{P}_{\vartheta_d(K)}$. Note that this is well defined since the only automorphism of stable bundles are given by multiplication by scalars.

We are now ready to state the following

Proposition 5.1. *Let $\boldsymbol{Q}_1^s$ be the system $\{Q_{2n+1}^s\}_{n\in\mathbb{N}}$. The sequence of morphisms*

$$j_{2n+1} : Q_{2n+1}^s \longrightarrow \mathbb{P}(E_{2n+1})$$

is a quasi-isomorphism. Moreover, there is also an asymptotic isomorphism of graded rings

$$H^*(N_1)[x] \longrightarrow H^*(\mathbb{P}(E_{2n+1}^{\cdot}))$$

(with x an independent variable of degree 2) and we have the identity of Poincaré series

$$P(N_1;t) = (1-t^2)\cdot P(\boldsymbol{Q}_1^s;t).$$

Proof. Consider the sequence of morphisms of graded rings

$$\varphi_{2n+1} : H^*(N_1)[x] \longrightarrow H^*(N_1)[\xi_{2n+1}]/(R_{2n+1}) \simeq H^*(\mathbb{P}(E_{2n+1}))$$

where

$$R_{2n+1} = \xi_{2n+1}^r - \pi_{2n+1}^*(c_1)\cdot \xi_{2n+1}^{r-1} + \ldots + (-1)^r \pi_{2n+1}^*(c_r),$$

the $c_i \in H^{2i}(N_1)$, $1 \leq i \leq r$, are the Chern classes of E_{2n+1} and $\varphi_{2n+1}(x) = \xi_{2n+1}$. Recall that the N_{2n+1}'s are all isomorphic to N_1. Since the rank r of E_{2n+1} grows indefinetely with n and N_1 is finite dimensional, it is clear that the sequence $\{\varphi_{2n+1}\}_{n\in\mathbb{N}}$ is an asymptotic isomorphism.

It is clear that the $\mathbb{P}(E_{2n+1})$, $n \in \mathbb{N}$, are smooth algebraic varieties and that the j_{2n+1} are open immersions. It remains for us to show that

$$\operatorname{codim} \mathbb{P}(E_{2n+1}) - j_{2n+1}(Q_{2n+1}^s)$$

goes to infinity as $n \to \infty$. By definition

$$j_{2n+1}(Q_{2n+1}^s) \cap \pi_{2n+1}^{-1}(s), \qquad s \in N_{2n+1}$$

coincides with the classes of injective homomorphisms

$$\operatorname{Hom}_{\mathcal{O}_X}^{\mathrm{Inj}}(\mathcal{O}_X^2, \mathcal{P}_s).$$

Fix a point $P \in X$, and consider $F = \mathcal{P}_s(-mP) = \mathcal{P}_s \otimes \mathcal{O}_X(-mP)$. If we take

$$m = n - 2g,$$

then

$$\frac{\deg F}{2} = \frac{4g+1}{2} > 2g - 2$$

and therefore

$$H^1(X, F) = 0.$$

Now we can apply the case $r = 2$ of the following

Lemma 5.1. *Fix a point $P \in X$ and let F be a locally free module of rank r such that $H^1(X,F) = 0$. If d_m, $m \geq 0$, denotes the codimension in* $\operatorname{Hom}_{\mathcal{O}_X}(\mathcal{O}_X^r, F(mP))$ *of the homomorphisms which are not injective, then $d_m \geq m$.*

Proof. Consider the short exact sequence

$$0 \to \mathcal{O}_X(-mP) \to \mathcal{O}_X \to \mathcal{O}_{mP} \to 0$$

where $\mathcal{O}_{mP} \cong \mathcal{O}_{X,P}/\mathfrak{m}_P^m$. Tensoring with $F(mP)$ and applying $\operatorname{Hom}_{\mathcal{O}_X}(\mathcal{O}_X^r, -)$ we obtain

$$\operatorname{Hom}_{\mathcal{O}_X}(\mathcal{O}_X^r, F(mP)) \to \operatorname{Hom}_{\mathcal{O}_{X,P}}(\mathcal{O}_{X,P}^r, F_P \otimes_{\mathcal{O}_{X,P}} \mathcal{O}_{mP}) \to 0$$

Any non-injective homomorphism $\mathcal{O}_X^r \to F(mP)$ is sent to a morphism that factors through a surjective $\mathcal{O}_{X,P}^r \to \mathcal{O}_{mP}^s$ for some $s < r$. Thus, in order to establish the Lemma, it suffices to prove that (for any $s < r$) the image of the composition map

$$\begin{array}{c} \operatorname{Hom}_{\mathcal{O}_{X,P}}^{\mathrm{Surj}}(\mathcal{O}_{X,P}^r, \mathcal{O}_{mP}^s) \times \operatorname{Hom}_{\mathcal{O}_{X,P}}(\mathcal{O}_{mP}^s, F_P \otimes_{\mathcal{O}_{X,P}} \mathcal{O}_{mP}) \\ \Phi^s \downarrow \\ \operatorname{Hom}_{\mathcal{O}_{X,P}}(\mathcal{O}_{X,P}^r, F_P \otimes_{\mathcal{O}_{X,P}} \mathcal{O}_{mP}) \end{array}$$

has codimension greater or equal to m. The group G of automorphisms of $\mathcal{O}_{mP}^s$ acts freely on the domain of Φ^s by $g \cdot (\alpha, \beta) = (g \circ \alpha, \beta \circ g^{-1})$. Since Φ^s is clearly constant along the orbits and G has dimension $s^2 \cdot m$, we have

$$\dim \operatorname{im} \Phi^s \leq r(sm) + (sr)m - s^2 m$$

and

$$\operatorname{codim} \operatorname{im} \Phi^s \geq r^2 \cdot m - 2rs \cdot m + s^2 \cdot m = (r-s)^2 \cdot m \geq m.$$

6. The Harder-Narasimhan-Shatz stratification

Recall that a vector bundle E of rank 2 is either semistable or has a unique sub line bundle $L_E \subset E$ such that $\deg L_E > \frac{1}{2}\deg E$. The filtration

$$0 \subset L_E \subset E$$

is called the Harder-Narasimhan filtration of E and we shall refer to the pair

$$(\deg L_E,\ \deg E/L_E)$$

as the Harder-Narasimhan type of E. If E is semistable we shall say that its type is

$$(\frac{\deg E}{2},\ \frac{\deg E}{2}).$$

Any parameter space for a family of bundles of rank 2 can be stratified according to its Harder-Narasimhan type (see [24].) In the case of the family $\mathcal{K}$, where $\mathcal{M} = \mathcal{O}^2_{X\times Q_{2n+1}}/\mathcal{K}$ is the universal quotient over $X \times Q_{2n+1}$, we define

$$S^{-1}_{2n+1} = Q^{ss}_{2n+1} = Q^{s}_{2n+1}$$

and for $r \geq 0$

$$S^r_{2n+1} = \{M \in Q_{2n+1} \mid M = \mathcal{O}^2_X/K \text{ with } \deg L_K = r - n\}\ .$$

Proposition 6.1. *We have a decomposition*

$$Q_{2n+1} = Q^s_{2n+1} \cup S^0_{2n+1} \cup S^1_{2n+1} \cup \ldots \cup S^n_{2n+1}$$

such that:

(1) *For each $k \geq -1$ the union of strata $U^k_{2n+1} = \bigcup_{-1\leq r\leq k} S^r_{2n+1}$ is a Zariski open subset of Q_{2n+1}.*
(2) *If $n \geq r + 2g$, then S^r_{2n+1} is smooth and has codimension $2r + g$ in Q_{2n+1}.*
(3) *If S^r_{2n+1} is smooth, then there is a subbundle $\mathcal{L}$ of the restriction of $\mathcal{K}$ to $X \times S^r_{2n+1}$ such that for every s in S^r_{2n+1}*

$$0 \subset \mathcal{L}_s \subset \mathcal{K}_s$$

is actually the Harder-Narasimhan filtration of $\mathcal{K}_s$.

Proof. These statements follow from general properties of the Shatz stratification of a family of vectors bundles over X (see [24, 25].) Nevertheless, for the convenience of the reader, we shall give a direct justification of (2).

Let $Z^r_{2n+1} \subset Q_{2n+1} \times \mathrm{Quot}^{[1,n-r]}_{\mathcal{O}^2_X/X/\mathbb{C}}$ be defined by

$$Z^r_{2n+1} = \{(M, A) \mid M = \mathcal{O}^2_X/K,\ \ A = \mathcal{O}^2_X/L \text{ and } L \subset K\}.$$

We embed S^r_{2n+1} in Z^r_{2n+1} by sending $M = \mathcal{O}^2_X/K$ to the pair $(M, \mathcal{O}^2_X/L_K)$, where

$$0 \subset L_K \subset K$$

is the Harder-Narasimhan filtration of K. This identifies S^r_{2n+1} with the open subset of Z^r_{2n+1} consisting of those pairs (M, A) such that K/L is locally free. By projection on the second factor, we obtain natural maps

$$\rho : S^r_{2n+1} \to \mathrm{Quot}^{[1,n-r]}_{\mathcal{O}^2_X/X/\mathbb{C}}\ .$$

The tangent space to Z^r_{2n+1} at (M, A) is the fibre product

$$\begin{array}{ccc}
\mathrm{T}_{(M,A)} Z^r_{2n+1} & \longrightarrow & \mathrm{T}_M Q_{2n+1} = \mathrm{Hom}(K, M) \\
\downarrow & & \downarrow{\scriptstyle \mathrm{Hom}(i,M)} \\
\mathrm{T}_A\, \mathrm{Quot}^{[1,n-r]}_{\mathcal{O}^2_X/X/\mathbb{C}} = \mathrm{Hom}(L, A) & \xrightarrow[\mathrm{Hom}(L,h)]{} & \mathrm{Hom}(L, M)
\end{array}$$

where $i : L \to K$ and $h : A \to M$ are the obvious morphisms. The fibre of ρ over some $A = \mathcal{O}_X^2/L$ can be naturally identified with an open subset of $\mathrm{Quot}_{A/X/\mathbf{C}}^{[0,2n+1]}$, namely the open subset of locally free submodules of A. It follows that the fibres are smooth of constant dimension

$$\dim \mathrm{Hom}_{\mathcal{O}_X}(K/L_K, \mathcal{O}_X^2/K) = 2n+1.$$

It is easy to see that, for (M, A) in S_{2n+1}^r, the vertical arrow on the left hand side of the diagram above can be identified with the tangent map $\mathrm{T}_M\rho$ of ρ at M. The vanishing of $\mathrm{Ext}^1(K/L, M)$ implies the surjectivity of $\mathrm{Hom}(i, M)$ and therefore, that of $\mathrm{T}_M\rho$.

Assume $n - r \geq 2g$. Every $A = \mathcal{O}_X^2/L$ in $\mathrm{Quot}_{\mathcal{O}_X^2/X/\mathbf{C}}^{[1,n-r]}$ is such that $\deg L^\vee \geq 2g-1$ and therefore $H^1(X, L^\vee)$ vanishes. On the other hand A is generated by global sections, so there is an exact sequence

$$0 \to \mathcal{O}_X \to A \to \mathcal{T} \to 0$$

with $\mathcal{T}$ a torsion sheaf. Since $H^1(X, L^\vee) = 0$, we have that

$$H^1(X, L^\vee \otimes_{\mathcal{O}_X} A) = 0.$$

It follows that $\mathrm{Quot}_{\mathcal{O}_X^2/X/\mathbf{C}}^{[1,n-r]}$ is smooth of dimension $2\cdot(n-r)+(1-g)$, and that S_{2n+1}^r is smooth of dimension $4n+2-(2r+g)$. Since Q_{2n+1} has dimension $4n+2$, the stratum S_{2n+1}^r has the expected codimension.

7. Cohomology of the strata

We compute the cohomology of the system $\mathbf{S}_1^r = \{S_{2n+1}^r\}_{n\in\mathbf{N}}$.

We have seen in the proof of Proposition 6.1 that $\mathrm{Quot}_{\mathcal{O}_X^2/X/\mathbf{C}}^{[1,d]}$ is smooth whenever $d \geq 2g$. In this case, we can apply the results of [3] and we obtain

Proposition 7.1. *Suppose $d \geq 2g$ and let $Q = \mathrm{Quot}_{\mathcal{O}_X^2/X/\mathbf{C}}^{[1,d]}$. There is a decomposition*

$$Q = W^d \cup C^d$$

where C^d is a closed subset isomorphic to $X^{(d)}$, and W^d is an open subset isomorphic to a vector bundle of rank $d-g+1$ over $X^{(d)}$. Moreover, the cohomology of Q is free of torsion and its Poincaré series is given by

$$P(Q;t) = (1 + t^{2\cdot(d-g+1)}) \cdot P(X^{(d)};t).$$

Proof. Recall from [3] that the fixed points, under the canonical action of $\mathbf{G}_m^2$, are precisely $X^{(d)} \times \mathrm{Hilb}_{X/\mathbf{C}}^{[1,0]}$ and $\mathrm{Hilb}_{X/\mathbf{C}}^{[1,0]} \times X^{(d)}$, embedded in Q by taking directs sums. Note that $\mathrm{Hilb}_{X/\mathbf{C}}^{[1,0]}$ consists of a single point (namely $\mathcal{O}_X = \mathcal{O}_X/0$) and that $\dim_{\mathbf{C}} \mathrm{Hom}_{\mathcal{O}_X}(\mathcal{O}_X(-D), \mathcal{O}_X) = \deg D - g + 1$ by Riemann-Roch. The result is now a straightforward application of the theorem in [3].

Proposition 7.2. *The sequence of restriction morphisms*

$$i_d^* : H^*(\mathrm{Quot}_{\mathcal{O}_X^2/X/\mathbf{C}}^{[1,d]}) \to H^*(C^d)$$

is an asymptotic isomorphism.

Proof. Consider the long exact sequence of cohomology groups with compact support

$$\ldots \to H_c^i(W^d) \to H^i(\mathrm{Quot}_{\mathcal{O}_X^2/X/\mathbf{C}}^{[1,d]}) \to H^i(C^d) \to H_c^{i+1}(W^d) \to \ldots$$

Since $H_j(W^d) = 0$ for $j \geq 2d+1$, we have by duality

$$H_c^i(W^d) \simeq H_{2(2d-g+1)-i}(W^d) = 0$$

for each $i \leq 2d - 2g + 1$. It follows that for a fixed $i \in \mathbf{N}$, the restriction map

$$H^i(\mathrm{Quot}_{\mathcal{O}_X^2/X/\mathbf{C}}^{[1,d]}) \to H^i(C^d)$$

is an isomorphism whenever $d \geq g + \frac{i}{2}$.

Note that the component $\Delta^r_{2n+1} = X^{(n-r)} \times X^{(n+r+1)}$ of the fixed points under the action of $\mathbb{G}_m^2$ on Q_{2n+1} is contained in S^r_{2n+1}.

Proposition 7.3. *The sequence of restriction morphisms*

$$\iota^*_{2n+1} : H^*(S^r_{2n+1}) \to H^*(\Delta^r_{2n+1})$$

induced by the inclusions $\iota_{2n+1} : \Delta^r_{2n+1} \hookrightarrow S^r_{2n+1}$, *is an asymptotic isomorphism. Moreover, there is also an asymptotic isomorphisms of graded rings*

$$H[x_1] \otimes H[x_2] \to H^*(\Delta^r_{2n+1}),$$

where $H = H^*(\mathrm{Pic}^0 X)$ *is the exterior algebra of a free* $\mathbb{Z}$*-module of rank* $2g$, *and* x_1, x_2 *are independent variables of degree* 2. *In particular, the Poincaré series of* S^r_1 *is given by*

$$P(S^r_1; t) = \frac{(1+t)^{4g}}{(1-t^2)^2}.$$

Proof. Let $\mathcal{L}$ be a Poincaré bundle over $X \times \mathrm{Pic}^{-(n+r+1)} X$. Let E^r_{2n+1} be the vector bundle over

$$\mathrm{Quot}^{[1,n-r]}_{\mathcal{O}^2_X/X/\mathbb{C}} \times \mathrm{Pic}^{-(n+r+1)} X$$

having fibre canonically isomorphic to $\mathrm{Hom}_{\mathcal{O}_X}(\mathcal{L}_{[F]}, A)$ over the point $(A, [F])$. Consider the maps

$$\alpha : S^r_{2n+1} \to \mathrm{Quot}^{[1,n-r]}_{\mathcal{O}^2_X/X/\mathbb{C}} \times \mathrm{Pic}^{-(n+r+1)} X$$

sending $M = \mathcal{O}^2_X/K$ to $(\mathcal{O}^2_X/L_K, [K/L_K])$ where $0 \subset L_K \subset K$ is the Harder-Narasimhan filtration of K. We can lift α to a map

$$j : S^r_{2n+1} \to \mathbb{P}(E^r_{2n+1})$$

by sending $M = \mathcal{O}^2_X/K$ to the class in $\mathbb{P}(\mathrm{Hom}_{\mathcal{O}_X}(\mathcal{L}_{[K/L_K]}, \mathcal{O}^2_X/L_K))$ of the homomorphism

$$\mathcal{L}_{[K/L_K]} \xrightarrow{\sigma} K/L_K \xrightarrow{i} \mathcal{O}^2_X/L_K$$

where σ denotes an isomorphism and i the inclusion. It is clear that this class is independent of the choice of σ.

Consider the commutative diagram

$$\begin{array}{ccc}
\Delta^r_{2n+1} = X^{(n-r)} \times X^{(n+r+1)} & \xrightarrow{\iota} & S^r_{2n+1} \\
\downarrow{\scriptstyle \tilde{j}} & & \downarrow{\scriptstyle j} \\
\mathbb{P}(\tilde{E}^r_{2n+1}) & \xrightarrow{\epsilon} & \mathbb{P}(E^r_{2n+1}) \\
\downarrow & & \downarrow \\
C^{n-r} \times \mathrm{Pic}^{-(n+r+1)} X & \xrightarrow{\delta} & \mathrm{Quot}^{[1,n-r]}_{\mathcal{O}^2_X/X/\mathbb{C}} \times \mathrm{Pic}^{-(n+r+1)} X
\end{array}$$

where $\tilde{E}^r_{2n+1}$ is the restriction of E^r_{2n+1} to $C^{n-r} \times \mathrm{Pic}^{-(n+r+1)} X$. We already know that δ, and therefore ϵ, induce asymptotic isomorphisms in cohomology. The claim would follow if we knew that j and $\tilde{j}$ do likewise. This is the content of the next two propositions.

Proposition 7.4. *The sequence of morphisms* $j : S^r_{2n+1} \to \mathbb{P}(E^r_{2n+1})$ *is a quasi-isomorphism.*

Proof. The morphisms are clearly open immersions. The intersection of the complement of the image of $\tilde{j}$ with the fibre of $\mathbb{P}(E^r_{2n+1})$ over $(A, [F])$ coincides with the classes of homomorphisms $\mathcal{L}_{[F]} \to A$ which are not injective. These are precisely the homomorphisms that factor through the inclusion $A_{\mathrm{tor}} \hookrightarrow A$ of the torsion submodule of A. The short exact sequence

$$0 \to \mathrm{Hom}_{\mathcal{O}_X}(\mathcal{L}_{[F]}, A_{\mathrm{tor}}) \to \mathrm{Hom}_{\mathcal{O}_X}(\mathcal{L}_{[F]}, A) \to \mathrm{Hom}_{\mathcal{O}_X}(\mathcal{L}_{[F]}, A/A_{\mathrm{tor}}) \to 0$$

shows that they form a subspace of dimension $\deg A_{\rm tor}$. Since $\deg A_{\rm tor} \leq n - r$ and $\dim_{\mathbb{C}} \mathrm{Hom}_{\mathcal{O}_X}(\mathcal{L}_{[F]}, A) = 2n-g+2$ whenever $n+r+1 \geq 2g-1$, we have that codim $\mathbb{P}(E^r_{2n+1}) - j(S^r_{2n+1})$ grows indefinitely as $n \to \infty$.

Proposition 7.5. *The sequence of morphisms* $\tilde{\jmath} : \Delta^r_{2n+1} \to \mathbb{P}(\tilde{E}^r_{2n+1})$ *is a quasi-isomorphism.*

Proof. For large n, $\tilde{\jmath}$ maps Δ^r_{2n+1} isomorphically onto the projectivization of the subbundle of $\tilde{E}^r_{2n+1}$ whose fibre over $(D, [F])$, $D \in X^{(n-r)}$, is

$$\mathrm{Hom}_{\mathcal{O}_X}(\mathcal{L}_{[F]}, \mathcal{O}_X) \subset \mathrm{Hom}_{\mathcal{O}_X}(\mathcal{L}_{[F]}, (\mathcal{O}_X/\mathcal{O}_X(-D)) \oplus \mathcal{O}_X)$$

The dimensions of the corresponding projective spaces are $n + r - g + 2$ and $2n - g + 2$ respectively. It follows that for large n the codimension of $\mathbb{P}(\tilde{E}^r_{2n+1}) - \tilde{\jmath}(\Delta^r_{2n+1})$ is $n - r$ which grows indefinitely as $n \to \infty$.

8. PERFECTION

Finally, we prove that the stratification is perfect.

By perfection we mean that the Thom-Gysin long exact sequences

$$\cdots \to H^{i-2d_r}(S^r_{2n+1}) \xrightarrow{\tau^{r,i}_{2n+1}} H^i(U^r_{2n+1}) \longrightarrow H^i(U^{r-1}_{2n+1}) \to \cdots$$

where $n \geq r+2g$ and $d_r = 2r+g$, are such that for every $i \in \mathbb{N}$ there is an n_i such that $\tau^{r,i}_{2n+1}$ is injective for every r and every $n \geq n_i$. Recall the fundamental observation of Atiyah and Bott [2] that, since the composite

$$\begin{array}{ccc} H^{i-2d_r}(S^r_{2n+1}) & \xrightarrow{\tau^{r,i}_{2n+1}} & H^i(U^r_{2n+1}) \\ & & J \big\downarrow \text{restriction} \\ & & H^i(S^r_{2n+1}) \end{array}$$

is simply cup product with the top Chern class e^r_{2n+1} of the normal bundle N^r_{2n+1} to S^r_{2n+1} in Q_{2n+1}, in order to check injectivity of $\tau^{r,i}_{2n+1}$ it suffices to check it for the multiplication map

$$- \cup e^r_{2n+1} : H^{i-2d_r}(S^r_{2n+1}) \longrightarrow H^i(S^r_{2n+1}).$$

Proposition 8.1. *Let X be a smooth projective algebraic curve of genus g. The stratification*

$$Q_1 = \bigcup_{r \in \{-1\} \cup \mathbb{N}} S^r_1$$

is perfect. In particular, we have the identity of Poincaré series:

$$P(Q_1; t) = P(Q^s_1; t) + \sum_{r \in \mathbb{N}} t^{2(2r+g)} \cdot P(S^r_1; t).$$

Proof. Consider the diagram

$$\begin{array}{ccccc} H^{i-2d_r}(S^r_{2n+1}) & \longrightarrow & H^{i-2d_r}(\Delta^r_{2n+1}) & \xleftarrow{\varphi^{i-2d_r}_{2n+1}} & H^{i-2d_r}(\mathbf{\Delta}^r) \\ {\scriptstyle -\cup e^r_{2n+1}} \big\downarrow & & {\scriptstyle -\cup \tilde{e}^r_{2n+1}} \big\downarrow & & \big\downarrow {\scriptstyle -\cup e^r} \\ H^i(S^r_{2n+1}) & \longrightarrow & H^i(\Delta^r_{2n+1}) & \xleftarrow[\varphi^i_{2n+1}]{} & H^i(\mathbf{\Delta}^r) \end{array}$$

where $H^*(\mathbf{\Delta}^r) = H \otimes H[x_1, x_2]$, n is large enough so that the horizontal arrows are isomorphisms, and $\varphi^{2d_r}_{2n+1}(e^r) = \tilde{e}^r_{2n+1}$. It is clear that if e^r is not a zero-divisor in $H^*(\mathbf{\Delta}^r)$, then all the vertical arrows are injective. We shall apply the following Lemma (see [2])

Lemma 8.1. *Let R^* be a graded ring such that R^0 is an integral domain and each R^i is a free R^0-module. If an element $e \in R^*[x]$ is sent, by the natural projection $R^*[x] \to R^0[x]$, into a nonzero element, then e cannot be a zero-divisor in $R^*[x]$.*

Fix a point $P \in X$ and define, for n large enough, the maps

$$\begin{array}{rcccl} \eta: & \mathbb{P}^{n-r-g} = |(n-r)P| & \longrightarrow & \Delta^r_{2n+1} = X^{(n-r)} \times X^{(n+r+1)} \\ & D & \longmapsto & (D,\ (n+r+1)P) \end{array}$$

We have a commutative diagram

$$\begin{array}{ccccc} H^{2d_r}(S^r_{2n+1}) & \longrightarrow & H^{2d_r}(\Delta^r_{2n+1}) & \xleftarrow{\varphi^{2d_r}_{2n+1}} & H^{2d_r}(\Delta^r) = R^*[x_2] \\ & & \downarrow \eta^* & & \downarrow \\ & & H^{2d_r}(\mathbb{P}^{n-r-g}) & \longleftarrow & H^{2d_r}(\mathbb{P}^\infty) = R^0[x_2] \end{array}$$

such that all the horizontal arrows are isomorphisms. Therefore, it suffices to show that the restriction $\eta^*\tilde{e}^r_{2n+1}$ is, for large n, different from zero. This is equivalent to showing that the top Chern class of the restriction to $|(n-r)P|$ of the normal bundle is non-zero.

Note that if $M = \mathcal{O}^2_X/K$ in S^r_{2n+1} has Harder-Narasimhan filtration $0 \subset L_K \subset K$, then the bottom sequence (where $A = \mathcal{O}^2_X/L_K$) of the following diagram

$$\begin{array}{ccccccc} & \mathrm{T}_M S^r_{2n+1} & \longrightarrow & \mathrm{T}_M Q_{2n+1} & \longrightarrow & \mathrm{Ext}^1(L_K, K/L_K) \to 0 \\ & \downarrow & & \downarrow \mathrm{Hom}(i,M) & & \| \\ 0 \to & \mathrm{T}_A\,\mathrm{Quot}^{[1,n-r]}_{\mathcal{O}^2_X/X/\mathbb{C}} & \xrightarrow[\mathrm{Hom}(L_K,h)]{} & \mathrm{Hom}(L_K, M) & \longrightarrow & \mathrm{Ext}^1(L_K, K/L_K) \to 0 \end{array}$$

is exact, and $\mathrm{Ext}^1(L_K, K/L_K)$ can be identified with the normal space to S^r_{2n+1} at M. Globally, if $\mathcal{K}$ still denotes its restriction to $X \times S^r_{2n+1}$, we have

$$N^r_{2n+1} \simeq \mathrm{R}^1 q_* \mathcal{H}om\,(\mathcal{L}, \mathcal{K}/\mathcal{L})$$

where $\mathcal{L} \subset \mathcal{K}$ is the subbundle over $X \times S^r_{2n+1}$ considered in Proposition 6.1, and $q: X \times S^r_{2n+1} \to S^r_{2n+1}$ is the canonical projection. Since $\mathcal{L}$ and $\mathcal{K}$ are flat over S^r_{2n+1}, and $\mathrm{R}^i q_*$ vanishes for $i \geq 2$, base change commutes with taking $\mathrm{R}^1 q_*$. Let $\mathcal{O}_{X \times X^{(d)}}/\mathcal{J}_d$ be the universal quotient over $X \times X^{(d)}$. The restriction of $\mathcal{H}om\,(\mathcal{L}, \mathcal{K}/\mathcal{L})$ to $X \times \Delta^r_{2n+1}$ is isomorphic to $\mathcal{H}om(q_1^*\mathcal{J}_{n-r}, q_2^*\mathcal{J}_{n+r+1})$ where $q_1 : X \times \Delta^r_{2n+1} \to X \times X^{(n-r)}$ and $q_2 : X \times \Delta^r_{2n+1} \to X \times X^{(n+r+1)}$ are the natural projections. Moreover, we have

$$\begin{aligned} &(1 \times \eta)^* \mathcal{H}om(q_1^*\mathcal{J}_{n-r}, q_2^*\mathcal{J}_{n+r+1}) = \\ &= \mathcal{H}om\,(p_2^*\mathcal{O}_{\mathbb{P}}(-1) \otimes p_1^*\mathcal{O}_X(-(n-r)P),\ p_1^*\mathcal{O}_X(-(n+r+1)P)) \end{aligned}$$

where p_1 and p_2 are the natural projections of $X \times \mathbb{P}^{n-r-g}$. Applying $\mathrm{R}^1 p_{2*}$ and using the projection formula we get:

$$\begin{aligned} &\mathrm{R}^1 p_{2*} \mathcal{H}om\,(p_2^*\mathcal{O}_{\mathbb{P}}(-1) \otimes p_1^*\mathcal{O}_X(-(n-r)P),\ p_1^*\mathcal{O}_X(-(n+r+1)P)) \simeq \\ &\simeq \mathrm{R}^1 p_{2*} p_2^*\mathcal{O}_{\mathbb{P}}(1) \otimes p_1^*\,(\mathcal{O}_X(-(n+r+1)P) \otimes \mathcal{O}_X((n-r)P)) \\ &\simeq \mathcal{O}_{\mathbb{P}}(1) \otimes \mathrm{R}^1 p_{2*} p_1^* \mathcal{O}_X(-(2r+1)P) \\ &\simeq \mathcal{O}_{\mathbb{P}}(1) \otimes H^1(X, \mathcal{O}_X(-(2r+1)P)) \\ &\simeq \mathcal{O}_{\mathbb{P}}(1)^{\oplus 2r+g}. \end{aligned}$$

Since we only consider strata of positive codimension $2r+g$, the top Chern class of this bundle is different from zero and we are done.

REFERENCES

1. E. Arbarello, M. Cornalba, P. A. Griffiths and J. Harris, *Geometry of Algebraic Curves I*, Grundlehren der math. Wissenschaften 267, Springer-Verlag, New York 1985.
2. M. F. Atiyah and R. Bott, The Yang-Mills equations over Riemann surfaces, *Phil. Trans. Roy. Soc. London, Ser. A* **308** (1982), 523–615.
3. E. Bifet, Sur les points fixes du schéma $\mathrm{Quot}_{\mathcal{O}^r_X/X/k}$ sous l'action du tore $G^r_{m,k}$, *C. R. Acad. Sci. Paris*, t. **309** (1989), 609–612.

4. G. D. Daskalopoulos, *The topology of the space of stable bundles over a compact Riemann surface*, Ph.D. Thesis, University of Chicago.
5. P. Deligne, Cohomologie à supports propres, Exposé XVII, in *Théorie des Topos et Cohomologie Etale des Schémas (SGA 4) Tome* 3, Lecture Notes in Math. 305, Springer-Verlag 1973.
6. U. V. Desale and S. Ramanan, Poincaré polynomials of the variety of stable bundles, *Math. Ann.*, **216** (1975), 233–244.
7. A. Dold and R. Thom, Une géneralisation de la notion d'espace fibré. Application aux produits symmétriques infinis, *C. R. Acad. Sci. Paris*, t. **242** (1956), 1680–1682.
8. A. Dold and R. Thom, Quasifaserungen und unendliche symmetrishe Produkte, *Ann. of Math.*, **67** (1958), 239–281.
9. F. Ghione and M. Letizia, Effective divisors of higher rank on a curve and the Siegel formula, preprint.
10. A. Grothendieck, Techniques de construction et théorèmes d'existence en géométrie algébrique IV: Les schémas de Hilbert, Séminaire Bourbaki 1960/61 no. 221.
11. G. Harder, Eine Bemerkung zu einer Arbeit von P. E. Newstead, *J. für Math.*, **242** (1970), 16–25.
12. G. Harder, Semisimple group schemes over curves and automorphic functions, *Actes, Congrès intern. Math. 1970, Tome 2*, 307–312.
13. G. Harder and M. S. Narasimhan, On the cohomology groups of moduli spaces of vector bundles on curves, *Math. Ann.*, **212** (1975), 215–248.
14. F. Kirwan, On spaces of maps from Riemann surfaces to Grassmannians and applications to the cohomology of moduli of vector bundles, *Ark. för Mat.*, **24** (1986), 221–275.
15. G. Laumon, Correspondence de Langlands géométrique pour les corps de fonctions, Duke Math. Jour., **54** (1987), 309–359.
16. H. B. Lawson, Algebraic cycles and homotopy theory, *Ann. of Math.*, **129** (1989), 253–291.
17. I. G. Macdonald, Symmetric products of an algebraic curve, *Topology*, **1** (1962), 319–343.
18. I. G. Macdonald, The Poincaré polynomial of a symmetric product, *Proc. Camb. Phil. Soc.*, **58** (1962), 563–568.
19. P. E. Newstead, Topological properties of some spaces of stable bundles, *Topology*, **6** (1967), 241–262.
20. P. E. Newstead, *Introduction to moduli problems and orbit spaces*, Tata Inst. Lecture Notes, Springer-Verlag 1978.
21. C. S. Seshadri, *Fibrés vectoriels sur les courbes algébriques*, Astérisque 96, 1982.
22. I. R. Shafarevich, On some infinite-dimensional groups, *Rend. Mat. Appl., V Ser.* **25** (1967), 208–212.
23. I. R. Shafarevich, On some infinite-dimensional groups II, *Math. USSR, Izv.*, **18** (1982), 185–194.
24. S. S. Shatz, The decomposition and specialization of algebraic families of vector bundles, *Compositio Math.*, **35** (1977), 163–187.
25. J.-L. Verdier and J. Le Potier, eds. *Module des Fibrés Stables sur les Courbes Algébriques*, Progress in Math. 54, Birkhäuser, Boston 1985.
26. A. Weil, Généralization des fonctions Abéliennes, *J. Math. Pures Appl.*, **17** (1938), 47–87.

MATHEMATICAL SCIENCES RESEARCH INSTITUTE, 1000 CENTENNIAL DRIVE, BERKELEY, CA 94720

Current address: DEPARTMENT OF MATHEMATICS AND INSTITUTE FOR MATHEMATICAL SCIENCES, STATE UNIVERSITY OF NEW YORK AT STONY BROOK, STONY BROOK, NY 11794-3651 (E. BIFET)
E-mail address: bifet@math.sunysb.edu

DIPARTIMENTO DI MATEMATICA, UNIVERSITÀ DI ROMA–TOR VERGATA, VIA DELLA RICERCA SCIENTIFICA, 00133 ROMA, ITALIA (F. GHIONE AND M. LETIZIA)

GAUSSIAN MAPS FOR CERTAIN FAMILIES OF CANONICAL CURVES

by

Ciro Ciliberto and Rick Miranda

Contents:

0. Introduction

In this paper we will make some computations and remarks concerning Gaussian maps for curves. Recall that given two line bundles L and M on a curve C, one has the multiplication map on global sections

$$\mu: H^0(C,L)\otimes H^0(C,M) \longrightarrow H^0(C,L\otimes M)$$

which has kernel denoted by $\mathcal{R}(L,M)$. The Gaussian map

$$\phi_{L,M}: \mathcal{R}(L,M) \longrightarrow H^0(C,\Omega^1_C\otimes L\otimes M)$$

is (essentially) defined by sending $\sigma\otimes\tau \in \mathcal{R}(L,M)$ to $d\sigma\otimes\tau - \sigma\otimes d\tau$, where differentiation is defined with respect to a local parameter. Note that if $L = M$, the map can be considered as going from $\Lambda^2H^0(L)$ to $H^0(\Omega^1\otimes L^2)$. Of course one can also consider the restriction to sublinear systems $V \subseteq H^0(L)$.

This map has a geometrical interpretation: it is the map on global sections for the Gauss map for the morphism of C into a projective space given by $H^0(L)$, if $|L|$ is base-point-free.

The Gaussian map we are primarily concerned with here is $\phi = \phi_{\omega,\omega}$, where ω is the canonical bundle of C; it maps $\Lambda^2H^0(\omega)$ to $H^0(\Omega^1\otimes\omega^{\otimes 2})$, and is perfectly well-defined in this form for a stable curve.

In a previous paper [C-M], we have analyzed the rank of this Gaussian map ϕ for curves of low genus, i.e., curves with genus g less than 12. Except for genus 9 and 11, we have found ϕ to be injective for a general curve. Together with J. Harris [C-H-M], we proved that for genus equal to 10 or at least 12, ϕ is surjective for the general curve. Precisely, these results are summarized below.

(0.1)<u>Table:</u> Generic behavior of the Gaussian map $\phi:\Lambda^2 H^0(\omega) \longrightarrow H^0(\omega^3)$.

g	$3\le g\le 8$	9	10	11	≥ 12
ϕ:	injective	1-dim'l kernel	isomorphism	corank 1	surjective

One can stratify the moduli space $\mathcal{M}_g$ for stable curves of genus g by the rank of the Gaussian map ϕ, which is well-behaved since this rank is semicontinuous. One of the motivations of this paper was to understand some of the features of this stratification. Specifically, we wanted to understand as far as possible the relationships between this stratification and other more classical loci (e.g., hyperelliptic, trigonal, etc.) in $\mathcal{M}_g$.

When the present paper was essentially completed, we became aware of Wahl's preprint [W2], which shares some of the same motivations of this note. Some results there are closely related, as we shall see, with the results here; however there is no real overlap, except perhaps in section 1.

For an excellent overview of the study of these Gaussian maps on curves, the reader may wish to consult [W3].

As noticed by J. Wahl [W1], these Gaussian maps are also related to the deformation theory of the cones over canonical curves. The way this comes about is because of the following two well-known results. The first relates the cokernels of these Gaussian maps to the spaces of global sections of twists of normal bundles to C.

(0.2)<u>Proposition:</u> Let C be a linearly normal curve in $\mathbb{P}^r$, let $L = \mathcal{O}_C(1)$, and let N_C be the normal bundle of C in $\mathbb{P}^r$. Then

(a) The sequence
$$0 \longrightarrow H^0(C,L)^\vee \longrightarrow H^0(C,N_C(-1)) \longrightarrow \operatorname{coker}(\phi_{\omega,L})^\vee \longrightarrow 0$$
is exact.

(b) If $k \ge 2$, then $H^0(C,N_C(-k)) \cong \operatorname{coker}(\phi_{\omega\otimes L^{k-1},L})^\vee$.

For a proof of this, one can consult [C-M, Proposition (1.2)]. The second fact is that these spaces of global sections are in turn related to the deformation theory of the cone over C. Specifically, one has the following (see [P]).

(0.3)<u>Proposition:</u> Let C be projectively normal in $\mathbb{P}^r$, and let X be the cone over C in $\mathbb{P}^{r+1}$. Then $H^0(X,N_X) \cong \bigoplus_{k\ge 0} H^0(C,N_C(-k))$.

In [C-M, Corollary (4.4)], we have explicitly computed $h^0(N_C(-k))$ for all nonnegative k, for the general curve of genus g; these results are reproduced below.

(0.4)Proposition: Let C be a general canonical curve of genus $g \geq 3$. Then we have the following dimensions for the global sections of nonpositive twists of the normal bundle to C.

$h^0(N_C(-k)) \setminus g$	3	4	5	6	7	8	9	10	11	≥ 12
$k = 0$:	$g^2 + 3g - 4$ for every g									
$k = 1$:	10	13	15	16	16	15	14	10	12	g
$k = 2$:	6	5	3	1	0	0	0	0	0	0
$k = 3$:	3	1	0	0	0	0	0	0	0	0
$k = 4$:	1	0	0	0	0	0	0	0	0	0
$k \leq 5$:	0 for every g									
$h^0(N_X)$:	$g^2 + 2g + 19$ for $g \leq 9$ or $= 11$; $g^2 + 4g - 4$ otherwise									

As a consequence of these computations, we could prove that the point ⟨X⟩ of the Hilbert scheme representing the cone X over the general canonical curve C is smooth. On the other hand, the numbers in the above table can only increase for special canonical curves, and they do if and only if the point ⟨X⟩ is a singular point of the Hilbert scheme. Therefore another interpretation of the strata of $\mathcal{M}_g$ defined by the rank of ϕ is in terms of extent of singularity of the point ⟨X⟩.

In section 1 of this paper we discuss the case of hyperelliptic curves, which is rather easy and well-known: the rank of ϕ is 2g-3 in this case. Essentially we only make certain remarks concerning the fact that this is a lower bound for the rank of ϕ, and that the only smooth curves for which it is attained are the hyperelliptic curves. This has been proved by J. Wahl in [W2, Theorem 4.6], but also follows from a classical theorem of B. Segre.

In section 2 we discuss the case of curves for which Petri's theorem fails, i.e., which are not cut out by quadrics: the trigonal curves and smooth plane quintics. We compute the cohomology of the normal bundle for the general such curve, and it turns out that corank of ϕ jumps by g + 5 for the trigonal curves; it jumps by 10 (= g+4) for the quintics. Related results about curves lying on scrolls are contained in [W2, section 4], where a lower bound for the dimension of the cokernel of ϕ is given for such curves. The interested reader should also consult [D-M] for a precise computation of the corank of ϕ for most curves on scrolls.

In section 3 we discuss elliptic-hyperelliptic curves and curves on Del Pezzo surfaces, and we obtain similar results using the same methods as in section 2. In section 4 we try to compute ϕ explicitly using equations, for certain cyclic covers of $\mathbb{P}^1$. We succeed in the triple cover case, and we find explicit curves with $\dim(\mathrm{coker}(\phi)) = g+5$. On the other hand, we can prove that for the cyclic covers studied here the Gaussian map ϕ is never surjective.

In the final section we make some speculations concerning the Gaussian map ϕ for general curves lying on a K3 surface. Our discussion suggests that for most large enough genera, the corank of ϕ for such curves should be one, although we do not feel confident enough to attempt a conjecture. For certain low genera we are able to apply the techniques developed here and in [C-M] to prove a smoothability statement concerning unions of planes to smooth K3 surfaces. For example we prove in Proposition (5.3) that the union of planes in $\mathbb{P}^{11}$ whose dual complex is the dodecahedron can be so deformed.

The first author would like to thank the Department of Mathematics at Colorado State University for an invitation to visit while this work was completed.

1. Hyperelliptic curves

Let C be a hyperelliptic curve of genus g, given as $y^2 = f(x)$, with f of degree 2g + 1, having simple roots. The space of holomorphic 1-forms on C has a basis of the form $\{x^i dx/y \mid 0 \le i \le g-1\}$. The image of the Gaussian map ϕ is then $\{g(x)\cdot(dx/y)^3 \mid \deg(g) \le 2g - 4\}$, so that:

(1.1)<u>Proposition:</u> For a hyperelliptic curve of genus $g \ge 2$, the rank of the Gaussian map ϕ is 2g - 3. In particular, the dimension of the cokernel of ϕ is 3g - 2.

This also has been noticed by J. Wahl [W, Remark (5.8.1)].

There are more general results which imply Proposition (1.1). In particular, firstly, one sees rather easily that 2g - 3 is a lower bound for the rank of ϕ; this follows from the general proposition below.

(1.2)<u>Proposition:</u>(G. Gherardelli [G]) Let L be a line bundle on a curve C, let V be a subspace of $H^0(C,L)$ of dimension at least 2, and let $\phi_V:\Lambda^2 V \longrightarrow H^0(L^{\otimes 2}\otimes \omega_C)$ be the associated Gaussian map. Then $\mathrm{rank}(\phi_V) \ge 2\dim(V) - 3$.

Proof: It is enough to notice that no indecomposable tensor $\sigma_1\wedge\sigma_2$ belongs to the kernel K of ϕ. Since the Grassmann of lines in $\mathbb{P}(V)$ has dimension $2\dim(V) - 4$, we must have that the codimension of $\mathbb{P}(K)$ in $\mathbb{P}(V)$ is at most $2\dim(V) - 3$, and this proves the result. ■

Note that this bound is achieved in the case that C is a smooth rational curve and V is complete. Note that if $f_V: C \longrightarrow \Gamma \subset \mathbb{P}(V^*)$ is the associated map to a linear system V on a curve C, then the Gaussian map for V on C factors through Γ. To be more precise, if $\tilde{\Gamma}$ is the normalization of Γ and M is the line bundle on $\tilde{\Gamma}$ which is the pull-back of $\mathcal{O}_\Gamma(1)$, and $W \subset H^0(\tilde{\Gamma}, M)$ is the subspace defining the map to Γ, then ϕ_V on C factors through ϕ_W on $\tilde{\Gamma}$. This shows that the bound is also attained for the canonical system on a hyperelliptic curve of genus at least 2.

In fact, a converse of this result is true, due to a theorem of B. Segre [S].

(1.3)<u>Theorem:</u> Suppose C is a smooth curve, L is a line bundle on C, and V is a base-point-free subspace of $H^0(C,L)$, with $\dim_{\mathbb{C}} V = r+1 \geq 5$. Then the rank of $\phi_V: \Lambda^2 V \longrightarrow H^0(L^2 \otimes \omega)$ is equal to $2r-1$ if and only if the image Γ of the map $\psi_V: C \longrightarrow \mathbb{P}(V^*)$ associated to V is a rational normal curve.

Proof: Assume first that Γ is a rational normal curve; in this case, if ψ_V is birational onto its image, the assertion is trivial. If ψ_V is not birational onto its image, the statement follows from the fact that the Gaussian map for V on C factors through Γ.

Our presentation of the proof of the converse is inspired by the argument by J. Wahl in [W2, Theorem 4.6]. By the same reasoning as above we may restrict ourselves to the case that ψ_V is birational onto its image. Let p be any point on C, and let $(0=a_0<a_1<\ldots<a_r)$ be the vanishing sequence of V at p; moreover let $(s_0,\ldots,s_r)$ be a basis of V which realizes this vanishing sequence. This means that if t is a local parameter at p, we may assume that a local description of s_i is given by $s_i = t^{a_i}$ + higher order terms. Then via the Gaussian map ϕ_V the pure tensor $s_i \wedge s_j$ is mapped to

$$[(a_j - a_i)t^{a_i+a_j-1} + \text{higher order terms}]dt.$$

Hence the dimension of the image of ϕ_V is at least the cardinality of the set $S_{\underline{a}} = \{a_i+a_j \mid 0 \leq i < j \leq r\}$; therefore $|S_{\underline{a}}| \leq 2r-1$, by our hypothesis.

Since the a_i's are strictly increasing, we see that $S_{\underline{a}}$ has at least $2r-1$ elements, namely $\{a_1,a_2,\ldots,a_r,a_1+a_r,a_2+a_r,\ldots,a_{r-1}+a_r\}$ (reproving Gherardelli's theorem). Therefore $|S_{\underline{a}}| = 2r-1$. It is now easy to see that this forces the numbers a_i+a_j to only depend on $i+j$, and if $r \geq 4$ then this forces $a_i = ia_1$ for $i \geq 1$. However since ψ_V is birational, we have that the greatest common divisor of the a_i's is one; hence $a_1 = 1$ and $a_i = i$ if $r \geq 4$. This means that p is not an inflectional point for the map ψ_V; since this is true for every p on C, we conclude that Γ is a rational normal curve, which is the only curve in a projective space without inflectional points. ■

As a consequence, we have the following.

(1.4)Corollary: Let C be a smooth curve of genus $g \geq 2$, and let ϕ be the Gaussian map for C. Suppose that the rank of ϕ is equal to $2g - 3$. Then C is hyperelliptic.

Segre's original proof of Theorem (1.3) is in the style of the old projective differential geometry which was common in his day.

For high genus, the rank of the Gaussian map ϕ is bounded only by the dimension of the target space (it is in fact generically surjective by [C-H-M]) so that we see that

$$2g - 3 \leq \text{the rank of } \phi \leq 5g - 5.$$

It is amusing to remark that there are exactly $3g - 3$ (which by chance is exactly the dimension of the moduli space for curves of genus g) numbers between the upper and lower bounds. Moreover note that both are achieved, the lower by the hyperelliptics and the upper by the general curve. One is led to wonder whether every number between the two is a rank for an actual (stable) curve of genus g.

For low genus (at most 11), the rank of ϕ is restricted by its domain not its range, (except in genus 9 and 11), because generically ϕ is injective in these cases (see [C-M]). Therefore for g between 2 and 8 we have that $2g - 3 \leq \text{rank}(\phi) \leq g(g-1)/2$. Note that for genus 2 and 3, this determines the rank of ϕ. For genus 4, both the lower bound of rank 5 and the upper bound of rank 6 are attained, by the hyperelliptics and the general curve respectively. The examples which are produced in later sections of this paper will demonstrate other intermediate ranks for most genera. This is essentially the only evidence we have for a positive answer to the question in the above paragraph.

2. Trigonal curves and plane quintics

There are two types of curves of genus $g \geq 4$ for which the canonical image is not cut out by hypersurfaces of degree 2, namely, the trigonal curves of any genus $g \geq 4$, and the smooth plane quintics (of genus 6). In this section we will compute the rank of the Gaussian map ϕ for these curves, and more generally we will calculate the cohomology of the twists of the normal bundle for the canonical image of C. We require a preliminary lemma.

(2.1) Lemma:

(a) Let S be the Veronese surface in $\mathbb{P}^{r=5}$. Then
$h^0(N_S(-1)) = 6$ and $h^0(N_S(-j)) = 0$ for $j \geq 2$.

(b) Let S be a rational normal scroll in $\mathbb{P}^{r \geq 4}$. Then
$h^0(N_S(-1)) = 2r-2$ and $h^0(N_S(-j)) = 0$ for $j \geq 2$.
Also h^1 and h^2 of $N_S(-1)$ vanish in this case.

Proof: Let $\mathbb{P} = \mathbb{P}^r$. Since in both cases the ideal of S is generated by forms of degree 2, the vanishing of the h^0's for $j \geq 3$ is clear. Using the Euler sequence for T_S and the defining sequence for N_S, one sees easily that $h^0(T_{\mathbb{P}}|_S(-1)) = r+1$ and $h^0(T_{\mathbb{P}}|_S(-2)) = 0$.

In the case of the Veronese, h^0 and h^1 of $T_S(-1)$ both vanish, so that $h^0(N_S(-1)) = 6$. Then $h^0(N_S(-2)) \leq h^1(T_S(-2)) = 0$, again using the Euler sequence for $\mathbb{P}^2$. This proves (a).

To make the computation for the scrolls, we are going to use the sequence introduced [C, (2.2)]. Let $\varphi: S \longrightarrow \mathbb{P}^2$ be a general projection; then we have the exact sequence

$$(2.2) \qquad 0 \longrightarrow \mathcal{O}_S(1)^{\oplus(r-2)} \longrightarrow N_S \longrightarrow N_\varphi \longrightarrow 0,$$

where N_φ is the normal sheaf to the map φ; it is supported on the ramification locus for φ, which is a curve. Therefore $h^2(N_\varphi(j)) = 0$ for every j, so that since $h^2(\mathcal{O}_S) = 0$, we have that $h^2(N_S(-1)) = 0$.

Since $h^1(\mathcal{O}_S)$ is also zero, we see that $h^1(N_S(-1)) = h^1(N_\varphi(-1))$. To calculate this final group, we refer to the notation and calculation given in [C, pp. 358-9]; the same argument there shows that $h^1(N_\varphi(-1)) = h^1(N'_\psi(-1))$, and this vanishes because $N'_\psi(-1)$ is a line bundle of degree r-1 on the ramification locus of φ, which is a smooth rational curve. Hence $h^1(N_S(-1)) = 0$.

The Euler-Poincare characteristic of $N_S(-1)$ is $2r - 2$, which is equal to h^0.

Using (2.2) again, we see that $H^0(N_S(-2))$ injects into $H^0(N_\varphi(-2))$, which has the same dimension as $H^0(N'_\psi(-2))$; this is zero by degree reasons. This finishes the proof of (b). ∎

Let us first turn our attention to the plane quintic curve.

(2.3)<u>Theorem:</u> Let C be a smooth plane quintic curve, so that the canonical image of C lies on the Veronese surface $S \subset \mathbb{P}^5$. Let N_C denote the normal bundle of C in $\mathbb{P}^5$. Then $h^0(N_C(k))$ (for $k \le 0$) is as in the following table.

k	0	-1	-2	$\le$-3
$h^0(C,N_C(k))$	50	16	3	0

Proof: Consider the following sequences:

(2.4) $$0 \longrightarrow N_{C/S} \longrightarrow N_C \longrightarrow N_{S|_C} \longrightarrow 0,$$

and

(2.5) $$0 \longrightarrow N_S(-C) \longrightarrow N_S \longrightarrow N_{S|_C} \longrightarrow 0,$$

which induce the sequences

(2.6) $$0 \longrightarrow N_{C/S}(-jH) \longrightarrow N_C(-jH) \longrightarrow N_{S|_C}(-jH) \longrightarrow 0$$

and

(2.7) $$0 \longrightarrow N_S(-C-jH) \longrightarrow N_S(-jH) \longrightarrow N_{S|_C}(-jH) \longrightarrow 0$$

for every j, where H is the hyperplane in $\mathbb{P}^5$.

It is the H^0 of the middle term of (2.6) that we are interested in. If we denote by Γ the curve C considered as a plane curve, then $N_{C/S}(-jH) \cong \mathcal{O}_\Gamma(5-2j)$, and so has cohomology

$$h^0(N_{C/S}(-jH)) = \begin{cases} 10 \text{ if } j = 1 \\ 3 \text{ if } j = 2, \\ 0 \text{ if } j \ge 3 \end{cases} \text{ and } h^1(N_{C/S}(-jH)) = \begin{cases} 0 & \text{if } j = 1 \\ 3 & \text{if } j = 2. \\ 10(j-2) & \text{if } j \ge 3 \end{cases}$$

Observe that h^0 and h^1 of $N_S(-C-jH)$ are both zero if $j \ge 1$; for h^0 this is a consequence of the fact that the ideal of S is cut out by quadrics, and for h^1 we refer to [C, Corollary (1.8)]. Therefore $h^0(N_{S|_C}(-jH)) = h^0(N_S(-jH))$ for $j \ge 1$, using (2.7).

If $j \ge 3$, we see that both $h^0(N_S(-jH))$ and $h^0(N_{C/S}(-jH))$ vanish, so that $h^0(N_C(-jH))$ is also zero by (2.6).

By Lemma (2.1), $h^0(N_S(-2H)) = 0$, and therefore $h^0(N_C(-2)) = 3$.

From (2.6) and Lemma (2.1) we see that

$$h^0(N_C(-1)) = h^0(N_{C/S}(-1)) + h^0(N_{S|_C}(-1)) = 10 + h^0(N_S(-1)) = 16. \quad \blacksquare$$

With the same methods as used above one can treat the case of the canonical image C of a smooth plane curve of degree at least 6. We obtain that $h^0(N_C(-1)) = 10 + (d-1)(d-2)/2$ and $h^0(N_C(-2)) = 0$ if $d \ge 7$; $h^0(N_C(-2)) = 1$ if $d = 6$.

Let us turn our attention to the case of trigonal curves. The analogue of Theorem (2.3) follows.

(2.8)Theorem: Let C be a general trigonal curve of genus $g \geq 4$. Then $h^0(N_C(k))$ (for $k \leq 0$) is given below.

k	g	0	-1	-2	≤-3
$h^0(C,N_C(k))$	≥10	(g-1)(g+4)	2g+5	0	0
	8,9	same	same	0	0
	7	same	same	1	0
	6	same	same	2	0
	5	same	same	3	0
	4	same	same	5	1 for k=-3, then 0

Proof: The line of argument is essentially equal to that for the plane quintic. The case of $k = 0$ has been remarked on above. Let S be the scroll in $\mathbb{P}^{r-g-1}$ in which the canonical image of C lies; note that if H is the hyperplane section of S and F is the fiber of the ruling, then the class of C is $3H - (r-3)F$, and $K_S = -2H + (r-3)F$. We have the same sequences (2.4) - (2.7) again, and the first step is to observe that $h^1(N_{C/S}(-1)) = 0$; putting this off for the moment, we have then that (2.6) is exact on global sections when $j = 1$, and moreover $h^0(N_{C/S}(-1)) = 9$ by Riemann-Roch on S. Since $h^1(N_S(-C-H)) = 0$ by the same result quoted above, we see that (2.7) is also exact on global sections when $j = 1$. Using Lemma (2.1), we see that $h^0(N_C(-1)) = 9 + (2r-2) - h^0(N_S(-C-H)) = 2g + 5 - h^0(N_S(-C-H))$.

In order to finish this part of the calculation, we must show that $h^0(N_S(-C-H)) = h^1(N_{C/S}(-1)) = 0$. As to the first group, note that since S is cut out by quadrics, N_S sits inside a direct sum of $\mathcal{O}_S(2)$'s; therefore, restricting to a general fiber F, we have that $N_{S|_F}$ sits inside a direct sum of $\mathcal{O}_F(2)$'s. Since $-C-H = -4H+(r-3)F$, $N_S(-C-H)|_F$ sits inside a direct sum of $\mathcal{O}_F(-2)$'s, and therefore has no nonzero sections; since F is general, $N_S(-C-H)$ can have no nonzero sections.

Note that $N_{C/S}(-1) \cong \mathcal{O}_C(2H-(r-3)F)$; therefore by Serre duality $h^1(N_{C/S}(-1)) = h^0(\mathcal{O}_C(-H+(r-3)F))$. We have the sequence

$$0 \longrightarrow \mathcal{O}_S(-4H+2(r-3)F) \longrightarrow \mathcal{O}_S(-H+(r-3)F) \longrightarrow \mathcal{O}_C(-H+(r-3)F) \longrightarrow 0$$

and since S is general, a standard cohomology computation on the scroll will show that $h^0(\mathcal{O}_S(-H+(r-3)F))$ and $h^1(\mathcal{O}_S(-4H+2(r-3)F))$ both vanish, proving what we want. This finishes the $k = -1$ part of the table.

Let us move to the $k = -2$ situation. We have the sequence

$$0 \longrightarrow N_{C/S}(-2) \longrightarrow N_C(-2) \longrightarrow N_{S|_C}(-2) \longrightarrow 0$$

and $h^0(N_{C/S}(-2)) = h^0(\mathcal{O}_C(H-(r-3)F)$, which equals zero for degree reasons as soon as $g \geq 10$ (in genus 10, the degree of the bundle is zero, but the bundle cannot be trivial because C is ample on S). In addition, we have the sequence

$$0 \longrightarrow \mathcal{O}_S(-2H) \longrightarrow \mathcal{O}_S(H-(r-3)F) \longrightarrow \mathcal{O}_C(H-(r-3)F) \longrightarrow 0,$$

and $h^0(\mathcal{O}_S(H-(r-3)F)$ vanishes if $g \geq 8$, as does $h^1(\mathcal{O}_S(-2H))$ in any case. Hence in these genera we also have $h^0(N_{C/S}(-2)) = 0$. Now from the sequence

$$0 \longrightarrow N_S(-2H-C) \longrightarrow N_S(-2H) \longrightarrow N_S(-2H)|_C \longrightarrow 0$$

we see that since $h^1(N_S(-2H-C)) = 0$ (using the result of [C]) and $h^0(N_S(-2H)) = 0$ as above, we have that $h^0(N_{S|_C}(-2)) = 0$. This then implies that $h^0(N_C(-2)) = 0$ for $g \geq 8$.

In lower genera this argument fails, but only because $h^0(N_{C/S}(-2))$ is not zero; in fact, $h^0(N_C(-2)) = h^0(N_{C/S}(-2)) = h^0(\mathcal{O}_C(H-(r-3)F))$ if $g \geq 5$. The case of genus 4 is trivial. The reader can easily check that the numbers written in the table are correct.

Finally, in case $k \leq -3$, the same sequences easily give the table answers. ■

Using Theorem (2.8) we can easily compute the dimensions of the kernel and cokernel of the Gaussian map ϕ.

(2.9) Corollary: Let C be a general trigonal curve of genus $g \geq 4$. Then

$$\dim(\ker(\phi)) = (g-4)(g-5)/2 \quad \text{and} \quad \dim(\mathrm{coker}(\phi)) = g + 5.$$

It is amusing to remark that any trigonal curve in genus 5 behaves like a general curve of genus 5 as far as the rank of ϕ is concerned.

The same methods that we use here should be applicable to a large class of curves on rational normal scrolls.

We should mention here that J. Wahl, in [W2, Theorem 4.8], shows that if C is a smooth curve of genus at least 5 on a rational normal scroll, then $\dim(\mathrm{coker}(\phi)) \geq 9$. This has been sharpened by J. Duflot and the second author in [D-M].

It is possible that by choosing special trigonal curves, one can achieve a variety of cokernel dimensions for the Gaussian map ϕ. In section 4 we will exhibit specific trigonal curves (whose genera are all equal to 1 modulo 3) such that the corank of ϕ is $g + 5$.

3. Elliptic-hyperelliptic curves and curves on Del Pezzo surfaces

In this section we will compute the rank of the Gaussian map ϕ for elliptic-hyperelliptic curves and for canonical curves on Del Pezzo surfaces.

Recall that normal linearly normal nondegenerate irreducible surfaces of degree r in $\mathbb{P}^r$ are of two types: Del Pezzo surfaces (up to $r = 9$) and elliptic normal cones. A general quadric section of these surfaces is a canonical curve of genus $g = r + 1$, and in particular the quadric sections of the elliptic cones are double coverings of elliptic curves and vice-versa.

If S is such a surface, and C is such a canonical curve, then the normal bundle for C in $\mathbb{P} = \mathbb{P}^r$ splits into the direct sum $\mathcal{O}_C(2)\oplus N_S|_C$, since C is a complete intersection on S. Therefore the computations concerning N_C reduce immediately to computations involving $N_S|_C$.

We begin with a lemma analogous to Lemma (2.1).

(3.1)<u>Lemma:</u>

(a) Let S be an elliptic cone in $\mathbb{P}^{r\geq 5}$. Then $h^0(N_S(-k)) = \begin{cases} 2r & \text{if } k = 1 \\ 0 & \text{if } k \geq 2 \end{cases}$.

In case $r = 4$, we have $h^0(N_S(-k)) = \begin{cases} 10 & \text{if } k = 1 \\ 2 & \text{if } k = 2. \\ 0 & \text{if } k \geq 3 \end{cases}$

In case $r = 3$, $h^0(N_S(-k)) = \begin{cases} 10 & \text{if } k = 1 \\ 4 & \text{if } k = 2 \\ 1 & \text{if } k = 3 \\ 0 & \text{if } k \geq 4 \end{cases}$.

(b) Let S be a smooth Del Pezzo surface in $\mathbb{P}^{r\geq 5}$. Then

$$h^0(N_S(-k)) = \begin{cases} 10 & \text{if } k = 1 \\ 0 & \text{if } k \geq 2 \end{cases}.$$

In cases $r = 3$ and 4, one has the same answers as in (a).

Proof: For the elliptic cones, recall that $H^0(N_S(-k)) \cong \bigoplus_{n\geq k} H^0(E,N_E(-n))$ (see [P]), where E is the elliptic curve in $\mathbb{P}^{r-1}$ over which S is the cone. On the other hand, we have $h^0(N_E(-n)) = \begin{cases} 2r & \text{if } n = 1 \\ 0 & \text{if } n \geq 2 \end{cases}$ (using Riemann-Roch for vector bundles on elliptic curves, see [P, Example (6.10)]); this proves (a), modulo the complete intersections which we leave to the reader.

For the Del Pezzos we have that $h^0(N_S(-k)) = 0$ for $k \geq 3$, since their ideal is generated by quadrics if $r \geq 4$. Moreover the fact that these surfaces are projectively Cohen-Macauley, with general hyperplane section an elliptic normal curve E, and the information that $h^0(N_E(-2)) = 0$ if $r \geq 5$, gives easily that $h^0(N_S(-2)) = 0$.

The dimension $h^0(N_S(-1))$ for general S is the number of parameters for Del Pezzo surfaces of degree r containing a fixed hyperplane section, which is easily seen to be 10. For a more formal argument, it is a standard computation with Euler-Poincare characteristics to see that for $N_S(-1)$, $h^0 - h^1 + h^2 = 10$. We need then to see that both h^1 and h^2 vanish. The vanishing of h^2 follows from the Euler sequence and Serre duality: $h^2(N_S(-1)) = h^0(N_S^*)$, and $H^0(N_S^*)$ injects into $H^0(\Omega^1_{\mathbb{P}|_S})$, which in turn injects into $H^0(\mathcal{O}_S(1)) \otimes H^0(\mathcal{O}_S(-1))$, which is zero. With an analogous argument $h^1(N_S(-1)) = h^1(N_S^*)$, and via the Euler sequence $H^1(N_S^*)$ injects into $H^1(\Omega^1_{\mathbb{P}|_S})$, which is isomorphic to $H^0(\mathcal{O}_S)$; in fact $H^1(N_S^*)$ is the kernel (after applying the isomorphism) of the natural map from $H^0(\mathcal{O}_S)$ to $H^1(\Omega^1_S)$, which is nonzero since the image is the 1-dimensional space generated by the polarization. Hence the kernel $H^1(N_S^*)$ is zero, and we finish.

The complete intersection are again left to the reader. ∎

Let us consider the case of the elliptic-hyperelliptic curves first.

(3.2)<u>Theorem:</u> Let C be an elliptic-hyperelliptic curve lying on an elliptic cone in $\mathbb{P}^r$, with $r \geq 5$. Then $h^0(N_C(-k))$ is given in the following table.

k	0	-1	-2	≤-3
$h^0(C,N_C(k))$	r^2+5r	$3r+1$	1	0

If $r = 4$, then we have

k	0	-1	-2	≤-3
$h^0(C,N_C(k))$	36	15	3	0

Finally, if $r = 3$, we have

k	0	-1	-2	-3	≤-4
$h^0(C,N_C(k))$	24	14	5	1	0

Proof: We leave the complete intersections to the reader, and concentrate on the cases when r is at least 5. In these cases the $k = 0$ and $k \leq -3$ computations are trivial. Let us turn to the case when $k = -1$.

First observe that the splitting of the normal bundle described above implies that $h^0(N_C(-1)) = (r+1) + h^0(N_{S|_C}(-1))$. Consider the sequence

$$0 \longrightarrow N_S(-3) \longrightarrow N_S(-1) \longrightarrow N_{S|_C}(-1) \longrightarrow 0.$$

Note that $h^0(N_S(-3)) = 0$ by Lemma (3.1), and also $h^1(N_S(-3)) = 0$ using a result of Pinkham (see [P, Lemma (5.2)]). Hence $h^0(N_{S|_C}(-1)) = h^0(N_S(-1))$, which equals 2r by Lemma (3.1). This verifies the assertion in this case.

The proof of the other term when $k = -2$ is analogous. ∎

Of course this gives us information about the rank of the Gaussian map ϕ for these curves.

(3.3)<u>Corollary:</u> Let C be an elliptic-hyperelliptic curve of genus g in $\mathbb{P}^{g-1}$, with $g \geq 6$. Then

$$\dim(\ker(\phi)) = (g-2)(g-3)/2 \quad \text{and} \quad \dim(\operatorname{coker}(\phi)) = 2g - 2.$$

Finally let us consider the curves on the Del Pezzo surfaces.

(3.4)<u>Theorem:</u> Let C be a canonical curve of genus $g \geq 6$ lying on a Del Pezzo surface in $\mathbb{P}^{r-g-1}$. Then $h^0(N_C(-k))$ is given in the following table.

k	0	-1	-2	≤-3
$h^0(C,N_C(k))$	r^2+5r	$\geq r+11$	≥ 1	0

In case $g = 4$ or 5, we have the same answers as given in Theorem (3.2).

Proof: Again we leave the cases $k = 0$, $k \leq -3$, and the complete intersections ($g = 4$ and 5) to the reader. The argument in case $k = -1$ is very similar to that given in Theorem (3.2), and we see that
$h^0(N_C(-1)) = (r+1) + h^0(N_S(-1)) + h^1(N_S(-3)) = (r+11) + h^1(N_S(-3)) \geq r+11$.
For $k = -2$, again the computation is analogous and in fact easier: we obtain that $h^0(N_C(-2)) = 1 + h^1(N_{S|_C}(-2)) \geq 1$. ∎

(3.5)<u>Remark:</u> We would like to have computed $h^1(N_S(-3))$ and $h^1(N_{S|_C}(-2))$. The first number is by Serre duality equal to $h^1(N_S^*(2))$, and using the Euler sequence is the cokernel of the natural map from $H^0(\Omega^1_{\mathbb{P}|_S}(2))$ to $H^0(\Omega^1_S(2))$. This is a surface version of the Gaussian map for the S. In fact, $H^1(\Omega^1_{\mathbb{P}|_S}(2))$ is indeed isomorphic to the kernel of the multiplication map from $H^0(\mathcal{O}_S(1))^{\oplus 2}$ to $H^0(\mathcal{O}_S(2))$. Of course for the complete intersections, one has that this H^1 vanishes, and this is the only evidence we have that it vanishes in general. A similar situation occurs in the computation of $h^1(N_{S|_C}(-2))$. In fact, this number is at most equal to $h^1(N_S(-4))$, which again can be interpreted as the dimension of the cokernel of an appropriate Gaussian map for S.

If C and S are general, then $h^0(N_C(-2))$ is one; in fact, C can be flatly degenerated to an elliptic-hyperelliptic curve (by degenerating S to the elliptic cone) and semicontinuity implies this.

In genus $g = 6$ the same degeneration to an elliptic-hyperelliptic curve shows that also for $h^0(N_C(-1))$ the equality holds (i.e., this $h^0 = 16$) if C is a general quadric section of a Del Pezzo. As mentioned above, this means something in terms of a Gaussian map for the general Del Pezzo.

(3.6)<u>Corollary:</u> Let C be a canonical curve of genus $g \geq 6$ lying on a Del Pezzo surface. Then

$$\dim(\ker(\phi)) \geq (g-3)(g+10)/2 \quad \text{and} \quad \dim(\operatorname{coker}(\phi)) \geq 2g + 10.$$

4. Gaussian maps for certain cyclic covers of the line

In this section we want to compute as explicitly as possible the Gaussian map ϕ for certain cyclic covers of $\mathbb{P}^1$, i.e., curves defined by an equation of the form

(4.1) $$y^N = f(x)$$

where f(x) is a polynomial in x with distinct roots. It is convenient to consider the special case when the degree of f(x) is exactly a multiple of the covering degree N: let us say that f(x) has degree LN, for some $L \geq 1$. We will see that in general the Gaussian map ϕ is never surjective for such curves, and in particular in the case of trigonal curves (when N is 3) that we are able to write down one such curve such that the corank of ϕ is exactly $g + 5$, which is the corank for a general trigonal curve by Corollary (2.9).

The affine curve described by (4.1) is smooth, but is however singular at infinity, in the plane $\mathbb{P}^2$. The assumption that f(x) has degree a multiple of N allows us to make a rather simple change of coordinates at infinity, and remove the singularity. To this end set $z = 1/x$, $w = y/x^L$, and $g(z) = z^{LN}f(1/z)$; then we have $w^N = (y/x^L)^N = (z^L y)^N = z^{LN}y^N = g(z)$, so that at infinity the model for the curve defined by (4.1) becomes

(4.2) $$w^N = g(z).$$

Note that g(z) has distinct roots also, since f(x) does. Let C be the complete curve described in the two charts given above.

This change of variable actually means that we are considering a smooth model for the curve on a surface $\mathbb{F}_L$, the rational scroll with a section for the ruling having self-intersection -L.

We are now in a position to explicitly write down all the holomorphic 1-forms on C. The first step is to compute the genus g of C; this can be done using the Hurwitz formula, since we have a degree N cover of $\mathbb{P}^1$ with exactly LN ramification points, each contributing N-1 to the ramification. Therefore

(4.3) $$g = 1 - N + LN(N-1)/2.$$

Let us first turn to the (x,y) chart, where equation (4.1) describes C. We have $Ny^{N-1}dy = f'(x)dx$, so that a holomorphic 1-form on C in this chart is given by

(4.4) $$Ndy/f'(x) = dx/y^{N-1},$$

which the reader can check is holomorphic everywhere in this chart. Therefore we have that any expression of the form

(4.5) $$\sigma_{ij} = x^i y^j dx/y^{N-1} = x^i y^{j-N+1} dx$$

is holomorphic in this chart, as long as i and j are nonnegative.

When is σ_{ij} holomorphic on all of C? To answer this, we write σ_{ij} in terms of the coordinates (z,w) at infinity, and we see that

(4.6) $$\sigma_{ij} = x^i y^j dx/y^{N-1} = z^{-i}(wz^{-L})^j(-z^{-2}dz)/(wz^{-L})^{N-1}$$
$$= -(z^{-i-Lj-2+L(N-1)}w^j)(dz/w^{N-1}),$$

which is holomorphic at infinity if and only if the powers of z and w appearing in (4.6) are non-negative, i.e., if

(4.7) $$i + Lj \le L(N-1)-2.$$

This leads to the following.

(4.8)<u>Proposition:</u> The set of 1-forms $\{\sigma_{ij} | i + Lj \le L(N-1)-2\}$ forms a basis for $H^0(C,\omega_C)$.

Proof: They are clearly linearly independent over $\mathbb{C}$, and by the analysis above they are all holomorphic. To finish, we need only check that there are g of them. Note that j can range over the integers from 0 through N-2, and for each j in this interval, i then ranges from 0 through L(N-1-j)-2. Thus we have a total of $\sum_{j=0}^{N-2}[L(N-1-j)-1]$ forms, which sums to $LN(N-1)/2 - (N-1)$, which is the genus g given in (4.3). ■

Recall that to compute the Gaussian map $\phi:\Lambda^2 H^0(\omega_C) \longrightarrow H^0(\omega_C^3)$, we have that $\phi(\omega_1 \wedge \omega_2) = (\alpha_1\alpha_2' - \alpha_2\alpha_1')\cdot(du)^3$, where u is a local parameter and $\omega_i = \alpha_i du$. By the previous Proposition, the image of ϕ is spanned by the elements $\phi(\sigma_{ij} \wedge \sigma_{kh})$, which we will now compute.

We have

$$\phi(\sigma_{ij} \wedge \sigma_{kh}) = \phi([x^i y^{j-N+1}dx] \wedge [x^k y^{h-N+1}dx])$$
$$= \Big[(x^i y^{j-N+1})(kx^{k-1}y^{h-N+1} + (h-N+1)x^k y^{h-N}(dy/dx))$$
$$- (x^k y^{h-N+1})(ix^{i-1}y^{j-N+1} + (j-N+1)x^i y^{j-N}(dy/dx))\Big]\cdot(dx)^3$$
$$= \Big[x^{i+k-1}y^{j+h-2N+2}(k-i) + x^{i+k}y^{j+h-2N+1}(dy/dx)(h-j)\Big]\cdot(dx)^3$$

$$= \left[x^{i+k-1}y^{j+h-2N+2}(k-i) + x^{i+k}y^{j+h-2N+1}(f'(x)/Ny^{N-1})(h-j)\right]\cdot(dx)^3,$$

which simplifies then to

$$(4.9)\quad \phi(\sigma_{ij}\wedge\sigma_{kh}) = \left[(k-i)x^{i+k-1}y^{j+h-2N+2} + \frac{(h-j)}{N}x^{i+k}y^{j+h-3N+2}f'(x)\right]\cdot(dx)^3$$

This general computation immediately gives the following.

(4.10)Proposition: Assume that L is at least 6. Then for C as above, the Gaussian map ϕ is not surjective.

Proof: The condition on L implies that $y^{-1}(dx)^3$ is in $H^0(\omega_C^3)$. However, it is easy to see that this is not in the image of ϕ; the maximum power of y which can be achieved is y^{-2}. ∎

Let us specialize the situation, and assume that $f(x) = x^{LN} - 1$. Then $f'(x) = LNx^{LN-1}$ so that the formula above specializes to

$$(4.11)\quad \phi(\sigma_{ij}\wedge\sigma_{kh}) = \left[(k-i)x^{i+k-1}y^{j+h-2N+2} + L(h-j)x^{i+k+LN-1}y^{j+h-3N+2}\right]\cdot(dx)^3$$

which is a bit more tractable. We want to investigate especially the case when N = 3, i.e., a cyclic trigonal curve; in particular, we want to compare the rank for this curve C to the rank for a general trigonal curve of this genus, which we have computed in section 2.

Hence now fix N = 3; in this case g = 3L-2, and the basis for $H^0(\omega_C)$ is $\{\sigma_{ij} | 0 \le j \le 1,\ 0 \le i \le L(2-j)-2\}$. Formula (4.11) now becomes

$$(4.12)\quad \phi(\sigma_{ij}\wedge\sigma_{kh}) = \left[(k-i)x^{i+k-1}y^{j+h-4} + L(h-j)x^{i+k+3L-1}y^{j+h-7}\right]\cdot(dx)^3.$$

Therefore the image of ϕ is spanned by essentially 3 types of elements:

$$\phi(\sigma_{i0}\wedge\sigma_{k0}) = \left[(k-i)x^{i+k-1}y^{-4}\right]\cdot(dx)^3 \text{ for } 0 \le i < k \le 2L-2,$$

$$\phi(\sigma_{i0}\wedge\sigma_{k1}) = \left[(k-i)x^{i+k-1}y^{-3} + Lx^{i+k+3L-1}y^{-6}\right]\cdot(dx)^3$$

for $0 \le i \le 2L-2$ and $0 \le k \le L-2$, and

$$\phi(\sigma_{i1}\wedge\sigma_{k1}) = \left[(k-i)x^{i+k-1}y^{-2}\right]\cdot(dx)^3 \text{ for } 0 \le i < k \le L-2.$$

Notice that for the elements $\phi(\sigma_{i0}\wedge\sigma_{k0})$, we obtain an image of dimension 4L-5 (the powers of x range from 0 to 4L-6), and because of the y^{-4} factor there is never any overlap with the image of the other two types. Similarly, for the elements $\phi(\sigma_{i1}\wedge\sigma_{k1})$, we obtain an image of dimension 2L-5, and again these are independent of the other types of elements.

Finally, for the elements $\phi(\sigma_{10}\wedge\sigma_{k1})$, we see that if γ can be represented in two ways as i+k, for allowable i and k, we can obtain in the image of ϕ both the terms $x^{\gamma-1}y^{-3}(dx)^3$ and $x^{\gamma+3L-1}y^{-6}(dx)^3$; otherwise we only can get one element in the image. Moreover, all these elements are independent. The possible γ's are those integers from 0 through 3L-4, and the extremes are the only γ's not obtained in two ways; therefore using these elements we obtain an image of dimension $1 + 2(3L-5) + 1 = 6L-8$. We have proved the following.

(4.13)<u>Theorem:</u> The rank of ϕ for the curve $y^3 = x^{3L} - 1$ is 12L-18. Therefore the dimension of the cokernel of ϕ is $g + 5$.

Thus these curves behave as "general" trigonal curves, with respect to the Gaussian map ϕ, by Corollary (2.9).

5. Speculations concerning curves on a K3 surface

In this section we will make some remarks concerning the Gaussian map for canonical curves lying on K3 surfaces, and the relationship between these matters and the Hilbert scheme for the K3 surfaces.

Let us begin by considering the case of genus g equal to 10 or at least 12. In this case it is known that the Gaussian map ϕ is surjective for the general curve (see [C-H-M]), and therefore the general curve cannot lie on a K3 surface (by the results of [W1]). We would like to know the dimension of the cokernel of ϕ for the general curve lying on a K3 surface: it must be strictly positive. We note here the recent work of C. Voisin, who has proved in [V] that the corank of ϕ is at most 3, for a generic curve of even genus $\geq$ 10 which satisfies the Brill-Noether-Petri conditions.

We are going to discuss a series of ideas which suggest to the authors that the corank of ϕ should be 1 for a general canonical curve on a K3 surface for most of these high genera.

Let X be the union of a set of 2-planes in $\mathbb{P}^g$ whose dual complex is a PL-decomposition of the 2-sphere; then X is numerically a K3 surface, and the Hilbert scheme for the smooth K3 surfaces of degree 2g-2 in $\mathbb{P}^g$ will contain the point <X> representing X (although it is not clear that <X> lies in the component containing the smooth K3's). The general hyperplane section of X is a "graph curve", namely, a union of $\mathbb{P}^1$'s whose dual graph G is planar. Moreover, this graph curve is canonically embedded if G is 3-edge-connected. Thus we see that this situation is a degenerate form of the situation above.

The second author has proved that under certain hypotheses, the Gaussian map ϕ for the graph curve has corank one (see [M]). Precisely, the result is:

(5.1)<u>Theorem:</u> Let G be a planar graph. Assume that

a) For every edge e of G, the faces adjacent to e have only e and the vertices of e in their common closure.

b) For every edge e of G, the faces neighboring e are different, and have only e in their common edge-neighborhoods.

Then the Gaussian map ϕ for the graph curve associated to G has corank 1.

The hypotheses on G force the graph curve to have genus at least 11, and in fact the first example of a graph satisfying these conditions is the dodecahedron graph, which gives a curve of genus 11. There is no such graph curve with genus 12, as the reader can rather easily check. However there are an infinite number of genera for which such graph curves exist.

We consider this some amount of weak evidence for the speculation that ϕ has corank one for a general curve lying on a K3 surface. First of all, what we described above is a degenerate form of the situation. Secondly, it possible that the union of planes, together with the hyperplane section which is the given graph curve, flatly deforms to a smooth K3 surface with a smooth hyperplane section. If this is true, then by the semicontinuity of the corank of ϕ, we would have corank one for the general hyperplane section of the smooth general K3.

We would like here to present some further motivation for this speculation. Let us make now a general remark. Let $\mathcal{X}_g$ be the locus inside the Hilbert scheme parameterizing cones over smooth canonical curves of genus g in $\mathbb{P}^{g-1}$. We recall from [C-M] that $\dim(\mathcal{X}_g) = g^2 + 4g - 4$, and that this is a generically smooth component of the Hilbert scheme if g is equal to 10 or is at least 12.

(5.2)<u>Lemma:</u> If C is a smooth canonical curve of genus g in $\mathbb{P}^{g-1}$, and the corank of ϕ is one, then $H^0(N_C(-2))$ vanishes. In particular, if g is equal to 10 or is at least 12, then the point $\langle X\rangle$ representing the cone X over C is a singular point of the Hilbert scheme, and its tangent space has dimension $g^2 + 4g - 3$, one more than dimension of $\mathcal{X}_g$.

Proof: Recall that $H^0(N_C(-2))$ is naturally isomorphic to the dual of the cokernel of the Gaussian map ϕ_{ω,ω^2}. We have the following commutative diagram:

$$\begin{array}{ccccccc}
H^0(\omega^3)\otimes H^0(\omega) & \xleftarrow{\phi\otimes id} & \Lambda^2H^0(\omega)\otimes H^0(\omega) & \xrightarrow{\beta} & (H^0(\omega)\otimes H^0(\omega))\otimes H^0(\omega) & & \\
\downarrow & & \downarrow\alpha & & \downarrow & & \\
H^0(\omega^4) & \xleftarrow[\phi_{\omega,\omega^2}]{} & \mathcal{R} & \longrightarrow & H^0(\omega^2)\otimes H^0(\omega) & \longrightarrow & H^0(\omega^3)
\end{array}$$

The maps which are not given names are either multiplication or inclusions. The map β send $f\wedge g\otimes h$ to $h\otimes(f\otimes g-g\otimes f)$, and the map α exists because then $\Lambda^2H^0(\omega)$ goes to zero in $H^0(\omega^2)$. It suffices therefore to show that the composite from $\Lambda^2H^0(\omega)\otimes H^0(\omega)$ to $H^0(\omega^4)$ is surjective. The image of $\Lambda^2H^0(\omega)$ in $H^0(\omega^3)$ has codimension one by hypothesis; therefore we can apply a lemma of M. Green [Gr], which says that multiplication by linear forms carries a codimension one subspace of the space of homogeneous polynomials of degree 3 onto the space of polynomials of degree 4. (Actually, Green's lemma is quite stronger than this.) Therefore $\mathrm{im}(\phi\otimes id)\otimes H^0(\omega)$ is all of $H^0(\omega^4)$, and we have proved the first statement.

Applying [C-M, Theorem (4.3) and Corollary (4.4)], we see that the dimension of the tangent space $H^0(N_X)$ to the Hilbert scheme at $\langle X\rangle$ is exactly one more than the dimension of the tangent space at the general point of $\mathcal{X}_g$. ∎

The vanishing of this $H^0(N_C(-2))$ has also been proved for most nonhyperelliptic curves by S. Tendian in [T].

Let us assume for the moment that one could prove that for a general hyperplane section of a K3 surface, the corank of ϕ is one. This should indicate two related results.

The first is that $\mathcal{X}_g$, at least infinitesimally, should cut out a hypersurface in the component $\mathcal{H}_g$ of the Hilbert scheme whose general point parameterizes a smooth K3 surface; moreover the normal vector corresponding to the cokernel of ϕ should point into the direction of smoothings to K3 surfaces.

On the other hand, note that the cones over hyperplane sections of K3 surfaces form a locus in the Hilbert scheme of dimension at most $g^2 + 2g + 18$, simply by counting parameters; we have equality if and only if the general such hyperplane section lies on only finitely many K3 surfaces up to projectivities. Moreover the dimension of $\mathcal{H}_g$ is $g^2 + 2g + 19$. The second result indicated above is that actually we should have this finiteness property as a consequence of the generic corank one statement. This is because if the finiteness fails, then the cones over hyperplane sections of K3's forms a locus of dimension strictly less that $g^2 + 2g + 18$, and is therefore of codimension at least two in the $\mathcal{H}_g$; however in this case we should have that there are two smoothing directions from such a cone toward the K3 locus, and so the corank of ϕ should be at least two.

The reader at this point might be tempted to conjecture that the corank of ϕ is one for the general hyperplane section of a K3 surface. However we feel compelled to warn you that the finiteness property mentioned in the last paragraph definitely fails in genus 12, by the work of S. Mori and S. Mukai in [M-M]. Therefore in this case the corank one statement is in serious doubt. And in fact recall that we are unable to build even a degenerate example using unions of planes and planar graph curves in this genus.

Now we wish to consider the case of low genus, in particular g at most 9 and genus 11. In these cases we know that $\mathcal{X}_g$ is contained in $\mathcal{H}_g$, and $\mathcal{H}_g$ is generically smooth at the points of $\mathcal{X}_g$ (see [C-M]). We would like to show that in these cases the union of planes of the type considered above can actually be smoothed to a K3. In high genus, it is exactly this point which is the stumbling block to providing a proof of the speculations given above.

As an example to be specific let us fix the genus to be 11. Let G be the planar graph of the dodecahedron, let S_G be the union of planes in $\mathbb{P}^{11}$ described by G, and let C_G be the graph curve associated to G, which is realized as the hyperplane section of S_G, canonically embedded in $\mathbb{P}^{10}$. We will denote by X the cone over C in $\mathbb{P}^{11}$: it is another union of planes.

(5.3)<u>Proposition:</u> The point $\langle S_G\rangle$ of the Hilbert scheme representing S_G is in $\mathcal{H}_{11}$. In other words, S_G can be flatly deformed to a smooth K3 surface in $\mathbb{P}^{11}$

Proof: We have computed the cohomology of the negative twists of the normal bundle of C_G in [C-M] and [M]: it turns out that $h^0(N_{C_G}(-1)) = 12$, and $h^0(N_{C_G}(-k)) = 0$ for $k \geq 2$. Therefore the tangent space $H^0(N_{X_G}) \cong \bigoplus_{k\geq 0} H^0(N_{C_G}(-k))$ at the point $\langle X_G\rangle$ has dimension 162. But now $\mathcal{X}_{11}$ is contained in $\mathcal{H}_{11}$ by the results of [M-M], and $\mathcal{H}_{11}$ also has dimension 162. Hence $\langle X_G\rangle$ is a smooth point of $\mathcal{H}_{11}$.

On the other hand, $\langle S_G\rangle$ lies in the same component of the Hilbert scheme as does $\langle X_G\rangle$, because S_G can be flatly degenerated to X_G (see [C-M], section 5]). By the smoothness of the Hilbert scheme at $\langle X_G\rangle$ we cannot have two different components passing through this point; hence the component containing $\langle S_G\rangle$ must be $\mathcal{H}_{11}$. ■

Of course a similar statement holds concerning degenerations of embedded K3 surfaces of degree 2g - 2 in $\mathbb{P}^g$ to unions of planes holds in the cases of complete intersections, namely for g = 3, 4, and 5. For genus 6, 7, 8, and 9, one can repeat the argument above, using the calculations of [C-M, section 3] for certain other planar graphs. (See [C-M, Remark 3.4] for the genus 8 and 9 cases.) The dimension counts all work as above, and we can finally state the following.

(5.4)<u>Theorem:</u> If $g \leq 9$ or $g = 11$, a general K3 surface of degree $2g - 2$ in $\mathbb{P}^g$ can be flatly degenerated to a reduced union of planes whose dual complex is a topological 2-sphere.

Results of this type are interesting by comparison to the situation for plane curves with nodes and Severi's problem. It is likely that these degenerations could be useful in studying general properties of K3 surfaces. We conjecture that a nice geometrical proof of the well-known fact that the general K3 surface in $\mathbb{P}^g$ has Picard number one could be proved by using such a degeneration, in the manner of Griffiths-Harris (see [G-H]).

Note further that the semi-stable degenerations of K3 surfaces are rather well understood now (see e.g. [K] and [P-P]), but the embedded degenerations are still somewhat mysterious.

Finally we would like tho mention that in genus 10 the locus in $\mathcal{M}_{10}$ of curves for which ϕ fails to be of maximal rank contains the K3 sections. In fact , F. Cukierman and D. Ulmer have recently announced a proof that in genus 10, the K3 locus in the moduli space is equal to the divisor where the Gaussian map is not an isomorphism.

References:

[C] C. Ciliberto:"On the Hilbert scheme of curves of maximal genus in a projective space", Math. Z. 194, (1987), 351-363.

[C-H-M] C. Ciliberto, J. Harris, and R. Miranda:"On the surjectivity of the Wahl map", Duke Math. Journal, 57 (1988), 829-858.

[C-M] C. Ciliberto and R. Miranda:"On the Gaussian map for canonical curves of low genus", Duke Math. Journal, 61 (1990), 417-443

[D-M] J. Duflot and R. Miranda:"The Gaussian map for rational ruled surfaces", to appear in the Transactions of the AMS.

[G] G. Gherardelli:"Un' osservazione sulla serie Jacobiana di una serie lineare", Rendic. R. Acc. Naz. dei Lincei, serie 4, 6 (1927), p.286

[G-H] P. Griffiths and J. Harris:"On the Noether-Lefschetz theorem and some remarks on codimension two cycles", Math. Ann. 271 (1985), 31-51.

[Gr] M. Green:"A new proof of the explicit Noether-Lefschetz theorem", to appear in the Journal of Diff. Geom.

[K] V. Kulikov:"Degeneration off K3 surfaces and Enriques surfaces", Math. USSR Izvestija, 11 (1977), 957-989.

[M] R. Miranda:"On the Wahl map for certain planar graph curves", in: Algebraic Geometry: Sundance 1988, edited by B. Harbourne and R. Speiser, Contemporary mathematics Vol. 116, AMS, 115-124.

[M-M] S. Mori and S. Mukai:"On the uniruledness of the moduli space of curves of genus 11", in Algebraic Geometry, Proceedings, Tokyo/Kyoto 1982, Springer LNM #1016 (1983), 334-353.

[P] H. Pinkham:"Deformations of algebraic varieties with $\mathbb{G}_m$-action", Thesis, Harvard University, 1974.

[P-P] U. Persson and H. Pinkham:"Degeneration of surfaces with trivial canonical bundle", Annals of Math., 113 (1981), 45-66.

[S] B. Segre:"Sulle curve algebriche le cui tangenti appartengono al massimo numero di complessi lineari indipendenti", Mem. Acc. Naz. dei Lincei 6 (1928), 578-592.

[T] S. Tendian:"Deformations of cones over curves of high degree", Ph.D. dissertation, University of North Carolina at Chapel Hill, July 1990.

[V] C. Voisin:"Sur l'application de Wahl des courbes satisfaisant la condition de Brill-Noether-Petri", preprint.

[W1] J. Wahl:"On the Jacobian algebra of a graded Gorenstein singularity", Duke Math. Journal, vol. 55, no. 4 (1987), 843-871.

[W2] J. Wahl:"Gaussian maps on algebraic curves", J. Diff. Geom. 32 (1990), 77-98

[W3] J. Wahl:"Introduction to Gaussian maps on an Algebraic Curve", notes prepared in connection with lectures at the Trieste Conference on Projective Varieties, June 1989.

Ciro Ciliberto
Dipartimento di Matematica
Universitá di Roma II
Via O. Raimondo
00173 Roma
ITALY

and

Rick Miranda
Department of Mathematics
Colorado State University
Fort Collins, CO 80523
USA

Geometry of the Horrocks bundle on $\mathbb{P}^5$

Wolfram Decker, Nicolae Manolache, Frank-Olaf Schreyer

Introduction

One of the few known vector bundles of rank r on a projective space $\mathbb{P}^n$ satisfying $2 \leq r \leq n-2$ is the Horrocks bundle $\mathcal{E}$ of rank 3 on $\mathbb{P}^5$. $\mathcal{E}$ is stable with Chern classes $c_1 = 0$, $c_2 = 3$, $c_3 = 0$ [Ho 2].

In this paper we study the geometry of sections $s \in H^0(\mathbb{P}^5, \mathcal{E}(2))$, 2 being the lowest twist with nonzero sections. As pointed out by Horrocks, the zero locus $Z(s)$, if of expected codimension, is a reducible (possibly nonreduced) surface of degree 14. Our main result describes the primary decomposition of $Z(s)$. For the general section s we obtain:

(a) $Z(s)$ has six components: two disjoint planes $\mathbb{P}(W)$, $\mathbb{P}(W^*)$, three smooth quadric surfaces Q_1, Q_2, Q_3 and a Del Pezzo surface D of degree 6.

(b) Each of the quadrics intersects each of the planes in a line, and the Del Pezzo surface in two opposite lines of its hexagon of lines. All intersections together form the 12 edges of an octahedron:

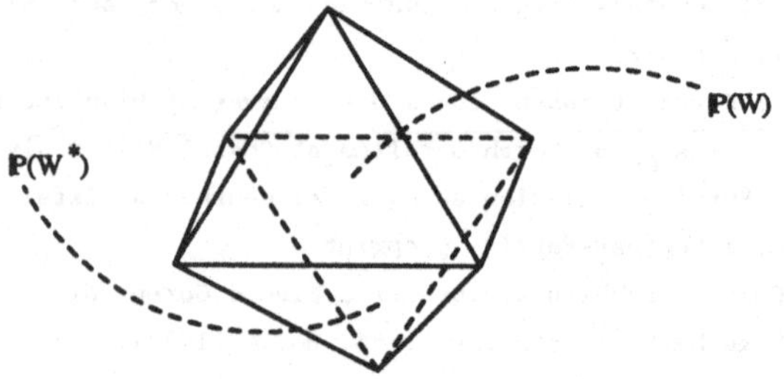

(c) The three quadrics Q_1, Q_2, Q_3 define a quadratic transformation

$$\xi : \mathbb{P}(W) - - \rightarrow \mathbb{P}(W^*)$$

as follows:

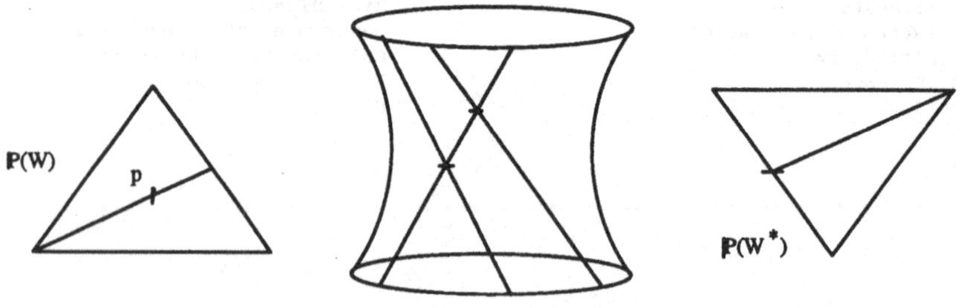

The projection of a point $p \in \mathbb{P}(W)$ from one of the vertices onto the opposite line of the distinguished triangle in $\mathbb{P}(W)$ gives a point on one of the quadrics Q_i, hence two lines in this quadric. One of them is the intersection line with $\mathbb{P}(W)$, the other intersects $\mathbb{P}(W^*)$ in a point of its triangle and thus defines a line in $\mathbb{P}(W^*)$ as indicated above. The three lines in $\mathbb{P}(W^*)$ obtained in this way intersect in a single point ! This point is $\xi(p)$.

(d) The projection $D - - \rightarrow \mathbb{P}(W)$ from $\mathbb{P}(W^*)$ realizes D as a triple cover of the plane $\mathbb{P}(W)$ branched totally along the triangle. (Similarly for $D - - \rightarrow \mathbb{P}(W^*)$.)

Abstractly the surface D is isomorphic to the graph of ξ. However, as we will prove, there is no canonical isomorphism.

That $Z(s)$ somewhat reflects the geometry of a triangle is not surprising: Horrocks proved that $H^0(\mathbb{P}^5, \mathcal{E}(2)) \cong W^* \otimes W/ < \mathbb{1} >$. Thus a section s corresponds to a traceless endomorphism

$$a(s) \in \mathrm{End}(W) \cong \mathrm{End}(W^*)$$

of W and W^* resp. The vertices of the triangle in $\mathbb{P}(W)$ and $\mathbb{P}(W^*)$ resp. are the eigenspaces of $a(s)$.
However the richness of the elementary projective geometry of triangles playing a role was a surprise for us.
It is not too hard to determine the sections of an explicitely given vector bundle. But one needs some computational skill. Of course we can improve our computational power tremendously by using representation theory or a computer [Mac]. We use both somewhat. We do not intend to give every detail of the necessary computations (though this could be done even without referring to a computer). Instead we give a detailed outline of the main steps of these computations.
In section 1 we review the construction of $\mathcal{E}$, i.e. its monad and its moduli scheme

$$M^0 = \mathrm{PGL}(5).[\mathcal{E}] \subset M_{\mathbb{P}^5}(3;0,3,0).$$

(There is a minor difference with the results stated by Horrocks: Over a field of characteristic 2, $\mathcal{E}$ is a *nonsplit* extension of Tango's bundle $\mathcal{F}$ [Ta 1]:

$$0 \longrightarrow \mathcal{F} \longrightarrow \mathcal{E} \longrightarrow \mathcal{O} \longrightarrow 0\,.)$$

As a pretty detail we mention that the morphism

$$M^0 \longrightarrow M^0\,, [\tilde{\mathcal{E}}] \longrightarrow [\tilde{\mathcal{E}}^*]$$

may be interpreted as a selfdual cubo-cubic Cremona transformation

$$\mathbb{P}^{13} - - \rightarrow \mathbb{P}^{13}.$$

In section 2 we compute the equations of the zero loci $Z(s)$, $s \in H^0(\mathbb{P}^5, \mathcal{E}(2))$, and obtain the results described above.
We are grateful to W. Barth, S. Ramanan and A. van de Ven for encouragement and helpful discussions. We also thank the DFG-Schwerpunktprogramm "Komplexe Mannigfaltigkeiten" for its financial support.

1. Monads for $\mathcal{E}$

Let V be a 6-dimensional vector space over an algebraically closed field k and $\mathbb{P}(V) \cong \mathbb{P}^5_k$ the projective space of lines in V. $\mathcal{T}_{\mathbb{P}(V)}$ denotes the tangent bundle, $\Omega^p(p) = \wedge^p \mathcal{T}^*_{\mathbb{P}(V)} \otimes \mathcal{O}(p)$.
(1.1) **Lemma** [Bei]. There are canonical isomorphisms

$$\mathrm{Hom}(\Omega^p(p), \Omega^q(q)) \cong \wedge^{p-q} V$$

defined by contraction. The composition of morphisms corresponds to the multiplication in $\wedge V$. □

Let $\omega \in \wedge^2 V$ be a nondegenerate symplectic form on V^*,

$$V^* \times V^* \longrightarrow k, \ (x,y) \longrightarrow x \wedge y \neg \omega$$

($\neg$ stands for contraction). Consider

$$\mathcal{B} = \operatorname{coker}(\Omega^4(4) \xrightarrow{\omega} \Omega^2(2)).$$

(1.2) Proposition.
(i) $\mathcal{B}$ is a rank 5 vector bundle on $\mathbb{P}(V)$ with Chern polynomial

$$c(\mathcal{B}) = (1-h^2)(1+3h^2)$$

(h denotes the positive generator of the Chow ring of $\mathbb{P}(V)$).
(ii) $\operatorname{Hom}(\mathcal{O}, \mathcal{B}) = 0$.
(iii) If $\operatorname{char}(k) \neq 2$ then $\mathcal{B}$ is self-dual. If $\operatorname{char}(k) = 2$ then $\operatorname{Hom}(\mathcal{B}, \mathcal{O})$ is 1-dimensional, in particular $\mathcal{B}$ is not self-dual.

Proof. (i) $\omega \in \Lambda^2 V$ defines a monomorphism $\Omega^4(4) \longrightarrow \Omega^2(2)$ of vector bundles iff ω is nondegenerate, which is satisfied by assumption. So $\mathcal{B}$ is a rank 5 vector bundle. Its Chern polynomial is

$$\begin{aligned} c(\mathcal{B}) &= c(\Omega^2(2)) \cdot c(\Omega^4(4))^{-1} = \tfrac{1+2h}{(1+h)^6} \cdot \tfrac{1-2h}{(1-h)^6} \\ &= (1-h^2)(1+3h^2) \in \mathbb{Z}[h]/(h^6) \end{aligned}$$

(use the Koszul complex on $\mathbb{P}^5$).
(ii) $H^0(\mathbb{P}^5, \Omega^2(2)) = H^1(\mathbb{P}^5, \Omega^4(4)) = 0$.
(iii) Let $\omega^* \in \Lambda^2 V^*$ be the symplectic form on V induced by ω, and let $\mathcal{N}$ be the null correlation bundle associated to ω^* (cf. [Ba]):

(1) $$0 \longrightarrow \mathcal{O}(-1) \xrightarrow{\omega^*} \Omega^1(1) \longrightarrow \mathcal{N} \longrightarrow 0.$$

The skewsymmetric pairing $V^* \times V^* \longrightarrow k$ induces a skewsymmetric pairing $\mathcal{N} \otimes \mathcal{N} \longrightarrow \mathcal{O}$. This determines a symmetric form on $\Lambda^2\mathcal{N}$, a distinguished section $\eta \in \operatorname{Hom}(\mathcal{O}, \Lambda^2\mathcal{N})$ and a distinguished cosection $\eta' \in \operatorname{Hom}(\Lambda^2\mathcal{N}, \mathcal{O})$. Consider the diagram

$$\begin{array}{ccccccccc} & & 0 & & & & & & \\ & & \downarrow & & & & & & \\ 0 & \longrightarrow & \mathcal{N}(-1) & \longrightarrow & \Omega^2(2) & \longrightarrow & \Lambda^2\mathcal{N} & \longrightarrow & 0 \\ & & \downarrow & & \| & & & & \\ 0 & \longrightarrow & \Omega^4(4) & \xrightarrow{\omega} & \Omega^2(2) & \longrightarrow & \mathcal{B} & \longrightarrow & 0 \\ & & \downarrow & & & & & & \\ & & \mathcal{O} & & & & & & \\ & & \downarrow & & & & & & \\ & & 0 & & & & & & \end{array}$$

whose first row is Λ^2 of (1) and whose first column is isomorphic to the dual of (1) twisted. Completing the diagram yields a short exact sequence

(2) $$0 \longrightarrow \mathcal{O} \xrightarrow{\eta} \Lambda^2\mathcal{N} \longrightarrow \mathcal{B} \longrightarrow 0.$$

The composition

$$\eta' \circ \eta \in \operatorname{Hom}(\mathcal{O}, \mathcal{O})$$

is multiplication by 2. Thus if $\operatorname{char}(k) \neq 2$ the sequence (2) splits and the nondegenerate symmetric form on $\Lambda^2\mathcal{N}$ induces one on $\mathcal{B}$. In particular $\mathcal{B} \cong \mathcal{B}^*$. If $\operatorname{char}(k) = 2$ then (2) does not split. Instead η' comes from an element of $\operatorname{Hom}(\mathcal{B}, \mathcal{O})$. Indeed, since

$$\omega \wedge \omega = 0 \quad \text{if} \quad \operatorname{char}(k) = 2,$$

$\omega \in \wedge^2 V \cong \mathrm{Hom}(\Omega^2(2), \mathcal{O})$ induces a nonzero cosection of $\mathcal{B}$. So $\mathrm{Hom}(\mathcal{B}, \mathcal{O})$ is 1-dimensional since

$$\mathrm{Hom}(\wedge^2 \mathcal{N}, \mathcal{O}) \cong \mathrm{Hom}(\mathcal{O}, \wedge^2 \mathcal{N}) \cong \mathrm{Hom}(\mathcal{O}, \mathcal{O}) \cong k.$$

□

Horrocks constructs his 3-bundle as the cohomology of a monad of type

$$\mathcal{O}(-1) \xrightarrow{\sigma} \mathcal{B} \xrightarrow{\tau} \mathcal{O}(1).$$

If $\mathrm{char}(k) \neq 2$, then $\mathcal{B} \cong \mathcal{B}^*$ and the description of all monads of this type is somewhat simpler, since we may regard σ and τ as elements of the same space. We treat this case first.

The Koszul presentation

$$\begin{array}{ccccccc} & \wedge^6 V^* \otimes \mathcal{O}(-2) & \xrightarrow{\neg\omega} & \wedge^4 V^* \otimes \mathcal{O}(-2) & & & \\ & \downarrow & & \downarrow & & & \\ & \wedge^5 V^* \otimes \mathcal{O}(-1) & \xrightarrow{\neg\omega} & \wedge^3 V^* \otimes \mathcal{O}(-1) & & & \\ & \downarrow & & \downarrow & & & \\ 0 \longrightarrow & \Omega^4(4) & \xrightarrow{\neg\omega} & \Omega^2(2) & \longrightarrow \mathcal{B} \longrightarrow 0 \\ & \downarrow & & \downarrow & & & \\ & 0 & & 0 & & & \end{array}$$

gives $\mathrm{Hom}(\mathcal{O}(-1), \mathcal{B}) \cong \wedge^3 V^* / (\wedge^5 V^* \neg \omega)$. Since $\mathrm{char}(k) \neq 2, \omega \wedge \omega \neq 0$ and

$$\wedge^5 V^* \xrightarrow{\neg(\omega\wedge\omega)} V^*$$

is an isomorphism. We can therefore identify

$$\mathrm{Hom}(\mathcal{O}(-1), \mathcal{B}) \cong \wedge^3 V^* / (\wedge^5 V^* \neg \omega) \cong \ker(\wedge^3 V^* \xrightarrow{\neg\omega} V^*) =: U.$$

(1.3) **Proposition.** ($\mathrm{char}(k) \neq 2$.)

(i) $I = \mathbb{P}(U) \cap G(3, V^*) \subset \mathbb{P}(\wedge^3 V^*)$
is the variety of maximal isotropic subspaces of V^* with respect to ω.

(ii) $\Gamma = \{(\sigma, \tau) \in \mathbb{P}(U) \times \mathbb{P}(U) \mid \tau \circ \sigma = 0\}$
is the graph of a selfdual cubo-cubic Cremona transformation $\gamma : \mathbb{P}(U) - - \rightarrow \mathbb{P}(U)$ whose base locus is the singular locus S of the tangent developable T of I, and whose exceptional divisor is the tangent developable T.

(iii) $(\sigma, \tau) \in \Gamma$ define a monad
$\mathcal{O}(-1) \xrightarrow{\sigma} \mathcal{B} \xrightarrow{\tau} \mathcal{O}(1)$
iff neither σ nor τ lies in T.

Proof. (i) is elementary: Let $x_1 \wedge x_2 \wedge x_3 \in \wedge^3 V^*$ be a nontrivial decomposable form and let $< x_1, x_2, x_3 > \subset V^*$ be the corresponding subspace. If $< x_1, x_2, x_3 >$ is isotropic then we can choose an isotropic complement $< x_4, x_5, x_6 > \subset V^*$ whose basis x_4, x_5, x_6 is dual to x_1, x_2, x_3 (with respect to ω). Then

$$\omega = e_1 \wedge e_4 + e_2 \wedge e_5 + e_3 \wedge e_6,$$

where $e_1, \cdots, e_6 \in V$ is the dual basis to $x_1, \cdots, x_6 \in V^*$, and

$$x_1 \wedge x_2 \wedge x_3 \neg \omega = 0.$$

Conversely, if $< x_1, x_2, x_3 >$ is not isotropic, say

$$x_2 \wedge x_3 \neg \omega = 1,$$

then we may choose $x_1 \in < x_1, x_2, x_3 > \cap < x_2.x_3 >^{\perp}$ and obtain

$$x_1 \wedge x_2 \wedge x_3 \neg \omega = x_1 \neq 0.$$

This proves that maximal isotropic subspaces correspond to decomposable forms $x_1 \wedge x_2 \wedge x_3$ with $x_1 \wedge x_2 \wedge x_3 \neg \omega = 0$.

(ii) To compute

$$\Gamma = \{(\sigma, \tau) \in \mathbb{P}(U) \times \mathbb{P}(U) \mid \tau \circ \sigma = 0\}$$

explicitely, we give yet another description of $\mathcal{B}$ and $\wedge^2 \mathcal{N}$ valid if $\mathrm{char}(k) \neq 2$. We have

$$\wedge^3 V^* = \wedge^5 V^* \neg \omega \oplus U$$

where $U = \ker(\wedge^3 V^* \xrightarrow{\neg\omega} V^*)$, and dually

$$\wedge^3 V = V \wedge \omega \oplus U^*$$

where $U^* = \ker(\wedge^3 V \xrightarrow{\wedge\omega} \wedge^5 V)$. $\wedge^2 \mathcal{N}$ is the image of the composition

$$b : \wedge^3 V^* \otimes \mathcal{O}(-1) \longrightarrow \wedge^2 V^* \otimes \mathcal{O} \cong \wedge^2 V \otimes \mathcal{O} \longrightarrow \wedge^3 V \otimes \mathcal{O}(1)$$

of Koszul maps with the isomorphism $\wedge^2 V^* \cong \wedge^2 V$ induced by ω. With respect to the decompositions of $\wedge^3 V^*$ and $\wedge^3 V$as above (and suitable bases chosen) b has block structure:

$$b = \begin{pmatrix} b_1 & 0 \\ 0 & b_2 \end{pmatrix}$$

with

$$b_1 = \begin{pmatrix} x_1^2 & x_1x_2 & x_1x_3 & x_1x_4 & x_1x_5 & x_1x_6 \\ x_2x_1 & x_2^2 & x_2x_3 & x_2x_4 & x_2x_5 & x_2x_6 \\ x_3x_1 & x_3x_2 & x_3^2 & x_3x_4 & x_3x_5 & x_3x_6 \\ x_4x_1 & x_4x_2 & x_4x_3 & x_4^2 & x_4x_5 & x_4x_6 \\ x_5x_1 & x_5x_2 & x_5x_3 & x_5x_4 & x_5^2 & x_5x_6 \\ x_6x_1 & x_6x_2 & x_6x_3 & x_6x_4 & x_6x_5 & x_6^2 \end{pmatrix}$$

and

$$b_2 = \begin{pmatrix} 0 & q & x_1^2 & x_1x_2 & x_1x_3 & x_2^2 & x_2x_3 \\ q & 0 & 0 & 0 & 0 & 0 & 0 \\ x_1^2 & 0 & 0 & 0 & 0 & x_6^2 & -x_5x_6 \\ x_1x_2 & 0 & 0 & -\frac{1}{2}x_6^2 & \frac{1}{2}x_5x_6 & 0 & \frac{1}{2}x_4x_6 \\ x_1x_3 & 0 & 0 & \frac{1}{2}x_5x_6 & -\frac{1}{2}x_5^2 & -x_4x_6 & \frac{1}{2}x_4x_5 \\ x_2^2 & 0 & x_6^2 & 0 & -x_4x_6 & 0 & 0 \\ x_2x_3 & 0 & -x_5x_6 & \frac{1}{2}x_4x_6 & \frac{1}{2}x_4x_5 & 0 & -\frac{1}{2}x_4^2 \\ x_3^2 & 0 & x_5^2 & -x_4x_5 & 0 & x_4^2 & 0 \\ 0 & x_4^2 & r_1 & x_2x_4 & x_3x_4 & 0 & 0 \\ 0 & x_4x_5 & x_1x_5 & -\frac{1}{2}x_3x_6 & \frac{1}{2}x_3x_5 & x_2x_4 & \frac{1}{2}x_3x_4 \\ 0 & x_4x_6 & x_1x_6 & \frac{1}{2}x_2x_6 & -\frac{1}{2}x_2x_5 & 0 & \frac{1}{2}x_2x_4 \\ 0 & x_5^2 & 0 & x_1x_5 & 0 & r_2 & x_3x_5 \\ 0 & x_5x_6 & 0 & \frac{1}{2}x_1x_6 & \frac{1}{2}x_1x_5 & x_2x_6 & -\frac{1}{2}x_1x_4 \\ 0 & x_6^2 & 0 & 0 & x_1x_6 & 0 & x_2x_6 \end{pmatrix}$$

$$\left.\begin{array}{ccccccc}
x_3^2 & 0 & 0 & 0 & 0 & 0 & 0 \\
0 & x_4^2 & x_4x_5 & x_4x_6 & x_5^2 & x_5x_6 & x_6^2 \\
x_5^2 & r_1 & x_1x_5 & x_1x_6 & 0 & 0 & 0 \\
-x_4x_5 & x_2x_4 & -\frac{1}{2}x_3x_6 & \frac{1}{2}x_2x_6 & x_1x_5 & \frac{1}{2}x_1x_6 & 0 \\
0 & x_3x_4 & \frac{1}{2}x_3x_5 & -\frac{1}{2}x_2x_5 & 0 & \frac{1}{2}x_1x_5 & x_1x_6 \\
x_4^2 & 0 & x_2x_4 & 0 & r_2 & x_2x_6 & 0 \\
0 & 0 & \frac{1}{2}x_3x_4 & \frac{1}{2}x_2x_4 & x_3x_5 & -\frac{1}{2}x_1x_4 & x_2x_6 \\
0 & 0 & 0 & x_3x_4 & 0 & x_3x_5 & r_3 \\
0 & 0 & 0 & 0 & x_3^2 & -x_2x_3 & x_2^2 \\
0 & 0 & -\frac{1}{2}x_3^2 & \frac{1}{2}x_2x_3 & 0 & \frac{1}{2}x_1x_3 & -x_1x_2 \\
x_3x_4 & 0 & \frac{1}{2}x_2x_3 & -\frac{1}{2}x_2^2 & -x_1x_3 & \frac{1}{2}x_1x_2 & 0 \\
0 & x_3^2 & 0 & -x_1x_3 & 0 & 0 & x_1^2 \\
x_3x_5 & -x_2x_3 & \frac{1}{2}x_1x_3 & \frac{1}{2}x_1x_2 & 0 & -\frac{1}{2}x_1^2 & 0 \\
r_3 & x_2^2 & -x_1x_2 & 0 & x_1^2 & 0 & 0
\end{array}\right)$$

where

$$q = x_1x_4 + x_2x_5 + x_3x_6$$

and

$$r_1 = x_1x_4 - x_2x_5 - x_3x_6, \quad r_2 = -x_1x_4 + x_2x_5 - x_3x_6, \quad r_3 = -x_1x_4 - x_2x_5 + x_3x_6.$$

Then

$$\mathcal{B} = \operatorname{Im}(U \otimes \mathcal{O}(-1) \xrightarrow{b_2} U^* \otimes \mathcal{O}(1))$$

and because of the symmetry of b_2 we see again that $\mathcal{B} \cong \mathcal{B}^*$. Now (ii) follows from this description of $\mathcal{B}$ by a straightforward calculation of which we give a rough outline: We consider

$$b_2 \in U^* \otimes U^* \otimes S_2V^*$$

as a homomorphism

$$\tilde{b}_2 : U \otimes \mathcal{O}_{\mathbb{P}(U)} \longrightarrow S_2V^* \otimes \mathcal{O}_{\mathbb{P}(U)}(1)$$

of vector bundles on $\mathbb{P}(U)$. One can check that

$$\ker(\tilde{b}_2) \cong \mathcal{O}_{\mathbb{P}(U)}(-3)$$

and that the 14 cubic forms defining $\mathcal{O}_{\mathbb{P}(U)}(-3) \longrightarrow U \otimes \mathcal{O}_{\mathbb{P}(U)}$ give a rational map

$$\gamma : \mathbb{P}(U) - - \rightarrow \mathbb{P}(U)$$

such that $\gamma^2 = \mathrm{id}_{\mathbb{P}(U)}$. The base locus of γ has codimension 4, degree 23 and is arithmetically Cohen-Macaulay. The (reduced) exceptional divisor of γ is a hypersurface of degree 4 whose singular locus coincides with the base locus. To indentify these varieties with the tangent developable T and its singular locus S resp. and to ease some of the calculations lateron we bring in the action of the symplectic group $\mathrm{Sp}\,\omega$ on $\mathbb{P}(U)$. γ is $\mathrm{Sp}\,\omega$-equivariant. $\mathrm{Sp}\,\omega$ has only 4 orbits on $\mathbb{P}(U)$: Consider the stratification

$$I \subset S \subset T \subset \mathbb{P}(U)$$

of $\mathbb{P}(U)$. The orbits of $\mathrm{Sp}\,\omega$ are the strata

$$I,\ S\setminus I,\ T\setminus S,\ \mathbb{P}(U)\setminus T$$

(as one can check). I is a manifold of dimension 6, T a hypersurface of degree 4. The singular locus S of T is defined by the partial derivatives of the equations of T, i.e. by 14 cubic forms. S coincides with the base locus of γ.

(iii) To determine which $\sigma \in \mathbb{P}(U)$ define a monomorphism

$$\mathcal{O}(-1) \xrightarrow{\sigma} \mathcal{B} \hookrightarrow U^* \otimes \mathcal{O}(1)$$

of vector bundles on $\mathbb{P}(V)$ it suffices to consider a representative of each of the 4 orbits of $\mathrm{Sp}\,\omega$. If

$$\omega = e_1 \wedge e_4 + e_2 \wedge e_5 + e_3 \wedge e_6$$

with associated symplectic basis $s_1, \dots, s_6$ of V^*, then we may consider

$$\begin{array}{ll} \sigma_1 = x_1 \wedge x_2 \wedge x_3 & \in I \\ \sigma_2 = \frac{1}{2}(x_1 \wedge x_2 \wedge x_5 - x_1 \wedge x_3 \wedge x_6) & \in S\setminus I \\ \sigma_3 = \frac{1}{2}(x_1 \wedge x_2 \wedge x_5 - x_1 \wedge x_3 \wedge x_6) + x_2 \wedge x_3 \wedge x_4 & \in T\setminus S \\ \sigma_4 = x_1 \wedge x_2 \wedge x_3 - x_4 \wedge x_5 \wedge x_6 & \in \mathbb{P}(U)\setminus T. \end{array}$$

The corresponding morphisms

$$\mathcal{O}(-1) \longrightarrow U^* \otimes \mathcal{O}(1)$$

are given by the 1st, 13th, sum of the 13th and 9th, sum of the 1st and 2nd row of b_2 resp. Only σ_4 defines a monomorphism of vector bundles, because only the entries of the generalized row corresponding to σ_4 generate an $(x_1, \dots, x_6)$-primary ideal.

$$\gamma(\sigma_4) = x_1 \wedge x_2 \wedge x_3 + x_4 \wedge x_5 \wedge x_6.$$

In particular we see that γ maps a secant line of $I \subset \mathbb{P}(U)$ onto itself. □

Let us fix dual bases $e_1, \dots, e_6$ of V and $x_1, \dots, x_6$ of V^* resp. and set

$$\begin{aligned} \omega &= e_1 \wedge e_4 + e_2 \wedge e_5 + e_3 \wedge e_6, \\ \sigma^{\pm} &= x_1 \wedge x_2 \wedge x_3 \pm x_4 \wedge x_5 \wedge x_6. \end{aligned}$$

The linear forms x_1, x_2, x_3 and x_4, x_5, x_6 define subspaces $W^* = V(x_1, x_2, x_3)$ and $W = V(x_4, x_5, x_6)$ of V resp. W^* is dual to W with respect to ω, $V = W \oplus W^*$. Let E be the cohomology of the monad

$$0 \longrightarrow \mathcal{O}(-1) \xrightarrow{\sigma^-} \mathcal{B} \xrightarrow{\sigma^+} \mathcal{O}(1) \longrightarrow 0$$

where $\mathcal{B}$ is defined by ω.

(1.4) **Proposition.**

(i) $\mathcal{E}$ is a stable rank 3 vector bundle with Chern polynomial $c(\mathcal{E}) = 1 + 3h^2$.

(ii) The orbit $\mathrm{PGL}(V).[\mathcal{E}] \subset M_{\mathbb{P}^5}(3;0,3,0)$ of $\mathcal{E}$ in the moduli scheme of stable rank 3 vector bundles on $\mathbb{P}^5$ with Chern classes $c_1 = 0$, $c_2 = 3$, $c_3 = 0$ is a smooth open part of a 27-dimensional irreducible component.

(iii) The symmetry group
$G = \{\Phi \in \mathrm{PGL}(V) \mid \Phi^*\mathcal{E} \cong \mathcal{E}\}$
is a semi-direct product
$G = \mathrm{SL}(W) \rtimes \mathbf{Z}_2$
where $\mathrm{SL}(W)$ acts on $V = W \oplus W^*$ naturally and the $\mathbf{Z}_2$-part is generated by the transformation
$R : e_i \longrightarrow e_{i+3}$ (indices mod 6).

Proof. (i) $H^0\mathcal{E} = H^0\mathcal{E}^* = 0$ follows from the monad. So $\mathcal{E}$ is stable.
(ii) By (1.3) the family of bundles constructed is 27-dimensional, it is an open subvariety M^0 of

$$M = \{(\tilde{\omega}, \sigma) \in \mathbb{P}(\wedge^2 V) \times \mathbb{P}(\wedge^3 V^*) \mid \sigma \neg \tilde{\omega} = 0\}.$$

More precisely:

$$\begin{array}{c} M^0 \\ \downarrow \\ \mathbb{P}(\wedge^2 V)\backslash H \end{array}$$

where $H \subset \mathbb{P}(\wedge^2 V)$ is the cubic hypersurface of degenerate symplectic forms and the fibres are isomorphic to $\mathbb{P}(U)\backslash T$ where T is a quartic hypersurface as in (1.3). $\mathrm{PGL}(V)$ operates transitively on $\mathbb{P}(\wedge^2 V)\backslash H$ and $\mathrm{PSp}(\tilde{\omega})$ acts transitively on the fibre $\mathbb{P}(U)\backslash T$ over a point $\tilde{\omega} \in \mathbb{P}(\wedge^2 V) \backslash H$.
It remains to check the tangent dimension of $M_{\mathbb{P}^5}(3;0,3,0)$ in the point $[\mathcal{E}]$.
From the monad, the resolution of $\mathcal{B}$

$$0 \longleftarrow \mathcal{B} \longleftarrow 14\mathcal{O}(-1) \longleftarrow 14\mathcal{O}(-2) \longleftarrow 6\mathcal{O}(-3) \longleftarrow \mathcal{O}(-4) \longleftarrow 0$$

and the cohomology table of $\mathcal{E}$ (compare (2.1.1)), we obtain

$$h^1\mathcal{E}^* \otimes \mathcal{E} \leq 27.$$

(iii) G is the stabilizer of $(\omega, \sigma^+) \in M^0$. □

(1.5) **Remark.** The Cremona transformation $\gamma : \mathbb{P}(U) - - \rightarrow \mathbb{P}(U)$ may be interpreted in terms of the moduli scheme as the transformation

$$M^0 \longrightarrow M^0,\ [\tilde{\mathcal{E}}] \longrightarrow [\tilde{\mathcal{E}}^*].\ \square$$

To see how to define $\mathcal{E}$ if $\mathrm{char}(k) = 2$ (then $\mathcal{B}$ is no longer self-dual), a less symmetric monad for $\mathcal{E}$ is more appropriate.
Write $\alpha = \frac{1}{2}\omega \wedge \omega = e_1 \wedge e_4 \wedge e_2 \wedge e_5 + e_1 \wedge e_4 \wedge e_3 \wedge e_6 + e_2 \wedge e_5 \wedge e_3 \wedge e_6$
and $\beta^{\pm} = e_1 \wedge e_2 \wedge e_3 \pm e_4 \wedge e_5 \wedge e_6$.

(1.6) **Proposition.** $\mathcal{E}$ is the cohomology of the monad

$$\Omega^5(5) \oplus \Omega^4(4) \xrightarrow{\varphi} \Omega^1(1) \oplus \Omega^2(2) \xrightarrow{\psi} V^* \otimes \mathcal{O}$$

with

$$\varphi = \begin{pmatrix} \alpha & \beta^+ \\ \beta^- & \omega \end{pmatrix}, \quad \psi = \begin{pmatrix} e_1 & e_2 & e_3 & e_4 & e_5 & e_6 \\ e_5 \wedge e_6 & e_6 \wedge e_4 & e_4 \wedge e_5 & e_2 \wedge e_3 & e_3 \wedge e_1 & e_1 \wedge e_2 \end{pmatrix}$$

Proof. The canonical diagram

$$
\begin{array}{ccccccccc}
 & & 0 & & 0 & & 0 & & \\
 & & \uparrow & & \uparrow & & \uparrow & & \\
0 & \longrightarrow & \mathcal{O}(-1) & \xrightarrow{\sigma^-} & \mathcal{B} & \xrightarrow{\sigma^+} & \mathcal{O}(1) & \longrightarrow & 0 \\
 & & \uparrow & & \uparrow & & \uparrow & & \\
0 & \longrightarrow & \Omega^5(5) \oplus \Omega^4(4) & \xrightarrow{\varphi} & \Omega^1(1) \oplus \Omega^2(2) & \xrightarrow{\psi} & V^* \otimes \mathcal{O} & \longrightarrow & 0 \\
 & & \uparrow & & \uparrow & & \uparrow & & \\
0 & \longrightarrow & \Omega^4(4) & \longrightarrow & \Omega^1(1) \oplus \Omega^4(4) & \longrightarrow & \Omega^1(1) & \longrightarrow & 0 \\
 & & \uparrow & & \uparrow & & \uparrow & & \\
 & & 0 & & 0 & & 0 & &
\end{array}
$$

is commutative. □

Now α is defined even if $\mathrm{char}(k) = 2$. So $\mathcal{E}$ is defined over the integers.

(1.7) **Remark.** The monad above is obtained by applying Beilinson's theorem [Bei] to $\mathcal{E}$. Killing cohomology [Ho 1] gives a monad of type

$$\mathcal{O}(-1) \longrightarrow \mathcal{B} \longrightarrow \mathcal{O}(1)$$

with

$$\mathcal{B} = \mathrm{coker}(\Omega^4(4) \xrightarrow{\omega} \Omega^2(2)). \quad \square$$

(1.8) **Proposition.** ($\mathrm{char}(k) = 2$.)
$\mathcal{E}$ is a nonsplit extension

$$0 \longrightarrow \mathcal{F} \longrightarrow \mathcal{E} \longrightarrow \mathcal{O} \longrightarrow 0$$

of $\mathcal{O}$ with Tango's bundle $\mathcal{F}$ on $\mathbb{P}^5$ [Ta1].

Proof The cosection $\eta' \in \mathrm{Hom}(\mathcal{B}, \mathcal{O})$ descends to a cosection of $\mathcal{E}$ defining the above sequence. The kernel $\mathcal{F}$ is the cohomology of the monad

$$0 \longrightarrow \Omega^5(5) \oplus \Omega^4(4) \longrightarrow \Omega^1(1) \oplus \Omega^2(2) \xrightarrow{(\psi | \frac{0}{\omega})} (V^* \otimes \mathcal{O}) \oplus \mathcal{O}.$$

Moreover, up to twist, $\mathcal{F}$ is Tango's bundle (see [Ta2, section 7]).
The extension does not split since $H^0\mathcal{E} = 0$. □

2. Sections of $\mathcal{E}$

In this section we work over an algebraically closed field k with $\mathrm{char}(k) \neq 2$. We want to describe the zero locus $Z(s)$ of a section $s \in H^0(\mathbb{P}^5, \mathcal{E}(2))$. Consider the display of the monad:

$$
\begin{array}{ccccc}
\mathcal{O}(-1) & \longrightarrow & \mathcal{K} & \longrightarrow & \mathcal{E} \\
\| & & \downarrow & & \downarrow \\
\mathcal{O}(-1) & \xrightarrow{\sigma^-} & \mathcal{B} & \longrightarrow & \mathcal{C} \\
 & & \downarrow{\scriptstyle \sigma^+} & & \downarrow \\
 & & \mathcal{O}(1) & = & \mathcal{O}(1)
\end{array}
$$

Let $S = \Gamma_*\mathcal{O} = \bigoplus_{n \in \mathbb{Z}} H^0(\mathbb{P}^5, \mathcal{O}(n))$ denote the polynomial ring, $E = \Gamma_*\mathcal{E}$, $B = \Gamma_*\mathcal{B}$ and $K = \Gamma_*\mathcal{K}$ the S-modules of global sections of $\mathcal{E}$, $\mathcal{B}$ and $\mathcal{K}$ resp. We have a short exact sequence

$$0 \longrightarrow S(-1) \longrightarrow K \longrightarrow E \longrightarrow 0.$$

Let $14S(-1)\xrightarrow{\sigma^+}S(1)$ be the generalized column of b_2 corresponding to σ^+. Notice that the symmetric form

$$q = x_1x_4 + x_2x_5 + x_3x_6$$

on $V = W \oplus W^*$ is one of the entries of σ^+.

(2.1) **Proposition.**

(i) The H^1-module

$$\bigoplus_{n\in\mathbb{Z}} H^1(\mathbb{P}^5, \mathcal{E}(n)) \cong [S/(q) + (x_1,x_2,x_3)^2 + (x_4,x_5,x_6)^2](1).$$

(ii) $K \cong \ker(14S(-1)\xrightarrow{\sigma^+}S(1))/\ker(14S(-1)\xrightarrow{b_2}14S(1))$

(iii) E has a minimal free resolution of type

$$0 \longleftarrow E \longleftarrow \begin{matrix} 8S(-2) \\ \oplus \\ 27S(-3)\end{matrix} \longleftarrow \begin{matrix} 6S(-3) \\ \oplus \\ 78S(-4)\end{matrix} \longleftarrow \begin{matrix} S(-4) \\ \oplus \\ 85S(-5)\end{matrix} \longleftarrow 42S(-6) \longleftarrow 8S(-7) \longleftarrow 0$$

Proof. (i) From the display of the monad we obtain

$$\begin{aligned}\bigoplus_{n\in\mathbb{Z}} H^1(\mathbb{P}^5, \mathcal{E}(n)) &\cong S(1)/\operatorname{Im}(B\xrightarrow{\sigma^+}S(1)) \\ &\cong \operatorname{coker}(14S(-1)\xrightarrow{\sigma^+}S(1)).\end{aligned}$$

The assertion follows by inspecting the entries of σ^+.

(ii) Complete the diagram

$$\begin{array}{ccccc} & & 0 & & \\ & & \downarrow & & \\ & & K & & \\ & & \downarrow & & \\ 14S(-1) & \longrightarrow & B & \longrightarrow & 0 \\ \downarrow{\scriptstyle\sigma^+} & & \downarrow{\scriptstyle\sigma^+} & & \\ S(1) & = & S(1) & & \\ \downarrow & & \downarrow & & \\ 0 & & 0 & & \end{array}$$

(iii) Comparing the syzygies of $\ker(14S(-1)\xrightarrow{\sigma^+}S(1))$,

$$\begin{matrix} S(-1) \\ \oplus \\ 22S(-2) \\ \oplus \\ 27S(-3)\end{matrix} \longleftarrow \begin{matrix} 12S(-3) \\ \oplus \\ 78S(-4)\end{matrix} \longleftarrow \begin{matrix} 2S(-4) \\ \oplus \\ 85S(-5)\end{matrix} \longleftarrow 42S(-6) \longleftarrow 8S(-7) \longleftarrow 0$$

with the syzygies of B,

$$14S(-1) \longleftarrow 14S(-2) \longleftarrow 6S(-3) \longleftarrow S(-4) \longleftarrow 0,$$

gives the result. □

(2.1.1) **Remark.**

(i) Notice that the part

$$\begin{matrix} S(-1) \\ \oplus \\ 22S(-2)\end{matrix} \longleftarrow 12S(-3) \longleftarrow 2S(-4) \longleftarrow 0$$

of the syzygies of $\ker(14S(-1)\xrightarrow{\sigma^+}S(1))$ is easy to compute. The number of terms in the remaining part follows by looking at the Hilbert function of $\bigoplus_{n\in\mathbb{Z}} H^1(\mathbb{P}^5,\mathcal{E}(n))$.

(ii) The resolution of E is defined over the ground ring $\mathbb{Z}[\frac{1}{2}]$. Thus the formulas for the cohomology groups of $\mathcal{E}$ as G-modules given by Horrocks for a ground field of characteristic zero remain valid for every field k with $\mathrm{char}(k) \neq 2$. □

(2.1.2) **Corollary.**
(i) $\mathcal{E}$ is 3-regular.
(ii) $H^0(\mathbb{P}^5,\mathcal{E}(2)) \cong W^*\otimes W/<\mathbb{1}>= \mathrm{Ad}(W)$ as G-modules.
(iii) $H^0(\mathbb{P}^5,\mathcal{E}(1)) = 0$.

Proof. For (ii) see [Ho2, section 2]. □

The zero locus of a general section $s \in H^0(\mathbb{P}^5,\mathcal{E}(n))$, $n \geq 3$, is a smooth surface [Ho2, section 3].
Here we want to describe the zero locus of a general section $s \in H^0(\mathbb{P}^5,\mathcal{E}(2))$.

(2.2) **Proposition.** If the zero locus $X = Z(s)$ of a section $s \in H^0(\mathbb{P}^5,\mathcal{E}(2))$ has expected codimension, then X is a (possibly reducible and nonreduced) locally complete intersection surface of degree 14, irregularity $q = h^1\mathcal{O}_X = 2$ and geometric genus $p_g = h^2\mathcal{O}_X = 1$.

Proof. If X has codimension 3 then

$$\deg X = c_2(\mathcal{E}(2)) = 14,$$

moreover the Koszul complex

$$0 \longrightarrow \mathcal{O}(-6) \longrightarrow \wedge^2\mathcal{E}^*(-4) \longrightarrow \mathcal{E}^*(-2)\xrightarrow{s}\mathcal{O} \longrightarrow \mathcal{O}_X \longrightarrow 0$$

associated to s is exact. We identify $\wedge^2\mathcal{E}^* \cong \mathcal{E}$ via $\wedge^3\mathcal{E} \cong \mathcal{O}$. Taking cohomology we obtain

$$H^2\mathcal{O}_X \cong H^5\mathcal{O}(-6) \cong k$$

and a short exact sequence

$$0 \longrightarrow H^2\mathcal{E}^*(-2) \longrightarrow H^1\mathcal{O}_X \longrightarrow H^3\mathcal{E}(-4) \longrightarrow 0.$$

$H^3\mathcal{E}(-4) \cong H^3\mathcal{K}(-4) \cong H^3\mathcal{B}(-4) \cong H^4\Omega^4 \cong k$ and similarly $H^2\mathcal{E}^*(-2) \cong k$. So $q = 2$. □

(2.2.1) **Remark.** Numerically X looks like an abelian surface with a (1,7)-polarization. However X is never irreducible: As pointed out by Horrocks [Ho2, section 3] $\mathbb{P}(W), \mathbb{P}(W^*) \subset \mathbb{P}(V)$ are always among the irreducible components of X. We will see this lateron once more. □

For the equations of X consider again the display

$$\begin{array}{ccccc}
 & & \mathcal{O}(-2) & & \\
 & & \swarrow s_{\mathcal{K}} \quad \searrow s & & \\
\mathcal{O}(-1) & \xrightarrow{\sigma^-} & \mathcal{K} & \longrightarrow & \mathcal{E} \\
\| & & \downarrow & & \downarrow \\
\mathcal{O}(-1) & \xrightarrow{\sigma^-} & \mathcal{B} & \longrightarrow & \mathcal{C} \\
 & & \downarrow \sigma^+ & & \downarrow \\
 & & \mathcal{O}(1) & = & \mathcal{O}(1)
\end{array}$$

and the inclusion $\mathcal{K} \subset \mathcal{B} \subset 14\mathcal{O}(1)$.

(2.3) **Proposition.**

(i) Each section $s \in H^0(\mathbb{P}^5, \mathcal{E}(2))$ has a unique lifting $s_{\mathcal{K}} \in H^0(\mathbb{P}^5, \mathcal{K}(2))$ whose zero locus $Z(s_{\mathcal{K}})$ contains $\mathbb{P}(W) \cup \mathbb{P}(W^*)$.

(ii) $X = Z(s)$ ist the locus where the map

$$\begin{pmatrix} \sigma^- \\ s_{\mathcal{K}} \end{pmatrix} \in \mathrm{Hom}(\mathcal{O}(-1) \oplus \mathcal{O}(-2), 13\mathcal{O}(1))$$

has rank≤ 1.

Proof. Since $\mathrm{Ext}^1(\mathcal{O}(-2), \mathcal{O}(-1)) = 0$ every section $s \in H^0\mathcal{E}(2)$ can be lifted to a section $s' \in H^0\mathcal{K}(2)$. Two liftings differ by an element of $\sigma^-(H^0\mathcal{O}(1)) \subset H^0\mathcal{K}(2)$. Moreover the dependancy locus of the map

$$\begin{pmatrix} \sigma^- \\ s' \end{pmatrix} \in \mathrm{Hom}(\mathcal{O}(-1) \oplus \mathcal{O}(-2), \mathcal{K})$$

is independent of the lifting and defines X. To identify $\binom{\sigma^-}{s'}$ with a 2×13 matrix with quadratic and cubic entries we compose with the inclusion $\mathcal{K} \subset 13\mathcal{O}(1)$ obtained from

$$\begin{array}{ccc} 0 & & 0 \\ \downarrow & & \downarrow \\ \mathcal{K} & \dashrightarrow & 13\mathcal{O}(1) \\ \downarrow & & \downarrow \\ \mathcal{B} & \hookrightarrow & 14\mathcal{O}(1) \\ \downarrow & & \downarrow \\ \mathcal{O}(1) & = & \mathcal{O}(1) \\ \downarrow & & \downarrow \\ 0 & & 0 \end{array}$$

This proves the second statement. For the first we have to find a distinguished lifting of s. This can be easily done by computing all 2×13 matrices as above explicitely:

By Proposition (2.1), (ii) any section of $\mathcal{K}(2)$ is represented by a linear syzygy among the entries of σ^+. Multiplying the corresponding 1×14-matrix of linear forms with b_2 yields a 1×14-matrix of cubic forms. The entry corresponding to the generalized column of σ^+ is equal to zero. Thus we obtain a 2×13-matrix

$$\begin{pmatrix} \sigma^- \\ s' \end{pmatrix} = \begin{pmatrix} q & x_1^2 & x_1x_2 & x_1x_3 & x_2^2 & x_2x_3 & x_3^2 & x_4^2 & x_4x_5 & x_4x_6 & x_5^2 & x_5x_6 & x_6^2 \\ f_0 & f_1 & f_2 & f_3 & f_4 & f_5 & f_6 & f_7 & f_8 & f_9 & f_{10} & f_{11} & f_{12} \end{pmatrix}$$

with $f_0, \ldots, f_{12} \in H^0\mathcal{O}(3)$. As it turns out f_0 is always a multiple of q.
$V(q)$ does not contain $\mathbb{P}(W) \cup \mathbb{P}(W^*)$. Let $s_{\mathcal{K}}$ be the distinguished lifting of s obtained by asking $f_0 = 0$. Inspecting $f_1, \ldots, f_{12}$ gives the result: If $f_0 = 0$ then

$$f_1, \ldots, f_{12} \in H^0\mathcal{O}(3) \cong \mathrm{S}_3 V^* \cong \mathrm{S}_3(W^* \oplus W)$$

are contained in the subspace

$$(\mathrm{S}_2 W^* \otimes W) \oplus (W^* \otimes \mathrm{S}_2 W) \subset \mathrm{S}_3 V^*$$

where we identify $W^* \cong \langle x_1, x_2, x_3 \rangle$ and $W \cong \langle x_4, x_5, x_6 \rangle$.
So the common zero locus of $f_1, \ldots, f_{12}$ contains $\mathbb{P}(W) \cup \mathbb{P}(W^*)$.

A somewhat different approach utilizes the G-module structure. As G-modules

$$(*) \qquad H^0\mathcal{K}(2) \cong H^0\mathcal{O}(1) \oplus H^0\mathcal{E}(2).$$

Hence there is a unique G-equivariant lifting

$$\begin{array}{ccc} H^0\mathcal{E}(2)\otimes\mathcal{O}(-2) & \dashrightarrow & \mathcal{K} \\ & \searrow & \downarrow \\ & & \mathcal{E}\,. \end{array}$$

To describe the matrix

$$\begin{pmatrix} \sigma^- \\ s_{\mathcal{K}} \end{pmatrix} \in \mathrm{Hom}(\mathcal{O}(-1)\oplus\mathcal{O}(-2),\mathcal{K}) \subset \mathrm{Hom}\,(\mathcal{K},\mathcal{O}(1))^* \otimes (\mathrm{S}_2V^* \oplus \mathrm{S}_3V^*)$$

in this setting we identify

$$\mathrm{Hom}\,(\mathcal{K},\mathcal{O}(1)) \subset \mathrm{Hom}(\mathcal{O}(-1),\mathcal{O}(1)) \cong \mathrm{S}_2V^*$$

with the summand

$$1\!\!1 \oplus \mathrm{S}_2W^* \oplus \mathrm{S}_2W \subset \mathrm{S}_2(W^*\oplus W) \cong \mathrm{S}_2V^*.$$

σ^- is then represented by the identity in

$$\mathrm{Hom}(\mathcal{K},\mathcal{O}(1))^* \otimes (1\!\!1 \oplus \mathrm{S}_2W^* \oplus \mathrm{S}_2W).$$

The splitting (*) defines a G-equivariant morphism

$$\mathrm{Hom}(\mathcal{K},\mathcal{O}(1)) \otimes H^0\mathcal{E}(2) \xrightarrow{\Phi} H^0\mathcal{O}(3) \cong \mathrm{S}_3V^*.$$

Φ may be interpreted as the map $s \longrightarrow s_{\mathcal{K}}$:
Since $\mathrm{S}_3V^* \cong (\mathrm{S}_3W^* \oplus \mathrm{S}_3W) \oplus ((\mathrm{S}_2W^* \otimes W) \oplus (W^* \otimes \mathrm{S}_2W))$ contains no copy of $H^0\mathcal{E}(2) \cong W^* \otimes W/ < 1\!\!1 >$. Φ is zero on the summand

$$1\!\!1 \otimes H^0\mathcal{E}(2) \longrightarrow \mathrm{S}_3V^*.$$

To compute the cubics $f_1, \ldots, f_{12}$ explicitely we have to consider the part

$$(\mathrm{S}_2W^* \oplus \mathrm{S}_2W) \oplus H^0\mathcal{E}(2) \longrightarrow \mathrm{S}_3V^*$$

of Φ:
Identify $W \cong \wedge^2 W^*$ and define a morphism

$$\mathrm{S}_2W^* \otimes W^* \otimes W \longrightarrow \mathrm{S}_3V^*$$

by composing

$$\begin{array}{ccc} S_2W^*\otimes W^*\otimes W & & \\ \| & & \\ S_2W^*\otimes W^*\otimes \Lambda^2 W^* & & g_1g_2\otimes f\otimes h_1\wedge h_2 \\ \downarrow & & \downarrow \\ W^*\otimes W^*\otimes W^*\otimes W^*\otimes W^* & & \frac{1}{2}(g_1\otimes g_2\oplus g_2\otimes g_1)\otimes f\otimes (h_1\otimes h_2 - h_2\otimes h_1) \end{array}$$

with

$$\begin{array}{ccc} W^*\otimes W^*\otimes W^*\otimes W^*\otimes W^* & & g_1\otimes g_2\otimes f\otimes h_1\otimes h_2 \\ \downarrow & & \downarrow \\ W^*\otimes\Lambda^2W^*\otimes\Lambda^2W^* & & h_2\otimes g_1\wedge f\otimes g_2\wedge h_1 \\ \downarrow & & \downarrow \\ W^*\otimes S_2\,\Lambda^2\,W^* & & h_2\otimes(g_1\wedge f)(g_2\wedge h) \\ \| & & \\ W^*\otimes S_2W & & \\ \cap & & \\ S_3V^* & & \end{array}$$

Similarly define

$$S_2W\otimes W^*\otimes W \longrightarrow S_2W^*\otimes W\subset S_3V^*.$$

The sum of the two morphisms is zero on the part

$$(S_2W^*\oplus S_2W)\otimes \mathbb{1} \longrightarrow S_3V^*$$

(compare (2.4) below), so it induces a G-morphism

$$(S_2W^*\oplus S_2W)\otimes(W^*\otimes W)/<\mathbb{1}>\longrightarrow S_3V^*$$

which is Φ. □

We are now ready to describe the zero locus of a general section $s\in H^0\mathcal{E}(2)\cong W^*\otimes W/ <\mathbb{1}>$. We identify s with a traceless endomorphism $a(s)$ of W. Since k is algebraically closed we may assume that our basis e_1,e_2,e_3 of W is chosen such that

$$a(s)=\begin{pmatrix} a_{11} & a_{12} & a_{13}\\ a_{21} & a_{22} & a_{23}\\ a_{31} & a_{32} & a_{33}\end{pmatrix}\in \mathrm{End}(W),\quad x=(x_1,x_2,x_3)\longrightarrow x\cdot a(s),$$

is in Jordan normal form, i.e. for a general section s we may assume that $a(s)$ is a diagonal matrix with three distinct eigenvalues a_{11},a_{22},a_{33}.

(2.4) **Proposition.** Let $X=Z(s)$ be the zero locus of the section corresponding to the diagonal matrix with three distinct eigenvalues a_{11},a_{22},a_{33}. Then X is the dependancy locus of the 2×13 matrix

$$\begin{pmatrix}\sigma^-\\ s_\mathcal{K}\end{pmatrix}=\begin{pmatrix} q & x_1^2 & x_1x_2 & x_1x_3 & x_2^2 & x_2x_3 & x_3^2 & x_4^2 & x_4x_5 & x_4x_6 & x_5^2 & x_5x_6 & x_6^2\\ 0 & f_1 & f_2 & f_3 & f_4 & f_5 & f_6 & f_7 & f_8 & f_9 & f_{10} & f_{11} & f_{12}\end{pmatrix}$$

with

$$\begin{pmatrix} f_1 \\ f_2 \\ f_3 \\ f_4 \\ f_5 \\ f_6 \\ f_7 \\ f_8 \\ f_9 \\ f_{10} \\ f_{11} \\ f_{12} \end{pmatrix} = a_{11} \begin{pmatrix} 0 \\ -\frac{1}{2}(x_2x_5x_6 + x_3x_6^2) \\ \frac{1}{2}(x_2x_5^2 + x_3x_5x_6) \\ x_2x_4x_6 \\ \frac{1}{2}(-x_2x_4x_5 + x_3x_4x_6) \\ -x_3x_4x_5 \\ 0 \\ -\frac{1}{2}(x_2x_3x_5 + x_3^2x_6) \\ \frac{1}{2}(x_2^2x_5 + x_2x_3x_6) \\ x_1x_3x_5 \\ \frac{1}{2}(-x_1x_2x_5 + x_1x_3x_6) \\ -x_1x_2x_6 \end{pmatrix} + a_{22} \begin{pmatrix} -x_1x_5x_6 \\ \frac{1}{2}(x_1x_4x_6 + x_3x_6^2) \\ \frac{1}{2}(x_1x_4x_5 - x_3x_5x_6) \\ 0 \\ -\frac{1}{2}(x_1x_4^2 + x_3x_4x_6) \\ x_3x_4x_5 \\ -x_2x_3x_4 \\ \frac{1}{2}(x_1x_3x_4 + x_3^2x_6) \\ \frac{1}{2}(x_1x_2x_4 - x_2x_3x_6) \\ 0 \\ -\frac{1}{2}(x_1^2x_4 + x_1x_3x_6) \\ x_1x_2x_6 \end{pmatrix} + a_{33} \begin{pmatrix} x_1x_5x_6 \\ \frac{1}{2}(-x_1x_4x_6 + x_2x_5x_6) \\ -\frac{1}{2}(x_1x_4x_5 + x_2x_5^2) \\ -x_2x_4x_6 \\ \frac{1}{2}(x_1x_4^2 + x_2x_4x_5) \\ 0 \\ x_2x_3x_4 \\ \frac{1}{2}(-x_1x_3x_4 + x_2x_3x_5) \\ -\frac{1}{2}(x_1x_2x_4 + x_2^2x_5) \\ -x_1x_3x_5 \\ \frac{1}{2}(x_1^2x_4 + x_1x_2x_5) \\ 0 \end{pmatrix}$$

Proof. Plug in the formula given in the proof of (2.3). □

Depending on a_{11}, a_{22}, a_{33} we set

$$\mu_i = \prod_{j \neq i} (a_{jj} - a_{ii}), \quad i = 1, 2, 3.$$

(2.5) **Proposition.** For s as above

$$Z(s_K) = \mathbb{P}(W) \cup \mathbb{P}(W^*) \cup Q_1 \cup Q_2 \cup Q_3$$

where

$$Q_i = V(x_i, x_{i+3}, q_i), \quad i = 1, 2, 3,$$

with

$$q_1 = \mu_2 x_3 x_6 - \mu_3 x_2 x_5, \quad q_2 = \mu_1 x_3 x_6 - \mu_3 x_1 x_4, \quad q_3 = \mu_2 x_1 x_4 - \mu_1 x_2 x_5.$$

Proof. An explicit computation shows:

$$I_{Z(s_K)} = I_{Q_1} \cap I_{Q_2} \cap I_{Q_3} \cap (x_1, x_2, x_3) \cap (x_4, x_5, x_6). \quad \square$$

(2.5.1) **Remark.** Each of the Q_i is a smooth quadric surface. □

(2.5.2) **Proposition**

(i) $Q_1 \cup Q_2 \cup Q_3$ intersects $\mathbb{P}(W)$ and $\mathbb{P}(W^*)$ resp. in a triangle whose vertices are the eigenspaces of $a(s) \in \mathrm{End}(W) \cong \mathrm{End}(W^*)$ in $\mathbb{P}(W)$ and $\mathbb{P}(W^*)$ resp.

(ii) Q_1, Q_2 and Q_3 define a quadratic transformation

$$\xi : \mathbb{P}(W) - - \rightarrow \mathbb{P}(W^*)$$

whose base locus consists of the eigenspaces of $a(s)$.

Proof.

(i) is clear.

(ii) The quadratic transformation is given by

$$(x_1 : x_2 : x_3) \longrightarrow (\mu_1 x_2 x_3 : \mu_2 x_1 x_3 : \mu_3 x_1 x_2).$$

Notice that the surprising fact that the three lines in $\mathbb{P}(W^*)$ constructed from a point $p \in \mathbb{P}(W)$ intersect in a single point follows from the fact that q_1, q_2, q_3 are linearly dependent.

□

(2.6) **Proposition.** Let s be as above. Then

(i) $Z(s) = \mathbb{P}(W) \cup \mathbb{P}(W^*) \cup Q_1 \cup Q_2 \cup Q_3 \cup D$ where D is a Del Pezzo surface of degree 6.

(ii) The singular locus of $Z(s)$ consists of the edges of the octahedron spanned by the eigenspaces $< e_1 >, \ldots, < e_6 > \in \mathbb{P}(W) \cup \mathbb{P}(W^*) \subset \mathbb{P}(V)$ of $a(s)$.

We first establish:

(2.6.1) **Proposition.**

(i) $Z(s) \setminus Z(s_\mathcal{K})$ is contained in $V(J)$ where

$$J = (x_1x_4 + x_2x_5 + x_3x_6, a_{11}x_1x_4 + a_{22}x_2x_5 + a_{33}x_3x_6, x_1x_2x_3 - x_4x_5x_6).$$

(ii) $V(J)$ is the union of a Del Pezzo surface D of degree 6 with 6 planes. These planes are the faces of the octahedron different from $\mathbb{P}(W)$ and $\mathbb{P}(W^*)$.

Proof.

(i) It suffices to prove

$$J \cdot I_{Z(s_\mathcal{K})} \subset I_{Z(s)}.$$

This follows by a computation.

(ii) J is generated by a regular sequence, hence $V(J)$ is an unmixed surface of degree $2 \cdot 2 \cdot 3 = 12$. Each of the 6 faces different from $\mathbb{P}(W)$ and $\mathbb{P}(W^*)$ is contained in $V(J)$. Let $p = (p_1 : \ldots : p_6) \in V(J) \setminus V(x_1x_2x_3x_4x_5x_6)$. (The existence of such a point is easily verified.) Then the image of $\mathbb{P}^2 = \mathbb{P}^2(u : v : w)$ under the rational map

$$\mathbb{P}^2 \dashrightarrow \mathbb{P}^5$$

$$(u : v : w) \longmapsto (p_1u^2v : p_2v^2w : p_3w^2u : p_4vw^2 : p_5wu^2 : p_6uv^2)$$

is contained in $V(J)$. This is a smooth Del Pezzo surface of degree 6. So the assertion follows by degree reasoning. □

Proof. of (2.6).

(i) Since

$$Z(s) \subset Z(s_\mathcal{K}) \cup V(J)$$

$Z(s)$ has the expected codimension. It remains to show that the 6 faces of the octahedron different from $\mathbb{P}(W)$ and $\mathbb{P}(W^*)$ are not among the components of $Z(s)$. This follows by inspecting the matrix $\begin{pmatrix} \sigma^- \\ s_\mathcal{K} \end{pmatrix}$.

(ii) Each irreducible component of $Z(s)$ is smooth. So its singular locus consists of the intersections of the various components. □

(2.7) **Remark.**

(i) The Del Pezzo surface D of degree 6 in the given embedding is hyperosculating in every point [To] (cf. [Sh] for a more recent treatment). The osculating hyperplane through a point $p \in D$ away from the hexagon of lines is spanned by the three conics which pass through p. To our knowledge D is (up to projective equivalence) the only known example of a nonruled hyperosculating surface in $\mathbb{P}^5$.

(ii) As an abstract surface D is isomorphic to the graph of the quadratic transformation

$$\xi : \mathbb{P}(W) \dashrightarrow \mathbb{P}(W^*)$$

defined by Q_1, Q_2 and Q_3.

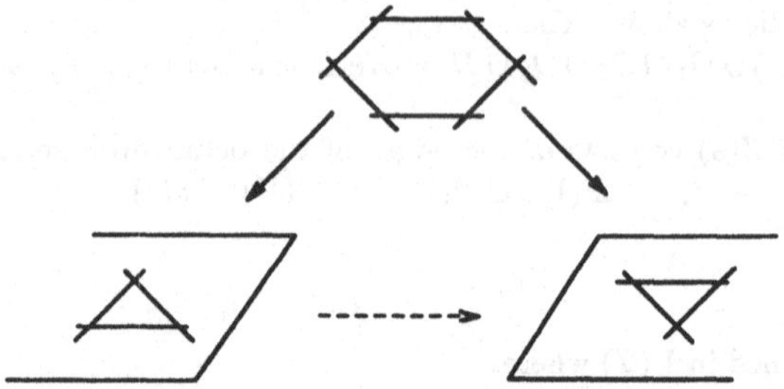

But D is not isomorphic to graph (ξ) in a canonical way. As we will prove later, there is no isomorphism $D \cong \text{graph}(\xi)$ which is rationally defined over $H^0(\mathbb{P}^5, \mathcal{E}(2))$.
(iii) There is however a connection between D and the plane triangles in $\mathbb{P}(W)$ and $\mathbb{P}(W^*)$ which is canonical. Consider the rational map

$$\pi : D - - \rightarrow \mathbb{P}(W)$$

induced by the projection $\mathbb{P}^5 \setminus \mathbb{P}(W^*) \longrightarrow \mathbb{P}(W))$. π factors as follows:
Let $D' = \{(w : x_1 : x_2 : x_3) \in \mathbb{P}^3 \mid w^3 = x_1 x_2 x_3\}$ be the triple cover $\pi_1 : D' \longrightarrow \mathbb{P}(W)$ of the plane branched totally over the triangle $x_1 x_2 x_3 = 0$. D' also contains a triangle of lines, the vertices are now singular points of type A_2. Let

$$\pi_2 : \tilde{D} \longrightarrow D'$$

be the minimal resolution of singularities of D'. $\tilde{D}$ contains a cycle of nine rational curves: A pair of (-2)-curves for each of the singularities and three (-1)-curves, which are the strict transforms of the lines in the triangle. Blowing down the (-1)-curves gives a map π_3 to a surface with a hexagon of (-1)-curves. This is D:

$$\pi_3 : \tilde{D} \longrightarrow D$$

and

$$\pi = \pi_1 \circ \pi_2 \circ \pi_3^{-1} : D - - \rightarrow \mathbb{P}(W).$$

Similarly we obtain a factorization of $D - - \rightarrow \mathbb{P}(W^*)$.

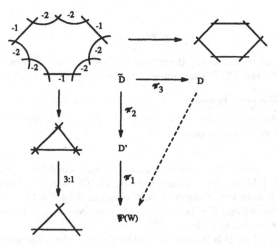

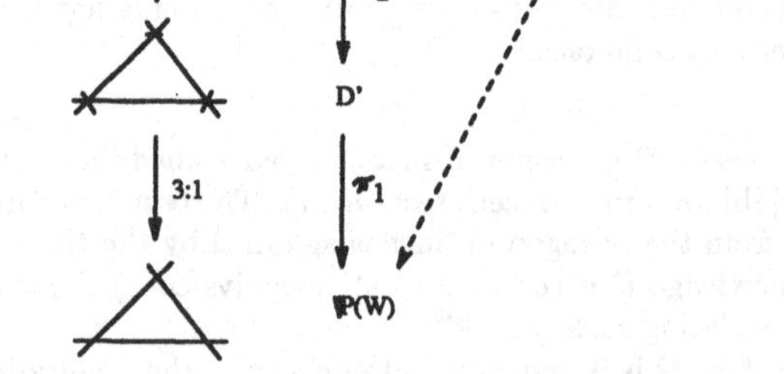

We now briefly discuss the geometry of more special nonzero sections $s \in H^0(\mathbb{P}^5, \mathcal{E}(2))$, i.e. those sections whose corresponding endomorphism $a(s) \in \mathrm{End}(W)$ has a multiple eigenvalue. Up to coordinate transformations and scalars there are only four more types depending on the Jordan normal form of $a(s)$:

$$\begin{pmatrix} 1 & 1 & 0 \\ 0 & 1 & 0 \\ 0 & 0 & -2 \end{pmatrix}, \begin{pmatrix} 1 & 0 & 0 \\ 0 & 1 & 0 \\ 0 & 0 & -2 \end{pmatrix}, \begin{pmatrix} 0 & 1 & 0 \\ 0 & 0 & 1 \\ 0 & 0 & 0 \end{pmatrix}, \begin{pmatrix} 0 & 1 & 0 \\ 0 & 0 & 0 \\ 0 & 0 & 0 \end{pmatrix}.$$

(2.8) **Proposition.**

(i) $Z(s)$ has codimension 3 iff the corresponding endomorphism $a(s) \in \mathrm{End}(W)$ is semistable.

(ii) In the semistable case $Z(s)$ consists of $\mathbb{P}(W)$, $\mathbb{P}(W^*)$, quadrics Q_λ for each eigenvalue λ of $a(s)$ with a multiple structure whose length is the multiplicity $m(\lambda)$ of the eigenvalue λ, and a surface $D(s)$ which is a component of the variety defined by the equations

$$x_1x_4 + x_2x_5 + x_3x_6, \quad \sum_{i,j=1}^{3} a_{ij}(s)x_ix_{j+3}$$

and

$$\prod_\lambda x_\lambda^{m(\lambda)} + \prod_\lambda (x_\lambda^*)^{m(\lambda)}$$

where

$$x_\lambda \in W^* \cong < x_1, x_2, x_3 >, \quad x_\lambda^* \in W \cong < x_4, x_5, x_6 >$$

are eigenvectors of $a(s)$ corresponding to λ.

Proof. As before this is easily checked by inspecting the corresponding 2×13 matrix defining $Z(s)$. □

Let $\Gamma^s \subset \Gamma^{ss} \subset \Gamma(\mathbb{P}^5, \mathcal{E}(2))$ denote the varieties of sections s whose corresponding endomorphism $a(s) \in \mathrm{End}(W)$ is stable and semistable resp.
Let $\mathbb{D} \subset \Gamma^s \times \mathbb{P}^5$ and $\mathbb{G} \subset \Gamma^s \times \mathbb{P}(W) \times \mathbb{P}(W^*)$ resp. be the smooth family whose fibre over $s \in \Gamma^s$ is the Del Pezzo component $D(s) \subset Z(s)$ and the graph $G(s)$ of the corresponding Cremona transformation

$$\xi(s) : \mathbb{P}(W) - - \to \mathbb{P}(W^*)$$

resp.

(2.9) **Proposition.** $\mathbb{D}$ and $\mathbb{G}$ are not isomorphic over Γ^s.

Proof. Suppose there is an isomorphism

$$\begin{array}{ccc} \mathbb{D} & \xrightarrow{\cong} & \mathbb{G} \\ & \searrow \quad \swarrow & \\ & \Gamma^s & \end{array}$$

The composition with the contraction

$$\begin{array}{ccc} \mathbb{G} & \longrightarrow & \Gamma^s \times \mathbb{P}(W) \\ & \searrow \quad \swarrow & \\ & \Gamma^s & \end{array}$$

identifies 3 lines of the hexagon in $D(s)$ with the exceptional locus of $D(s) - - \to \mathbb{P}(W)$. This allows to define an "orientation" of the triangle of lines in $\mathbb{P}(W)$, say (e_1, e_2, e_3) if the lines $\overline{e_1 e_6}, \overline{e_2 e_4}, \overline{e_3 e_5}$ are contracted:

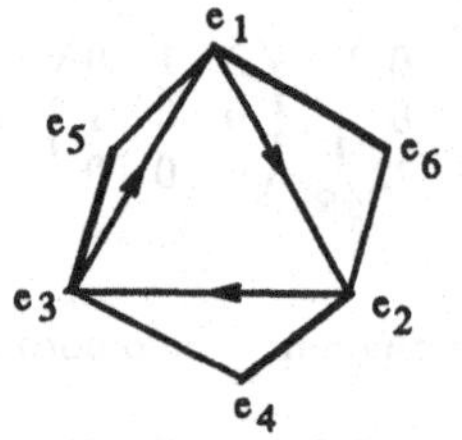

But this contradicts the fact that the characteristic polynomial

$$\chi(t) = \det(t \cdot I - a) \in k(a_{ij})[t]$$

has the full symmetric group S_3 as Galois group. □

Finally we consider the zero locus of $s_1 \wedge s_2 \in H^0(\mathbb{P}^5, \wedge^2 \mathcal{E}(4))$ for general sections $s_1, s_2 \in H^0(\mathbb{P}^5, \mathcal{E}(2))$. Set theoretically

$$Z(s_1 \wedge s_2) = \bigcup_{(\lambda:\mu) \in \mathbb{P}^1} Z(\lambda s_1 + \mu s_2) \subset \mathbb{P}^5.$$

Thus we expect that $Z(s_1 \wedge s_2)$ is a 3-fold fibred over $\mathbb{P}^1$.
To describe this 3-fold we introduce some notation. We write $a = a(s_1) = (a_{ij})$, $b = a(s_2) = (b_{ij}) \in \mathrm{End}(W)$ for the endomorphisms corresponding to s_1 and s_2 resp.
Let

$$q_a = \sum a_{ij} x_i x_{j+3}, \qquad q_b = \sum b_{ij} x_i x_{j+3}$$

be the quadrics as in Proposition (2.6.1). We associate to (a, b) a cubic

$$f^*_{(a,b)} - f_{(a,b)} \in \mathrm{S}_3 W^* \oplus \mathrm{S}_3 W \subset \mathrm{S}_3 V^*$$

by applying the following morphism to $a \otimes b$:

$$\begin{array}{ccc}
(W^* \otimes W) \otimes (W^* \otimes W) & & \\
\| & & \\
(W^* \otimes W^*) \otimes (W \otimes W) & & (g_1 \otimes g_2) \otimes (h_1 \otimes h_2) \\
\downarrow & & \downarrow \\
(\mathrm{S}_2 W^* \otimes \wedge^2 W^*) \oplus (\wedge^2 W^* \otimes \mathrm{S}_2 W) & & (g_1 g_2 \otimes h_1 \wedge h_2) - (g_1 \wedge g_2 \otimes h_1 h_2) \\
\| & & \\
(\mathrm{S}_2 W^* \otimes W^*) \oplus (W \otimes \mathrm{S}_2 W) & & \\
\downarrow & & \\
\mathrm{S}_3 W^* \oplus \mathrm{S}_3 W & &
\end{array}$$

The vanishing locus of $f^*_{(a,b)}$ in $\mathbb{P}(W)$ is isomorphic to the spectral curve

$$E = V(\det(tI - \lambda a - \mu b)) \subset \mathbb{P}^2(t : \lambda : \mu)$$

of a and b. The graph of this isomorphism in $\mathbb{P}^2(t:\lambda:\mu)\times\mathbb{P}(W)$ is given by

$$(tI-\lambda a-\mu b)\cdot\begin{pmatrix}x_1\\x_2\\x_3\end{pmatrix}=0.$$

Similarly for $f_{(a,b)}$.

(2.10) **Proposition.** For general $s_1, s_2, \in H^0(\mathbb{P}^5, \mathcal{E}(2))$ the zero locus of $s_1 \wedge s_2$ has two irreducible components

$$Z(s_1\wedge s_2)=X_6\cup X_9$$

of dimension 3 and degree 6 and 9 resp.
X_6 is the complete intersection

$$X_6=V(q,f^*_{(a,b)}-f_{(a,b)}).$$

It contains a pencil of Del Pezzo surfaces and is singular in the 18 intersection points of any two of these. $X_9 = Z(s_{1K}\wedge s_{2K})$ is not normal. Its normalization is a $\mathbb{P}^1\times\mathbb{P}^1$-bundle over the spectral curve E.
X_6 and X_9 intersect in a surface Y obtained via linkage as follows:

$$\mathbb{P}(W)\cup\mathbb{P}(W^*)\ \underset{q,q_a,q_b}{\sim}\ Y'\ \underset{q,f^*_{(a,b)},f_{(a,b)}}{\sim}\ Y.$$

Proof. Equations for $Z(s_1\wedge s_2)$ are obtained by taking the 3×3-minors of the 3×13-matrix corresponding to s_1 and s_2. The assertion follows by a computation. (Most of the geometry is clear from our description of $Z(\lambda s_1+\mu s_2)$). □

References

[Ba] Barth, W.: Some properties of stable rank-2 vector bundles on $\mathbb{P}_n$. Math. Ann. **226**, 125–150 (1977).

[Bei] Beilinson, A.: Coherent sheaves on $\mathbb{P}^N$ and problems of linear algebra. Functional Anal. Appl. **12**, 214–216 (1978).

[Ho 1] Horrocks, G.: Construction of bundles on $\mathbb{P}^n$. In: Les équations de Yang-Mills, A. Douady, J.-L. Verdier, séminaire E.N.S. 1977-78. Astérisque **71–72**, 197–203 (1980)

[Ho 2] Horrocks, G.: Examples of rank three vector bundles on five-dimensional projective space. J. London Math. Soc. (2),**18**, 15–27 (1978)

[Mac] Bayer, D., Stillman, M.: Macaulay, a computer algebra system for algebraic geometry.

[Sh] Shifrin, T.: The osculatory behavior of surfaces in $\mathbb{P}^n$. Pac. J. Math. **123**, 227–256 (1986).

[Ta 1] Tango, H.: On morphisms from projective space $\mathbb{P}^n$ to the Grassmann variety $\mathrm{Gr}(n,d)$. J. Math. Kyoto Univ. **16**, 201–207 (1976).

[Ta 2] Tango, H.: On vector bundles on $\mathbb{P}^n$ which have σ-transition matrices. Preprint

[To] Togliatti, E.: Alcuni esempi di superficie algebriche degli iperspazi che rappresentano un' equazione di Laplace. Comm. Math. Helv. **1**, 255–272 (1929).

Wolfram Decker
Fachbereich 9 Mathematik
Universität des Saarlandes
D–6600 Saarbrücken
Germany

Nicolae Manolache
Institute of Mathematics
of the Romanian Academy
Str. Academiei 14
Bucharest
Romania

Frank-Olaf Schreyer
Fakultät für Mathematik
und Physik
Universität Bayreuth
Postfach 101251
D–8580 Bayreuth
Germany

STABILITY AND RESTRICTIONS OF PICARD BUNDLES, WITH AN APPLICATIONS TO THE NORMAL BUNDLES OF ELLIPTIC CURVES

by

Lawrence Ein* and Robert Lazarsfeld**

Introduction.

Let C be a smooth irreducible projective curve of genus $g \geq 1$, and for each integer d let $J_d(C)$ be the Jacobian of C, which we view as parametrizing all line bundles on C of degree d. Denote by L_t the bundle on C corresponding to the point $t \in J_d(C)$. Provided that $d \geq 2g-1$, the vector spaces $H^0(C,L_t)$ fit together to form the fibres of a vector bundle P_d on $J_d(C)$, of rank d+1-g, called the degree d *Picard bundle* (defined by this description up tp tensoring by line bundles on $J_d(C)$). These bundles have been the focus of considerable study in recent years, notably by Kempf and Mukai ([K1], [K2], [K3], [M]). To better understand their geometry, it is natural to ask whether P_d is stable with respect to the canonical principal polarization of $J_d(C)$. Kempf [K1] shows that this is indeed the case for the first bundle P_{2g-1}. The main purpose of this note is to complete Kempf's result by proving the following

Theorem. *For every* $d \geq 2g$, *the Picard bundle* P_d *is stable with respect to the polarization on* $J_d(C)$ *defined by the theta divisor* $\Theta_C \subset J_d(C)$.

For g = 2, this was established by Umemura [U]. As in [K1], the proof depends on analyzing the restriction of P_d to C. We show that the restriction of P_d to both $C \subset J_d(C)$ and $(-C) \subset J_d(C)$ are stable; either of these statements implies the result. In the hope that the techniques involved may find other uses in the future, we give rather different arguments for the stability of each of these restrictions.

The Theorem leads to a quick proof of the semi-stability of the normal bundles to an elliptic curve embedded by a complete linear series. More precisely, suppose that X is a compact Riemann surface of genus 1. Let L be a line bundle of degree d on X, and denote by $P^i(L)$ the bundle of ith order principal parts of L, so that $P^i(L)$ has rank i+1. The global sections of L lift canonically to sections of $P^i(L)$, and they surject when $i \leq d-2$. In this case we define a vector bundle $R^i(L)$ by the exact sequence

$$0 \longrightarrow R^i(L) \longrightarrow H^0(L) \otimes_{\mathbb{C}} \mathcal{O}_X \longrightarrow P^i(L) \longrightarrow 0.$$

Thus $R^1(L) = N^* \otimes L$, where N is the normal bundle to X in $\mathbb{P}H^0(L)$, and in general we think of the $R^i(L)$ as higher-order conormal bundles of X. Observing that $R^i(L)$ is essentially the pull-back of a Picard bundle under an étale morphism $X \to X = J_{d-i-1}(X)$, we deduce in §4 the

Corollary. *Provided that* $\deg(L) \geq i+2$, *the bundle* $R^i(L)$ *is semi-stable.*

When i=1 the result is due to Ellingsrud (although by a more involved argument). The general case answers a question of Dolgachev.

The theorem and its corollary give rise to some interesting open problems. First, it follows by well known results of Donaldson and Uhlenbeck-Yau that P_d, like any stable bundle, carries a Hermitian-Einstein metric. The question, suggested by Narasimhan, is whether one can construct

---

* Partially Supported by a Sloan Fellowship and N.S.F. Grant DMS 89-04243
** Partially Supported by N.S.F. Grant DMS 89-02551

these metrics explicity. The second problem concerns a characterization of the Picard bundles. Mukai [M] proves that when g=2, P_d is (up to twists and translations) the only stable bundle on $J_d(C)$ with the appropriate Chern classes. Is there an analogous result in higher genus? Mukai [M] and Kempf [K3] have shown that if C is non-hyperelliptic, then in any event a small deformation of P_d is again (a twist of a translate of) a Picard bundle. Kempf [K2] has also given some other characterizations of Picard bundles. Finally, if L is a line bundle of degree $d \geq 2g+i$ on any curve C of genus g, one may define higher conormal bundles $R^i(L)$ as above. We conjecture that $R^i(L)$ is always semi-stable for $d >> 0$. Some evidence in this direction appears in §4.

We wish to take this opportunity to thank D.Butler, G.Kempf, J.Li and S.Mukai for valuable discussions.

§ 1. Restrictions of Picard Bundles

We start with some notation. Throughout, C denotes a smooth irreducible projective curve of genus $g \geq 1$ defined over an algebraically closed field of arbitrary characteristic, and $x_0 \in C$ is a fixed base-point. We denote by $J_d(C)$ the Picard variety parametrizing line bundles of degree d on C, and we write $[L] \in J_d(C)$ for the point corresponding to a bundle L. Finally, let U_d be the universal bundle on $P_d(C) \times C$, normalized so that U_d is trivial on $J_d(C) \times \{x_0\}$. Thus if $\pi : J_d(C) \times C \to J_d(C)$ is the projection, and if $[L] \in J_d(C)$ is an arbitrary point, then $U_d | \pi^{-1}([L]) \cong L$. The degree d Picard sheaf on $J_d(C)$ is defined by

$$P_d = \pi_*(U_d).$$

It follows from the base-change theorem and Riemann Roch that if $d \geq 2g-1$, then P_d is actually a vector bundle on $J_d(C)$, with $rk(P_d) = d+1-g$.

Next, suppose given line bundles

$$A \in J_{d-1}(C) \quad \text{and} \quad B \in J_{d+1}(C).$$

Define embeddings

$$u_A : C \longrightarrow J_d(C) \quad \text{and} \quad v_B : C \longrightarrow J_d(C)$$

via

$$u_A(x) = [A(x)], \qquad v_B(x) = [B(-x)],$$

where as customary $A(x) = A \otimes \mathcal{O}_C(x)$ and $B(-x) = B \otimes \mathcal{O}_C(-x)$. We denote by $C_A \subset J_d(C)$ and $C_B \subset J_d(C)$ the images of u_A and v_B respectively. Observe that if $A, A' \in J_{d-1}(C)$ are two line bundles of degree d-1, then $C_{A'}$ is a translate of C_A (and similarly for C_B and $C_{B'}$).

Lemma 1.1. *If* $d \geq 2g-1$, *then one has canonical isomorphisms*

$$(u_A)^*(P_d) = p_*(q^*A \otimes \mathcal{O}_{C\times C}(\Delta)) \otimes \mathcal{O}_C(-x_0)$$

and

$$(v_B)^*(P_d) = p_*(q^*B \otimes \mathcal{O}_{C\times C}(-\Delta)) \otimes \mathcal{O}_C(x_0),$$

where $p : C \times C \longrightarrow C$ *and* $q : C \times C \longrightarrow C$ *denote the first and second projection respectively, and* $\Delta \subset C \times C$ *is the diagonal.*

Proof. Taking into account the normalization of U_d, one sees fibrewise that

$$(u_A \times 1_C)^*(U_d) = q^*(A) \otimes \mathcal{O}_{C\times C}(\Delta) \otimes p^*(\mathcal{O}_C(-x_0))$$

and

$$(v_B \times 1_C)^*(U_d) = q^*(B) \otimes \mathcal{O}_{C\times C}(-\Delta) \otimes p^*(\mathcal{O}_C(x_0)).$$

The lemma then follows from the theorem on cohomology and base-change. ■

Let $\Theta \subset J_d(C)$ denote the canonical principal polarization. Recall that the *slope* (with respect to Θ) of a torsion-free sheaf F on $J_d(C)$ is the rational number

$$\mu(F) = \frac{c_1(F).[\Theta]^{g-1}}{rk(F)} .$$

By definition, a vector bundle P is *stable* [resp. *semi-stable*] with respect to Θ if $\mu(F)<\mu(P)$ [resp. $\mu(F)\le\mu(P)$] for every non-zero torsion free subsheaf $F \subset P$ with rank(F)<rank(P). Similarly, if V is a bundle on C, then $\mu(V)=\deg(V)/rk(V)$, and V is stable [resp. semi-stable] if $\mu(W)<\mu(V)$ [resp. $\mu(W)\le\mu(V)$] for all sub-bundles $W \subset V$ with rk(W)<rk(V). (It is equivalent to demand the reverse inequalities on quotients.) As in [K1], the next point to observe is

Lemma 1.2. *Fix* $d \ge 2g$. *Then the stability of* P_d *is implied by either the stability of* $u_A^*(P_d)$ *for general* $[A] \in J_{d-1}(C)$, *or by the stability of* $v_B^*(P_d)$ *for general* $[B] \in J_{d+1}(C)$.

Proof. (Compare [K1]). Working in the ring $Num(J_d(C))$ of cycles on $J_d(C)$ modulo numerical equivalence, recall that $[C_A]=[\Theta]^{g-1}/(g-1)!$ c.f. [F, pp. 256-257] or [ACGH]. Hence the stability of P_d is equivalent to the assertion that

$$(*) \qquad \frac{[C_A].c_1(F)}{rk\ F} < \frac{[C_A].c_1(P_d)}{rk\ P_d}$$

for every torsion free $F \subset P_d$ with $rk(F)<rk(P_d)$. On the other hand, F is locally free outside a set $Z \subset J_d(C)$ of codimension ≥ 2, and we may assume that in fact F is sub-bundle of P_d outside Z. It follows by a dimension count that F is locally free in a neighborhood of $C_A \subset J_d(C)$ for sufficienty general $[A] \in J_{d-1}(C)$, and that $F|C_A$ sits as a sub-bundle of $P_d|C_A$. But this being so, (*) is implied by the stability of $u_A^*(P_d)$ for general A. The same argument proves the statement for $v_B^*(P_d)$ upon observing that if $-1 : J_d(C) \to J_d(C)$ denotes multiplication by -1, then $(-1)^*[\Theta]=[\Theta]$ in $Num(J_d(C))$, and hence $[C_B]=(-1)^*[C_A]=[\Theta]^{g-1}/(g-1)!$. ■

In view of Lemma 1.2, the issue is to understand something about the bundles appearing in Lemma 1.1. To this end, suppose that L is a non-special line bundle on C, generated by its global sections. Define bundles M_L and E_L on C by

$$E_L = p_*(q^*L \otimes \mathcal{O}_{C\times C}(\Delta))$$

and

$$M_L = p_*(q^*L \otimes \mathcal{O}_{C\times C}(-\Delta)).$$

Starting with the sequence $0 \to q^*L \otimes \mathcal{O}_{C\times C}(-\Delta) \to q^*L \to L \otimes \mathcal{O}_\Delta \to 0$ and taking direct imags, one finds that M_L sits in an exact sequence

$$(1.3) \qquad 0 \longrightarrow M_L \longrightarrow H^0(L) \otimes_{\mathbb{C}} \mathcal{O}_C \longrightarrow L \longrightarrow 0,$$

the homomorphism on the right being the canonical evaluation map. This bundle -- which controls the syzygies of L -- is quite well understood (c.f. [GL], [PR], or [L, §1]). As for E_L, we obtain analogously the exact sequence

$$0 \longrightarrow H^0(L)\otimes\mathcal{O}_C \longrightarrow E_L \longrightarrow L\otimes\theta_C \longrightarrow 0, \tag{1.4}$$

where θ_C denotes the tangent bundle to C. The extension class of (1.4) is given by an element $e_L \in H^0(L)\otimes H^1(L^*\otimes\omega_C) \cong H^0(L)\otimes H^0(L)^*$, and we will see in §2 that up to scalars e_L=id. In any event, putting together Lemmas 1.1 and 1.2, and noting that tensoring by a line bundle does not affect stability, we see that the Theorem stated in the introduction follows from

Proposition 1.5. *If* $\deg(L)\geq 2g-1$, *then* E_L *is stable, and if* $\deg(L)\geq 2g+1$, *then* M_L *is stable.*

We prove the first statement in §2, while the stability of M_L occupies §3.

§ 2. Stability of E_L

Throughout this section, L denotes a non-special line bundle of degree d on C, generated by its global sections. As above we put $E_L=p_*(q^*L\otimes\mathcal{O}_{C\times C}(\Delta))$, where $p:C\times C\to C$ and $q:C\times C\to C$ are the two projections. Taking direct images of $0 \to q^*L \to q^*L\otimes\mathcal{O}_{C\times C}(\Delta) \to L\otimes\mathcal{O}_\Delta(\Delta) \to 0$ yields the basic exact sequence

$$0 \longrightarrow H^0(L)\otimes\mathcal{O}_C \longrightarrow E_L \longrightarrow L\otimes\theta_C \longrightarrow 0, \tag{2.1}$$

θ_C being the tangent bundle to C. Our purpose it to prove

Proposition 2.2. *If* $d>2g-2$ [*resp. if* $d\geq 2g-2$] *then* E_L *is stable* [*resp. is semi-stable*].

We start with several lemmas.

Lemma 2.3. *Using Serre duality to make the identification* $H^1(L^*\otimes\omega_C)=H^0(L)^*$, *the extension class* $e\in H^0(L)\otimes H^1(L^*\otimes\omega_C)$ *defining* (2.1) *is given by a non-zero scalar multiple of the identity* $\mathrm{id}\in H^0(L)\otimes H^0(L)^*$. *In particular,* $H^0((E_L)^*)=0$.

Proof. The second statement follows easily from the first. Consider the sequence $0\to q^*L\otimes p^*(\omega_C\otimes L^*) \to q^*L\otimes p^*(\omega_C\otimes L^*)(\Delta) \to \mathcal{O}_\Delta \to 0$. Then e is the image of $1\in H^0(\mathcal{O}_\Delta)$ in $H^0(L)\otimes H^1(\omega_C\otimes L^*)$, i.e. the kernel of $H^0(L)\otimes H^1(\omega_C\otimes L^*) \to H^1(q^*L\otimes p^*(\omega_C\otimes L^*)(\Delta))$. Now compute this latter map by taking direct images under q: using duality for q it follows that e spans the kernel of the map induced on global sections by $q_*(p^*L)^*\otimes L = =H^0(L)^*\otimes L \to q_*(p^*L(-\Delta))^*\otimes L=(M_L)^*\otimes L$. But we recognize this homomorphism as a piece of the Euler sequence, and the assertion follows. ∎

Lemma 2.4. (Compare [PR] and [B]). *Let* V *be a globally generated vector bundle on* C, *with no trivial summands (i.e. with* $h^0(V^*)=0$). *Then* $\mu(V)>1$.

Proof. Suppose that V has rank r and degree n. Choosing (r+1) general sections of V, we construct an exact sequence

$$0 \longrightarrow V^* \longrightarrow \mathcal{O}^{r+1} \longrightarrow \det V \longrightarrow 0, \tag{*}$$

and since $h^0(V^*)=0$ it follows that $h^0(\det V)\geq r+1$. If det V is special, then Clifford's theorem applies to yield $n=\deg(\det(V))\geq 2(h^0(\det V)-1)\geq 2r$, and so $\mu(V)=n/r\geq 2$ in this case. On the other hand, if det V is non-special then $r \leq h^0(\det V)-1=n-g$ by Riemann Roch, and hence $\mu(V) \geq 1+(g/r)$. ■

Lemma 2.5. *Consider an exact sequence*

$$0 \longrightarrow T \longrightarrow V \longrightarrow \tau \longrightarrow 0$$

of sheaves on C, *where* $T=\mathcal{O}^r$ *is a trivial bundle of rank* r, *and* τ *is a torsion sheaf supported on a finite set. If* $\text{length}(\tau)<r$, *then* $h^0(V^*) \neq 0$, *i.e.* V *has a trivial summand.*

Proof. Dualizing the given sequence yields $0 \to V^* \to T^* \to \mathcal{E}xt^1(\tau,\mathcal{O}_C) \to 0$, and $\text{length}(\mathcal{E}xt^1(\tau,\mathcal{O}_C))=\text{length}(\tau)$. The assertion follows. ■

Proof of Proposition 2.2. When $d=\deg(L)=2g-2$ the semi-stability of E_L is clear from (2.1), so we assume $d \geq 2g-1$. Then $\mu(E_L)<1$. Suppose now that E_L fails to be stable. Then there exists a stable quotient sheaf G of E_L with $\mu(G) \leq \mu(E_L)<1$. Letting F be the image of the composition $H^0(L)\otimes\mathcal{O}_C \to E_L \to G$, the situation is summarized in the following diagram, which defines a sheaf τ:

$$\begin{array}{ccccccccc}
0 & \longrightarrow & H^0(L)\otimes\mathcal{O}_C & \longrightarrow & E_L & \longrightarrow & L\otimes\mathcal{O}_C & \longrightarrow & 0 \\
 & & \downarrow & & \downarrow & & \downarrow & & \\
0 & \longrightarrow & F & \longrightarrow & G & \longrightarrow & \tau & \longrightarrow & 0 \\
 & & \downarrow & & \downarrow & & \downarrow & & \\
 & & 0 & & 0 & & 0 & &
\end{array}$$

Note that τ -- being a quotient of $L\otimes\theta_C$ -- is either a torsion sheaf or isomorphic to $L\otimes\theta_C$. In particular, $F \neq 0$: for otherwise $G=\tau=L\otimes\theta_C$, but $L\otimes\theta_C$ doesn't destabilize E_L when $d \geq 2g-1$.

We assert that F is trivial. In fact, since F is generated by its global sections one can write $F=F_1\oplus F_2$, where F_1 is trivial and F_2 has no trivial summands. Thus F_2 is a sub-sheaf of G. But if $F_2 \neq 0$, then $\mu(F_2)>1$ by Lemma 2.4. This contradicts the stability of G and hence $F=F_1$ is trivial as claimed.

If $\tau=L\otimes\theta_C$ then $\mu(G)>\mu(E_L)$ by a direct computation, so we may assume that τ is a torsion sheaf. If $\text{length}(\tau) \geq \text{rank}(G)$, then again $\mu(G) \geq 1>\mu(E_L)$. So there remains only the possibility that $\text{length}(\tau)<\text{rank}(G)$. But then $h^0((E_L)^*) \neq 0$ thanks to Lemma 2.5, and this contradicts Lemma 2.3. This complete the proof of the Proposition. ■

§ 3. Cohomological Stability of M_L

Let L be a globally generated line bundle on C, and define M_L as at the end of §1. The stability of M_L when $\deg(L) \geq 2g+1$ follows almost immediately from the proof of Lemma 2.4. Indeed, an argument along these lines was given with M. Green some years ago, and Paranjape and Ramanan [PR] independently used such an approach to prove the stability of M_Ω when C is non-hyperelliptic. However, in response to a question of Kempf, we will give an alternative cohomological proof. We start with a

Definition 3.1. Let V be a vector bundle on C. We say that V is *cohomologically stable* [resp. *cohomologically semistable*] if for every line bundle A of degree a, and for every integer $t<\mathrm{rk}(V)$, one has

$$H^0(\wedge^t V\otimes A^*)=0 \qquad \text{whenever} \qquad a\geq t.\mu(V) \qquad [\text{resp. when } a>t.\mu(V)\ ].$$

Note that cohomological stability indeed implies stability in the usual sense. In fact, a proper subbundle $T\subset V$ of degree a and rank t determines an inclusion $A=_{\mathrm{def}}\wedge^t T\subset\wedge^t V$, and hence a non-zero section of $\wedge^t V\otimes A^*$. The condition in the definition then implies that $\mu(T)<\mu(V)$. In characteristic zero any exterior power of a semistable bundle is semistable, and it follows that in this case cohomological semistability is equivalent to semistability.

Proposition 3.2. *If* $\deg(L)\geq 2g+1$ [*resp.* $\deg(L)\geq 2g$] *then* M_L *is cohomologically stable* [*resp. cohomologically semistable*].

Proof. We assume $d\geq 2g+1$, the other case being almost identical. Keeping notation as in the definition, we must prove that $H^0(\wedge^t M_L\otimes A^*)=0$ whenever

$$(*) \qquad \frac{a}{t}\geq\mu(M_L)=-1-\frac{g}{d-g}>-2.$$

We use what is by now a standard filtration argument, as in [GLP, p.498], [GL], or [L]. Specifically, set $r=r(L)=d-g$, and choose general points $x_1,\dots,x_{r-1}\in C$. Then (c.f. [L,§1.4]) there is an exact sequence

$$0\longrightarrow L^*(x_1+\dots+x_{r-1})\longrightarrow M_L\longrightarrow\bigoplus^{r-1}\mathcal{O}_C(-x_i)\longrightarrow 0.$$

Put $D=D_{r-1}=x_1+\dots+x_{r-1}$. Taking exterior powers yields

$$0\longrightarrow\wedge^{t-1}\{\bigoplus^{r-1}\mathcal{O}_C(-x_i)\}\otimes L^*(D)\longrightarrow\wedge^t M_L\longrightarrow\wedge^t\bigoplus^{r-1}\mathcal{O}_C(-x_i)\longrightarrow 0.$$

One deduces from this that $H^0(\wedge^t M_L\otimes A^*)=0$ so long as:

(i). $H^0(A^*(-D_r))=0$ for a general effective divisor D_r of degree t,

and

(ii). $H^0(A^*\otimes L^*(D_{r-t}))=0$ for a general effective divisor D_{r-t} of degree $r-t=d-g-t$.

The line bundle appearing in (i) has degree $-a-t$, and we have t degrees of freedom in choosing it. So provided that $-a-t<g$, the desired vanishing will follow if $t>-(a+t)$. But both of these inequalities are consequences of (*). Similarly, for (ii) it is enough that $\deg(A^*\otimes L^*(D_{r-t}))=-a-t-g<0$. ■

Remark. If E is a globally generated vector bundle on C, one can use the canonical sequence $0\to M_E\to H^0(E)\otimes\mathcal{O}_C\to E\to 0$ to define a bundle M_E on C. Butler [B] has generalized Proposition 3.2 by proving that M_E is stable provided that E is stable and $\mu(E)>2g$. He applies this to obtain interesting surjectivity theorems for the multiplication maps $H^0(E)\otimes H^0(F)\to H^0(E\otimes F)$ on sections of stable bundles, and to prove a conjecture of Kempf concerning the syzygies of the homogeneous coordinate rings of curves. He also studies the stability of M_L for line bundles L with $\deg(L)\leq 2g$. The referee informs us that the stability of M_L has also been investigated by Paranjape in his 1989 thesis.

§ 4. Poly-stability of Normal Bundles to Complete Linear Series on an Elliptic Curve

A number of authors have considered the stability of the normal bundles to space cures (c.f. [GS], [EV], [EL],[Hu], or [Ha]), but in general the situation seems rather complicated. However as a very simple application of our main theorem, we show that for linearly normal embeddings of elliptic curves in characteristic zero, one obtains a fairly clean picture.

Let X be a compact Riemann surface of genus 1, and let L be a line bundle of degree d on X. Fix $i \leq d-2$, and let $R^i(L)$ be the rank d-i-1 vector bundle on X defined by

$$R^i(L)=p_*(q^*L\otimes\mathcal{O}_{X\times X}(-(i+1)\Delta)),$$

where as above $p,q:X\times X \to X$ denote the two projections. As noted in the introduction, these higher conormal bundles fit into exact sequences

$$0 \longrightarrow R^i(L) \longrightarrow H^0(L)\otimes_{\mathbf{C}}\mathcal{O}_X \longrightarrow P^i(L) \longrightarrow 0.$$

Theorem 4.1. *The bundle* $R^i(L)$ *is poly-stable, i.e. it is a direct sum of stable bundles of the same slope. In particular,* $R^i(L)$ *is semi-stable.*

Proof. The line bundle $\mathcal{L}=q^*L\otimes\mathcal{O}_{X\times X}(-(i+1)\Delta)$ defines a family of degree d-i-1 line bundles on X parametrized by X. This induces a finite surjective (and hence étale) classifying morphism

$$f:X \longrightarrow J_{d-i-1}(X)=X$$

with the property that $\mathcal{L}=(1\times f)^*(U_{d-i-1})\otimes p^*\eta$ for some bundle η on X, where as in §1 U_d denotes the Poincaré bundle on X×X. It follows from the base-change theorem that $R^i(L)=f^*(P_{d-i-1})\otimes\eta$, and hence $R^i(L)$ is a twist of the pull-back of a stable bundle P under an étale covering. But such a pull-back is automatically polystable. (The semi-stability of f^*P follows by a standard descent argument from the uniqueness of a maximal destabilizing sub-bundle. The stronger assertion that f^*P is actually poly-stable ia a consequence of the characterization of such bundles as those having an Hermitian-Einstein connection: alternatively, in the case at hand one could give a more direct elementary argument.) ■

Theorem 4.1 suggests that unlike the situation for incomplete linear series, the normal bundles of curves embedded by a complete linear series of sufficiently large degree behave in a uniform manner:

Conjecture 4.2. *There is an integer* d(g,i) *such that if* C *is any curve of genus* g *(say in characteristic zero), then the conormal bundle* $R^i(L)$ *defined as above is semi-stable for any line bundle* L *of degree* $d\geq d(g,i)$.

One can use Proposition 3.2 to show that in any event $R^i(L)$ cannot be "too unstable" for d>>0. In fact, to fix ideas let L be a line bundle of degree $d\geq 2g+1$, and consider $R(L^2)=R^1(L^2)$, which has slope $-2-4g/(4d-g-1)$. One may identify the fibre of M_L at a point $p\in C$ with the vector space $H^0(L(-p))$, and similarly the fibre of $R(L^2)$ at p is $H^0(L^2(-2p))$. Then the canonical map

$$H^0(L(-p))\otimes H^0(L(-p)) \longrightarrow H^0(L^2(-2p))$$

globalizes to a vector bundle homomorphism $M_L\otimes M_L \to R(L^2)$ which is surjective for $d\geq 2g+2$. But in characteristic zero the tensor product of two stable bundles is semi-stable, and hence $M_L\otimes M_L$ is semi-stable, of slope $2\mu(M_L)=-2-4g/(4d-2g)$. In particular, any quotient of $R(L^2)$ has slope ≥-2-4g/(4d-2g). Unfortunately, when g≥2 this falls slightly short of proving the semi-stability of $R(L^2)$.

References

[ACGH] E.Arbarello, M.Cornalba, P.Griffiths and J.Harris, *Geometry of Algebraic Curves*, Springer-Verlag (1985).

[B] D.Butler, Normal generation of vector bundles over a curve, to appear in J. Diff. Geom.

[EL] G.Ellingsrud and D.Laksov, The normal bundle of elliptic space curves of degree 5, in Proc. 18th Scand. Congr. Math. Prog. in Math. vol. 11, Birkhauser (1981), pp. 285-287.

[EV] D.Eisenbud and A.Van de Ven, On the normal bundles of smooth rational space curves, Math. Ann. 256 (1981), pp. 453-463.

[F] W.Fulton, *Intersection Theory*, Springer-Verlag (1983).

[GS] F.Ghione and G.Sacchiero, Normal bundles of rational curves in $\mathbb{P}^3$, Manuscripta Math. 33 (1980), pp. 111-128.

[GL] M.Green and R.Lazarsfeld, Some results on the syzygies of finite sets and algebraic curves, Compos. Math. 67 (1988), pp. 301-314.

[GLP] L.Gruson, R.Lazarsfeld and C.Peskine, On a theorem of Castelnuovo and the equations defining space curves, Inv. Math. 72 (1983), pp. 491-506.

[K1] G.Kempf, Rank g Picard bundles are stable, to appear in the Am. J. Math.

[K2] G.Kempf, Notes on the inversion of integrals, I, II, to appear.

[K3] G.Kempf, Towards the inversion of abelian integrals, I, Ann. Math. 110 (1979), pp. 243-247.

[Ha] R.Hartshorne, Classification of algebraic space curves, Proc. Sym. Pure Math. 46 (1985), pp. 145-164.

[Hu] K.Hulek, Projective Geometry of Elliptic Curves, Asterisque 137 (1986).

[L] R.Lazarsfeld, A sampling of vector bundle techniques in the study of linear serie, in *Riemann Surfaces* (Proceedings of the 1988 ICTP College on Riemann surfaces), World Scientific Press (1989).

[M] S.Mukai, Duality between D(X) and $D(\hat{X})$ with an application to Picard sheaves, Nagoya Math. J. 81 (1981), pp. 153-175.

[PR] K.Paranjape and S.Ramanan, On the canonical ring of a curve, in *Algebraic Geometry and Commutative Algebra* (in honor of M.Nagata), Kinokaniya (1988).

[U] H.Umemura, On a property of symmetric products of a curve of genus 2, Proc. Int. Symp. on Algebraic Geom., Kyoto 1977, pp. 709-721.

Department of Mathematics
University of Illinois at Chicago
Chigago, IL 60680

Department of Mathematics
University of California, Los Angeles
Los Angeles, CA 90024

Sections planes et majoration du genre des courbes gauches

Philippe Ellia - Rosario Strano

0. Introduction.

Dans cet article l'on donne une contribution ultérieure au problème d'Halphen (cf.[Ha]): déterminer G(d,s), le genre maximal des courbes intègres de degré d non contenues dans une surface de degré <s. Ce problème classique a déjà fait l'objet de quelques travaux ([Ha], [GP1], [GP2], [H1], [H2], [HH], [E]) et sa résolution apporterait une première esquisse complète de la géografie des courbes gauches. Rappelons que ce problème dépend fortement du rapport d/s et l'on distingue habituellement trois domains: $(s^2+4s+6)/6 \leq d < (s^2+4s+6)/3$ (A); $(s^2+4s+6)/3 \leq d \leq s(s-1)$ (B); $d > s(s-1)$ (C).

La valeur de G(d,s) dans le domain C est donnée par Halphen ([Ha]) mais ce n'est que dans [GP1] que l'on trouve une démonstration complète de cet énoncé. Dans le domaine A, le théorème de Clifford fournit $G(d,s) \leq G_A(d,s) := d(s-1)+1-\binom{s+2}{3}$ (cf.[H1]) et l'on conjecture l'égalité dans ce domaine (si $d \geq (3G_A(d,s)+12)/4$, l'égalité est une conséquence du théorème du rang maximum [BE3,4]).

Ces premiers résultats montrent que la subdivision ci-dessus n'est pas arbitraire mais reflete les faits suivants: une courbe, X, du domaine A a une spécialité moderée ($e(X) \leq s-2$) donc un genre relativement petit et, en général, $\sigma(X)$ est bien plus petit que s(X) (cf.0.4 pour les notations). Dans ce domaine la majoration est aisée, la difficulté est la minoration. Au contraire dans le domaine C, la spécialité et le genre sont grands et $\sigma(X) = s(X)$ grâce au lemme de Laudal [L]. Comme l'ont montré Gruson-Peskine on en déduit alors, par l'étude de la section plane, que toute courbe réalisant G(d,s) dans ce domaine est arithmétiquement Cohen-Macaulay. Grâce à une amélioration du lemme de Laudal, Gruson et Peskine ont montré ([GP2]) que ce comportement vaut encore si $d > s^2-2s+2$ mais, pour $d = s^2-2s+2$, ils prouvent que les courbes réalisant G(d,s) sont sections de fibrés de corrélation nulle, donc $\sigma = s-1$.

Ainsi le domain B est un glissement progressif de C vers A dans lequel les difficultés se somment. D'une part le théorème de Clifford ne donne rien (en général $e \geq s-1$) et d'autre part une majoration à la Castelnuovo, par la section plane, se heurte au problème d'estimer la fonction de Hilbert de cette section plane (que vaut σ?). Finalement, en l'absence de toute conjecture sérieuse, le problème de la minoration n'a plus de sens. Telle était la situation jusqu'à la parution des travaux [H2,3], [Hi], [HH]. Dans [H3] (cf.aussi [GP2]) Hartshorne, par la biais des faisceaux réflexifs, donne une majoration de $E(d,s) = \max\{e(X) \mid X$ intègre, de degré d, avec $h^0(\mathcal{I}_X(s-1)) = 0\}$ dans le domain B. Dans [H2],[HH] Hartshorne et Hartshorne-Hirschowitz énoncent la conjecture suivante:

0.1. ***Conjecture*** ([H2], [HH]). Pour s, f des entiers on pose :

$A(s,f) = \lceil \frac{1}{3}(s^2-sf+f^2-2s+7f+12) \rceil$ (resp. le même +1 si f=2s-7, 2s-9)

$B(s,f) = \lceil \frac{1}{3}(s^2-sf+f^2+6f+11) \rceil$ (resp. le même +1 si $f=2s-8$, $2s-10$).

Si $s \geq 5$, $s-1 \leq f \leq 2s-6$ et $A(s,f) \leq d < A(s,f+1)$ on définit

$$G_H(d,s) := d(s-1)+1-\binom{s+2}{3}+\binom{f-s+4}{3}+h$$

où

$$h = \begin{cases} 0 & \text{si } A(s,f) \leq d \leq B(s,f) \\ (d-B)(d-B+1)/2 & \text{si } B(s,f) \leq d < A(s,f+1). \end{cases}$$

Alors, avec les notations précédentes: $G(d,s) = G_H(d,s)$.

Cette conjecture est fondée sur le principe que G(d,s) est réalisé par des courbes de spécialité maximale, E(d,s).Pour la commodité du lecteur rappelons comment l'on arrive à cette conjecture. Soit X une courbe du domaine B et posons e=e(X). Si $e<s-1$ on a immediatement $g(X) \leq G_A(d,s)$. Si $e \geq s-1$ alors $e \leq 2s-6$ et $d \geq A(s,e)$ (cf.[H2],3.2). Ceci dit remarquons que pour s fixé et $s-1 \leq f \leq 2s-6$, A(s,f) est une fonction croissante de f et les intervalles $[A(s,f), A(s,f+1)[$ recouvrent le domaine B. Ainsi pour d, s donnés on définit $f(d,s) := \max\{f \mid d \geq A(s,f)\}$. La majoration de E(d,s) ([H2], [GP2]) est alors $E(d,s) \leq f(d,s)$.

En fait ([HH] Thm.5.4) il s'agit d'une égalité. Revenons à notre courbe X. Une section de $\omega_X(-e)$ fournit: $0 \to \mathcal{O}_P \to \mathcal{F} \to \mathcal{J}_X(e+4) \to 0$. Si $\mathcal{E} := \mathcal{F}(-e-5+s)$ on a $h^0(\mathcal{E})=0$ et $h^2(\mathcal{E}(e-s+2))=0$. Les hypothèses et la suite $0 \to \mathcal{O}_P(s-e-5) \to \mathcal{E} \to \mathcal{J}_X(s-1) \to 0$ donnent: $g(X) \leq d(s-1)+1-\binom{s+2}{3}+h^0(\mathcal{O}_P(e-s+1))+h^2(\mathcal{E})$. Il reste à majorer $h^2(\mathcal{E})$ ce qui peut être fait en appliquant [H2],2.1 avec $b=e-s+2$. Si $e=f(d,s)$ on trouve $g(X) \leq G_H(d,s)$. Ainsi si $e=E(d,s)$, $g(X) \leq G_H(d,s)$ et, d'après le principe énoncé plus haut, on peut conjecturer que dans le domaine B, $G(d,s) \leq G_H(d,s)$.

D'autre part dans [HH] et en utilisant [Hi],les auteurs prouvent:

0.2. Théorème. ([HH] Thm.5.4). Dans le domaine B, $G(d,s) \geq G_H(d,s)$.

Par consequent, modulo le principe initial, la conjecture est complètement justifiée. Ajoutons que la conjecture est vérifiée pour les courbes avec $e \geq E(d,s)-1$ ([H2], 3.6,3.7) et qu'elle est démontrée pour $f=2s-6$ ([GP2]), $f=2s-7$ ([E]) et $f \leq s$ ([H2]). Elle est donc vraie aux bords du domaine B. Ici nous démontrons:

0.3. Théorème. Pour $s \geq 8$ et $2s-8 \geq f \geq 2s-9$ la conjecture est vraie. De plus, sous ces hypothèses, toute courbe intègre C réalisant G(d, s) est de rang maximum avec $h^1(\mathcal{J}_C(s-1))=0$ et $e(C)=f$.

En outre nous décrivons le caractère des courbes réalisant G(d,s). Notre démonstration utilise 0.2 puisque nous limitons à majorer G(d, s).

Pour résoudre 0.1 il est indispensable de disposer de résultats adéquats sur le problème suivant: quels liens existe-t-il entre la postulation d'une courbe intègre et celle de sa section plane générale? Les premiers résultats sur cette question (Laudal [L], Gruson-Peskine [GP2]) ne font intervenir que d, σ (cf. 0.4 pour les notations).

Il semble clair maintenant que l'on peut obtenir des résultats plus fins en faisant intervenir la résolution minimale de la section plane générale (cf. [S1], [S2]). En reprenant ce point de vue et en utilisant des résultats de [S1], [S2], [H2], nous obtenons de nouveaux résultats sur la question précédente.

Ces résultats sont utilisés au paragraphe III pour majorer G(d, s) lorsque $f=2s-8$, $f=2s-9$. Pour cela on procéde comme dans [E]: d'après [GP1], [S2] nous savons que, à une exception près, dans le domaine considéré $\sigma = s-1$. Les résultats du paragraphe II nous permettent d'écarter certains caractères de degrè d, longueur σ et donc (cf.II.6) de majorer $G(d,s)$ par la valeur maximale de $g(\chi)-1-\#\{n_i \in \chi \mid n_i = \sigma\}$ pour les caractères restants. Les résultats nécessaires pour estimer ce maximum se trouvent au paragraphe I. Signalons en outre le lemme II.6 qui nous pemet de montrer que toute courbe C réalisant $G(d,s)$ est de rang maximum avec $h^1(\mathcal{I}_C(s-1))=0$ et $e(C)=f$ (cf. aussi III.7).

Notre propos aurait été pratiquement impossible sans la conjecture 0.1. Cependant cette conjecture, fondée sur le principe que $G(d,s)$ est réalisé par des courbes d'indice de spécialité maximal, est difficile à interpréter en termes de caractères numériques. Il est donc difficile d'imaginer (et à fortiori de démontrer!) les énoncés de type II.2, II.4, II.7, II.10, nécessaires pour couvrir le domaine B tout entier.

0.4. ***Notations.*** On travaille sur un corps **k** algébriquement clos, de caractéristique nulle. Sauf indications contraires une courbe $C \subset \mathbf{P}^3$ est un sous-schéma fermé, de dimension un, équidimensionnel, localement Cohen-Macaulay.
On note
$s(C)=\min\{k \mid h^0(\mathcal{I}_C(k))\neq 0\}$; $\sigma(C)=\min\{k \mid h^0(\mathcal{I}_{C\cap H}(k))\neq 0\}$ pour H un plan général, $e(C)=\max\{k \mid h^2(\mathcal{I}_C(k))\neq 0\}$. Si aucune confusion n'est à craindre on écrit s, σ, e.
Une courbe C est de rang maximum si $h^0(\mathcal{I}_C(k))\cdot h^1(\mathcal{I}_C(k))=0$, $k\in\mathbf{Z}$. Finalement, sauf mention expresse du contraire, tout caractère numérique (cf.I.1) est supposé connexe. Ceci est justifié par [GP1], 3.2.

I. Groupes de points de $\mathbf{P}^2$.

I.1. Caractères. Soit $E\subset\mathbf{P}^2$ un groupe de points de degré d. On note $\sigma(E)$ (ou σ si aucune confusion n'est à craindre) le plus petit degré d'une courbe contenant E. On note $\chi(E)=(n_0,\dots,n_{\sigma-1})$, $n_0\geq \dots \geq n_{\sigma-1}\geq\sigma$, le caractère numérique de E. Nous renvoyons à [GP1] pour la définition du caractère. Rappelons simplement que le caractère détermine la fonction de Hilbert h_E de E et réciproquement. En fait:

(1) $$h_E(n)=\sum_{i=0}^{\sigma-1}[(n-i+1)_+-(n-n_i+1)_+].$$

Notons aussi la relation suivante:

(2) $$h^1(\mathcal{I}_E(n))=\sum_{i=0}^{\sigma-1}[(n_i-n-1)_+-(i-n-1)_+].$$

Le genre du caractère $\chi(E)$ (ou le genre virtuel de E) est défini par :

(3) $$g(\chi(E))=\sum_{n\geq 1}h^1(\mathcal{I}_E(n)).$$

I.1.1. Proposition. Soit $E\subset\mathbf{P}^2$ un groupe de points vérifiant le principe de position uniforme; alors $\chi(E)$ est connexe (i.e. $n_{i+1}\geq n_i-1$).

Démonstration. [MR1] Cor. 2; cf. aussi [EP]. $\boxdot$

Ce résultat est une conséquence du "lemme de décomposabilité" suivant :

I.1.2. Proposition. Soit $E\subset\mathbf{P}^2$ un groupe de points de caractère $(n_0, \dots, n_{\sigma-1})$.

i) Soit $n_i > n_{i+1}$ et supposons que toutes les formes de degré $n_i - 1$ passant par E ont un plus grand commun diviseur, F, de degré i+1. Alors $F \cap E = R$ est un sous-groupe de points avec $\chi(R) = (n_0, \ldots, n_i)$.
ii) Si $n_i > n_{i+1}+1$ alors les conditions précédentes sont verifiées.

Démonstration. [D] Thm.4.1, cf.aussi [EP]. □

I.2. Fonction de Hilbert et résolution libre minimale. La résolution libre minimale de $I(E) := H^0_*(\mathcal{I}_E)$ induit une suite exacte :

$$(*) \qquad 0 \to \bigoplus \mathcal{O}_{\mathbf{P}^2}(-m_{2j}) \to \bigoplus \mathcal{O}_{\mathbf{P}^2}(-m_{1j}) \to \mathcal{I}_E \to 0.$$

On pose $m_1^+ = \max\{m_{1j}\}$, $m_1^- = \min\{m_{1j}\}$ et l'on définit m_2^+, m_2^- de la même façon.

Pour tout entier n on pose $\alpha(n) = \#\{m_{1j} \mid m_{1j} = n\}$, $\beta(n) = \#\{m_{2j} \mid m_{2j} = n\}$. Finalement si $\chi(E) = (n_0, \ldots, n_{\sigma-1})$ on pose $c(n) = \#\{n_i \mid n_i = n\}$.

Remarquons que la donnée de (*) détermine la fonction de Hilbert. La réciproque n'est pas vraie mais nous avons:

I.2.1. Proposition. Avec les notations précédentes:
(1) $m_1^- < m_2^-$, $m_1^+ < m_2^+ = n_0+1 = \tau+3$ $(\tau := \max\{n \mid h^1(\mathcal{I}_E(n)) \neq 0\})$.

De plus $h^1(\mathcal{I}_E(\tau)) = c(n_0)$.
(2) $\alpha(\sigma) = c(\sigma)+1$; si $\alpha(\sigma) = 1$ alors $\alpha(\sigma+1) = c(\sigma+1)$.
(3) $\beta(n) = \alpha(n) - c(n) + c(n-1)$ $(n > \sigma)$
(4) $\max\{0, c(n)-c(n-1)\} \leq \alpha(n) \leq c(n)$ $(n > \sigma)$
(5) si de plus E vérifie le principe de position uniforme alors:
$\max\{0, c(n)-c(n-1)\} \leq \alpha(n) \leq c(n)-1$ $(n > \sigma+1$ et $n = \sigma+1$ si $\alpha(\sigma) > 1)$.

Démonstration. (1) et (2) sont bien connus. Pour (3), (4), (5) cf. [C], [MR2] Th.1.5. □

I.2.2. Proposition. Soit $T \subset \mathbf{P}^2$ un groupe de points de degré d vérifiant le principe de position uniforme. On suppose $d \geq \sigma^2 - 2\sigma + 6$, $\sigma \geq 5$ et $h^0(\mathcal{I}_T(\sigma)) \geq 2$. Alors $h^0(\mathcal{I}_T(\sigma)) \leq 3$. Supposons en outre $m_2^- < \sigma+3$; alors l'un des cas suivants a lieu:

(1) $\alpha(\sigma) = 3$, $\alpha(\sigma+1) = 0$, $\beta(\sigma+1) = 1$ et $\beta(\sigma+2) = 0$
(2) $\alpha(\sigma) = 2$, $\alpha(\sigma+1) = 1$, $\beta(\sigma+2) = 1$ et $\beta(\sigma+3) = 0$
(3) $d = \sigma^2 - 2\sigma + 6$ et $\chi(T) = (2\sigma-3, \ldots, \sigma+2, \sigma+2, \sigma+1, \sigma+1, \sigma)$.

Dans les cas (2) et (3), si R est lié à T par une intersection complète $(\sigma+1, \sigma+1)$, alors toutes les formes de degré $\sigma-1$ passant par R ont un PGCD, F, de degré 3. Par conséquent $F \cap R = R'$ est un sous-groupe de points de R avec $\chi(R') = (\sigma+1, \sigma+1, \sigma)$.

Démonstration. Si $h^0(\mathcal{I}_T(\sigma)) \geq 4$ alors $(\sigma, \sigma, \sigma) \subset \chi(T)$ (cf. I.2.1 (2)). Par connexité (I.1.1) l'on déduit $d \leq d_0$ où $d_0 = \deg(\chi_0)$, $\chi_0 = (2\sigma-3, \ldots, \sigma, \sigma, \sigma)$. Ceci est impossible car $d_0 = \sigma^2 - 2\sigma + 3$.

On a alors $c(\sigma) \leq 2$. Si $c(\sigma) = 2$ l'on montre, comme ci-dessus, que $c(\sigma+1) = c(\sigma+2) = 1$. Il s'ensuit (par I.2.1) que nous sommes dans le cas (1).

Supposons alors que $c(\sigma) = 1$. Comme ci-dessus, l'on montre que $c(\sigma+1) \leq 2$.

i) Si $c(\sigma)=c(\sigma+1)=1$, d'après I.2.1.(5), $\alpha(\sigma+1)=0$. Vu l'hypothèse $m_2^- < \sigma+3$, il existe une relation de degré au plus $\sigma+2$ entre les deux générateurs de degré σ. Ceci est impossible parce que T vérifie le principe de position uniforme.

ii) Supposons enfin que $c(\sigma)=1$, $c(\sigma+1)=2$. En raisonnant comme dans les cas précédents l'on montre que $c(\sigma+2)=1$ sauf si $d=\sigma^2-2\sigma+6$ et si $\chi(T)=(2\sigma-3, \ldots, \sigma+2, \sigma+2, \sigma+1, \sigma+1, \sigma)$.

a) Supposons $c(\sigma+2)=1$: en utilisant I.2.1 on obtient le cas (2). Le caractère $\chi(T)$ a la forme $(\ldots, \sigma+3, \sigma+2, \sigma+1, \sigma+1, \sigma)$. Si R est lié à T par $(\sigma+1,\sigma+1)$, posons $\chi(R)=(n'_i)$ et l'on obtient facilement, par liaison: $n'_0=\sigma+1$, $n'_1=\sigma+1$, $n'_2=\sigma$, $n'_3<n'_2-1$. On conclut avec I.1.2.

b) Soit maintenant $c(\sigma+2)>1$. Comme déjà dit plus haut, les hypothèses impliquent $d=\sigma^2-2\sigma+6$ et $\chi(T)=(2\sigma-3, \ldots, \sigma+2, \sigma+2, \sigma+1, \sigma+1, \sigma)$. En utilisant I.2.1 l'on voit que la résolution minimale de I(T) a la forme suivante:

$$0 \to \mathcal{O}(-2\sigma+2)\oplus\mathcal{O}(-\sigma-3)\oplus\varepsilon\mathcal{O}(-\sigma-2) \to \varepsilon\mathcal{O}(-\sigma-2)\oplus\mathcal{O}(-\sigma-1)\oplus 2\mathcal{O}(-\sigma) \to \mathcal{I}_T \to 0$$

avec $\varepsilon=0,1$. Si $\varepsilon=0$, lorsque $\sigma\geq 5$, on a $m_2^- \geq \sigma+3$. Supposons donc $\varepsilon=1$. Si R est lié à T par $(\sigma+1,\sigma+1)$, par mapping cone:

$$0 \to 2\mathcal{O}(-\sigma-2)\oplus\mathcal{O}(-\sigma) \to \mathcal{O}(-\sigma-1)\oplus\mathcal{O}(-\sigma)\oplus\mathcal{O}(-\sigma+1)\oplus\mathcal{O}(-4) \to \mathcal{I}_R \to 0 .$$

Soient F_4 et $F_{\sigma-1}$ les générateurs de degrés $4, \sigma-1$. La relation de degré σ s'écrit : $LF_{\sigma-1}+P_{\sigma-4}=0$. Bien entendu L ne divise pas $P_{\sigma-4}$ donc $F_4=LF_3$, $F_{\sigma-1}=-P_{\sigma-4}F_3$.
On conclut avec I.1.2. □

I.2.3.Proposition. Soit $T\subset \mathbf{P}^2$ un groupe de points de degré d vérifiant le principe de position uniforme. On suppose $d=\sigma^2-3\sigma+6+\nu$, avec $\sigma\geq 5$ et $2\leq\nu\leq\sigma-3$. Alors $h^0(\mathcal{I}_T(\sigma))\leq 4$. Supposons $h^0(\mathcal{I}_T(\sigma))=4$, alors:

a) si $\nu=2$ la résolution minimale de I(T) vérifie: $\alpha(\sigma)=4$, $\alpha(\sigma+1)=0$, $\beta(\sigma+1)=2$;

b) si $\nu\geq 3$ on a en outre $\beta(\sigma+2)=\beta(\sigma+3)=0$.

Démonstration. Si $h^0(\mathcal{I}_T(\sigma))\geq 5$ alors $d\leq d_0=\deg(2\sigma-4, \ldots, \sigma, \sigma, \sigma, \sigma)$. Ceci est impossible car $d_0=\sigma^2-3\sigma+6$.
Supposons $h^0(\mathcal{I}_T(\sigma))=4$ et montrons $c(\sigma+1)=1$: dans le cas contraire $d\leq d_1=\deg(2\sigma-4, \ldots, \sigma+1,\sigma+1, \sigma, \sigma, \sigma)$. Or $d_1=d_0+1$ donc, pour $\nu\geq 2$, $c(\sigma+1)=1$ et a) suit de I.2.1. De la même façon on obtient $c(\sigma+2)=1$ si $\nu\geq 3$ et on conclut avec I.2.1. □

I.3. Sur le genre des caractères. Le lemme ci-dessous est bien utile pour calculer le genre des caractères. En effet le caractère maximal, φ, de degré d, longueur σ, est connu ainsi que son genre qui vaut $G_{CM}(d,\sigma)$ (cf.[GP1], Thm. 2.7 et sa démonstration). Si χ est un caractère donné on peut calculer son genre à partir de $g(\varphi)$ par applications répétées de I.3.1.

I.3.1. Lemme. Soit $\psi=(n_i)$ un caractère de degré d, longueur σ. Soit $\chi=(m_i)$ un caractère tel que $m_j=n_j-1$, $m_{j+k}=n_{j+k}+1$, $m_i=n_i$ sinon. Alors $g(\chi)=g(\psi)-(n_j-n_{j+k}-1)$.

Démonstration. D'après I.1 (3) :

$$g(\psi)-g(\chi)=\sum_{n\geq 1}[(n_j-n-1)_+ + (n_{j+k}-n-1)_+ - (n_j-n-2)_+ - (n_{j+k}-n)_+]$$ d'où le résultat. □

I.3.2. Lemme. Soient $\sigma \geq 5$ un entier et $d=\sigma^2-2\sigma+5$. Posons $\psi=(2\sigma-3, \ldots, \sigma+1, \sigma+1, \sigma+1, \sigma)$. Soit $\chi=(n_i)$ un caractère de degré d, longueur σ et $x:=\#\{n_i | n_i=\sigma\}$. Si $x \leq 1$ alors $g(\chi)-(x+1) \leq g(\psi)-2$ avec égalité si et seulement si $\chi=\psi$.

Démonstration. Supposon d'abord $x=1$. Soit $T(\chi)$ le caractère tronqué: $(n_0, \ldots, n_{\sigma-2})$. Il est clair que $T(\chi) \leq T(\psi)$ (ordre lexicographique). D'après [E] I.5 $g(T(\chi)) \leq g(T(\psi))$ avec égalité si et seulement si $T(\chi)=T(\psi)$ d'où le lemme si $x=1$.

Supposons $x=0$ et soit $\varphi'=(2\sigma-4, 2\sigma-4, \ldots, \sigma+1, \sigma+1, \sigma+1, \sigma+1)$. On a $\chi \leq \varphi'$ et $g(\chi) \leq g(\varphi')$ ([E] I.5). On conclut en observant que $g(\varphi')=g(\psi)-\sigma+4$ (cf.I.3.1). □

I.3.3. Lemme. Soient $\sigma \geq 2$ un entier et $d=\sigma^2-2\sigma+4$. Posons $\theta=(2\sigma-4, 2\sigma-4, \ldots, \sigma+1, \sigma+1, \sigma+1, \sigma)$. Soit $\chi=(n_i)$ un caractère de degré d, longueur σ et $x:=\#\{n_i | n_i=\sigma\}$. Si $x \leq 1$ alors $g(\chi)-(x+1) \leq g(\theta)-2$ avec égalité si et seulement si $\chi=\theta$.

Démonstration. Elle est identique à la précédente: si $x=1$ on tronque et si $x=0$ on considére $\varphi'=(2\sigma-4, 2\sigma-5, 2\sigma-5, \ldots, \sigma+1, \sigma+1, \sigma+1, \sigma+1)$. □

I.3.4. Lemme. Soient $\sigma \geq 7$, $2 \leq \nu \leq \sigma-3$ des entiers et $d=\sigma^2-3\sigma+6+\nu$. Posons $\psi=(2\sigma-4, \ldots, \sigma+\nu-1, \sigma+\nu-1, \ldots, \sigma+1, \sigma+1, \sigma, \sigma)$. Soit $\chi=(n_i)$ un caractère de degré d, longueur σ et $x:=\#\{n_i | n_i=\sigma\}$. Si $x \leq 2$ alors $g(\chi)-(x+1) \leq g(\psi)-3$ (*). De plus si l'on a égalité dans (*) alors $\chi=\psi$ si $\nu \neq 3$ et $\chi=\psi$ ou $\chi=\psi'=(2\sigma-4, \ldots, \sigma+1, \sigma+1, \sigma+1, \sigma+1, \sigma)$ si $\nu=3$.

Démonstration. Elle est similaire aux précédentes: on exhibe les caractères de genre maximal à x fixé.

Supposons $x=2$. En tronquant et en utilisant [E] I.5 on obtient $g(\chi) \leq g(\psi)$ avec égalité si et seulement si $\chi=\psi$.

Si $x=1$ et $\nu \geq 3$ on pose $\psi'=(2\sigma-4, \ldots, \sigma+\nu-2, \sigma+\nu-2, \ldots, \sigma+1, \sigma+1, \sigma+1, \sigma)$; si $\nu=2$, $\psi'=(2\sigma-5, 2\sigma-5, \ldots, \sigma+1, \sigma+1, \sigma+1, \sigma+1, \sigma)$. Comme précedemment on vérifie $g(\chi) \leq g(\psi')$ avec égalité si et seulement si $\chi=\psi'$. Si $\nu \geq 3$ on a $g(\psi')=g(\psi)-\nu+2$ et donc $g(\psi')-2 \leq g(\psi)-3$ avec égalité si et seulement si $\nu=3$. Si $\nu=2$, $g(\psi')=g(\psi)-\sigma+5$.
Si $x=0$ et $\nu \geq 4$ on pose $\psi''=(2\sigma-4, \ldots, \sigma+\nu-3, \sigma+\nu-3, \ldots, \sigma+1, \sigma+1, \sigma+1, \sigma+1)$. On a $g(\chi) \leq g(\psi'')=g(\psi)-2\nu+5$, ce qui permet de conclure.

Si $x=0$ et $\nu=3$ on pose $\psi''=(2\sigma-5, 2\sigma-5, \ldots, \sigma+1, \sigma+1, \sigma+1, \sigma+1, \sigma+1)$; on a $g(\chi) \leq g(\psi'')=g(\psi)-\sigma+4$. De même si $x=0$ et $\nu=2$ on pose $\psi''=(2\sigma-5, 2\sigma-6, 2\sigma-6, \ldots, \sigma+1, \sigma+1, \sigma+1, \sigma+1, \sigma+1)$; on a $g(\psi'')=g(\psi)-2\sigma+11$. Ceci achéve la démonstration car $\sigma \geq 7$ par hypothèse. □

I.3.5. Lemme. Soit $\sigma \geq 7$ un entier et $d=\sigma^2-3\sigma+7$. Soit $\varphi=(2\sigma-4, \ldots, \sigma+1, \sigma+1, \sigma, \sigma, \sigma)$ et soit $\chi=(n_i)$ un caractère de degré d, longueur σ et $x:=\#\{n_i | n_i=\sigma\}$. Alors $g(\chi)-(x+1) \leq g(\varphi)-4$ avec égalité si et seulement si $\chi=\varphi$.

Démonstration. Elle est identique à celles de I.3.2, I.3.3, I.3.4. □

II. Postulation et section plane.

Soit $C \subset \mathbf{P}^3$ une courbe (localement Cohen-Macaulay et équidimensionelle) et Γ sa section plane générale. Dans ce paragraphe nous considérons le problème suivant: connaissant la postulation de Γ peut on en déduire des information sur celle de C ?

Les premiers résultats sur ce sujet (cf. II.1, II.2) ne font intervenir que d et $\sigma(\Gamma)$. En fait pour aller au delà il semble qu'il faille considérer non seulement σ mais toute la résolution minimale de Γ (cf. II.3, II.8).

II.1. Lemme. Si C est intègre et si $d > \sigma^2+1$ alors $h^0(\mathcal{I}_C(\sigma)) \neq 0$.

Démonstration. [L], [GP2]. □

II.2. Lemme. Si C est intègre et si $d > \sigma^2-\sigma+4$ alors $h^0(\mathcal{I}_C(\sigma+1)) \neq 0$.

Démonstration. [S2], Th. 3. □

II.3. Lemme. Si $m_2^-(\Gamma) \geq t+3$ alors $H^0(\mathcal{I}_C(t)) \rightarrow H^0(\mathcal{I}_\Gamma(t))$ est surjective.

Démonstration. [S1], Th.4. □

II.4. Lemme. Soit C intègre. On suppose $\sigma \geq 5$. Si $h^0(\mathcal{I}_\Gamma(\sigma)) \geq 2$ et si $d > \sigma^2-2\sigma+6$ alors $h^0(\mathcal{I}_C(\sigma)) \neq 0$.

Démonstration. D'après I.2.2 on a les possibilités suivantes:

(0) $m_2^-(\Gamma) \geq \sigma+3$

(1) $\alpha(\sigma)=3$, $\alpha(\sigma+1)=0$, $\beta(\sigma+1)=1$ et $\beta(\sigma+2)=0$

(2) $\alpha(\sigma)=2$, $\alpha(\sigma+1)=1$, $\beta(\sigma+2)=1$ et $\beta(\sigma+3)=0$. De plus si R est lié à Γ par une intersection complete $(\sigma+1, \sigma+1)$, alors toutes les formes de degré $\sigma-1$ passant par R ont un PGCD , F, de degré 3. Par conséquent $F \cap R = R'$ est un sous-groupe de points de R avec $\chi(R') = (\sigma+1, \sigma+1, \sigma)$.

(0) D'après II.3, $h^0(\mathcal{I}_C(\sigma)) \neq 0$.

(1) On conclut avec [S2], §3, iii).

(2) D'après [S2] Thm. 4, $h^0(\mathcal{I}_C(\sigma+1)) \geq 3$. On peut donc lier C à une courbe C' par une intersection complète $(\sigma+1, \sigma+1)$. Comme $h^0(\mathcal{I}_C(\sigma))=0$, les deux surfaces liantes, G, G' sont intègres. Notons R la section plane générale de C'. D'après [S2], lemme 2, C' contient une courbe (loc.C.M. et équidimensionnelle) X ayant R' comme section plane générale. On a $\deg(X) = 3\sigma-1$ et $h^0(\mathcal{I}_X(3)) \neq 0$: en effet, si $\sigma \geq 5$, on a $m_2^-(R') \geq 6$ et l'on utilise II.3.

Notons S la surface cubique contenant X. On a $C' = X \cup C''$, C'' la courbe résiduelle de X dans C' par rapport à S, et, par construction $X = C' \cap S$ (à des composantes 0-dimensionnelles près). Comme X est tracée sur une surface intègre, G, de degré $\sigma+1$, on peut lier X, par (G,S), à une courbe (loc.C.M.), Z, de degré 4.

Remarquons que C' est obtenue par liaison géométrique (G, G') à partir de C. D'après [PS] lemme 3.5, il s'ensuit que Z est génériquement localement intersection complète.

Si $h^0(\mathcal{I}_Z(2)) \neq 0$ par liaison on a $h^0(\omega_X(2-\sigma)) \neq 0$: ceci implique $h^0(\omega_{C'}(2-\sigma)) \neq 0$ et, de nouveau par liaison, $h^0(\mathcal{I}_C(\sigma)) \neq 0$.

Si $h^0(\mathcal{I}_Z(2))=0$, d'après II.4.1 ci-dessous, toute section de $\omega_Z(1)$ s'annule sur une composante irréductible, Z_0, de Z. La suite de liaison $0 \to \mathcal{I}_U(\sigma+1) \to \mathcal{I}_X(\sigma+1) \to \omega_Z(1) \to 0$ montre que toute courbe liée à X dans une intersection complète $(\sigma+1, \sigma+1)$ contient Z_0. Montrons maintenant qu'il exist X', contenant strictement X et tel que toute surface de degré $\sigma+1$ contenant X contienne aussi X'. L'existence de X' permet de conclure. En effet en considérant les surfaces de la forme $S\cdot P_{\sigma-2}$ on deduit $X' \subset S$. D'autre part $G \cap G'$ contient X et donc X'. Comme C est intègre, C ne contient aucune composante de X', donc $X' \subset C'$ mais ceci contredit $S \cap C' = X$.

Si $Z_0=Z$, la suite de liaison fournit $H^0(\mathcal{I}_U(\sigma+1)) \cong H^0(\mathcal{I}_X(\sigma+1))$ et l'on pose X'=U. Supposons $Z_0 \neq Z$. On peut supposer Z_0 intègre et donc loc.C.M..Comme Z est généri-quement localement intersection complète on peut définir une résiduelle, Z', à Z_0 dans Z: en chaque point générique, η, Z est intersection complète et Z' est la liée à Z_0 dans Z (i.e. $\mathcal{I}_{Z',Z,\eta} = \mathcal{H}om_{\mathcal{O}_{Z,\eta}}(\mathcal{O}_{Z_0,\eta}, \mathcal{O}_{Z,\eta})$).On obtien un sous-schéma fermé en prenant l'adhérence dans Z; Z' est le plus grand sous-schéma loc.C.M.contenu dans cette adhérence. Finalement X' est le lieé a Z' dans l'intersection complètè U. Par construction $X' \supset X$ et $X' \neq X$. Il ne reste plus qu'à montrer que toute surface, T , de degré $\sigma+1$ contenant X contient aussi X'. Pour cela on se restreint à un plan général, H. On a un diagramme commutatif:

$$\begin{array}{ccccccccc}
 & & & & 0 & & 0 & & \\
 & & & & \downarrow & & \downarrow & & \\
0 & \to & H^0(\mathcal{I}_{U\cap H}(\sigma+1)) & \to & H^0(\mathcal{I}_{X'\cap H}(\sigma+1)) & \to & H^0(\omega_{Z'\cap H}(1)) & \to & H^1(\mathcal{I}_{U\cap H}(\sigma+1)) \\
 & & \downarrow\approx & & \downarrow & & \downarrow & & \downarrow\approx \\
0 & \to & H^0(\mathcal{I}_{U\cap H}(\sigma+1)) & \to & H^0(\mathcal{I}_{X\cap H}(\sigma+1)) & \to & H^0(\omega_{Z\cap H}(1)) & \to & H^1(\mathcal{I}_{U\cap H}(\sigma+1)) \\
 & & & & & & \downarrow & & \\
 & & & & & & H^0(\omega_{Z_0\cap H}(1)) & &
\end{array}$$

Soit $\overline{T}$ l'image de T dans $H^0(\mathcal{I}_{X\cap H}(\sigma+1))$. Comme toute surface de degré $\sigma+1$ contenant X contient aussi Z_0 ,l'image de $\overline{T}$ dans $H^0(\omega_{Z_0\cap H}(1)) \cong H^0(\mathcal{O}_{Z_0\cap H})$ est nulle.Il s'ensuit que $\overline{T}$ provient de $H^0(\mathcal{I}_{X'\cap H}(\sigma+1))$, donc T contient X'. ⊡

II.4.1. Sous-Lemme. Soit $Z \subset \mathbf{P}^3$ une courbe localement Cohen-Macaulay et génériquement localement intersection complète, de degré 4. L'un des cas suivants a lieu:

(1) $h^0(\mathcal{I}_Z(2)) \neq 0$

(2) toute section de $\omega_Z(1)$ s'annule sur une composante irréductible, Z_0, de Z.

Démonstration. Supposons $h^0(\mathcal{I}_Z(2))=0$. Si (2) n'est pas vrai alors $\omega_Z(1)$ a une section non nulle qui engendre presque partout. Cette section permet de construire une suite exacte:

$$0 \to \mathcal{O}_{\mathbf{P}} \to \mathcal{F}(2) \to \mathcal{I}_Z(3) \to 0$$

où $\mathcal{F}$ est un faisceau réflexif. Par hypothèse $h^0(\mathcal{F}(1))=0$. Les classe de Chern de $\mathcal{F}(1)$ sont $c_1=1$, $c_2=2$. Ceci contredit [H2], Thm.1.1. ⊡

II.4.2. *Remarque*. Le lemme II.4 n'est pas loin d'etre optimal. En effet, pour $\sigma \geq 3$, il existe une courbe lisse, connexe, C, vérifiant: $\deg(C) = \sigma^2 - 2\sigma + 5$, $h^0(\mathcal{I}_\Gamma(\sigma))=2$, $h^0(\mathcal{I}_C(\sigma))=0$.

Pour construire une telle courbe on considère une courbe lisse,connexe, Y, de degré 8 et genre 5 et de rang maximum. On a donc $h^0(\mathcal{I}_Y(3))=0$, $\mathcal{I}_Y(k)$ est engendré par ses sections si $k\geq 4$ et la résolution minimale de $\mathcal{I}_{Y\cap H}$ pour H général est:

$$0 \to 2\mathcal{O}(-5) \to \mathcal{O}(-4)\oplus 2\mathcal{O}(-3) \to \mathcal{I}_{Y\cap H} \to 0.$$

En faisant une liaison $(4,\sigma+1)$ on obtient une courbe, T, de degré $4\sigma-4$. Comme $\sigma+1\geq 4$, on peut supposer T et les surfaces liantes lisses. La suite de liaison:

$$0 \to \mathcal{I}_U(\sigma+1) \to \mathcal{I}_T(\sigma+1) \to \omega_Y \to 0$$

montre que $\mathcal{I}_T(\sigma+1)$ est engendré par ses sections. On peut donc lier T à une courbe lisse C par $(\sigma+1,\sigma+1)$. On vérifie facilement que C est connexe et $h^0(\mathcal{I}_C(\sigma))=0$. Finalement par mapping cone on obtien le caractère $\chi(C)=(2\sigma-3, \ldots, \sigma+1, \sigma+1, \sigma+1, \sigma)$ et la résolution libre minimale:

$$0 \to \mathcal{O}(-2\sigma+2)\oplus 2\mathcal{O}(-\sigma-2) \to 2\mathcal{O}(-\sigma)\oplus 2\mathcal{O}(-\sigma-1) \to \mathcal{I}_\Gamma \to 0.$$

II.4.3. ***Remarque.*** Comme nous allons le voir l'énoncé II.4 s'étend aux courbes de degré $\sigma^2-2\sigma+6$ dont le caractère est différent de $\varphi'=(2\sigma-3, \ldots, \sigma+2, \sigma+2, \sigma+1, \sigma+1, \sigma)$ (cf.la démonstration de II.5). D'autre part si $\chi(C)=\varphi'$ on a un résultat partiel (cf.II.5,II.6) en faisant intervenir le genre arithmétique.

II.5. Lemme. Soit $C\subset \mathbf{P}^3$ une courbe intègre de degré $\sigma^2-2\sigma+6$. Si $\sigma\geq 5$ et $h^0(\mathcal{I}_C(\sigma+1))\geq 2$ alors $h^0(\mathcal{I}_C(\sigma))\neq 0$.

Démonstration. On peut supposer que C est tracée sur deux surfaces intègres de degré $\sigma+1$ et donc que $C\cap H$ est contenu dans deux courbes intègres de degré $\sigma+1$. Nous sommes donc sous les hypothèses de I.2.2. Si $m_2^-(\Gamma)\geq \sigma+3$ on conclut avec II.3. Soit donc $m_2^-(\Gamma)<\sigma+3$. D'après la démonstration précédente, le seul cas à considérer est I.2.2.(3).On procéde alors comme dans II.4 (2). ⊡

Pour faire intervenir le genre arithmétique de C nous utiliseront (ii) du lemme suivant:

II.6. Lemme. Soient $C\subset \mathbf{P}^3$ une courbe intègre et $\mathbf{M}=H^1_*(\mathcal{I}_C)$ son module de Hartshorne-Rao. Si H est un plan général soient N et R les modules définis par la suite exacte:

(*) $$0 \to \mathbf{N} \to \mathbf{M}(-1) \xrightarrow{\cdot H} \mathbf{M} \to \mathbf{R} \to 0.$$

On pose $r(k)=\dim R_k$, $n(k)=\dim N_k$ et on désigne par $\chi=(n_i)$ le caractère de C.

(i) $g(\chi)-p_a(C)=\sum_{k\geq 1} r(k) =\sum_{k\geq 1} n(k)$

(ii) pour tout entier $k\geq 1$: $h^1(\mathcal{I}_C(k))\leq \sum_{j=1}^{k} r(j)\leq g(\chi)-p_a(C)$

(iii) s'il existe $t<s(C)$ tel que $h^0(\mathcal{I}_{C\cap H}(t))=g(\chi)-p_a(C)$ alors $h^1(\mathcal{I}_C(l))=0$ pour $l\geq t$, et C est de rang maximum.

(iv) si $h^1(\mathcal{I}_C(n_0-2))=0$ alors $e(C)=n_0-3$.

Démonstration. (i). En restreignant à H on obtient la suite exacte:

$$\ldots \to H^1(\mathcal{I}_C(k)) \to H^1(\mathcal{I}_{C\cap H}(k)) \to H^2(\mathcal{I}_C(k-1)) \to H^2(\mathcal{I}_C(k)) \to 0\,.$$

Soit Δ_k le conoyau de $H^1(\mathcal{I}_C(k)) \to H^1(\mathcal{I}_{C\cap H}(k))$ et $\delta_k=\dim\Delta_k$. On a $p_a(C)=\sum_{k\geq 1}\delta(k)$,

$g(\chi)=\sum_{k\geq 1} h^1(\mathcal{I}_{C\cap H}(k))$ et donc $g(\chi)-p_a(C)=\sum_{k\geq 1} r(k)$.

La suite exacte (*) montre $\sum_{k\geq 1} r(k)=\sum_{k\geq 1} n(k)$.

(ii). On a $h^1(\mathcal{I}_C(1))=r(1)$ car C est intègre. La suite de restriction à H montre: $h^1(\mathcal{I}_C(k))\leq r(k)+h^1(\mathcal{I}_C(k-1))$ et l'on obtient la première inégalité par récurrence.

(iii). Pour $t<s$, $h^0(\mathcal{I}_{C\cap H}(t))\leq h^1(\mathcal{I}_C(t-1))$. L'hypothèse et (ii) impliquent: $h^1(\mathcal{I}_C(t-1))=\sum_{j=1}^{t-1} r(j)=g(\chi)-p_a(C)$. D'après (i), $r(j)=0$ si $j\geq t$. Comme $h^0(\mathcal{I}_{C\cap H}(t))=h^1(\mathcal{I}_C(t-1))$ il vient $h^1(\mathcal{I}_C(t))=r(t)=0$. On termine par récurrence.

(iv). Si $h^1(\mathcal{I}_C(n_0-2))=0$ alors $0<h^1(\mathcal{I}_{C\cap H}(n_0-2))\leq h^2(\mathcal{I}_C(n_0-3))$ (cf.I.1 (2) pour la première inégalité). D'autre part, comme $h^1(\mathcal{I}_{C\cap H}(l))=0$ si $l\geq n_0-1$, $e(C)\leq n_0-3$. □

II.6.1. ***Remarque.*** L'énoncé II.6 (i) se trouve dans [W] et II.6 (ii) améliore III.3 de [E].

II.7. Corollaire. Soit $C\subset \mathbf{P}^3$ une courbe intègre de degré $\sigma^2-2\sigma+6$ avec $\sigma\geq 5$. Si $h^0(\mathcal{I}_{C\cap H}(\sigma))>1$ et si $p_a(C)\geq G_{CM}(d,\sigma)-7$ alors $h^0(\mathcal{I}_C(\sigma))\neq 0$.

Démonstration. D'après les démonstrations de II.4, II.5 le seul cas à considérer est celui où $\chi(C)=(2\sigma-3,\ \ldots,\ \sigma+2,\ \sigma+2,\ \sigma+1,\ \sigma+1,\ \sigma)$. On a $g(\chi(C))=G_{CM}(d,\sigma)-2$, $h^0(\mathcal{I}_{C\cap H}(\sigma+1))=7$. Si $p_a(C)\geq g(\chi(C))-5$ alors $h^1(\mathcal{I}_C(\sigma))\leq 5$ (II.6 (ii)) et $h^0(\mathcal{I}_C(\sigma+1))\geq 2$. On conclut avec II.5. □

II.7.1. ***Remarque.*** L'obstacle pour étendre II.4 au cas $d=\sigma^2-2\sigma+6$ est donc de montrer $h^0(\mathcal{I}_C(\sigma+1))\geq 2$. Les méthodes de [S2] semblent insuffisantes pour cela.

II.8. Proposition. Soit $C\subset \mathbf{P}^3$ une courbe intègre et soit Γ sa section plane générale. Avec les notations de I.2 supposons $\alpha_\sigma=4$, $\beta_{\sigma+1}=2$, $\beta_{\sigma+2}=\beta_{\sigma+3}=0$. Alors $h^0(\mathcal{I}_C(\sigma+1))\geq 6$.

Démonstration. L'hypothèse implique $h^0(\mathcal{I}_\Gamma(\sigma+1))\geq 10$. D'après la suite exacte

$$0 \to H^0(\mathcal{I}_C(\sigma)) \to H^0(\mathcal{I}_C(\sigma+1)) \to H^0(\mathcal{I}_\Gamma(\sigma+1)) \to H^1(\mathcal{I}_C(\sigma)) \to H^1(\mathcal{I}_C(\sigma+1))$$

il nous suffit de montrer que le noyau de $H^1(\mathcal{I}_C(\sigma)) \to H^1(\mathcal{I}_C(\sigma+1))$ a dimension au plus quatre. Pour cela on procéde comme dans [S2] §3 i).

La résolution libre minimale de $\mathcal{I}_\Gamma$ est de la forme:

$$0 \to \bigoplus_{n_{2i}\geq\sigma+4} \mathcal{O}(-n_{2i})\oplus 2\mathcal{O}(-\sigma-1) \to \bigoplus_{n_{1i}\geq\sigma} \mathcal{O}(-n_{1i})\oplus 4\mathcal{O}(-\sigma) \to \mathcal{I}_\Gamma \to 0.$$

En tensorisant par $\omega_h(\sigma+3)$ (cf. [S2] pour la définition de ω_h)

$$(f'_h)_0 : (2\omega_h(2))_0 \to \Big(\bigoplus_{n_{1i}\geq\sigma} \omega_h(-n_{1i}+\sigma+3)\Big)_0 \oplus (4\omega_h(3))_0$$

Il nous faut maintenant majorer la dimension du noyau de $(f'_h)_0$. Pour cela il suffit de majorer la dimension du noyau de l'application induite:

$$(f_h)_0 : (2\omega_h(2))_0 \to (4\omega_h(3))_0 .$$

Posons $k(h) := \dim\ker(f_h)_0$, $h \geq 1$. Observons pour commencer que toute courbe de $H^0(\mathcal{I}_\Gamma(\sigma))$ est irréductible en vertu du principe de position uniforme. Ceci dit si $G_1, \ldots, G_4$ est une base de $H^0(\mathcal{I}_\Gamma(\sigma))$, on peut supposer que les deux relations s'écrivent:

$$\begin{cases} G_1x_1+G_2x_2+G_3x_3=0 \\ G_1l_1+G_2l_2+G_3l_3+G_4l_4=0 \end{cases}$$

On a $l_4 \neq 0$ car autrement $G_2(l_2x_1-x_2l_1)+G_3(l_3x_1-x_3l_1)=0$, ce qui est absurde. On peut donc supposer l_2, l_3, l_4 linéairement indépendants. Une fois posé $l_1=\sum_{n=2}^{4} a_il_i$ on peut écrire les deux relations sous la forme

$$\begin{cases} G_1x_1+G_2x_2+G_3x_3=0 & \text{(R1)} \\ G_2l_2+G_3l_3+G_4l_4=0 & \text{(R2)} \end{cases}$$

avec l_2, l_3, l_4 linéairement indépendants. Posons $l_i=\sum_{j=1}^{3} c_{ij}x_j$, $2\leq i\leq 4$. Considérons $k(h)$ pour $h\geq 1$. Il résulte immédiatement des définitions que $k(h)=0$, $1\leq h\leq 2$, $k(3)=2$.

Pour $h=4$ on a $(f_4)_0 : \mathbf{k}^3\oplus\mathbf{k}^3 \to \mathbf{k}^4$ donnée par

$(f_4)_0(a_1, a_2, a_3; b_1, b_2, b_3)=(a_1, a_2+Bl_2, a_3+Bl_3, Bl_4)$ où $Bl_i=\sum_{j=1}^{3} c_{ij}b_j$. On vérifie facilement que $(f_4)_0$ est surjective donc $k(4)=2$.

Pour conclure la démonstration de la proposition il nous suffit (cf. [S2] §3 et Thm.2) de montrer que $(f_h)_0$ est injective pour $h\geq 5$. Ceci découle de II.8.1 et II.9 ci-dessous. □

II.8.1. Sous-Lemme. Avec les notations de la démonstration précédente, $(f_5)_0 : \mathbf{k}^6\oplus\mathbf{k}^6 \to \mathbf{k}^3\oplus\mathbf{k}^3\oplus\mathbf{k}^3\oplus\mathbf{k}^3$ est injective.

Démonstration. Soient (e_{ij}), $i\leq j$, $1\leq i, j\leq 3$ une base de $\mathbf{k}^6$ et (e_i), $1\leq i\leq 3$, une base de $\mathbf{k}^3$. Si $A=(\ldots, a_{ij}, \ldots)$, $B=(\ldots, b_{ij}, \ldots)$ sont deux vecteurs de $\mathbf{k}^6$ tels que $(f_5)_0(A;B)=(0;0)$ alors:

$$\begin{cases} a_{11}e_1+a_{12}e_2+a_{13}e_3=0 & (0) \\ a_{12}e_1+a_{22}e_2+a_{23}e_3+Bl_2=0 & (1) \\ a_{13}e_1+a_{23}e_2+a_{33}e_3+Bl_3=0 & (2) \\ Bl_4=0 & (3) \end{cases}$$

où pour $i=2,3,4$: $Bl_i=\left(\sum_{j=1}^{3} b_{1j}c_{ij}\right)e_1+\left(\sum_{j=1}^{3} b_{2j}c_{ij}\right)e_2+\left(\sum_{j=1}^{3} b_{3j}c_{ij}\right)e_3$ (on a posé $b_{ij}=b_{ji}$).

De (0) il vient $a_{11}=a_{12}=a_{13}=0$ et en considérant le coefficient de e_1 dans (1), (2), (3): $b_{11}=b_{12}=b_{13}=0$ (car l_2, l_3, l_4 sont linéairement indépendants).

En considérant les coefficients de e_2, e_3 dans (1), (2), (3) on obtient:

$$(*)\quad \begin{cases} a_{22}+b_{22}c_{22}+b_{23}c_{23}=0\\ a_{23}+b_{23}c_{22}+b_{33}c_{23}=0\\ a_{23}+b_{22}c_{32}+b_{23}c_{33}=0\\ a_{33}+b_{32}c_{32}+b_{33}c_{33}=0\\ b_{22}c_{42}+b_{23}c_{43}=0\\ b_{32}c_{42}+b_{33}c_{43}=0 \end{cases}$$

Les relations $(*)$ sont indépendantes si et seulements si $\Delta\neq 0$ où $\Delta:=c_{43}^2c_{32}-c_{42}^2c_{23}+(c_{22}-c_{33})c_{42}c_{43}$. Si $\Delta\neq 0$, $a_{ij}=b_{ij}=0$ pour tout i,j et $(A;B)=(0;0)$. Supposons donc $\Delta=0$ et distinguons plusieurs cas:
(a) $c_{42}=c_{43}=0$.
Les deux sygygies s'écrivent

$$\begin{cases} G_1x_1+G_2x_2+G_3x_3=0 & (R1)\\ G_2l_2+G_3l_3+G_4x_1=0 & (R2) \end{cases}$$

On obtient: $G_3(l_2x_3-x_2l_3)+(G_1l_2-G_4x_2)x_1=0$. On voit facilement que cette relation est absurde.
(b) c_{42} et c_{43} non tous les deux nuls. Sans nuire à la généralité on peut supposer $c_{43}\neq 0$. Considérons la deuxième relation (R2) et écrivons la sous la forme:

$$G_2(l_2-(c_{23}l_4)/c_{43})+G_3l_3+(G_4+(c_{23}G_2)/c_{43})l_4=0\ .$$

Quitte à poser $l'_2=l_2-(c_{23}l_4)/c_{43}$, $G'_4=G_4+(c_{23}G_2)/c_{43}$, on peut supposer $c_{23}=0$.
Soit $\gamma:=-c_{42}/c_{43}$. La relation $\Delta=0$ s'écrit: $(**)$ $c_{32}=\gamma(c_{22}-c_{33})$. Quitte à écrire (R1) et (R2) sous la forme:

$$\begin{cases} x_1G_1+x_2(G_2+\gamma G_3)+G_3(x_3-\gamma x_2)=0\\ l_2(G_2+\gamma G_3)+(l_3-\gamma l_2-(c_{33}l_4)/c_{43})G_3+l_4(G_4+(c_{33}G_3)/c_{43})=0 \end{cases}$$

on peut supposer $c_{32}=0$ par la suite. Ainsi $(**)$ devient $\gamma(c_{22}-c_{33})=0$.
(b.1) $\gamma=0$ (i.e. $c_{42}=0$). La relation (R2) s'écrit :
$G_2l_2+G_3(l_3-(c_{33}l_4)/c_{43})+(G_4+(c_{33}G_3)/c_{43})l_4=0$ ou encore (+) $G_2l_2+G_3cx_1+G'_4l_4=0$ avec $l_2=c_{21}x_1+c_{22}x_2$, $l_4=c_{41}x_1+c_{43}x_3$. En substituant $G_2x_2=-G_1x_1-G_3x_3$ (cf. (R1)) dans (+) on obtient une contradiction.
(b.2) $c_{33}=c_{22}$. De (R1) il vient $G_2x_2+G_3x_3=-G_1x_1$. En substituant dans (R2) et vu que $l_2=c_{21}x_1+c_{22}x_2$, $l_3=c_{31}x_1+c_{22}x_3$ on obtient encore une contradiction. □

II.9. Lemme. Soit $E\subset\mathbf{P}^2$ un groupe de points de résolution libre minimale:

$$0\to\bigoplus\mathcal{O}_{\mathbf{P}^2}(-b_j)\to\bigoplus\mathcal{O}_{\mathbf{P}^2}(-a_i)\to\mathcal{I}_E\to 0.$$

En tensorisant par $\omega_h(a)$ on obtient f_h: $\bigoplus\omega_h(a-b_j)\to\bigoplus\omega_h(a-a_i)$ qui, en degré zéro, induit $(f_h)_0:\bigoplus_{b_j\leq a}(\omega_h)_{a-b_j}\to\bigoplus_{a_i\leq a}(\omega_h)_{a-a_i}$. Si $(f_h)_0$ est injective et si $h-(a-b_j)>0$ pour tout j tel que $a\geq b_j$ alors $(f_t)_0$ est injective pour $t\geq h$.

Démonstration. Nous renvoyons à [S2] §2, lemma 1 pour plus de détails sur les modules ω_h. Rappelons simplement que l'on a une injection i: $\omega_h \to \omega_{h+1}(1)$ (qui provient de la surjection naturelle $R/m^{h+1} \to R/m^h$); pour s, $1 \le s \le 3$, un morphisme x_s: $\omega_{h+1} \to \omega_h$ (qui provient de $R/m^h \xrightarrow{x_s} R/m^{h+1}(1)$) et la multiplication $\cdot x_s$: $\omega_{h+1} \to \omega_{h+1}(1)$. Ce dernier morphisme est la composition $i \circ x_s$. Il s'ensuit que pour tout $h \ge 1$ on a un diagramme commutatif:

$$\begin{array}{ccc} \bigoplus(\omega_{h+1})_{a-b_j} & \xrightarrow{(f_{h+1})_0} & \bigoplus(\omega_{h+1})_{a-a_i} \\ \downarrow x_s & & \downarrow x_s \\ \bigoplus(\omega_h)_{a-b_j} & \xrightarrow{(f_h)_0} & \bigoplus(\omega_h)_{a-a_i} \end{array}$$

Si $(f_{h+1})_0(\alpha)=0$ et si $(f_h)_0$ est injective alors $x_s \cdot \alpha = 0$, $1 \le s \le 3$. Vu la description des ω_i ([S2] loc. cit.) et comme $h-(a-b_j)>0$ pour tout j concerné, ceci implique $\alpha=0$. □

II.10. Proposition. Soient $\sigma \ge 6$ un entier et $C \subset \mathbf{P}^3$ une courbe intègre de degré $d=\sigma^2-3\sigma+6+v$, $2 \le v \le \sigma-3$. Si $v \ge 3$ on suppose que la résolution minimale de la section plane générale Γ de C vérifie: $\alpha_\sigma=4$, $\alpha_{\sigma+1}=0$, $\beta_{\sigma+1}=2$, $\beta_{\sigma+2}=\beta_{\sigma+3}=0$. Si $v=2$ on suppose $\alpha_\sigma=4$, $\alpha_{\sigma+1}=0$, $\beta_{\sigma+1}=2$ et $p_a(C) \ge g(\chi(C))-8$. Alors $h^0(\mathcal{I}_C(\sigma)) \ne 0$.

Démonstration. Si $v \ge 3$, d'après II.8, C est contenue dans une intersection complète $(\sigma+1,\sigma+1)$. Si $v=2$ on a $h^0(\mathcal{I}_\Gamma(\sigma+1))=10$ et (cf. II.6. (ii)) $h^1(\mathcal{I}_C(\sigma)) \le 8$ donc $h^0(\mathcal{I}_C(\sigma+1)) \ge 2$ et l'on peut supposer C contenue dans une intersection complète $(\sigma+1,\sigma+1)$. Soit T la liée à C dans cette intersection complète. Le caractère $\chi(C)$ est de la forme $(\ldots, \sigma+2, \sigma+1, \sigma, \sigma, \sigma)$ et par liaison on a que, si $\chi(T)=(p_j)$, $p_0=\ldots=p_3=\sigma+1$, $p_4<\sigma$. Ainsi $\chi(T)$ n'est pas connexe et, d'après I.1.2, $T \cap H$ contient R avec $\chi(R)=(\sigma+1, \sigma+1, \sigma+1, \sigma+1)$. Il s'ensuit ([S2] lemma 2) que T contient une courbe T' telle que $T' \cap H = R$. D'après I.2 on a la résolution :

$$0 \to 4\mathcal{O}(-\sigma-2) \to 4\mathcal{O}(-\sigma-1) \oplus \mathcal{O}(-4) \to \mathcal{I}_R \to 0.$$

Si $\sigma \ge 5$, $h^0(\mathcal{I}_{T'}(4)) \ne 0$ (cf.II.3). Maintenant on procéde comme dans la démonstration de II.4. On lie T' à une courbe génériquement localement intersection complète, Z, par une intersection complète $(4,\sigma+1)$. On a $\deg(Z)=6$. Si $h^0(\mathcal{I}_Z(3)) \ne 0$ alors $h^0(\mathcal{I}_C(\sigma)) \ne 0$. Si $h^0(\mathcal{I}_Z(3))=0$, d'après II.11 ci-dessous, toute section de ω_Z s'annule sur une composante irréductible, Z_0, de Z et par conséquent C contient Z_0 ce qui est absurde. □

II.11. Lemme. Soit $Z \subset \mathbf{P}^3$ une courbe localement Cohen-Macaulay et génériquement localement intersection complète, de degré 6. L'un des cas suivants a lieu:

(1) $h^0(\mathcal{I}_Z(3)) \ne 0$

(2) toute section de ω_Z s'annule sur une composante irréductible, Z_0, de Z.

Démonstration. Identique à celle de II.4.1. □

III. Majoration de $G(d,s)$.

Dans ce paragraphe l'on démontre le théorème 0.3 de l'introduction. En outre l'on décrit le caractère des courbes réalisant $G(d,s)$ (i.e. des courbes intègres de degré d, genre arithmétique $G(d,s)$, non contenus dans une surface de degré $s-1$).

Pour $f=2s-8$ l'intervalle pour le degré est:

$A(s,2s-8)=s^2-4s+7\leq d<s^2-3s+5=A(s,2s-7)$ et l'on a $B(s,2s-8)=s^2-4s+10$.

III.1. Proposition. Soient d, s des entiers vérifiant $s^2-3s+4\geq d\geq s^2-4s+10$, $s\geq 6$. Alors:

(1) $G(d,s)=G_H(d,s)$

(2) Soit C une courbe intègre réalisant $G(d,s)$.

Si $\sigma(C)=s$ alors $d=s^2-4s+10$, C est arithmétiquement Cohen-Macaulay, de caractère le caractère maximal de degré d, longueur s. En particulier $e(C)=2s-8$.

Autrement $\sigma(C)=s-1$ et le caractère de C est:

$$\psi=(2\sigma-3,\ \dots,\ \sigma+\nu-2,\ \sigma+\nu-2,\ \dots,\ \sigma+1,\ \sigma+1,\ \sigma+1)$$

où l'on a posé $d=\sigma^2-2\sigma+3+\nu$, $4\leq\nu\leq\sigma-1$. De plus $e(C)=2s-8$, C est de rang maximum avec $h^1(\mathcal{I}_C(k))=0$ si $k\geq s-1$ et $h^1(\mathcal{I}_C(k))\leq 1$ sinon.

Démonstration. (1) Soit C une courbe intègre de degré d vérifiant $h^0(\mathcal{I}_C(s-1))=0$. D'après II.1 on a $\sigma\geq s-1$. Si $\sigma=s$ alors $p_a(C)\leq G_{CM}(d,s)$ (cf.[GP1]). On vérifie que $G_{CM}(d,s)<G_H(d,s)$ sauf si $d=s^2-4s+10$ auquel cas $G_{CM}(d,s)=G_H(d,s)$. Sous ces conditions il s'ensuit que si $p_a(C)=G_H(d,s)$ alors C est a.C.M. et $\chi(C)=\varphi$ le caractère maximal de degré d, longueur s. On a $\varphi=(2s-5,\ \dots\dots)$ (cf.[GP1]) et donc $e(C)=2s-8$.

Supposons désormais $\sigma=s-1$. D'aprés II.4, $h^0(\mathcal{I}_{C\cap H}(\sigma))=1$. Il s'ensuit que $p_a(C)\leq G-1$ où G est le genre maximal des caractères $\chi=(n_i)$, de degré d, longueur σ avec $n_{\sigma-1}\geq\sigma+1$. Posons $d=\sigma^2-2\sigma+3+\nu$, $4\leq\nu\leq\sigma-1$, et soit ψ le caractère:

$(2\sigma-3,\ \dots,\ \sigma+\nu-2,\ \sigma+\nu-2,\ \dots,\ \sigma+1,\ \sigma+1,\ \sigma+1)$. Il est clair que si χ' est un caractère de degré d, longueur σ alors $\chi'>\psi$ implique $n_{\sigma-1}<\sigma+1$. D'après [E], I.5, si χ est un caractère de degré d,longueur σ avec $n_{\sigma-1}\geq\sigma+1$ alors $g(\chi)\leq g(\psi)$ avec égalité si et seulement si $\chi=\psi$. Par conséquent $G=g(\psi)$ et un simple calcul montre que $g(\psi)=G_H(d,s)+1$. On conclut avec II.6. ⊡

III.2. Proposition. Soit $s\geq 6$ un entier et posons $d=s^2-4s+9$. Alors:

(1) $G(d,s)=G_H(d,s)$

(2) Soit C une courbe intègre réalisant $G(d,s)$. Alors $\sigma(C)=s-1$ et $\chi(C)=\psi$ où $\psi=(2\sigma-3,\ \dots,\ \sigma+1,\ \sigma+1,\ \sigma+1,\ \sigma+1)$. De plus $e(C)=2s-8$, C est de rang maximum avec $h^1(\mathcal{I}_C(k))=0$ si $k\geq s-1$, et $h^1(\mathcal{I}_C(k))\leq 1$ sinon.

Démonstration. Soit C intègre avec $h^0(\mathcal{I}_C(s-1))=0$ et $p_a(C)\geq G_H(d,s)$. D'après II.1 $\sigma(C)\geq s-1$. Comme $G_{CM}(d,s)<G_H(d,s)$ on a $\sigma(C)=s-1$. Si $h^0(\mathcal{I}_{C\cap H}(\sigma))\geq 2$, comme $G_{CM}(d,\sigma)=G_H(d,s)+4$, d'après II.7 on a $h^0(\mathcal{I}_C(\sigma))\neq 0$. Il s'ensuit que $p_a(C)\leq G-1$ où G est le genre maximal des caractères $\chi=(n_i)$, de degré d, longueur σ avec $n_{\sigma-1}\geq\sigma+1$. Il est clair que si χ est un tel caractère alors $\chi\leq\psi$. D'après [E], I.5, $g(\chi)\leq g(\psi)$ avec égalité si et seulement si $\chi=\psi$. On conclut avec II.6 car $g(\psi)=G_H(d,s)+1$. ⊡

III.3. Proposition. Soit $s\geq 6$ un entier et posons $d=s^2-4s+8$. Alors:

(1) $G(d,s)=G_H(d,s)$

(2) Soit C une courbe intègre réalisant G(d,s). Alors $\sigma(C)=s-1$ et $\chi(C)=\psi$ où $\psi=(2\sigma-3, \ldots, \sigma+1, \sigma+1, \sigma+1, \sigma)$. De plus $e(C)=2s-8$, C est de rang maximum avec $h^1(\mathcal{J}_C(k))=0$ si $k\geq s-1$, et $h^1(\mathcal{J}_C(k))\leq 2$ sinon.

Démonstration. Soit C intègre avec $h^0(\mathcal{J}_C(s-1))=0$ et $p_a(C)\geq G_H(d,s)$. D'après II.1 $\sigma(C)\geq s-1$. Comme $G_{CM}(d,s)<G_H(d,s)$ on a $\sigma(C)=s-1$. Le caractère maximale pour (d,σ) est: $\varphi=(2\sigma-3, \ldots, \sigma+2, \sigma+2, \sigma+1, \sigma, \sigma)$ et $g(\varphi)=G_H(d,s)+3$. Posons $\chi(C)=(n_i)$ et $x:=\#\{n_i| n_i=\sigma\}$. On a $p_a(C)\leq g(\chi(C))-(x+1)$ (cf.II.6.(i)).

Si x=2 alors $p_a(C)\leq G_H(d,s)$ avec égalité si et seulement si $\chi(C)=\varphi$. Ceci est impossible car I.2.1 et [S2] §3, iii) montrent que si $\chi(C)=\varphi$ alors $h^0(\mathcal{J}_C(\sigma))\neq 0$. Donc $x\leq 1$ et $g(\chi(C))-(x+1)\leq g(\psi)-2$ avec égalité si et seulement si $\chi(C)=\psi$ (cf.I.3.2). On termine avec II.6 et en observant que $g(\psi)=G_H(d,s)+2$. □

III.4. Proposition. Soit $s\geq 8$ un entier et posons $d=s^2-4s+7$. Alors:
(1) $G(d,s)=G_H(d,s)$
(2) Soit C une courbe intègre réalisant G(d,s). Alors $\sigma(C)=s-1$ et $\chi(C)=\varphi$ où $\varphi=(2\sigma-3, \ldots, \sigma+1, \sigma+1, \sigma, \sigma)$. De plus $e(C)=2s-8$, C est de rang maximum avec $h^1(\mathcal{J}_C(k))=0$ si $k\geq s-1$, et $h^1(\mathcal{J}_C(k))\leq 3$ sinon.

Démonstration. Soit C intègre avec $h^0(\mathcal{J}_C(s-1))=0$ et $p_a(C)\geq G_H(d,s)$. D'après II.1 $\sigma(C)\geq s-1$. Comme $G_{CM}(d,s)<G_H(d,s)$ on a $\sigma(C)=s-1$. Le caractère maximale pour (d,σ) est φ et $g(\varphi)=G_H(d,s)+3$. Posons $\chi(C)=(n_i)$ et $x:=\#\{n_i| n_i=\sigma\}$. On a $p_a(C)\leq g(\chi(C))-(x+1)$ (cf.II.6.(i)). Si x=2 alors $p_a(C)\leq G_H(d,s)$ avec égalité si et seulement si $\chi(C)=\varphi$.

Si $x\leq 1$, $p_a(C)\leq g(\theta)-2$ où $\theta=(2\sigma-4, 2\sigma-4, \ldots, \sigma+1, \sigma+1, \sigma+1, \sigma)$ (cf.I.3.3). Comme $g(\theta)\leq g(\varphi)-\sigma+4$, on voit que ceci est impossible. On termine avec II.6. □

Les propositions III.1,..., III.4 démontrent le théorème pour $f=2s-8$.
Si $f=2s-9$, l'intervalle pour le degré est:
$A(s,2s-9)=s^2-5s+11\leq d<s^2-4s+7=A(s,2s-8)$ et l'on a $B(s,2s-9)=s^2-5s+13$.

III.5. Proposition. Soient d, s des entiers vérifiant $s^2-5s+12\leq d\leq s^2-4s+6$, $s\geq 8$. Alors:
(1) $G(d,s)=G_H(d,s)$
(2) Posons $d=\sigma^2-3\sigma+6+\nu$, $2\leq\nu\leq\sigma-3$ et

$$\psi=(2\sigma-4, \ldots, \sigma+\nu-1, \sigma+\nu-1, \ldots, \sigma+1, \sigma+1, \sigma, \sigma)$$
$$\psi'=(2\sigma-4, \ldots, \sigma+1, \sigma+1, \sigma+1, \sigma+1, \sigma) \text{ si } \nu=3.$$

Si C réalise G(d,s) alors $\sigma(C)=s-1$. De plus $\chi(C)=\psi$ si $\nu\neq 3$ et $\chi(C)=\psi$ ou ψ' si $\nu=3$. En outre $e(C)=2s-9$, C est de rang maximum avec $h^1(\mathcal{J}_C(k))=0$ si $k\geq s-1$, et $h^1(\mathcal{J}_C(k))\leq 3$ (resp. 2 si $\chi(C)=\psi'$) sinon.

Démonstration. Soit C une courbe intègre de degré d vérifiant $h^0(\mathcal{J}_C(s-1))=0$ et $p_a(C)\geq G_H(d,s)$. D'après II.2 $\sigma(C)\geq s-1$. Comme $G_{CM}(d,s)<G_H(d,s)$ on a $\sigma(C)=s-1$. Posons $d=\sigma^2-3\sigma+6+\nu$, $2\leq\nu\leq\sigma-3$, $\chi(C)=(n_i)$ et $x:=\#\{n_i| n_i=\sigma\}$. On a $x\leq 3$ (I.2.2).

Supposons pour commencer x=3. Ceci implique (cf.II.6) $p_a(C)\leq g(\chi(C))-4$.
Si $\nu=2$ le caractère maximale pour (d,σ) est $\varphi=(2\sigma-4, \ldots, \sigma+2, \sigma+2, \sigma+1, \sigma, \sigma, \sigma)$. On a $g(\varphi)=G_H(d,s)+4$. Sous nos hypothèses on déduit: $p_a(C)=G_H(d,s)$ et $\chi(C)=\varphi$. Ceci est

impossible car I.2.3 et II.10 impliquent $h^0(\mathcal{I}_C(\sigma))\neq 0$. De même si $\nu\geq 3$, I.2.3 et II.10 impliquent $h^0(\mathcal{I}_C(\sigma))\neq 0$.

Par conséquent $x\leq 2$ et de $p_a(C)\leq g(\chi(C))-(x+1)$ on déduit $p_a(C)\leq g(\psi)-3$ (ou $g(\psi')-2$ si $\nu=3$) (cf.I.3.4). On verifie que $g(\psi)=G_H(d,s)+3$ (resp. $g(\psi')=G_H(d,s)+2$) et on conclut avec II.6. □

III.5.1. ***Remarque.*** Notons que, lorsque $\nu=3$, les deux possibilités $\chi(C)=\psi$ et $\chi(C)=\psi'$ ont effectivement lieu.

(1) $\chi(C)=\psi'$. Pour construire une telle courbe lisse connexe, C, on considère une courbe lisse connexe, Y, de degré 13, genre 18 et de rang maximum (cf.[HH] Th.5.2). On a $h^0(\mathcal{I}_Y(4))=0$, $\mathcal{I}_Y(k)$ est engendré par ses section si $k\geq 5$ et $\chi(Y)=(5,5,5,4)$. Soit S une surface lisse de degré $\sigma+1$, $\sigma\geq 4$, contenant Y et considerons sur S le système linéaire complet $|Y+(\sigma-4)\Pi|$, où $\Pi=S\cap H$. D'après la suite exacte:

$$0\to \mathcal{O}_S((\sigma-4)\Pi)\to \mathcal{O}_S(Y+(\sigma-4)\Pi)\to \omega_Y(-1)\to 0$$

comme $H^1(S,\mathcal{O}_S((\sigma-4)\Pi))=0$ et $H^0(Y,\omega_Y(-1))\neq 0$, on voit que Y n'est pas une composante fixe de $|Y+(\sigma-4)\Pi|$. En outre, comme $|(\sigma-4)\Pi|$ n'a pas des points fixes et Y est lisse, l'on déduit que l'élément général, C, de $|Y+(\sigma-4)\Pi|$ est lisse et connexe. On vérifie que C a le degré et le genre voulu. En outre $H^0(\mathcal{I}_C(\sigma))=H^0(S,\mathcal{O}_S(\sigma\Pi-C))= H^0(S,\mathcal{O}_S(4\Pi-Y))=H^0(\mathcal{I}_Y(4))=0$ et $h^0(\mathcal{I}_{C\cap H}(\sigma))= h^0(\mathcal{I}_{Y\cap H}(4))=2$. On conclut que C est la courbe cherchée.

(1) $\chi(C)=\psi$. Pour construire une telle courbe lisse connexe, C, on considère une courbe nodale, connexe, Y, de degré 13, genre 18 comme suit: $Y=X\cup P$ où X est une courbe lisse connexe de degré 7, genre 2 et de rang maximum, P est une courbe plane lisse de degré 6; on suppose que X et P se coupent quasi-transversalement en 7 points. On a $h^0(\mathcal{I}_Y(4))=0$, $\mathcal{I}_Y(k)$ est engendré par ses section si $k\geq 6$ et $\chi(Y)=(6,5,4,4)$. Soit S une surface lisse de degré $\sigma+1$, $\sigma\geq 5$, (cf.[E], lemme III.5) contenant Y et considérons sur S le système linéaire complet $|P+(\sigma-4)\Pi|$.D'après la suite exacte:

$$0\to \mathcal{O}_S((\sigma-4)\Pi)\to \mathcal{O}_S(P+(\sigma-4)\Pi)\to \omega_P(-1)\to 0,$$

comme $\omega_P(-1)$ est engendré par ses sections,on voit que $|P+(\sigma-4)\Pi|$ n'a pas des points fixes; l'on déduit que l'élément général, T, de $|P+(\sigma-4)\Pi|$ est lisse et connexe. L'on considère alors le système linéaire complet $|X+T|$ et l'on continue comme ci-dessus en notant que dans la suite exacte:

$$0\to \mathcal{O}_S(T)\to \mathcal{O}_S(X+T)\to \omega_X\to 0$$

on a $H^1(S,\mathcal{O}_S(T))=0$ car T est a.CM, et $H^0(X,\omega_X)\neq 0$. L'élément général, C, de $|X+T|$ est la courbe cherchée.

III.6. Proposition. Soit $s\geq 8$ un entier et posons $d=s^2-5s+11$. Alors:

(1) $G(d,s)=G_H(d,s)$

(2) Soit C une courbe intègre réalisant $G(d,s)$. Alors $\sigma(C)=s-1$ et $\chi(C)=\varphi$ où $\varphi=(2\sigma-4, \ldots,\sigma+1, \sigma+1, \sigma, \sigma, \sigma)$. De plus $e(C)=2s-9$, C est de rang maximum avec $h^1(\mathcal{I}_C(k))=0$ si $k\geq s-1$, et $h^1(\mathcal{I}_C(k))\leq 4$ sinon.

Démonstration. Soit C intègre avec $h^0(\mathcal{J}_C(s-1))=0$ et $p_a(C)\geq G_H(d,s)$. D'après II.2 $\sigma(C)\geq s-1$. Comme $G_{CM}(d,s)<G_H(d,s)$ on a $\sigma(C)=s-1$.

D'après II.6 $p_a(C)\leq g(\chi(C))-(x+1)$ où $x:=\#\{n_i | n_i=\sigma\}$. Finalement de I.3.5 on déduit $p_a(C)\leq g(\varphi)-4$ avec égalité si et seulement si $\chi(C)=\varphi$. On conclut avec II.6, en observant que $g(\varphi)=G_H(d,s)+4$. ☐

III.7. ***Remarque.*** D'après [GP2], [E], et II.6 on obtient, avec les notations de 0.1: pour $f\geq 2s-9$ toute courbe réalisant G(d,s) dans le domaine B est de rang maximun avec $h^1(\mathcal{J}_C(s-1))=0$ et e=f.
En ce qui concerne le rang maximum ceci conforte une conjecture de [BE1] (cf.aussi [BE2]).

BIBLIOGRAPHIE

[BE1] E. Ballico - Ph. Ellia: "A program for space curves" Rend. Sem. Mat. Torino, **44** (1986) 25-42.

[BE2] E. Ballico - Ph. Ellia: "On space curves with maximal genus in the range A" Revue Roumaine de Math. Pures et appliquées, **XXXIII**, n° 3 (1988) 165-174.

[BE3] E. Ballico - Ph. Ellia:"The maximal rank conjecture for non special curves in $\mathbf{P}^3$ " Invent.Math.,**79** (1985) 541-555.

[BE4] E. Ballico - Ph. Ellia: "Beyond the maximal rank conjecture for corves in $\mathbf{P}^3$ " Space Curves,Rocca di Papa, 1985, LNM **1266**,1987,1-23.

[C] G. Campanella: "Standard bases of perfect homogeneous polynomial ideals of height 2 " J. of Algebra, **100** (1986) 47-60.

[D] E. Davis: "Complete intersections of codimension 2 in $\mathbf{P}^r$: the Bezout-Jacobi-Segre theorem revisited" Rend. Sem. Mat. Torino, **43 (2)** (1985) 333-353.

[E] Ph. Ellia: "Sur le genre maximal des courbes gauches de degré d non sur une surface de degré s-1" à paraître sur Crelle Journal.

[EP] Ph. Ellia - C. Peskine: "Groupes de points de $\mathbf{P}^2$: caractère et position uniforme" LNM **1417**, 1990, 111-116.

[GP1] L. Gruson - C. Peskine: "Genre des courbes de l'espace projectif" in Algebraic geometry, Tromsø 1977, LNM **687**, 1978, 31-60.

[GP2] L. Gruson - C. Peskine: "Postulation des courbes gauches" in Open problems, Ravello 1982, LNM **997**, 1983, 218-227.

[Ha] G. Halphen: "Mémoire sur la classification des courbes gauches algébriques" in Oeuvres complètes t. III, 261-455.

[H1] R. Hartshorne: "On the classification of algebraic space curves" in Vector Bundles and differential equations, Nice 1979,Progr.Math.**7**,1980,83-112.

[H2] R. Hartshorne: "Stable reflexive sheaves III" Math. Ann. **279** (1988) 517-534.

[H3] R. Hartshorne: "Stable reflexive sheaves II" Invent.Math. **66** (1982) 165-190.

[HH] R. Hartshorne - A. Hirschowitz: "Nouvelles courbes de bon genre dans l'espace projectif" Math. Ann. **280** (1988) 353-367.

[Hi] A. Hirschowitz:"Existence de faisceaux réflexifs de rang deux sur $\mathbf{P}^3$ à bonne cohomologie" Publ.Math.IHES **66** (1987).

[L] O. A. Laudal: "A generalized trisecant lemma" in Algebraic geometry, Tromsø 1977, LNM **687**, 1978, 112-149.

[MR1] R. Maggioni - A. Ragusa: "The Hilbert function of generic plane section of curves in $\mathbf{P}^3$" Inv. Math. **91** (1988) 253-258.

[MR2] R. Maggioni - A. Ragusa: "Construction of smooth curves of $\mathbf{P}^3$ with assigned Hilbert function and generators' degrees" Le Matematiche **42** (1987) 195-210.

[PS] C. Peskine - L. Szpiro: "Liaison des varietés algébriques" Inv. Math. **26** (1974), 271-302.

[S1] R. Strano: "Sulle sezioni iperpiane delle curve" Rend. Sem. Mat. e Fis. Milano **57** (1987) 125-134.

[S2] R. Strano: "On generalized Laudal's lemma" à paraître in Proc. Trieste 1989.

[W] Walter: "Maximal rank curves in $\mathbf{P}^3$ with σ=s" Preprint.

Adresses des auteurs:

Philippe ELLIA
CNRS L.A. 168
Département de Mathématiques
Université de Nice
Parc Valrose 06034 - Nice Cedex (France)

Rosario STRANO
Dipartimento di Matematica
Università di Catania
Viale A.Doria, 6
95125 - Catania (Italia)

A tribute to Corrado Segre

Franco Ghione and Giorgio Ottaviani

In this work we review some papers by Corrado Segre published during the eighties of the XIX century, when he was just above twenty[1]. We believe that doing so may be interesting from the historical point of view as well as helpful to recognizing a link between methods of research used in those years (scarcely present in contemporary literature) and a number of results rediscovered (often without knowing it) in the current century. We thus try to reconstruct the origin of the path that has led to the modern theory of vector bundles on an algebraic curve.

§1. Corrado Segre's programme

To appreciate the innovative character of the ideas put forward by the very young Segre, it is convenient to recall that in those years it was harshly debated upon the usefulness of studying Hyperspace Geometry. Some authors mantained that addressing the geometry of the hyperspaces was an unfruitful intellectual game not certainly helpful to understand the "real" geometry in two or three dimensions. On the other side, Veronese and Bertini at first, and then C.Segre were perfectly aware that not only the study of the geometry of hyperspaces would shed new light on the geometry of curves and surfaces of ordinary space, but also that these latter could be viewed - and this is certainly innovative - as points (defined by a number of parameters) belonging to new algebraic varieties that could not be placed in ordinary space.

The first and probably most inspiring result in this trend of ideas is due to Veronese([V] p.208):

Every rational curve of degree n is a projection of a unique curve C_n of the n-dimensional space $\mathbb{P}^n$ whose (affine) coordinate functions, in terms of a parameter t, are simply:

$$C_n: \begin{cases} x_i = t^i \\ i=1,2,\ldots,n \end{cases}$$

Thus, for example, if we consider the rational cubics in the plane, we see that each of so many different patterns is but a different "shadow" of the twisted[2] cubic in $\mathbb{P}^3$. Having in mind Plato's myth of the cavern, we could think of the plane or the three-space as the wall of the cavern on which the shadows are cast of objects "living" in hyperspaces.

This result can easily be extended by Veronese:

[1]C.Segre was born in 1863

[2]"twisted" means "not contained in a hyperplane".

Every rational curve C of degree n, twisted in $\mathbf{P}^r$, *is a projection of the curve* $C_n \subset \mathbf{P}^n$.

In particular it is $r \leq n$. Veronese called C_n the *normal model* of C. The fact that every projective variety $X \subset \mathbf{P}^r$ of degree d admits a normal model (now called linearly normal), that is the fact that X is a projection of a well-defined variety $Y \subset \mathbf{P}^N$ of degree d not obtainable as a projection of a variety of the same degree placed in a higher dimensional space, was altogether clear to Segre, as it was clear the importance of calculating this maximum dimension N. This fact leads to the problem of calculating the dimension of $H^0(X,\mathcal{O}(1))$, i.e. to the Riemann-Roch problem.

Once he has determined the number N, Segre sets himself the task of finding all the normal models of the varieties of a given type up to projective transformations of the space $\mathbf{P}^n$. After this second step has been carried out, he can conclude that every given variety is the shadow of some of these normal models. The properties of the variety can be inferred by the geometry of the normal model and the way it is projected into $\mathbf{P}^r$.

§2. Rational ruled surfaces

Segre tries out at once (1884) his difficult programme in the case of the rational ruled surfaces, where he obtains a complete result [S1]:

Every rational ruled surface, not a cone, of degree n, twisted in $\mathbf{P}^3$ *(or* $\mathbf{P}^r$*) has, as its normal model, one of the surfaces* $F_m \subset \mathbf{P}^{n+1}$, $m=1,2,\ldots,[\frac{n}{2}]$, *where* F_m *is defined by the equations:*

$$F_m: \begin{cases} x_i=t^i & x_{m+1+j}=ut^j \\ i=1,2,\ldots,m & j=0,1,\ldots,n-m \end{cases}$$

The surfaces F_m, now called Hirzebruch surfaces, can be projectively characterized as the surfaces made of the lines joining two corresponding curves C_n and C_{n-m}, placed in two spaces $\mathbf{P}^n$ and $\mathbf{P}^{n-m}$, which in turn are skewly embedded in $\mathbf{P}^{n+1}$.

We now wish to sketch the proof that Segre gives of the above result, which also contains a proof, for the rank two case, of the celebrated theorem - now known as *Grothendieck's theorem*[3] [Gr] - according to which every algebraic vector bundle over $\mathbf{P}^1$ splits as a direct sum of line bundles.

Let $S \subset \mathbf{P}^r$ be a rational ruled surface of degree n. Then its hyperplane section $C=S \cap H$ will be a rational curve of degree n, twisted in H. Otherwise one could find a hyperplane H' containing C and a point $p \in S\setminus C$, in which case H' would contain the fiber r through p (since it would contain p and $x=C \cap r$). $H' \cap S$ would then have degree

[3]this theorem has a long history (see e.g.[OSS]).In an algebraic form it had also been proved by Dedekind and Weber [DW].

at least n+1, and thus $H' \supset S$, a contradiction. From Veronese's theorem it then follows $n \geq r\text{-}1$. The opposite inequality, i.e. the fact that every rational ruled surface in $\mathbf{P}^r$ of degree n ($r < n+1$) can always be obtained as a projection of a similar ruled surface in $\mathbf{P}^{n+1}$, is thought by Segre to be evident. In our opinion the assertion should have been given some justification. In fact Bertini in his comprehensive book about Hyperspace Geometry[Ber] gives of this assertion a rather complicated proof (which is actually a proof by Segre himself relative to the case of ruled surfaces on an elliptic curve adapted to the rational case (see Appendix I).

In any event, Segre reduces himself to classifying the rational ruled surfaces S, not cones, of degree n in $\mathbf{P}^{n+1}$(that otherwise wolud be cones over a C_n). Now, if C is an irreducible unisecant curve of degree $m \leq n$, then C is normal, i.e. it generates a $\mathbf{P}^m$. Indeed, if it were in a space $\mathbf{P}^{\mu}$ with $\mu < m$ *"siccome tutte le generatrici dovrebbero tagliare quella curva, si potrebbe per lo spazio stesso e per n−μ punti della superficie posti fuori di esso e su generatrici diverse far passare un iperpiano il quale conterrebbe le n−μ generatrici passanti per quei punti ed inoltre la curva di ordine m e quindi taglierebbe la superficie in una curva composta di ordine n+m−μ > n il che non può essere se quell'iperpiano non contiene tutta la superficie"*[4].

Once this has been established, Segre considers the unisecant curve C_m of least possible degree m contained in S. Since there always exists a hyperplane containing $[\frac{n+1}{2}]$ generatrices and S is not a cone, such hyperplane will also cut S along a unisecant curve of degree $\leq n-[\frac{n+1}{2}]$. *It follows* $m \leq [\frac{n}{2}]$. He then considers a hyperplane H passing through m distinct generatrices of S and n−2m+1 other points located on generatrices different from the previous ones. It is easy to see that, being $m \leq [\frac{n}{2}]$, such a hyperplane actually exists and moreover *does not contain generatrices other than the m already given.* Otherwise H would cut the curve C_m in at least m+1 points and would thus contain the whole C_m. But then, H would also contain the generatrices through the additional n−2m+1 points. It follows that H would contain C_m and m+(n−2m+1)=n−m+1 generatrices and hence would intersect S in a curve of degree n+1, contrary to S being twisted in $\mathbf{P}^{n+1}$. Thus H cuts S along the m generatrices as well aas along a further irreducible curve C_{n-m} of degree n−m. From this one can easily conclude that $C_m \cap C_{n-m} = \emptyset$, hence *S is projectively equivalent to a* F_m.

The fact that the above theorem is equivalent to the splitting theorem of vector bundles is immediately clear if we set up a "dictionary" that allows us to translate the

[4] [S1] p.267, *"since all the generatrices should cut that curve, we could have a hyperplane containing that same space and n−μ points of the surface placed outside it, and on different generatrices. Such hyperplane would contain the n−μ generatrices passing through those points and, moreover, the curve of degree m; hence it would cut the surface along a reducible curve of degree n+m−μ > n, which is impossible because the hyperplanes does not contain the whole surface."*

language of projective geometry into that of vector bundles.

§3. A useful dictionary

A ruled variety $S \subset \mathbf{P}^n$ with fibers $\mathbf{P}^s$ of genus g will be thought of as being defined by a morphism from a curve X of genus g into the grassmannian of the subspaces of $\mathbf{P}^n$ of dimension s:

$$\phi: X \to Gr(\mathbf{P}^s, \mathbf{P}^n)$$

such that ϕ is birational onto its image and $S = \bigcup_{x \in X} S_x$ where $S_x = \phi(x)$ is a vector subspace of $\mathbf{P}^n$ of dimension s which we shall call the *fiber* over x (or, with classic terminology, the *generatrix* over x).

We can canonically associate to S a holomorphic (algebraic) bundle over X of rank s+1, given by the preimage under ϕ, of the tautological bundle on the grassmannian. More specifically, such bundle, denoted by E_S, is given as a point set by

$$E_S = \{(x,y) \mid y \in E_x\} \subset X x C^{n+1}$$

where, after a choice of coordinates, we have set $\mathbf{P}^n = \mathbf{P}(C^{n+1})$ e $S_x = \mathbf{P}(E_x)$, E_x being a vector subspace of C^{n+1} of dimension s+1.

Moreover, a hyperplane $\mathbf{P}^{n-1} = \mathbf{P}(H)$ of $\mathbf{P}^n$ defines (noncanonically) a holomorphic section of E_S^* and (canonically) an element of $\mathbf{P}(\Gamma(X, E_S^*))$. Indeed, fixing a basis in C^{n+1}/H which is of dimension one, determines the projection $p: C^{n+1} \to C^{n+1}/H \simeq C$ and hence the global section $s_H: X \to E_S^*$ defined by

$$s_H(x)(y) = p(y) \quad x \in X, \ y \in E_x \subset C^{n+1}$$

that is zero at all points $x \in X$ such that $E_x \subset H$. It is also immediately seen that such sections generate the fibers at all points. Of course, if we change the basis of C^{n+1}/H, the section will be multiplied by a non-zero constant, and we will have a natural inclusion

$$\mathbf{P}^{n*} \to \mathbf{P}(\Gamma(X, E_S^*)) \tag{1}$$

It is now easy to verify that if S is a projection of S′ from a point outside S′, then E_S will be isomorphic to $E_{S'}$ and moreover if E^* is a bundle generated by its sections, i.e. if there is a morphism of bundles

$$X x C^{n+1} \to E^*$$

which is surjective on each fiber, then $\bigcup_{x \in X} \mathbf{P}(E_x) \subset \mathbf{P}(C^{n+1})$ defines a ruled variety S whose associated bundle is E. Thus S is linearly normal if and only if tha map (1) is an isomorphism and, moreover, denoted by N the dimension of the space where the normal model "lives", $N = \dim H^0(X, E_S^*) - 1$. We note also that the sections s_H allow to compute the *Chern classes* of E_S^*; indeed the locus of points where s+1 generic sections of E_S^*

become dependent dually corresponds to the set of fibers S_x that meet the space $\mathbf{P}^{n-s-1}$, intersection of the s+1 hyperplanes corresponding to these sections, and thus

$$\deg S = \deg E_S^* \tag{2}$$

Ruled subvarieties of S will then correspond to fiber subbundles of E_S (and viceversa) whose degrees will still verify (2) and, in particular, ***unisecant curves*** will correspond to line subbundles of E, generic hyperplane sections - not containing whole fibers - will correspond to subbundles of E with trivial quotient, and finally, cones in $\mathbf{P}^n$ having a $\mathbf{P}^k$ as vertex and a ruled variety S in $\mathbf{P}^{n-k-1}$ as basis will correspond to bundles of the form $1^{k+1} \oplus E_S$ (denoting by 1=XxC the trivial bundle). Two ruled subvarieties S_1 and S_2 added in $\mathbf{C}^{n+1}$ fiber by fiber will give rise to a map

$$E_{S_1} \oplus E_{S_2} \to E_S \tag{3}$$

which will be injective (as a map between sheaves) if the fibers S_{1_x} and S_{2_x} do not meet genericallyu, whereas it will be an isomorphism if S_{1_x} and S_{2_x} generate S_x for every x and $S_1 \cap S_2 = \emptyset$.

In particular, if S is a ruled surface, C_1 and C_2 distinct unisecant of S and L_1, L_2 the associated line bundles, then

$$0 \to L_1 \oplus L_2 \to E_S \to T \to 0$$

where we have denoted by T a torsion sheaf with support in $C_1 \cap C_2$. We thus find, upon computing the Chern classes, the simple intersection formula (already known to Segre):

$$\deg C_1 \cap C_2 = (\deg C_1 + \deg C_2) - \deg S \tag{4}$$

We finally observe that if we project $S \subset \mathbf{P}^n$ in $S' \subset \mathbf{P}^{n-1}$ from a point $p \in S_{x_0}$, the projection E_x in E'_x will be an isomorphism for $x \neq x_0$ having in x_0 cokernel of dimension one. We thus obtain the exact sequence (of sheaves):

$$0 \to E_S \to E_{S'} \to \mathcal{O}_{x_0} \to 0 \tag{5}$$

where $\mathcal{O}_{x_0}$ is the skyscraper sheaf in x_0 with fiber C. Moreover, if we perform a projection with center a fiber S_{x_0} of S we obtain

$$E_{S'} \simeq E_S(x_0)$$

The transformation (5) is called by many authors (e.g.. [M],[MN],[T]) an *elementary transformation.*

§4. Bundles on $\mathbf{P}^1$ of higher rank

In a paper of 1885-86[S2], Segre considers the case of bundles of $\mathbf{P}^1$ of rank i+1, with special attention to the case i=2. He obtains the Riemann-Roch formula, which can be stated by saying that if E is generated by its global sections and has degree n and rank i+1, then

$$h^0(\mathbf{P}^1, E) - 1 = n + i$$

Using the same methods and arguments given for the i=1 case, he studies, when i=2, the rational ruled surfaces and the unisecant curves of minimal degree contained in the surface S being able to prove that there always three curves of degrees m_1,m_2,m_3 generating the plane S_x for every x. From this we deduce that

$$E=L_1 \oplus L_2 \oplus L_3$$

The case of arbitrary rank, which was to be addressed in the dissertation of his disciple A.Bellatalla[Bel], is summarized thus:: "*i ragionamenti qui fatti pel caso i=2 si estenderanno facilmente ad i qualunque e l'analogia permetterà di prevederne senz'altro i risultati, sicchè non ne farò più oggetto di un nuovo lavoro*"[5].

§5. Rank two bundles on an elliptic curve

The methods used in the rational case still apply, with little change, to the elliptic case, essentially because if $X \subset \mathbf{P}^n$is an elliptic curve, then X is non-special. It is easy to see that if $S \subset \mathbf{P}^N$ is a twisted elliptic ruled surface of degree n, then its hyperplane section C will also be twisted in $\mathbf{P}^{N-1}$ and thus, being non-special it will result $n-1 \geq N-1$. On the other hand, if S is not a cone, it cannot be n=N; otherwise a generic hyperplane H of $\mathbf{P}^N$ containing a generatrix would cut S along a further unisecant elliptic curve C_{n-1} of degree n−1, which would generate a space L of dimension at most n−2. Now the linear system of hyperplanes through L would cut S along the fixed C_{n-1} and a variable line so that S would turn out to be a rational ruled surface, against the assumption. It is thus $N \leq n-1$. Also, Segre succeeds in constructing explicitly (see Appendix I) starting with S in $\mathbf{P}^N$ (if $N < n-1$) another elliptic ruled surface $S' \subset \mathbf{P}^{n-1}$ of degree n projecting itself onto S.

Having solved in this way the Riemann-Roch problem, Segre begins to classify the elliptic ruled surfaces of degree n in $\mathbf{P}^{n-1}$, not cones, investigating first the possible degrees of the unisecants, and then, from these possible numerical invariants, obtaining explicit examples of elliptic ruled surfaces with the given numerical invariants.

It is worth noting that this analysis leads to a new aspect (not present in the rational case): the existence of *indecomposable* ruled surfaces, that is, such that every two distinct unisecant curves have nonempty intersection. In other words, we believe for the first time, *indecomposable bundles of rank two make their appearance*. In order to construct the easiest (but easily extendible) such example let us consider a smooth plane cubic Γ on which we set an inflection point O, zero element of Γ thought of as an algebraic group. Let $A \in \Gamma, A \neq O$ be a point and $\tau_A:\Gamma \to \Gamma$ la *translation* taking a point P to the point $P \oplus A$ (where $\oplus$ denotes addition in Γ). Embedding two copies of Γ in $\mathbf{P}^5$ in such a way as to make them skew, we can construct a ruled surface $S_A \subset \mathbf{P}^5$ by joining

[5][S2], "*The argument given for the case i=2 will easily extend to the case of arbitrary i, where, by analogy, we can anticipate the results. Hence I shall not address this case in future works*".

with a line the images in $\mathbf{P}^5$of every pair of points P and $\tau(P)$. It is easy to see that the bundle E_{S_A} associated to such surface splits, and in addition

$$E_{S_A}=\mathcal{O}(-3O)\oplus\mathcal{O}(-3A)$$

Moreover the surface S_A contains only two cubic unisecants and ∞^2 quartic unisecants forming an algebraic system of degree 2. In other words two generic curves of the system meet transversally at two points. Projecting now in $\mathbf{P}^4$ the surface S_A from a point P_O chosen on the generatrix through O, outside the two planes, we obtain a surface $S'_A\subset\mathbf{P}^4$ of degree 5 containing an algebraic system of ∞^1 unisecants of minimal degree 3 obtained by projecting the quartics of S_A passing through P_O. The surface S'_A does not contain conics, since otherwise these would be projections of either a conic of S_A (there is no such a conic) or a cubic of S_A passing through P_O (there is no such a cubic by the choice of P_O). The surface S'_A , being smooth and normal, is thus indecomposable since otherwise if C_1 and c_2 were unisecants of degree n_1 and n_2 in S'_A , with $C_1\cap C_2=\emptyset$, then by (4) $n_1+n_2=5$ and hence $n_1\leq 2$ or $n_2\leq 2$.

It is also easy to see that once the points O,P_0 and the embeddings of the two planes in $\mathbf{P}^5$ are fixed, the surfaces S'_A , obtained by different values of A, are not projectively equivalent in $\mathbf{P}^4$. On the other hand, we can choose coordinates in order that we can set arbitrarily on a given surface the points O,P_0 and the two planes. In other words there is, in $\mathbf{P}^4$, a family S'_A $(A\in\Gamma\backslash\{0\})$ of indecomposable ruled surfaces of degree 5. For A=O the above construction "degenerates" in a decomposable surface generated by a plane cubic and a double line. However, making a different choice for the zero element of Γ, we obtain that the indecomposable surfaces are parametrized by the points of Γ, and any other surface is projectively equivalent to one of those. Therefore, counting the moduli, we see that all corresponding bundles can be obtained from a fixed one by tensoring by a line bundle of degree zero.

Translating this into modern language the indecomposable bundle $E^*_A:=E_{S'_A}$ is given by the sequence

$$0\to\mathcal{O}(-3O)\oplus\mathcal{O}(-3A)\to E^*_A\to\mathcal{O}_O\to 0$$

from which we derive, with easy calculations, the extension

$$0\to\mathcal{O}\to E_A(-2O)\to\mathcal{O}(3A-2O)\to 0$$

Hence we see that the indecomposable bundles of rank 2 and degree 1 can all be obtained as extensions of the type

$$0\to\mathcal{O}\to ?\to\mathcal{O}(P)\to 0$$

for some $P\in C$. It is now easy to find out (see [Ha] p.377) can all be gotten from the particular extension with P=O by tensoring by a suitable line bundle. This is in accordance with the corollary on p.434 of [A], where the case of rank > 2 is also treated.

§6. The ruled surfaces of arbitrary genus

The Riemann-Roch problem for the vector bundles of arbitrary rank on a curve of theorem on the base curve X. (A modern account of this method is given in [Gh]).

The computation of the index 2^p is based, instead, on a counting formula of Castelnuovo ([C]) whose correctness Segre himself doubted: "*La démonstration ingénieuse, que ce géomètre y donne de cette importante formule, pourrait laisser sur sa validité absolue des doutes, qui se réfléchiraient sur le n° présent et plus loin sur les n°s 20 et 21 de ces Recherches; cependant les confirmations qu'on trove de ces résultats me portent à penser qu'ils sont absolument vrais*"[6].

It is surprising that these results have remained unknown for nearly a hundred years until - starting in 1950 - various authors, such as Gunning, Nagata, Maruyama, Atiyah and others, have rediscovered, without being aware of, Segre's results (or, rather, some of them, not even the deepest ones). Thus, for example, (7) implies that, on a ruled surface S of genus p, the minimal self-intersection of one of its unisecants C_0 is never greater than p:

$$C_0^2 \leq p$$

as it is easily seen by noting that the self-intersection of a unisecant curve C of degree m on the surface S of degree n is given by

$$C^2 = 2m - n.$$

This is the form in which Nagata[N] gives this theorem without quoting Segre's paper. In the language of vector bundles, an equivalent formulation of the above fact can be given by saying that if E is a bundle of rank 2 and degree n on a curve X of genus p and L_0 is the line subbundle of E of maximal degree, then

$$\deg L_0 \geq \left[\frac{n-p+1}{2}\right].$$

This fact is an immediate consequence of (7) if E is generated by its global sections, case to which we can however reduce ourselves by tensoring by a suitable line bundle. It is in this form that we find Segre's results in Gunning[G] and others, among whom Stuhler[St], Lange-Narasimhan[LN], Lange[L1], at the end of the seventies. A study of the family of unisecants $\mathfrak{C}_m$, together with a critical revision of Segre's results and a clarification of the assumptions of generality that he makes, can be found in Ghione[Gh], where the number 2^p is also computed (see also [GhLa]) without using Castelnuovo's counting formula. This number has also been computed by Hirschowitz[Hi] in 1984 and by Lange[L2] in 1985. We also wish to note that the general ruled surfaces, in the sense of Segre, give rise to vector bundles having the property that the subbundle of maximal degree has degree $\left[\frac{n-p+1}{2}\right]$, therefore being particular instances of *stable bundles in the*

[6][S6], *footnote n.16, "The ingenious proof that this geometer[*Castelnuovo*] has given of such important formula, can leave some doubts about its validity. These will be reflected on the results of the present section and of §§ 20-21 of this paper. However, the confirmations that one has found of these results lead me to think that they are absolutetely true".*

sense of Mumford.

A bundle E for which the inequality

$$\deg L \leq \frac{\deg E-p+1}{2}$$

holds for every line subbundle $L \subset E$ is called ***strongly stable*** [Hi],[T]. The fact that the generic bundle is strongly stable and an extended version to the case of arbitrary rank have been profitably used by some authors.

We remark also that the family of unisecants $\mathfrak{C}_m$ of the surface S corresponds, using our dictionary, to (Grothendieck) scheme of the quotients of E_S^* of rank 1 and degree m. Of these schemes Segre had in a sense given the dimension d_m and - for $d_m=0$ - the number of points.

To conclude we would like to observe that Segre's interest for the theory of ruled varieties was also motivated by the study of algebraic curves from a projective point of view. Indeed, a linear series g_1^k on a curve $X \subset \mathbf{P}^n$ defines a rational ruled variety, made of the linear spaces generated by the divisors of the series, containing X as k-secant. Conversely, the existence of such variety S guarantees the existence of the g_1^k on X. Starting from this observation E.Mezzetti and G.Sacchiero ([MS]) could study, in recent times, those components of the Hilbert scheme of the curves of $\mathbf{P}^n$ arising from the consideration of the k-gonal curves.

In view of the previous considerations we would like to reflect on how the irregular flow of the mathematical (and cultural) fashions, with its whirlpools, can bury, for decades, entire theories, outstanding results and methods, to see them come up again out of necessity but without apparent link with their origin.

Appendix I: Riemann-Roch for rational and elliptic ruled surfaces

In this appendix we want to give the proof of Riemann-Roch theorem for an elliptic ruled surface following Segre. Really the construction is relative to the rational case such as bertini([Ber] pag.356) proposes it, but it is essentially the same(with minor changes) as the proof of Segre in the elliptic case[S3].

We need the following lemma: *Let* $C_n \subset \mathbf{P}^n$ *be the Veronese curve, set a point* $O \notin C_n$ *and let* Γ *be the cone over* C_n *with vertex O. If* $P_1,\ldots,P_{n+1}$ *are independent points of* Γ*, there exists a rational normal curve contained in* Γ *which contains them.* Indeed, let H be the hyperplane generated by $P_1,\ldots,P_n$ and set $C:=\Gamma \cap H$. WE embed the ambient space in $\mathbf{P}^{n+1}$ in such a away that H is contained in a hyperplane H' of $\mathbf{P}^{n+1}$ not containing O but containing the normal model C' of C. C is the projection of C' from a point $O' \in H'$. Let Γ' be the cone over C' with vertex O. Γ' projects itself from O' onto Γ

and the points $P_1,\ldots,P_{n+1}$ are the projections of the independent points $P_1',\ldots P_{n+1}'$ of Γ' which generate the hyperplane H'' of $\mathbf{P}^{n+1}$. The projection of the rational curve $H''\cap\Gamma'$ of degree n is the curve that we looked for.

Now, let S be a rational ruled surface of degree n generating a space $\mathbf{P}^{n+1-k}$. We want to explicitly construct a new rational ruled surface of degree n generating a space $\mathbf{P}^{n+1}$ and projecting itself onto S. For the convenience of the reader we join a picture (fig. 1) in which the indexes denote the sequence of the various steps.

1. In the space $H_0=\mathbf{P}^{n+1-k}\supset S$ we choose a generic hyperplane $H_1=\mathbf{P}^{n-k}$ such that $C_1:=H_1\cap S$ is an irreducible, rational curve of degree n.
2. We can obtain C_1 projecting a rational normal curve $V_2\subset\mathbf{P}^n=:H_2$ from a space $O_2=\mathbf{P}^{k-1}$. Let $K_2=\mathbf{P}^{n+1}$ be the space generated by H_0 and O_2, and let Γ_2 be the projecting cone.
3. An hyperplane $H_3=\mathbf{P}^{n-1}$ such that $O_2\in H_3\subset H_2$ cuts V_2 in the independent points $P_1,P_2,\ldots,P_n$.
4. H_3 projects itself onto the hyperplane $H_4=\mathbf{P}^{n-k-1}$ of H_1 which cuts C_1 in the points $P_1',P_2',\ldots,P_n'$ obtained by projecting $P_1,P_2,\ldots P_n$
5. Let now $H_5=\mathbf{P}^{n-k}$ be an hyperplane of H_0 containing H_4, it will meet the surface in the curve C_5 which have in common with C_1 the points $P_1',P_2',\ldots,P_n'$.
6. Let $\Gamma_6\subset K_2$ be the cone over C_5 with vertex O_2, this cone of degree n generates a space $H_6=\mathbf{P}^n$ and contains the n independent points $P_1,\ldots,P_n$. We choose on Γ_6 another generic point P_{n+1}. By the previous lemma, there exists a rational curve V_6 of degree n containing $P_1,\ldots,P_{n+1}$ which projects itself from O_2 onto C_5.
7. Let A be a point on V_2 and A' its projection on C_1. Let B' be the corresponding point on C_5 by means of the generatrix of S through A', and B be the point which projects itself on B' from O_2. Joining A and B with a line in $K_2=\mathbf{P}^{n+1}$ we get a rational ruled surface of degree n which generates K_2 and projects itself onto S.

The construction works with minor changes in the case of genus 1([S3] n.4).

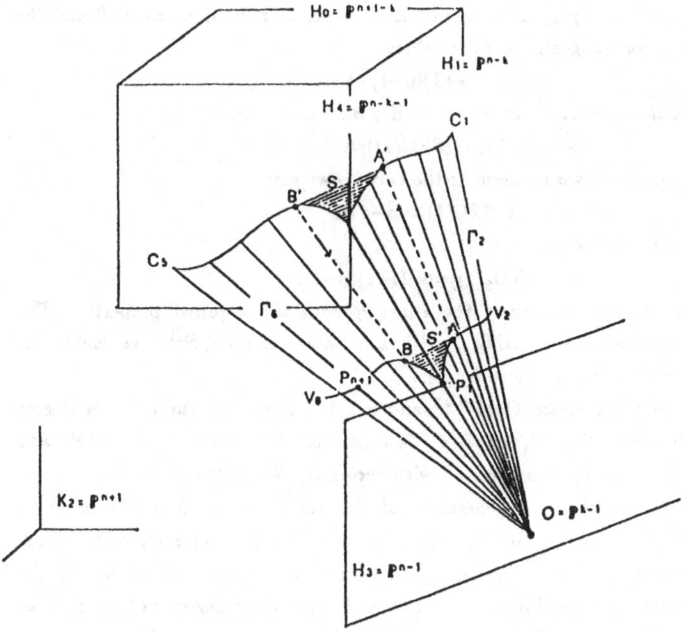

Fig. 1

Appendix II: Riemann-Roch for ruled varieties

The proof of Segre of the Riemann-Roch theorem for ruled varieties in projective spaces over a curve X of genus p is based on a counting formula that Schubert communicated to him in a letter. A modern proof with an analysis of the assumptions under which the result holds is in [GS].

Thus, let $S \subset \mathbf{P}^N$ be a variety of degree n ruled in projective spaces of dimension s on a curve X of genus p. Let $\gamma \subset S$ be a curve of degree m k-secant($k \geq s+1$), i.e. such that the morphism

$$\pi \mid_\gamma : \gamma \to X$$

is a k-fold covering of the base X. Now the genus of the curve γ can be computed in terms of these numerical invariants and of a further invariant z which measures essentially the number of divisors of degree s+1 contained in $\gamma_x = \gamma \cap S_x$ and which do not generate the fiber S_x. Thus, under suitable and natural assumptions, the genus g of γ is given by:

$$g = \frac{k-1}{s(s+1)}[(s+1)(m-s)-kn]+kp-\frac{z}{\binom{k-2}{s-1}}$$

We suppose now that there exists on S a curve γ (s+1)-secant such that for every

$x \in X$ the $s+1$ points belonging to $\gamma \cap S_x = \mathbf{P}^s$ are independent. In this case $z=0$ and the above formula gives for the genus of γ the value:

$$g = m + (s+1)(p-1) + 1 - n$$

It is easy to check that S is normal if and only if γ is normal, that is

$$\chi(\gamma, \mathcal{O}_\gamma(1)) = \chi(S, \mathcal{O}_S(1)) = \chi(X, E_S^*).$$

If we apply the Riemann-Roch theorem to the curve γ we get:

$$\chi(S, \mathcal{O}_S(1)) = m - g + 1$$

and substituting the value of g:

$$\chi(S, \mathcal{O}_S(1)) = n - (s+1)(p-1).$$

Then the key point is to construct the curve γ with the required properties. The technique of Segre consists in cutting S with a suitable cone $\Gamma \subset \mathbf{P}^N$. We make this construction in the following steps(see fig. 2):

1. We choose in $\mathbf{P}^N$ a space $O_1 = \mathbf{P}^{N-s-2}$ and a rational normal curve C_1 of degree $s+1$ in such a way that $O_1 \cap S = \emptyset$, C_1 is contained in a space $H_1 = \mathbf{P}^{s+1} \subset \mathbf{P}^N$ and $H_1 \cap O_1 = \emptyset$. Let Γ be the cone over C_1 with vertex O_1. We have dim $\Gamma = N - s$.
2. Let $S_x = \mathbf{P}^s$ ($x \in X$) be any generatrix of the ruled variety S and consider a space $H_2 = \mathbf{P}^{s+1}$ which contains S_x but does not meet the vertex O_1 of the cone. Such a space exists because $O_1 \cap S_x = \emptyset$ and then O_1 and S_x generate an hyperplane of $\mathbf{P}^N$. H_2 cuts the cone Γ in a rational normal curve C_2 of degree $s+1$. as S_x is an hyperplane of H_2, it cut C_2 in the $s+1$ independent pointsP_x, P'_x, P''_x. indipendenti. It follows that $\Gamma \cap S_x$ consists of $s+1$ independent points for every x and then $\gamma := \Gamma \cap S$ is the curve that we looked for.

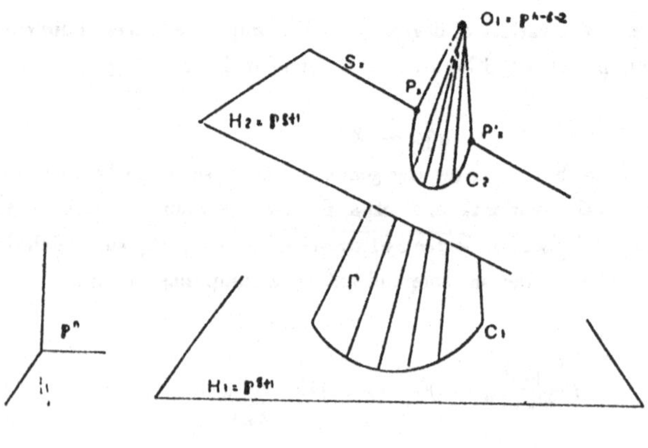

Fig. 2

References

[A] *M.F.Atiyah,* Vector bundles over an elliptic curve, Proc. London Math. Soc. (3)VII 27(1957), 414-452

[Bel] *A.Bellatalla,* Sulle varietà razionali normali composte di ∞^1 spazi lineari, Atti R. Accad. Torino, 36(1901), 803-833

[Ber] *E.Bertini,* Introduzione alla geometria proiettiva degli iperspazi, Messina 1923

[C] *G.Castelnuovo,* Una applicazione della geometria enumerativa alle curve algebriche, Rend. Circ. Mat. Palermo,III n.4(1889)

[DW] *J.W.R.Dedekind, Weber,* Theorie der algebraischen Funktionen einer Veränderlichen,Crelle J. 92(1882),181-290

[Gh] *F.Ghione,* Quelques résultats de Corrado Segre sur les surfaces réglées, Math. Ann. 255(1981), 77-95

[GhLa] *F.Ghione, A.Lascu,* Unisecant curves on a general ruled surface, Manuscr. Math. 26(1978), 169-177

[GS] *F.Ghione, G.Sacchiero,* Genre d'une courbe lisse traceé sur une varieté réglée, Lect. Notes Math 1260(1987), 97-107

[Gr] *A.Grothendieck,* Sur la classification de fibré holomorphe sur la sphère de Riemann, Amer. J. Math. 79(1957), 121-138

[Gu] *R.C.Gunning,* On the divisor order of vector bundles of rank 2 on a Riemann surface, Bull.Inst.Math.Acad.Sinica 6(1978), 295-302

[Ha] *R.Hartshorne,* Algebraic geometry, New York 1977

[Hi] *A.Hirschowitz,* Rank techniques and jump stratifications, 1984

[L1] *H.Lange,* Higher secant varieties of curves and the Theorem of Nagata on ruled surfaces, Manuscripta Math. 47(1984), 263-269

[L2] *H.Lange,* Höhere sekantvarietäten und Vektorbundel auf Kurven, Manuscr. Math. 52(1985), 63-80

[LN] *H.Lange, M.S.Narasimhan,* Maximal subbundles of rank two vector bundles on curves, Math. Ann. 266(1983), 55-72

[M] *M.Maruyama,* Elementary transformations in the theory of algebraic vector bundles, Lect. Notes Math. 961(1982), 241-266

[MN] *M.Maruyama, M.Nagata,* Note on the structure of a ruled surface, J.reine angew.Math., 239-240(1969), 68-73

[MS] *E.Mezzetti, G.Sacchiero,* Gonality and Hilbert schemes of smooth curves, Lect. Notes Math. 1389, 183-194, Berlin 1989

[N] *M.Nagata,* On self-intersection number of a section on a ruled surface, Nagoya Math. J. 37(1970), 191-196

[OSS] *C.Okonek. M.Schneider, H.Spindler,* Vector bundles on complex projective spaces, Progress in Math. 3(1980), Boston 1980

[S1]* *C.Segre,* Sulle rigate razionali in uno spazio lineare qualunque, Atti R. Accad. Torino, 19(1883-84), 265-282

[S2]* *C.Segre,* Sulle varietà normali a tre dimensioni composte di serie semplici razionali di piani, Atti R. Accad. Torino, 21(1885-86),95-115

[S3]* *C.Segre,* Ricerche sulle rigate ellittiche di qualunque ordine, Atti Accad. Tor., 21(1885-86), 628-651

[S4]* *C.Segre,* Recherches générales sur les courbes et les surfaces réglées algébriques I, Math. Ann. 30(1887), 203-226

[S5]* *C.Segre,* Sulle varietà algebriche composte di una serie sempicemente infinita di spazi, Atti Ac. Lincei, (IV),3(1887), 149-153

[S6]* *C.Segre,* Recherches générales sur les courbes et les surfaces réglées algébriques II, Math. Ann. 34(1889), 1-25

[St] *U.Stuhler,* Unterbündel maximalen Grades von Vektor-bündeln auf algebraischen Kurven, Manuscripta Math. 27(1979), 313-321

[T] *A.N.Tyurin,* The geometry of moduli of vector bundles, Russian Math.Surveys 29:6(1974), 57-88

[V] *G.Veronese,* Behandlung der projectivischen Verhältnisse der Räume von verschiedenen Dimensionen durch das Princip des Projicirens und Schneidens, Math. Ann. 19(1882), 161-234

The works of C.Segre marked in this list by a "" are reprinted in the volume "Opere scelte", Ed. Cremonese, Roma 1957.*

Franco Ghione and Giorgio Ottaviani
Dipartimento di Matematica
II Università di Roma, Tor Vergata, I-00133 Roma

UN APERÇU DES TRAVAUX MATHÉMATIQUES DE G.H. HALPHEN (1844-1889)

par Laurent Gruson

Introduction.

La notice qui suit a été rédigée pour le centenaire de la mort (le 21 mai 1889) du mathématicien Georges-Henri Halphen.

La réputation d'Halphen est surtout fondée aujourd'hui sur deux séries de mémoires de géométrie algébrique. Aux débuts de la géométrie énumérative (entre Chasles et Schubert) une première série propose une théorie exhaustive des "familles" de coniques du plan (complètement justifiée dans [2]); c'est le premier exemple d'étude énumérative des espaces homogènes symétriques, thème de recherches fort actif actuellement. Une seconde série aboutit au mémoire de 1882 sur la classification des courbes gauches algébriques (dont une première version fut écrite en 1869), d'une ampleur de vues exceptionnelle. Les questions que ce mémoire a soulevées sont encore loin d'être résolues (cf. [4]). L'essentiel de ces travaux semble antérieur à 1875.

De 1875 à 1885, Halphen suit une ligne de recherche très précise mais difficilement caractérisable, partant de l'étude énumérative des points d'une courbe algébrique plane qui sont solution d'une équation différentielle donnée (à coefficients fonctions rationnelles des coordonnées), aboutissant à une série de mémoires sur la classification des équations différentielles linéaires et la "théorie de Galois" de celles-ci. L'étude critique de ces travaux n'a pas été faite. Après 1885 Halphen se consacre à la rédaction de son livre [6], synthèse très élaborée de la théorie classique des fonctions elliptiques, malheureusement interrompue avant l'entrée en scène des applications algébriques et arithmétiques, mais gardant sans doute un intérêt actuel sur de nombreux points (cf. 3.6).

Dans ce "premier aperçu", j'ai laissé de côté les travaux dont il est rendu compte dans [2] et [4] ainsi que la partie "géométrie projective et équations différentielles" (sur laquelle j'espère revenir un jour); il reste une vue d'ensemble extrêmement schématique et quelques détails sur le fonctions elliptiques ...

Il me reste à remercier les organisateurs de la conférence de Trieste de leur hospitalité et à me souvenir des recherches entreprises avec C.Peskine en 1974-76, qui ont fondé notre intérêt commun pour les travaux d'Halphen.

1 On trouve dans [5] quatre notices, dues respectivement à Picard, Poincaré, Jordan et Brioschi, dont j'extrais les renseignements biographiques suivants. Halphen est né à Rouen le 30 octobre 1844; entré à l'Ecole polytechnique en 1862, il a suivi une carrière d'officier d'artillerie (et a participé comme tel à la guerre de 1870). De 1872 à 1886, il a enseigné à l'Ecole polytechnique comme répétiteur d'analyse (les professeurs étaient alors Hermite, jusqu'en 1876, puis Jordan); il a été également examinateur (du concours d'entrée) de 1884 à 1887.

En 1886 il revint au "service actif" et fut affecté comme commandant de batteries à Versailles: "c'était une lourde tâche qui venait s'ajouter à l'effort considérable que lui demandait en ce moment même la préparation de son Traité sur les fonctions elliptiques" (E.Picard) - il rassemblait alors le matériel nécessaire à l'exposé des aspects algébriques et arithmétiques de la théorie. Selon les notices citées, l'excès de travail aurait alors gravement compromis la santé d'Halphen: d'après Hermite ([8] I p. 448) sa mort serait due à une pneumonie, complication d'un rhumatisme articulaire dont il souffrait au printemps 1889.

2 Je hasarde quelques remarques sur l'environnement mathématique d'Halphen. Lorsqu'en 1872 Halphen devient un mathématicien professionel, il a déjà rédigé un mémoire important sur les courbes gauches algébriques, où les influences d'Hermite et de Cayley sont évidentes; ses méthodes et ses goûts sont affirmés. Son horizon s'élargit lorqu'il approfondit les travaux de l'école allemande de géométrie algébrique (sans doute ceux de Clebsch et de Noether en premier lieu) - la première allusion qu'il fasse aux notions de correspondance birationnelle et de genre date, sauf erreur, de 1874 ([5] I pp. 312-316) - et ceux de Jordan sur la théorie des groupes. On admet en général qu'Halphen a influencé les débuts de Schubert, mais en 1879 les deux hommes étaient en controverse sur la métaphysique de la géométrie énumérative, si l'on en croit les extraits des lettres d'Halphen à Zeuthen réunis dans ([5] IV pp. 629-644).

3 Je présente maintenant une description succincte de la chronologie des travaux d' Halphen; il convient auparavant de renvoyer au très bel exposé qu'il en a fait lui-même en 1886 ([5] I pp. 1-47, écrit à l'occasion de sa candidature (heureuse) à l'Académie des sciences) et à la liste thématique de ses articles ([5] IV pp. 645-653).

3.1 ***Géométrie réglée*** - On peut citer une note (classique) décrivant la théorie d'intersection de la variété des droites de l'espace ([5] I pp. 75-79, complété pp. 83-90, 1869), l'application au nombre de tétraèdres dont les arêtes appartiennent à des congruences de droites données ([5] I pp. 167-170, 1873), la description de la surface réglée (quartique) lieu d'une droite se déplaçant de manière que quatre de ses points restent dans des plans fixes ([5] I pp. 94-97 (1873) complété en II pp. 345-347), la détermination des lignes asymptotiques (algébriques) d'une surface réglée ayant deux directrices rectilignes ([5] II pp. 86-88, 1876).

3.2 ***Classification des courbes gauches algébriques*** - Halphen a rédigé en 1869 (à l'occasion d'un congé semestriel) la première version d'un mémoire sur ce sujet (résumée en [5] I pp. 80-82 (1870)), dont la version définitive a paru en 1882 ([5] III pp. 261-455); cf. [4]: L'analyse tentée dans [4] reste fort incomplète. Sans chercher à la mettre à jour ici, je hasarderai une remarque sur la ligne de démarcation entre les deux versions du mémoire (bien que je n'aie pas lu la première!): en 1869, la partie "interprétation géométrique de la théorie de l'élimination" (i.e. le chap. III de la version définitive) semble achevée; ensuite, le mouvement de réflexion d'Halphen semble centré sur l'idée de "schéma de Hilbert", et la formulation la plus précise de son aboutissement semble l'introduction du mémoire. Entre-temps on peut citer deux notes ([5] I pp. 91-93 et 203-207) sur le même sujet, ainsi que le mémoire "Recherches de géométrie à n dimensions" ([5] I pp. 171-192, 1873) qui relève du même ordre d'idées: application de la "construction monoidale" de Cayley à l'étude des sous-variétés de codimension > 1 de l'espace projectif de dimension n (ce mémoire contient notamment une tentative (incomplète) de démonstration du "théorème de Bezout" dans ce contexte).

3.3 ***Géométrie énumérative de la variété des coniques du plan*** - La première publication d'Halphen sur ce sujet date de 1873 ([5] I pp. 98-157); il l'a reprise, en en bouleversant les conclusions, en 1876-77 ([5] I pp. 543-545, 553-556, II pp. 1-85, 275-289); cf. [2].

3.4 ***Points singuliers des courbes algébriques planes*** - Ce sujet (déjà traité par Newton) était alors très étudié; Halphen y a contribué à partir de 1874. Il disposait du théorème de Noether (1871) ("réduction des singularités": en appliquant à une courbe plane une suite convenable de transformations "quadratiques" de son plan, on parvient à une courbe dont les seuls points singuliers sont "ordinaires": au voisinage de chacun d'eux, la courbe est réunion de "branches" lisses se coupant deux à deux transversalement) et de la définition d'invariants numériques (les "exposants caractéristiques") attachés à la donnée d'un germe irréductible de courbe analytique plane (ou "branche"), dont on sait aujourd'hui qu'ils séparent les composantes irréductibles de la variété qui paramètre ces germes (cf. [13] p. 16).

Halphen a donné trois exposés de synthèse sur ce sujet ([5] I pp. 216-311, 420-474, IV pp. 1-93) (la définition qu'il donne des exposants caractéristiques diffère légèrement de celle de [13]: elle est adaptée à l'étude de la dualité mais n'est pas invariante par déformation). Les deux derniers décrivent un procédé de "réduction des singularités" différent de celui de Noether: on introduit une transformation des courbes du plan qui généralise (de manière "non-euclidienne" selon le modèle de Cayley) le passage à la développée (dans le plan dual) - plus précisément, ayant fixé une conique C, on transforme un point M "doublé" selon une tangente L en le point d'intersection de L et de la polaire de M relative à C. Halphen montre que les transformées successives, d'ordre assez grand, d'une courbe arbitraire n'ont que des points singuliers ordinaires, et (surtout) que leurs degrés (donc aussi leurs classes) sont finalement en progression arithmétique - ce dernier résultat est extrêmement remarquable à mes yeux.

Un autre procédé de réduction des singularités est décrit dans [5] I pp. 358-361.

On trouve dans [5] II pp. 108-120 une étude des points singuliers des courbes gauches et surtout, pp. 154-196, une étude des courbes singulières d'une surface algébrique (1878) dont le but est d'élucider les singularités de la surface duale qui en provienent: dans cette étude Halphen fait intervenir les deux premiers exposants caractériques (en son sens) de chaque branche, centrée sur la ligne singulière, de la section plane générale de la surface; on retrouve cette idée dans son étude des familles de coniques, cf. [2].

3.5 ***Géométrie projective et équations différentielles*** - Dans cette rubrique je range les articles sur le nombre de points d'une courbe algébrique, plane ou gauche, où est vérifiée une équation différentielle (1876), la thèse de doctorat sur les invariants différentiels (1878), le mémoire sur la réduction des équations différentielles linéaires aux formes intégrables (1880) et divers travaux voisins.

3.6. ***Fonctions elliptiques*** - Ce thème fait son entrée en 1878, lorsque Halphen, dans sa thèse, utilise le mode elliptique de représentation paramétrique des cubiques planes pour définir le premier "invariant différentiel absolu" (d'ordre 7) des germes analytiques de courbes du plan.

L'article suivant ("Sur diverses formules récurrentes concernant les diviseurs des nombres entiers", [5] II pp. 256-271, 1878) contient une démonstration arithmétique élémentaire de la formule du triple produit de Jacobi (on trouve dans [6] I pp. 405-410 une autre démonstration originale de cette formule).

Peu après, quatre notes et deux articles ([5] II pp. 292-344, 1878-79) traitent du "problème de la division" (des périodes d'une intégrale elliptique de première espèce par un entier $n > 2$); cette étude est amplifiée dans [6] (surtout dans le tome II, chap. IX-XI et XIV, mais non dans le tome III); elle va faire l'objet d'une

Digression (3.6.1) - Soit $E = \mathbf{C}/\Omega$ une courbe elliptique. Le problème de la division dans E par un entier $n > 2$ est la recherche effective d'une fonction méromorphe sur E dont le diviseur des zéros soit l'ensemble des points d'ordre exact n, avec multiplicité 1. Traditionnellement, on introduit d'abord la fonction $u \longrightarrow \psi_n u = \sigma\, nu/\sigma^{n^2} u$ (où σ est la fonction sigma de Weierstrass relative au réseau Ω): elle est méromorphe sur E, homogène de degré n^2-1 en (u,Ω), et son diviseur de zéros est l'ensemble des points non nuls de E d'ordre divisant n; on trouve dans [12] (pp. 33-46) une étude algébrique de cette fonction lorsque E est la courbe d'équation $y^2 = x^3 + ax + b$, où a et b sont des indéterminées sur $\mathbf{Z}$. Dans [6] I, pp. 102 sq., Halphen ébauche une étude de ce type de l'expression $\gamma_n = \sigma_n . \sigma_2^{-(n^2-1)/3}$ (irrationnelle) qui est homogène de degré 0 en (u,Ω). On obtient des relations algébriques entre les fonctions γ_n (ou ψ_n) en utilisant "l'équation à trois termes" vérifiée par la fonction sigma:

(1) $\quad (ab).(cd)-(ac).(bd)+(ad).(bc) = 0$, où $(ab) = \sigma(a+b)\,\sigma(a-b)$, d'où:

(2) $\quad \gamma_{m+n}\,\gamma_{m-n}\,\gamma_{p+q}\,\gamma_{p-q} - \gamma_{m+p}\,\gamma_{m-p}\,\gamma_{n+q}\,\gamma_{n-q} + \gamma_{m+q}\,\gamma_{m-q}\,\gamma_{n+p}\,\gamma_{n-p} = 0$

Halphen pose alors $X = \gamma_3^3$ et $Y = \gamma_4$, et déduit de (2) que si n est un entier non multiple, resp. multiple, de 3, l'expression γ_n, resp. $X^{-1/3}\gamma_n$, appartient à $\mathbf{Z}[X,Y]$: ainsi $\gamma_5 = X-Y$, $\gamma_6 = X^{1/3}(Y-X-Y^2)$, $\gamma_7 = X(Y-X)-Y^3$, $\gamma_8 = Y((Y-X)(2X-Y)-XY^2)$. En fait, les fonctions X et Y sont déjà définies dans la thèse d'Halphen: soit $u \to M(u)$ une représentation elliptique propre d'une cubique lisse C du plan projectif $\mathbf{P}_2$, normalisée de sorte que M(0) soit un point d'inflexion de C; avec les notations de [5] II p. 237, on a $\eta(M(u)) = X(3u)$, $\xi(M(u)) = Y(3u)$, où η et ξ sont les deux premiers invariants différentiels absolus des germes de courbes analytiques de $\mathbf{P}_2$. D'un autre côté, les cubiques de $\mathbf{P}_2$ vérifient une équation différentielle d'ordre 9, invariante par automorphismes de $\mathbf{P}_2$, linéaire en les dérivées neuvièmes (c'est la restriction à $\mathbf{P}_2$, identifié, par le plongement de Veronese, à une surface de degré 9 de $\mathbf{P}_9$, du "wronskien", équation différentielle des courbes de $\mathbf{P}_9$ situées dans un hyperplan). Halphen écrit cette équation sous forme "autonome" en (η, ξ) ou (ce qui revient au même d'après ce que nous venons de voir) en (X, Y):

$$\text{(3)} \qquad \frac{dY}{dX} = \frac{3Y(Y+1)-4X}{X(8Y-1)}$$

([5] II pp. 242-247, voir aussi pp. 440-445).

Sachant que $X(u) = \wp'^{-2}2u.(\wp u-\wp 2u)^3$, $Y(u) = \wp'^{-1}u.\wp' 2u$, Halphen montre directement que ces fonction vérifient (3) ([5] II pp. 292-295) et obtient les identités

$$\frac{g_2}{12\wp^2 u} = \frac{18X^2+8X(Y+1)(Y-2)+(Y+1)^4}{(4X+(Y+1)^2)^2},$$

$$\frac{g_3}{8\wp^3 u} = \frac{64X^3+24X^2(2Y^2-2Y+5)+12X(Y+1)^3(Y-2)+(Y+1)^6}{(4X+(Y+1)^2)^3}$$

(et $j = -27\dfrac{(16X^2+8X(Y+1)(Y-2)+(Y+1)^4)^3}{(X(Y+1))^3((8Y-1)^3+(32X+8Y^3-20Y-1)^2)}$, sauf erreur) ([5] II pp. 296-299, modifié du fait qu'Halphen utilisait alors les notations de Jacobi). Il en résulte que $\mathbf{C}(X, Y)$ est le corps des fonctions méromorphes (y compris "à l'infini"), homogènes de degré 0, en (u, Ω), qui sont Ω-périodiques en u; et aussi que l'expression $\delta_n := \prod_{m/n} \gamma_{n/m}^{\mu(m)}$ appartient à $\mathbf{Z}[\frac{1}{6}][X,Y]$ si $n>3$ (μ est la fonction de Möbius) (ceci n'est pas explicitement mentionné par Halphen).

Ces calculs ont des applications géométriques, par exemple dans l'étude des covariants absolus des courbes de genre un munies d'un plongement "normal" de degré n dans $\mathbf{P}_{n-1}$ (sous le groupe d'automorphismes de ce dernier). Halphen s'est intéressé aux cas n = 2,3; je vais donner quelques détails sur l'exemple des "polygones de Poncelet" (n = 2) en suivant [6] II, chap.X, pp. 367-412.

Le début est célèbre: soient C_1, C_2 deux coniques de $\mathbf{P}_2$, Γ le lieu des couples $(M_1,M_2) \in C_1 \times C_2$ tels que la tangente en M_2 à C_2 passe par M_1: c'est une courbe de genre un; sa "jacobienne" (de degré 0) est une courbe elliptique C, non canoniquement isomorphe à Γ; elle est munie d'un point U, la différence des deux modules inversibles sur Γ provenant par image réciproque des modules inversibles de degré 1 (dits "tautologiques") sur C_1 et C_2. Jacobi a montré que si n est un entier ≥ 3, la condition $nU = 0$ est équivalente à l'existence d'un (ou, ce qui revient au même - c'est le "grand théorème de Poncelet" - d'une infinité de) polygone (s) à n côtés inscrits dans C_1 et circonscrits à C_2. Cayley a donné de cette condition une expression rationnelle en les "invariants fondamentaux" $(s_i)_{0 \leq i \leq 3}$ du couple de coniques (C_1, C_2) (ces derniers sont définis comme suit: on choisit une équation de C_i, $q_i(X) := {}^tXM_iX = 0$, où M_i est une matrice symétrique 3×3, et l'on pose $\text{dét}(tM_1+M_2) = s_0t^3+s_1t^2+s_2t+s_3 = P(t)$). On trouve une démonstration moderne de ce résultat dans [3].

La suite est moins connue. C'est d'abord la "formule de duplication" de Gundelfinger; je transcris Halphen en langage moderne. On identifie $\mathcal{O}_{C_1}(-1)$ et le module des différentielles ω_{C_1}: si $t \to X(t)$ est une représentation paramétrique locale de C_1, exprimée en coordonnées homogènes, on vérifie que le rapport des vecteurs (proportionnels) $dX(t) \wedge X(t)$ et $M_1X(t)$ définit une section de $\omega_{C_1}(1)$ indépendante des choix intermédiaires ([6] II p. 391). La section q_2 de $\mathcal{O}_{C_1}(2)$ définit alors le carré d'une forme différentielle irrationnelle α, singulière aux points de $C_1 \cap C_2$. D'autre part on introduit un covariant q de (C_1,C_2): l'équation, convenablement normalisée, de la conique passant par les huit points de contact des tangentes communes à C_1 et C_2. Soit t la fraction rationnelle q/s_0q_2, définie sur C_1. La formule de Gundelfinger s'écrit $\alpha=\pm dt/2\sqrt{P}(t)$; sa source est bien sûr la formule de duplication d'Hermite (1856, cf. [9] p. 16 et [6] II pp. 360-362), qui en est aussi le cas particulier $s_1 = 0$.

Soient maintenant $(m_i)_{1\leq i\leq 4}$ quatre points de $\mathbf{P}_2$ en position générale; Halphen leur associe une "représentation elliptique" de $\mathbf{P}_2$: en langage moderne, il introduit la désingularisée S du revêtement double de $\mathbf{P}_2$ ramifié le long des côtés du quadrilatère complet de sommets (m_i). Si M est un point général de S d'image m dans $\mathbf{P}_2$, on note C_M la conique de $\mathbf{P}_2$ passant par m et les m_i, Γ_M son image réciproque dans S, qui est une courbe de genre un ramifiée en les m_i au-dessus de C_M. Le choix d'un des m_i pour origine fait de Γ_M une courbe elliptique et permet de définir les fonctions méromorphes X, Y, γ_n, δ_n de la variable M, qui ne dépendent que de m i.e. sont définies sur $\mathbf{P}_2$. Halphen calcule $X(M)$ et $Y(M)$ ([6] II p. 407). Soit $D_{i,n}$ la courbe de $\mathbf{P}_2$ lieu des zéros de δ_n: avec les notations du problème de Poncelet pour le couple (C_M,C'_M) où C'_M est la conique passant par les m_j et tangente en m_i à m_im, la courbe Γ s'identifie à Γ_M et le point U à M, donc la condition $m \in D_{i,n}$ signifie que U est d'ordre exact n; le degré de $D_{i,n}$ est $\frac{1}{4}n^2 \prod_{p/n \text{ p premier}} (1-p^{-2})$. D'un autre côté, on voit aisément que les fonctions $D_{i,2n}$ ne dépendent pas du choix de i et sont invariantes sous l'action sur $\mathbf{P}_2$ du groupe S_4 de permutation des m_i. Utilisant la formule de duplication, Halphen donne l'expression des fonctions $X(2M)$ et $Y(2M)$ comme "combinants" du pinceau des coniques contenant les m_i, i.e. comme fonctions rationnelles sur $\mathbf{P}_2$ invariantes sous S_4 ([6] II p. 403). On voit de plus que si $n = 2m$ et m est impair, $D_{i,n}$ est réunion des $D_{j,m}$ $(j\neq i)$ tandis que si n est multiple de 4, $D_{i,n}$ est indépendant de i et réunion de trois courbes permutées entre elles par S_4.

L'article "Recherches sur les courbes planes du troisième degré" ([5] II pp. 319-344, 1878) et le chap. XI de [6] II décrivent la "représentation elliptique" du plan par un pinceau de cubiques de points d'inflexion fixés; le développement est proche de celui que je viens de résumer, le rôle de S_4 étant tenu par le "groupe de Hesse" des automorphismes du plan permutant les points d'inflexion. Je signale deux articles voisins, "Problème concernant les courbes planes du troisième degré" ([5] II pp. 450-466, 1880-81) (trouver une courbe de degré 3 et une courbe de classe 3 du plan, dont l'intersection sur chacune soit le double d'un diviseur), "Sur les courbes du sixième degré à neuf points doubles" ([5] II pp. 547-557, 1881-82) (conditions que doivent vérifier neuf points d'un plan pour qu'il existe une infinité de courbes de degré 3n ayant multiplicité n en chacun d'eux: ces courbes forment alors un "pinceau d'Halphen").

Le chap. XIV de [6] II étudie le développement en fraction continue, dit "régulier",

$$(4) \qquad \frac{\sqrt{P(x)}-\sqrt{P(y)}}{x-y} = A_0 + \frac{B_1h^2}{A_1} + \frac{B_2h^2}{A_2}$$

où x est une variable voisine de x_0, y est un paramètre, P est un polynôme de degré 4 sans facteur carré; rappelons que le second membre est une expression formelle (prolongée indéfiniment) dans laquelle h est l'accroissement $x - x_0$, les A_i sont linéaires en x et les B_i sont indépendants de x, qui est caractérisée par la propriété que pour tout m, la "réduite" R_m, fraction rationnelle déduite de la fraction continue en supprimant les termes d'indice > m, diffère de $\frac{\sqrt{P(x)} - \sqrt{P(y)}}{x-y}$ de $0(h^{2m+2})$ quand h tend vers 0.

Halphen introduit l'argument elliptique $u = \int_x^\infty P^{-1/2}(x)\,dx$, défini modulo le réseau Ω des périodes de cette intégrale, et regarde x et $x' = P(x)$ comme fonctions de u ; il note a l'argument initial $(x(a) = x_0,\ x'(a) = \sqrt{P(x_0)}\ )$, v l'argument tel que $x(v/2)$ soit une racine de P pour chaque détermination de v/2 mod Ω, c_1 (noté aussi -v-w) l'argument tel que $x(c_1) = y$, $x'(c_1) = \sqrt{P(y)}$, $c_m = c_1 + (m-1)(2a+v)$; $y_m = x(c_m)$. Alors le développement en fraction continue de $\frac{\sqrt{P(x)}-x'(c_m)}{x-y_m}$ s'écrit $A + \frac{KB_mh^2}{A_m} + \frac{KB_{m+1}h^2}{A_{m+1}} + \ldots$(avec les notations de (4), A étant linéaire en x et K indépendant de x). Le développement régulier (4) est interrompu lorsque $B_m = 0$, ce qui a lieu si et seulement si $a = c_{m+2}$; on doit alors le modifier, par exemple le cas $m = 0$ conduit au développement $\sqrt{P(x)} = (q_0 + q_1 h + q_2 h^2) + \frac{q_3h^3}{1-\varphi_2h} + \frac{\theta_3h^3}{1-\varphi_3h^2} + \ldots$ (dont le début est atypique). Ce dernier développement a été étudié par Jacobi, qui a donné des expressions elliptiques de φ_i et θ_i, reprises par Halphen (loc. cit. pp. 607-608, cf. aussi Préambule, p. 576).

Pour étudier la convergence de la fraction continue (4), Halphen part de l'expression $R_m = \frac{f_m\sqrt{P(x)}-\sqrt{P(y)}}{x-y}$ de la m -ème réduite, où $f_m = (1+\lambda h_m)/(1-\lambda h_m)$ avec λ indépendant de m et $h_m = \sigma^{2m-1}(a-u)\,\sigma(c_{m+1}+u+v)/\sigma^{2m-1}(a+u+v)\,\sigma(c_{m+1}-u)$ (p.615). En étudiant le comportement de h_m pour $m \to +\infty$, il montre que si P admet quatre racines réelles et x_0 est réel, la fraction continue (4) converge, en tout point de la sphère $\mathbf{C} \cup \{\infty\}$ privée des deux segments de droite joignant les racines de P et contigus à celui qui contient x_0, vers l'unique détermination de $\frac{\sqrt{P(x)}-\sqrt{P(y)}}{x-y}$ prenant la valeur proposée en x_0. La convergence est uniforme en tout point n'appartenant pas à une composante spécifiée de la courbe de représentation paramétrique $t \to \wp(c_1+t)$, $t \in \mathbf{R}$. Par contre, la fraction continue (4) diverge quel que soit x si l'on a $2a+v = b_1\omega_1 + b_2\omega_2$, avec b_1 et b_2 réels linéairement indépendants sur $\mathbf{Q}$.

J'interromps ici ma digression.

3.6.2 Le "Traité des fonctions elliptiques et de leurs applications" [6], auquel ont été incorporés les articles postérieurs à 1884 sur le même sujet, est une oeuvre très complexe, dont je ne peux tenter ici qu'une caractérisation grossière.

Le tome I est intitulé "Théorie des fonctions elliptiques et de leurs développements en série"; paru en 1886, il est légèrement postérieur à l'exposé de Jordan ([10] II chap. VII): sachant qu'Halphen fut un collaborateur de Jordan à l'Ecole polytechnique de 1876 à 1886 (et manifestement lié d'amitié avec lui) on peut poser la question de l'interaction de leurs réflexions sur le sujet. Cependant les deux projets étaient fort différents; celui d'Halphen, clairement posé dans sa préface, était d'écrire une somme qui soit accessible aux "utilisateurs" (rappelons qu'une normalisation des notations relatives aux fonctions elliptiques, indispensable à un tel projet, avait été introduite par Weierstrass dans un cours de 1861, dont la diffusion en France date du début des années 1880; cette normalisation est universellement utilisée aujourd'hui). Halphen se prive jusqu'au chap. XIV et dernier du tome I de l'outil des fonctions holomorphes. Partant des fonctions sn, cn, dn de Jacobi, il leur substitue rapidement la fonction $\wp$ de Weierstrass: si g_2 et g_3 sont deux paramètres réels liés par l'inégalité $\Delta := g_2^3 - 27g_3^2 > 0$, le polynôme $P(x) = 4x^3 - g_2x - g_3$ a trois racines réelles $e_1 > e_2 > e_3$; la fonction $\wp$ est définie dans le domaine réel comme la réciproque de la fonction $x \rightarrow -\int_x^{+\infty} P^{-1/2}(t)\,dt$ sur $[e_1, +\infty[$. Halphen passe au domaine complexe en suivant l'ancien chemin des découvreurs de la double périodicité: il démontre le théorème d'addition, passe aux arguments purement imaginaires en changeant le signe de g_3 puis aux arguments complexes en posant le théorème d'addition. Il traite ensuite le cas où (g_2 et g_3 sont réels et) Δ est négatif. Il introduit les fonctions ζ (p. 134) et σ (p. 168) de Weierstrass selon la même démarche: définition dans le domaine réel, par intégrations successives, puis passage au domaine complexe en posant le théorème d'addition. Il classifie les intégrales elliptiques en trois "espèces" (chap. VII). Le chap. VIII, consacré aux fonctions thêta, marque l'apparition des développements en série: jusque là, les développements de Taylor n'étaient utilisés que sous forme de développement limité (dans le domaine complexe). C'est alors seulement que Halphen peut, p. 284, introduire les fonctions elliptiques d'invariants complexes (le développement en série de la fonction thêta gardant un sens dans ce contexte). Halphen ne reviendra pas là-dessus (sauf malentendu de ma part), en particulier l'idée que $\wp$ est fonction réciproque de l'intégrale complexe indéfinie $-\int^{\infty} P^{-1/2}(z)\,dz$ et l'existence de fonctions elliptiques d'invariants (g_2, g_3) donnés (tels que $\Delta \neq 0$) ne sont pas abordées (en contraste avec [10] pp. 413-423). Je note aussi que parmi les identités classiques liant les fonctions thêta, Halphen cite "l'équation à trois termes" (formule (1) de 3.6.1) et la formule du triple produit, mais non la relation quartique générale de Riemann (qu'il étudie, en plusiers variables, dans [5] IV pp. 545-569) .

Les chap. IX et X étudient les dérivées des fonctions de Weierstrass par rapport aux périodes et aux invariants, calculées (comme dans [10] pp. 518-552) au moyen des opérateurs $\omega_1 \frac{\partial}{\partial\omega_1} + \omega_2 \frac{\partial}{\partial\omega_2}$ (homogénéité) et $D = -2(\eta_1 \frac{\partial}{\partial\omega_1} + \eta_2 \frac{\partial}{\partial\omega_2}) = 12g_3\frac{\partial}{\partial g_2} + \frac{1}{3} g_2^2\frac{\partial}{\partial g_3}$ ("the Halphen-Fricke operator" de [11]). Je cite seulement le système différentiel de la fin du chap. IX: $x_i' + x_j' = x_i x_j$ $(1 \leq i < j \leq 3)$ (dont une solution est $x_i = \omega_1^2 \wp(\omega_i/2) - \omega_1 \eta_1$ $(1 \leq i \leq 3)$, la variable étant $2i\pi\omega_2/\omega_1$); il est repris à la fin du chap. X (système différentiel vérifié par les intégrales complètes de Legendre) et surtout généralisé dans deux intéressantes notes ([5] II pp. 475-481).

Les chap. XI-XIII contiennent les développements en série double, produit infini et série trigonométrique des fonctions de Weierstrass; en l'absence de la théorie de Cauchy, les méthodes sont parfois différentes de celles que nous connaissons, ainsi la série double qui sert aujourd'hui à définir $\wp$ est ici déduite, p. 366, de la décomposition en éléments simples de la fraction rationnelle exprimant $\wp nu$ en $\wp u$, pour $n = +\infty$. Les interactions entre ces développements et les fonctions arithmétiques ou les équations modulaires devaient faire partie du tome III; il n'y en a pas trace ici. Enfin, comme signalé plus haut, le chap. XIV est une brève récapitulation, fondée sur l'emploi des fonctions holomorphes.

Le tome II, paru en 1888, est intitulé "Applications à la mécanique, à la physique, à la géodésie, à la géométrie et au calcul intégral".

Les quatre premiers chapitres étudient la cinématique du solide; après l'exposé d'une "représentation elliptique" de SO(3) (apparentée à celle de l'espace euclidien qui est adaptée à l'equation de Lamé) on y trouve:

(i) le mouvement d'un solide pesant suspendu par un point, dans les deux cas où l'on sait l'intégrer par les fonctions elliptiques (ou plutôt leurs généralisations dites "fonctions doublement périodiques de seconde espèce"): le cas d'Euler-Poinsot, où le point fixe est le centre de gravité; le cas de la toupie de Lagrange, où l'ellipsoïde d'inertie est de révolution autour d'un axe passant par le point fixe. Sur ce sujet très classique, l'apport d'Halphen fut, selon ses dires, de développer un travail de Jacobi exprimant le mouvement d'une toupie de Lagrange comme composé de deux mouvements à la Poinsot "concordants" (ce qui fut également réalisé par Darboux);

(ii) le mouvement d'une solide dans un liquide dans un cas, signalé par Clebsch à la suite d'un travail de Kirchhoff, caractérisé par des conditions sur la forme quadratique d'inertie, où l'intégration par les fonctions de seconde espèce est possible; à nouveau Halphen décompose le mouvement en deux mouvements à la Poinsot concordants et une rotation autour d'un axe fixe.

Le chap. V est une étude exhaustive des figures d'équilibre d'une barre élastique plane soumise à une pression normale uniforme et des forces et couples en ses extrémités, problème traité antérieurement par M. Lévy: dans un système approprié de coordonnées polaires, la courbe cherchée vérifie l'équation différentielle $\gamma = Ar^2 + B$ où A et B sont des constantes, γ est la courbure et r est le rayon vecteur.

Les chap. VI et VII étudient les géodésiques des quadriques de révolution (qui admettent des représentation paramétriques de seconde espèce) - Halphen ne traite pas les géodésiques des quadriques quelconques (qui demandent l'emploi des fonctions de genre 2). Citons un résultat original sous la forme que lui a ensuite donnée Darboux ([6] II p. 270): soient C et C' deux géodésiques d'une quadrique Q, tangentes à une même ligne de courbure de Q (d'après Chasles, cette condition définit une intégrale première de l'équation différentielle des géodésiques de Q): si deux points variables m de C et m' de C' sont liés par la condition que les tangentes en m à C et en m' à C' se coupent en un point n, on a $|dm| - |dm'| = d|mn| \pm d|m'n|$.

Le chap. VIII développe le calcul, fait par Gauss, de l'attraction newtonienne en un point m de l'espace euclidien par une ellipse E munie d'une répartition de masse proportionelle à l'aire du secteur angulaire limité par E et issu d'un des foyers: la solution de Gauss utilise les périodes d'une intégrale elliptique de première espèce attachée (comme dans le "problème de Poncelet", cf. 3.6.1) au couple de coniques du plan de l'infini formé de l'ombilicale et de la projection de E vue de m. Halphen modifie cette solution en exprimant l'invariant absolu en fonction rationnelle de m.

Les chap. IX-XI et XIV contiennent les applications "à la géométrie et au calcul intégral", cf. 3.6.1.

Le chap. XII étudie l'équation de Lamé. On sait que celle-ci est le résultat de la méthode de "séparation des variables" à l'équation de Laplace dans l'espace euclidien rapporté au "système triple orthogonal" des quadriques homofocales à un ellipsoïde donné. La situation est clarifiée par l'introduction de la courbe elliptique C, jacobienne de la biquadratique de l'espace (projectif) dual, enveloppe des plans tangents communs à ces quadriques: on obtient une "représentation elliptique" de l'espace à laquelle Halphen donne la forme $x_\alpha = U_\alpha^2 \sigma_\alpha u \sigma_\alpha v \sigma_\alpha w$ (en coordonnées homogènes): ici (u, v, w) est une triplet non ordonné dans $\mathbf{C}$ vu comme revêtement universel de C, α est 0 ou une demi-période (modulo le réseau des périodes), U_α est une constante appropriée (cf. aussi, p. 455, la description projective de cette représentation comme un revêtement ramifié (de degré 64) de l'espace par la variété des diviseurs de degré 3 sur C).

Avec ces notations, on a les expressions $(\wp u - \wp v)(\wp u - \wp w)\, du^2 + (\wp v - \wp u)(\wp v - \wp w)\, dv^2 + (\wp w - \wp u)(\wp w - \wp v) dw^2$ du ds^2 de l'espace euclidien, $(\wp v - \wp w)\, \partial^2/\partial u^2 + (\wp w - \wp u)\, \partial^2/\partial v^2 + (\wp u - \wp v)\, \partial^2/\partial w^2$ du laplacien. La première donne (par une méthode de Jacobi systématisée par Liouville) les géodésiques de l'ellipsoïde. Si d'autre part on cherche, avec Lamé et Hermite, les solutions de l'équation de Laplace de la forme $x(u)\, y(v)\, z(w)$ ("méthode de séparation des variables") on est conduit à une équation différentielle du type $x'' = (n(n+1)\, \wp u + B)x$, où n est un entier et B une constante: c'est l'équation de Lamé, qui sous ces hypothèses peut être intégrée par les fonctions doublement périodiques de seconde espèce.

Le chap. XIII, intitulé "Equations différentielles linéaires", se rattache aux travaux d'Halphen sur ce sujet.

Le tome III devait contenir les théories de l'équation modulaire et de la multiplication complexe, les applications arithmétiques et un aperçu historique. Après la mort d'Halphen, la section de géométrie de l'Académie des sciences a publié deux manuscrits sur la division des périodes par 5 et 7 (le second contient une forme originale de la résolvante de degré 7 pour la division des périodes par 7), un article paru antérieurement sur la multiplication complexe par $\sqrt{-23}$, et quelques fragments (pp. 194-269); je n'essaierai pas d'en parler ici.

3.7 ***Questions diverses d'analyse*** - Je mentionne d'abord une note ([5] IV pp. 471-483), non publiée du vivant d'Halphen et datée anterieurement à 1870 par ses éditeurs) qui "somme" la fraction continue $\frac{c+d}{a+b} + \frac{2c+d}{2a+b} + \frac{3c+d}{3a+b} + \ldots$; il me semble probable qu'il y ait eu, pour Halphen, interaction entre le thème des fractions continues (très développé à l'époque) et celui des courbes gauches (par l'intermédiaire du formalisme de l'élimination) ce qui donne à cette note quelqu'importance à mes yeux. En 1881-82 Halphen a publié une série d'articles sur des développements en série variés de fonctions entières ([5] II pp. 485-543). Le mémoire "Sur quelques séries pour le développement des fonctions à une seule variable" étudie les séries formées à partir de "polynômes d'Appell" i.e. les développements tayloriens généralisés de [1] VI § 1 (aujourd'hui surtout envisagés comme développements asymptotiques). Etant donnée une fonction "génératrice" de développement de Taylor $u(\xi) = \sum_{n\geq 0} u_n \frac{\xi^n}{n!}$, on lui associe l'opérateur $U = u\left(\frac{d}{dx}\right)$

et la suite $P_n(x) = U(x^n) = \sum_{0\leq k\leq n} \binom{n}{k} u_k\, x^{n-k}$ des polynômes d'Appell. Si $u_0 \neq 0$, on peut former l'opérateur inverse V de U ; si f est une fonction et si l'on pose $g = Vf$, la série dont il s'agit est $f(x+y) = \sum_{n\geq 0} \frac{1}{n!} P_n(x)\, g^{(n)}(y)$. C'est une identité si f est un polynôme; pour définir commodément g dans des cas plus généraux, Halphen suppose que $v = 1/u$ s'écrit $v(\xi)=\int_\Gamma \theta(x)e^{x\xi}\, dx$, où θ est une fonction et Γ un chemin de $\mathbf{C}$, d'où $g(y) = \int_\Gamma \theta(x)\, f(x+y)\, dx$ lorsque le second membre existe; il montre que le développement taylorien généralisé de f converge vers f si $\limsup_{m\to+\infty} |f^{(m)}(x)|^{1/m} < R$ (rayon de convergence de u) pour au moins un x: il étudie ensuite l'exemple $u(\xi) = e^{(-1)^n a\xi^{2n}}$ (n entier, $a > 0$, $\Gamma = \mathbf{R}$) où R est infini.

Un mémoire assez connu, "Sur une série d'Abel", étudie le développement formel $f(x) = \sum_{n\geq 0} \frac{1}{n!} f^{(n)}(na)\, x(x-na)^{n-1}$ qu'Abel avait déduit de la substitution à u de l'opérateur de dérivation, dans la série de Lagrange exprimant e^{xu} en ue^{au}; la validité de ce développement est équivalente à la condition $\limsup_{m\to+\infty} | f^{(m)}(x) |^{1/m} < \alpha/|a|$, où α est la racine positive de l'équation $ze^{1+z} = 1$, pour au moins un x.

Une note, "Sur l'approximation des sommes de fonctions numériques ([5] IV pp. 95-98, 1883), donne une démonstration d'un théorème de Mertens: la "probabilité" que deux entiers soient premiers entre eux est $6/\pi^2$. Lipschitz a donné le développement en série de Dirichlet, $(\zeta(s-1)/\zeta(s)) = \sum_{n\geq 1} \varphi(n).n^{-s}$, où φ est l'indicateur d'Euler et ζ est la fonction zêta de Riemann. Halphen identifie la probabilité cherchée, $\lim_{n \to +\infty} \frac{2}{n^2}(\varphi(1) + \ldots + \varphi(n))$, au résidu du premier membre en son pôle $s = 2$: le raisonnement rappelle la "première approximation" de la démonstration du théorème des nombres premiers, selon [7] p. 30. Ce théorème est lui-même mentionné comme devant faire l'objet d'une note ultérieure (qui ne fut jamais publiée: rappelons que ses premières démonstrations datent de 1896). Un peu plus tard ([5] I p. 42) Halphen parle "d'obstacles imprévus" à cette tentative, et cite l'absence d'une démonstration de la décomposition de ξ en produit de "facteurs primaires" annoncée par Riemann. Indépendamment, et à peu près en même temps, Stieltjes a discuté de problème (lettres à Mittag-Leffler, [8] II pp. 445-457), mentionnant la même difficulté (p. 449) et mettant en lumière, contrairement à Halphen, le rôle que doit jouer la localisation des zéros de ζ.

On peut noter que chacun de ces travaux illustre un aspect formel de variantes de la transformation de Laplace.

Bibliographie

[1] N.Bourbaki, *Fonctions d'une variable réelle*. Hermann, Paris.

[2] E.Casas-Alveró et S.Xambó-Descamps, *The enumerative theory of conics after Halphen*. Lecture notes 1192 (1986), Springer.

[3] P.Griffiths and J.Harris, *On Cayley's explicit solution to Poncelet's porism*. L'enseignement mathématique 24 (1978), pp. 31-40.

[4] L.Gruson et C.Peskine, *Théorème de spécialité*, exposé 13 de: Les équations de Yang-Mills, Séminaire ENS 1977-78, Astérisque 71-72, SMF.

[5] G.H.Halphen, *Oeuvres*, Gauthier-Villard, Paris (1916-1924).

[6] G.H.Halphen, *Traité des fonctions elliptiques et de leurs applications*, Gauthier-Villars, Paris (1886-1891).

[7] G.H.Hardy, *Ramanujan*, Chelsea, New York.

[8] C.Hermite et T.J.Stieltjes, *Correspondance*, Gauthier-Villars, Paris.

[9] C.Houzel, *Fonctions elliptiques et intégrales abéliennes*, chap. VII de: Abrégé d'histoire des mathématiques (sous la direction de J.Dieudonné), Hermann , Paris.

[10] C.Jordan, *Cours d'analyse à l'école polytechnique*, Gauthier-Villars, Paris (nouveau tirage, 1959).

[11] N.Katz, *p -adic interpolation of real analytic Eisenstein series*, Ann. math. 104 (1976), pp. 459-571.

[12] S.Lang, *Elliptic curves: diophantine geometry*, Grundlehren des math. Wissenschaften n. 231, Springer.

[13] O.Zariski, *Le problème des modules pour les branches planes*, Hermann, Paris.

Departement de Mathématiques
Université de Lille - USTL
59655 Villeneuve d'Ascq
France

THE SOURCE DOUBLE-POINT CYCLE OF A FINITE MAP OF CODIMENSION ONE

STEVEN KLEIMAN,[1] JOSEPH LIPMAN,[2] AND BERND ULRICH[3]

Abstract. Let X, Y be smooth varieties of dimensions n, $n+1$ over an algebraically closed field, and $f: X \to Y$ a finite map, birational onto its image Z. The source double-point set supports two natural positive cycles: (1) the fundamental cycle of the divisor M_2 defined by the conductor of X/Z, and (2) the direct image of the fundamental cycle of the residual scheme X_2 of the diagonal in the product $X \times_Y X$. Over thirteen years ago, it was conjectured that the two cycles are equal if the characteristic is 0 or if f is "appropriately generic." That conjecture will be established in a more general form.

1. Introduction

Let X and Y be smooth varieties over an algebraically closed field, and assume that $\dim Y - \dim X = 1$. Let $f: X \to Y$ be a finite map that is birational onto its image Z. For example, X might be a projective variety, and f a general central projection onto a hypersurface Z in $Y := \mathbf{P}^{n+1}$. Consider the *source double-point scheme* M_2 of f. By definition, M_2 is the effective divisor whose ideal is the conductor C_X of X/Z. Its underlying set consists of the points x of X whose fiber $f^{-1}f(x)$ is a scheme of length at least 2. Consider also the residual scheme X_2 of the fiber product $X \times_Y X$ with respect to the diagonal. By definition, $X_2 := \mathbf{P}(\mathcal{I}(\Delta))$ where $\mathcal{I}(\Delta)$ is the ideal of the diagonal. Consider finally the map $f_1: X_2 \to X$ induced by the second projection. Its image $f_1 X_2$ too consists of the x whose fiber $f^{-1}f(x)$ has length at least 2. Thus there are two natural source double-point cycles: the fundamental cycle $[M_2]$, and the direct image $f_{1*}[X_2]$. Are the two cycles equal? For over thirteen years, the equation

$$[M_2] = f_{1*}[X_2] \tag{1.1}$$

has been known if $\dim X = 1$, and conjectured if $\dim X$ is arbitrary provided also the characteristic is zero or f is "appropriately generic" [**19**, p. 383; **9**, p. 95]. This article will establish that conjecture in a more general form.

In arbitrary characteristic, Equation (1.1) holds if and only if

$$\operatorname{cod}(\overline{\Sigma}_2, X) \geq 2, \tag{1.2}$$

where $\overline{\Sigma}_2$ is the locus of points x in X such that $\dim \Omega^1_f(x) \geq 2$; see (3.11). (In other words, $\overline{\Sigma}_2$ is the "Thom–Boardman" locus of points where the "kernel rank," or the "differential corank," of f is at least 2. It is also the locus where the fibers of f are not "curvilinear.") If

---

AMS(MOS) subject classifications (1985). Primary 14C25; Secondary 14O20, 14N10.

Acknowledgements. It is a pleasure to thank Fabrizio Catanese, David Eisenbud, and Christian Peskine for fruitful discussions. Eisenbud explained his work with Buchsbaum [**4**], and discussed other points of commutative algebra. Peskine explained at length his work with Gruson [**15**], and called attention to Catanese's work [**6**, **7**]. Catanese described that work, and called attention to Mond and Pellikaan's work [**27**] and to van Straten and de Jong's work [**32**].

[1] Partially supported by NSF grant DMS-8801743.

[2] Partially supported by NSF grant DMS-8803054, and at MIT 21–30 May 1989 by Sloan Foundation grant 88-10-1.

[3] Partially supported by NSF grant DMS-8803383.

f is a "generic" map, such as a generic projection [**19**, pp. 365–366], then the ramification locus $\overline{\Sigma}_1$, the locus where $\dim \Omega^1_f(x) \geq 1$, is of codimension 2 or empty, and $\overline{\Sigma}_2$ is of codimension 6 or empty. In many important cases in practice, $\overline{\Sigma}_1$ is, however, of codimension 1, but (1.2) holds nevertheless. For example, f might be a central projection of a smooth curve X onto a plane curve Z with cusps. Indeed, Condition (1.2) holds automatically if $\dim X = 1$ or if the characteristic is zero, but not always if the characteristic is positive; see (2.6) and (2.7). In any event,

$$[M_2] = f_{1*}[X_2] + D, \tag{1.3}$$

where $D \geq 0$; moreover, the components of D are exactly the components of codimension 1 of $\overline{\Sigma}_2$; see (3.10).

From the point of view of the enumerative theory of singularities of mappings, $[M_2]$ is the right double-point cycle whether or not Condition (1.2) is satisfied. Indeed, its rational equivalence class is given by the double-point formula,

$$f^*f_*[X] - c_1(f)[X], \tag{1.4}$$

where $c_1(f)$ is the first Chern class of the virtual normal sheaf $\mathcal{V}_f = f^*\mathcal{T}_Y - \mathcal{T}_X$. That statement follows from Grothendieck duality theory; see (2.3). However, if f is not finite or if f is of codimension s greater than 1, then the cycle class defined by the conductor need not be given by the general double-point formula $f^*f_*[X] - c_s(f)[X]$; Fulton [**9**, 2.4, 2.5, pp. 95–96] gave examples. For an introduction to some classical instances of the double-point formula, see [**19**, pp. 312–315 and 366–368] and [**10**, pp. 167–170].

Suppose that f is *appropriately generic* in the sense that $\dim X - \dim X_2 = 1$. For example, f is appropriately generic if it is a general central projection [**19**, pp. 388]. Then Condition (1.2) holds; see the proof of (3.12). Moreover, the class of $f_{1*}[X_2]$ too is given by the double-point formula (1.4); that statement follows from residual-intersection theory, and the proof works whenever the map f_1 is of the same codimension s as f [**19**, pp. 377–384; **20**, pp. 46]. Since the cycles $f_{1*}[X_2]$ and $[M_2]$ have the same class, it was reasonable to conjecture that they are equal.

In practice, there are three important cases where it is too restrictive to assume that X and Y are smooth over a field: (1) iterative multiple-point theory [**19**, pp. 384-391; **20**], where rth-order theory for $f: X \to Y$ is derived from $(r-1)$th-order theory for the "iteration" map $f_1: X_2 \to X$; (2) Catanese's theory of "quasi-generic canonical projections" [**6**, **7**], where Y is a (singular) weighted projective space; and (3) van Straten and de Jong's deformation theory of "normalizations" [**32**, §3], where the base is an Artin ring. However, that restrictive assumption can be suitably relaxed. In fact, a priori, it is natural to assume instead that f is Gorenstein and Y is (S_2). For example, Y could be a normal scheme or the flat deformation of a normal scheme. (Coincidentally, Avramov and Foxby [**2**] are now developing a local algebraic theory of Gorenstein maps.) On the other hand, multiple-point theory of higher order or of higher codimension requires an assumption of intermediate strength, namely, that X is a local complete intersection in a smooth Y-scheme and Y is Cohen-Macaulay.

The theory in this article is part of a larger body of theory, which has had a remarkable history over the last fifteen to twenty years. On the very day (in June 1976) that Equation (1.1) was conjectured, Fulton solved the first case, where X is a smooth curve and Y a smooth surface. He proceeded by analyzing the effect of blowing up a singular point of the image Z of X in Y. Fulton's proof appears in [**9**, pp. 98–99]. Two months later, Teissier told Fulton that, the previous year, he [**30**, pp. 118–121] had been led to discover virtually the same equality and proof, while studying the equisingularity of curves over the complex numbers. At the same time, Teissier [**30**, pp. 121–123] gave a second proof, based on deforming Z.

It was a theorem whose time had come; indeed, closely related work had already been done independently. In 1974, Gusein-Zade [**16**, p. 23], as part of a study of vanishing cycles, proved Equation (1.1) for a smooth curve mapping into a smooth surface over the complex numbers; he used blowups in about the same way as Fulton and Teissier. In 1973, Fischer [**11**] studied the module of jets of a unibranched map from a smooth curve into a smooth surface by considering the same blowups. In that case, he obtained a length formula that is equivalent to Equation (1.1). The equivalence holds because the module of jets and the structure sheaf of X_2 are locally isomorphic as $\mathcal{O}_X$-modules, for example, because of (3.2)(2). Later, in 1976, Brown [**3**] generalized Fischer's work, eliminating the hypothesis of unibranchedness. Brown also found that Fisher's proofs were mildly incomplete in the case of positive characteristic, leading Fischer to publish an improved version [**12**] in 1978. Of those five authors, only Fulton mentioned the case of higher dimensional X and Y.

In 1972 Artin and Nagata [**1**, (5.8), pp. 322], inspired by some unpublished results and questions of Mumford, proved a version of Equation (1.1) in the case that X is a smooth curve, Y is a smooth surface, $f: X \to Y$ is any map birational onto its image, and the base is a field of any characteristic. Their version, like Fischer's and Brown's, is a statement about the ideal of the diagonal of $X \times_Y X$. Their proof, like Teissier's second, involves deforming f into a map whose image has simple nodes at worst. Artin and Nagata also gave an example that shows that their version of Equation (1.1) does not generalize to the case that X is a smooth surface and $Y := \mathbf{P}^4$.

In a nutshell, the proof of (1.3) runs as follows; see (3.9). First, it is shown that the direct image of $[X_2]$ on $X \times_Y X$ is equal to the fundamental cycle of the ideal of the diagonal, $[\mathcal{I}(\Delta)]$, diminished by a positive cycle C whose components lie in the diagonal and correspond precisely to the components of codimension 1 of $\overline{\Sigma}_2$. In fact, off the image of $\overline{\Sigma}_2$ in the diagonal subscheme, the structure map $p: X_2 \to X \times_Y X$ is a closed embedding, and its ideal is $\mathcal{A}nn(\mathcal{I}(\Delta))$; see (3.4)(2). Moreover, off the image of $\overline{\Sigma}_2$, locally $\mathcal{I}(\Delta)$ is generated by one element, and hence is isomorphic to $p_*\mathcal{O}_{X_2}$; see (3.3). (So, in particular, X_2 is equal, off the image of $\overline{\Sigma}_2$, to the double-point scheme X_2' considered by Mond [**26**, § 3, pp. 368–371] and Marar and Mond [**25**, 1.1, pp. 554-555], which is defined by $\mathcal{A}nn(\mathcal{I}(\Delta))$. Moreover, as $\mathcal{I}(\Delta)$ is locally generated by one element,

$$\mathcal{A}nn_{X\times_Y X}(\mathcal{I}(\Delta)) = \mathcal{F}itt^0_{X\times_Y X}(\mathcal{I}(\Delta)).$$

Mond [**26**, 3.2(i), p. 369] made a note of that equation because the Fitting ideal is "more readily calculable" [**25**, bottom p. 554]. Furthermore [**25**, bottom p. 554; **31**], if X is a local complete intersection in a smooth Y-scheme and Y is Cohen–Macaulay — for example, if X and Y are smooth over a field — then that equation continues to hold, and X_2' is Cohen–Macaulay and is of finite flat dimension over Y; see (3.13).)

On the other hand, at any point w of the diagonal in the image of $\overline{\Sigma}_2$, the fiber $p^{-1}w$ has dimension at least 1; see (3.2)(1). Therefore, the components of X_2 lying over the image of $\overline{\Sigma}_2$ do not contribute to $p_*[X_2]$, and the other components of X_2 contribute with the same multiplicity to both $p_*[X_2]$ and $[\mathcal{I}(\Delta)]$. (However, Ulrich [**31**] has proved that, if X is a local complete intersection in a smooth Y-scheme and Y is (S_2), then $[X_2'] = [\mathcal{I}(\Delta)]$; see (3.13).) The preceding considerations (including those in parentheses) are valid in great generality; in particular, f may have any codimension s (provided that, in the more sophisticated statements, it is assumed that M_2 is of codimension at least s in X — whence it follows that M_2 and X_2' are of codimension exactly s; see (3.13)).

It now suffices to prove that the direct image of $[\mathcal{I}(\Delta)]$ is equal to $[M_2]$; in other words, at each generic point ξ of M_2, the length of $\mathcal{I}(\Delta)$ is equal to the colength of $\mathcal{C}_X$. The latter statement follows from these equations:

$$\mathcal{F}itt^0_X(\mathcal{I}(\Delta)) = \mathcal{F}itt^0_Y(f_*\mathcal{O}_X/\mathcal{O}_Z)\mathcal{O}_X = \mathcal{C}_X. \tag{1.5}$$

Indeed, it will be proved that C_X is invertible as f is Gorenstein; see (2.3). Hence, $\mathcal{I}(\Delta)$ is of flat dimension 1 over X, and so the desired length-colength equation holds.

The first equation in (1.5) is proved via rather simple and general considerations, which require no special hypotheses; see (3.4)(1). The key lemma (3.3) was apparently known to Artin and Nagata, and perhaps to Mumford; see the statement in parentheses on line 6 of p. 322 in [**1**]. The second equation in (1.5) is an immediate consequence of the following equation on Y:

$$\mathcal{F}itt^0_Y(f_*\mathcal{O}_X/\mathcal{O}_Z) = \mathcal{A}nn_Y(f_*\mathcal{O}_X/\mathcal{O}_Z). \tag{1.6}$$

Equation (1.6) follows from a general theorem of Buchsbaum and Eisenbud [**4**, p. 232]; see (3.7) and (3.5). However, only a special case of the general theorem is needed here, and in that case, the theorem's proof simplifies to a few lines involving the Hilbert–Burch theorem; Eisenbud showed that short proof to the authors (on 26 May 1989), and it too is given in (3.5).

The *target double-point scheme* N_2 is, by definition, the subscheme of Y of the adjoint ideal $\mathcal{A}nn_Y(f_*\mathcal{O}_X/\mathcal{O}_Z)$. So, Equation (1.6) says, in other words, that N_2 is determinantal, cut out locally by the maximal minors of any matrix presenting $f_*\mathcal{O}_X/\mathcal{O}_Z$. Now, the proof in (3.5) of (1.6) also shows that N_2 is of flat dimension 2 in Y; hence, it is of pure codimension 2 in Y by the Intersection Theorem of Peskine–Szpiro and Roberts, and it is Cohen–Macaulay if Y is by the Auslander-Buchsbaum Theorem.

Equation (1.6) was already known, however. Mond and Pellikaan, in their March 1988 preprint [**27**, p. 121], had obtained it independently and also from Buchsbaum and Eisenbud's theorem. They prove (1.6) in the course of proving the equation,

$$\mathcal{F}itt^0_Y(f_*\mathcal{O}_X/\mathcal{O}_Z) = \mathcal{F}itt^1_Y(X). \tag{1.7}$$

That equation interested them because it, together with (1.6), says that the target double-point scheme N_2 is also defined by the Fitting ideal $\mathcal{F}itt^1_Y(X)$.

Another proof of (1.7) is found, as Mond and Pellikaan indicated, in a June 1988 preprint of van Straten and de Jong, who used the Hilbert–Burch theorem directly, [**32**; combine (4.8), (4.12), and (4.13)]. They used (1.7) to compare the deformation theory of the pair (X, Z) with that of (N_2, Z). Mond and Pellikaan, and van Straten and de Jong gave credit to Catanese [**6**] (see [**7**] also) for introducing the key ideas in 1982. (In turn, Catanese said that he drew inspiration from work of Arbarello, Sernesi, and Ciliberto.) Catanese's purpose was to study "pluriregular varieties of free general type" via "quasi-generic canonical projections."

Independently, in 1981, Gruson and Peskine [**15**] were led to Equations (1.6) and (1.7) while studying the scheme of r-secants of a smooth space curve C. They viewed the secant scheme as the target r-fold locus of the map $f: X \to Y$, where Y is the Grassmannian of lines L and where X is the variety of pairs (P, L) with $P \in L \cap C$. They did not prove (1.6) directly, but first proved a form [**15**, 1.5, p. 5] of the equation,

$$\mathcal{A}nn_Y(f_*\mathcal{O}_X/\mathcal{O}_Z) = \mathcal{F}itt^1_Y(X). \tag{1.8}$$

Their proof is simple and direct, and does not involve the Hilbert–Burch theorem or anything like it. They do prove a form [**15**, 1.3, p. 4] of (1.7), but their proof needs an additional hypothesis, which, as it turns out, amounts to the assumption that $\overline{\Sigma}_2$ is empty. On the other hand, under that assumption, they prove a more general statement, involving the higher-order Fitting ideals.

The higher-order multiple-point loci of $f: X \to Y$ are also of some interest. The first job is to find a reasonable scheme-theoretic definition of them. Assume, as always, that $f: X \to Y$ is finite and birational onto its image Z. Assume also that $\overline{\Sigma}_2$ is empty. This hypothesis is not that much of a restriction in the rth-order theory for $r \leq 6$, because the expected codimension in X of $\overline{\Sigma}_2$ is 6. Moreover, there are many applications where, in fact, $\overline{\Sigma}_2$ is empty. Assume finally that X is a local complete intersection in a smooth Y-scheme and that Y is Cohen-Macaulay. Under roughly those hypotheses, Gruson and Peskine [**15**] and Mond and Pellikaan [**27**] independently strove to show that the Fitting ideal $\mathcal{F}itt_Y^{r-1}(X)$ defines a reasonable scheme N_r of target r-fold points. For example, for $r=1$, that Fitting ideal defines the scheme-theoretic image Z; see (2.2). For $r=2$, the Fitting ideal is equal to the adjoint ideal by (1.8); so the new definition of N_2 agrees with the old. In the work [**23**] under preparation, the present authors will develop the following additional evidence for the reasonableness of this definition of N_r.

Following the iterative approach to multiple-point theory of [**20**], define the scheme M_r of source r-fold points of f as the scheme of target $(r-1)$-fold points of f_1; in other words, define M_r as the scheme with ideal $\mathcal{F}itt_X^{r-2}(X_2)$. For example, for $r=2$, that Fitting ideal is equal to the conductor $\mathcal{C}_X$; see (3.4) and (3.7). So the new definition of M_2 agrees with the old. Now, if the definitions of M_r and N_r are indeed reasonable, then these schemes should be compatible under pullback:

$$M_r = f^{-1}N_r.$$

That compatibility equation will be proved in [**23**]; the proof is similar to the proof of (3.4).

Assume that M_r and N_r have the expected codimensions, $r-1$ and r, everywhere. Then a general point of N_r has an inverse image of length r; so the cycle relation

$$f_*[M_r] = r[N_r] \tag{1.9}$$

should hold. For example, for $r=2$, this relation is equivalent to the usual Gorenstein formula, because then M_2 and N_2 are defined by the conductors on X and Z. Relation (1.9) will be proved for arbitrary r in [**23**].

There is another generalization of the usual Gorenstein formula, due to Gruson and Peskine [**15**, Prop. 2.6, p. 13]. For $r=2$, it reduces to the other form of the usual formula: the colength of the conductor in a 1-dimensional Gorenstein domain is equal to the colength of the domain in its normalization. For arbitrary r, the generalization says intuitively that a general $(r+1)$-fold point counts as $r+1$ r-fold points. In [**23**], following the approach to multiple-point theory based on the Hilbert scheme, which is developed in [**22**], it will be shown how to interpret Gruson and Peskine's generalization of the Gorenstein formula as a statement about the Hilbert scheme Hilb_f^r, and how to derive it from Relation (1.9). The key step is to prove that Hilb_f^r is equal to the blowup of N_r along N_{r+1}.

2. The double-point schemes

(2.1) *Setup.* Let $f: X \to Y$ be a finite map of locally Noetherian schemes. Assume that f is birational onto its image; more precisely, assume that there is an open subset U of Y such that its preimage $f^{-1}U$ is dense in X and the restriction $f^{-1}U \to U$ is an embedding. Assume that f is of pure codimension 1; that is, if ξ is the generic point of an arbitrary component of X, then $\dim \mathcal{O}_{Y,f\xi} = 1$. Assume that f is of flat dimension 1. Finally, assume that Y satisfies Serre's condition (S_2) [**14**, (5.7.2), p. 103]: for every $y \in Y$,

$$\mathrm{depth}(\mathcal{O}_{Y,y}) \geq \inf(2, \dim \mathcal{O}_{Y,y}).$$

These conditions will be assumed without further mention throughout Section 2.

If, in the derived category, $f^!\mathcal{O}_Y$ is isomorphic to a (shifted) invertible sheaf ω_f, then f is called *Gorenstein* [**17**, p. 144]. If f is Gorenstein, define its first Chern class as that of ω_f:

$$c_1(f) := c_1(\omega_f).$$

For example, if there is a factorization $f = \pi i$ where $i: X \hookrightarrow P$ is a regular embedding and $\pi: P \to Y$ is smooth, then f is Gorenstein and

$$\omega_f = \det(\mathcal{V}_i) \otimes \det(\mathcal{T}_\pi)^{-1}$$

where $\mathcal{V}_i$ is the normal sheaf and $\mathcal{T}_\pi$ is the tangent sheaf. For instance, if X and Y are smooth over some base scheme S, then the product $P := X \times_S Y$, the graph map $i: X \to P$, and the projection $\pi: X \times_S Y \to Y$ will work; in this case,

$$\omega_f = \det(\mathcal{T}_{X/Y})^{-1} \otimes \det(\mathcal{T}_{Y/S}).$$

The ideal $\mathcal{A}nn_Y(f_*\mathcal{O}_X/\mathcal{I}m\,\mathcal{O}_Y)$ is called the *adjoint ideal.* The scheme it defines is denoted by N_2 and called the *target double-point scheme.* Its underlying set consists of the points y of Y whose fiber $f^{-1}(y)$ is a scheme of length at least 2 over $k(y)$. The adjoint ideal is an $f_*\mathcal{O}_X$-module, and the associated sheaf on X

$$\mathcal{C}_X := \widetilde{\mathcal{A}nn_Y(f_*\mathcal{O}_X/\mathcal{I}m\,\mathcal{O}_Y)}$$

is an ideal, called the *conductor on* X. The corresponding scheme is denoted by M_2 and called the *source double-point scheme.* Obviously, $M_2 = f^{-1}N_2$ as schemes, and the restriction $M_2 \to N_2$ is finite and surjective.

The Fitting ideal $\mathcal{F}itt_X^{r-1}(\Omega_f^1)$ defines a scheme, denoted $\overline{\Sigma}_r$. Its underlying set consists of the points x of X such that $\dim \Omega_f^1(x) \geq r$. Obviously, $\overline{\Sigma}_0 = X$, and

$$M_2 \supseteq \overline{\Sigma}_1 \supseteq \overline{\Sigma}_2 \supseteq \cdots .$$

The formation of $\overline{\Sigma}_r$ commutes with base change as the formation of Ω_f^1 does and the formation of a Fitting ideal does.

Denote by Z the scheme-theoretic image of X in Y. By definition [**13**, (6.10.1), p. 324], Z is the smallest closed subscheme of Y through which f factors. Because f is quasi-compact and quasi-separated, Z exists and is defined by the ideal $\mathcal{A}nn_Y(f_*\mathcal{O}_X)$. Obviously, N_2 is a closed subscheme of Z; its ideal is the sheaf

$$\mathcal{C}_Z := \mathcal{A}nn_Z(f_*\mathcal{O}_X/\mathcal{O}_Z),$$

and $\mathcal{C}_Z$ is called the *conductor on* Z.

Proposition (2.2) *The scheme-theoretic image Z of X in Y is a divisor, and its ideal is equal to the Fitting ideal $\mathcal{F}itt_Y^0(X)$. In other words, locally Z is defined by the determinant of any square matrix presenting $\mathcal{O}_X$ over $\mathcal{O}_Y$; such matrices exist, and their determinants are regular elements. Moreover, the formation of Z commutes with base change.*

Proof. Because f is finite, the Fitting ideal $\mathcal{F}itt_Y^0(X)$ exists. Because f is of flat dimension 1 and of codimension 1, locally $\mathcal{O}_X$ is presented over $\mathcal{O}_Y$ by a square matrix whose determinant is regular and generates $\mathcal{F}itt_Y^0(X)$. Let W denote the corresponding divisor. Then W has no embedded components because Y satisfies (S_2).

The schemes W and Z have the same support, and $Z \subseteq W$ because

$$\mathcal{A}nn_X(Y)^n \subseteq \mathcal{F}itt_Y^0(X) \subseteq \mathcal{A}nn_X(Y)$$

for some integer n. Moreover, W and Z are generically equal because f is generically an embedding. Therefore, W and Z are equal because W has no embedded components.

The formation of Z commutes with base change because the formation of a Fitting ideal does.

Theorem (2.3) *The conductor $\mathcal{C}_X$ is an invertible sheaf if and only if f is a Gorenstein map. In either case, the double-point cycle $[M_2]$ is given by the double-point formula,*

$$[M_2] = f^*f_*[X] - c_1(f)[X],$$

which holds modulo rational equivalence.

Proof. The proof is a version of that in [**19**, pp. 365–366]; this version uses more abstract, but nevertheless standard Grothendieck duality theory [**17**, **24**].

Say $f = jg$ where $g: X \to Z$ and $j: Z \hookrightarrow Y$. Work in the derived category of quasi-coherent sheaves. Trivially

$$f^!\mathcal{O}_Y = \mathbf{R}\mathcal{H}om_X(\mathcal{O}_X, f^!\mathcal{O}_Y).$$

Since g and j are finite, $\mathbf{R}g_* = g_*$ and $\mathbf{R}j_* = j_*$. Also, $f^! = g^!j^!$ and $f_* = j_*g_*$. Hence, duality yields the equations,

$$(g_*)f^!\mathcal{O}_Y = \mathbf{R}\mathcal{H}om_Z(g_*\mathcal{O}_X, j^!\mathcal{O}_Y),$$
$$j_*\mathbf{R}\mathcal{H}om_Z(g_*\mathcal{O}_X, j^!\mathcal{O}_Y) = \mathbf{R}\mathcal{H}om_Z(f_*\mathcal{O}_X, \mathcal{O}_Y).$$

The latter complex has all its cohomology concentrated in degree 1, because f is finite and of flat dimension 1 and because Z is nowhere dense in Y and Y has no embedded components as it satisfies (S_2). Hence, $f^!\mathcal{O}_Y$ does too. Say $f^!\mathcal{O}_Y[1]$ is isomorphic to the quasi-coherent sheaf ω_f.

By (2.2), Z is a divisor in Y. So $j^!\mathcal{O}_Y = \mathcal{O}_Z(Z)[-1]$. Hence

$$g_*\omega_f = \mathcal{H}om_Z(g_*\mathcal{O}_X, \mathcal{O}_Z(Z)) = \mathcal{H}om_Z(g_*\mathcal{O}_X, \mathcal{O}_Z) \otimes \mathcal{O}_Z(Z).$$

Now, X has no embedded component because f is of flat dimension 1 and Y satisfies (S_2); hence, $g_*\mathcal{O}_X$ is contained in the sheaf of total quotient rings of $\mathcal{O}_Z$ (that condition is not implied by the definition of birationality adopted in (2.1)). Therefore, standard elementary considerations show that evaluation at 1 defines an isomorphism,

$$\mathcal{H}om_Z(g_*\mathcal{O}_X, \mathcal{O}_Z) = \mathcal{A}nn_Z(g_*\mathcal{O}_X/\mathcal{O}_Z);$$

its inverse sends a local section to multiplication by that section. Therefore, taking associated sheaves yields the following equation on X:

$$\omega_f = \mathcal{C}_X \otimes g^*\mathcal{O}_Z(Z).$$

The assertions follow immediately.

Proposition (2.4) *The source double-point scheme M_2 is of pure codimension 1 in X, and the target double-point scheme N_2 is of pure codimension 1 in Z and of pure codimension 2 in Y.*

Proof. Since $X \to Z$ is finite and birational and since $M_2 = f^{-1}N_2$, it suffices to treat N_2. By (2.1), Z is a divisor in Y. So it suffices to prove that N_2 is of pure codimension 2 in Y. Now, by definition, N_2 is the support of the $\mathcal{O}_Y$-module $f_*\mathcal{O}_X/\mathcal{O}_Z$. That module is of flat dimension at most 2 because $\mathcal{O}_X$ and $\mathcal{O}_Z$ are both of flat dimension 1. Hence N_2 is everywhere of codimension at most 2 by virtue of the Intersection Theorem; it is well known that the case needed here may be derived easily from the work of Peskine and Szpiro [**28**], but the general case was proved by P. Roberts [**29**]. On the other hand, N_2 is of codimension at least 2, because $X \to Z$ is birational. Therefore, N_2 is of pure codimension 2.

The Intersection Theorem is not needed here if f is Gorenstein, for then $\mathcal{C}_X$ is invertible by (2.3).

Lemma (2.5) *Let ξ be a generic point of a component of M_2. Suppose that X is regular at ξ and that the field extension $k(\xi)/k(f\xi)$ is separable. Then $\xi \notin \overline{\Sigma}_2$.*

Proof. Denote the reduced scheme $(M_2)_{\mathrm{red}}$ by D. Then D is of pure codimension 1 in X by (2.4). So D is a divisor in X at ξ because X is regular there. At ξ, consider the standard exact sequence,

$$\mathcal{O}_D(-D) \longrightarrow \Omega^1_f|D \longrightarrow \Omega^1_{D/Y} \longrightarrow 0.$$

The first term is invertible, and the third term vanishes because $k(\xi)/k(f\xi)$ is finite and separable. Hence, $\dim_{k(\xi)} \Omega^1_f(\xi) \le 1$. In other words, $\xi \notin \overline{\Sigma}_2$.

Proposition (2.6) *Suppose that X and Y are of finite type over a field k, and that X is regular in codimension 1 (for example, normal). Suppose either* (a) $\dim X = 1$ *and k is perfect or* (b) *k is of characteristic zero. Then* $\operatorname{cod}(\overline{\Sigma}_2, X) \ge 2$.

Proof. Let ξ be a generic point of a component of M_2. Then X is regular at ξ because M_2 is of pure codimension 1 in X by (2.4). If $\dim X = 1$ and k is perfect, then $k(\xi)/k$ is finite and separable; whence, then $k(\xi)/k(f\xi)$ is separable. Of course, $k(\xi)/k(f\xi)$ is separable if the characteristic of k is 0. Therefore, $\xi \notin \overline{\Sigma}_2$ by (2.5). Thus the assertion holds.

(2.7) *Example.* Here is an example where $\overline{\Sigma}_2$ has codimension 1. Fix an algebraically closed field of positive characteristic p. Let X be a closed, reduced, and irreducible surface in $\mathbf{P}^3$, and consider its Gauss map $f: X_0 \to Y$, where X_0 is the smooth locus of X, and Y is the dual $\mathbf{P}^3$. Fix a point P of X_0, and choose affine coordinates x, y, z such that x, y are regular parameters of X at P. Then

$$\dim \Omega^1_f(P) = 2 - \operatorname{rank} \frac{\partial^2 z}{\partial(x, y)^2}$$

by [**18**, (2.6.1), p. 153; **21**, § I-5, 175–177]. Therefore, the point P lies in $\overline{\Sigma}_2$ if and only if the Hessian $\frac{\partial^2 z}{\partial(x,y)^2}$ vanishes at P.

Suppose $p = 2$. Then $\frac{\partial^2 z}{\partial x^2}$ and $\frac{\partial^2 z}{\partial y^2}$ vanish identically near P. Hence $\overline{\Sigma}_2$ is defined near P by the vanishing of $\frac{\partial^2 z}{\partial y \partial x}$. Suppose that X is smooth of degree d at least 2. Then $f: X \to Y$ is finite. That fact is well known, and holds in any characteristic. It holds, for example, because $f^*\mathcal{O}_Y(1)$ is ample, as it is equal to $\mathcal{O}_X(d-1)$ [**19**, middle of p. 360; **21**, § II-2, p.190]. Suppose that X is general of its degree. Then f is birational onto its image [**18**, (5.6), p. 176; **21**, (21), p. 180]. In particular, $\overline{\Sigma}_2 \ne X$. Therefore, $\operatorname{cod}(\overline{\Sigma}_2, X) = 1$.

Suppose $p \ge 3$. Then $\frac{\partial^2 z}{\partial x^2}$ and $\frac{\partial^2 z}{\partial y^2}$ vanish identically if, for instance,

$$z = xy(x+y)^p + x^{p+1} + y^{p+1}.$$

Then, moreover, $\frac{\partial^2 z}{\partial y \partial x} = (x+y)^p$. So, if X is the surface with that equation, then its Gauss map f is birational onto its image by the Hessian Criterion [**18**, (3.3), p. 155; **21**, (12), p. 176]. Unfortunately, X has a (unique) singular point at infinity, (0,0,1,0). However, that point corresponds to a curve C in Y, and a computation shows that the (reducible) curve D of X corresponding to C contains the entire curve at infinity of X, but does not contain the locus $\{x + y = 0\}$. Hence, the restriction $(X - D) \to (Y - C)$ is finite, and its $\overline{\Sigma}_2$ is of codimension 1.

3. The residual double-point cycle

Definition (3.1) Let $f: X \to Y$ be a separated map of schemes. Form the residual scheme of the diagonal and the corresponding map,

$$X_2 := \mathbf{P}(\mathcal{I}(\Delta)) \quad \text{and} \quad f_1: X_2 \xrightarrow{p} X \times_Y X \xrightarrow{p_2} X,$$

where $\mathcal{I}(\Delta)$ is the ideal of the diagonal, p is the structure map, and p_2 is the second projection. Then X_2 is called the *iteration*, or *derived*, scheme, and f_1 is called the *iteration*, or *derived*, map [**20**, 4.1, pp. 36–37; **22**, (2.10)]. If f is proper, then $f_{1*}[X_2]$ is defined and will be called the *residual double-point cycle* of f.

Lemma (3.2) *Let $f: X \to Y$ be a separated map locally of finite type between locally Noetherian schemes. Let w be a point of $X \times_Y X$.*

(1) *The following four conditions are equivalent:*

(a) *The structure map $p: X_2 \to X \times_Y X$ is a closed embedding at w.*

(b) *The fiber $p^{-1}w$ is empty or of dimension 0.*

(c) *Either w lies off the diagonal, or w lies on the diagonal and $\dim \Omega^1_f(p_2 w) \le 1$.*

(d) *The ideal $\mathcal{I}(\Delta)$ of the diagonal is generated by one element at w.*

(2) *Let U be an open subset of $X \times_Y X$ on which $\mathcal{I}(\Delta)$ is generated by a single element of $\Gamma(U, \mathcal{I}(\Delta))$. Then the restriction $p^{-1}U \to U$ is a closed embedding, its ideal is $\mathcal{A}nn(\mathcal{I}(\Delta))|U$, and there is an isomorphism of $\mathcal{O}_U$-modules,*

$$p_*\mathcal{O}_{X_2}|U \simeq \mathcal{I}(\Delta)|U.$$

Proof. Trivially (a) implies (b). For convenience, set $\mathcal{I} := \mathcal{I}(\Delta)$. Then $X_2 = \mathbf{P}(\mathcal{I})$ by (3.1). So the fiber $p^{-1}w$ is equal to $\mathbf{P}((\mathcal{I}/\mathcal{I}^2)(w))$. Hence (b) implies (c) because $\mathcal{I}/\mathcal{I}^2$ is isomorphic to the direct image under the diagonal map of Ω^1_f. Also because of that isomorphism and by Nakayama's lemma, (c) implies (d).

Let U be an open subset of $X \times X$ on which $\mathcal{I}$ is generated by a single section. That section defines a surjection on U from $\mathcal{O}_{X\times X}$ to $\mathcal{I}$, and its kernel is obviously $\mathcal{A}nn(\mathcal{I})$; in other words, there is an exact sequence,

$$0 \longrightarrow \mathcal{A}nn(\mathcal{I})|U \longrightarrow \mathcal{O}_{X\times X}|U \longrightarrow \mathcal{I}|U \longrightarrow 0. \tag{3.2.1}$$

In general, if $\mathcal{E} \to \mathcal{F} \to \mathcal{G} \to 0$ is an exact sequence of quasi-coherent sheaves on an arbitrary scheme, then the ideal of $\mathbf{P}(\mathcal{G})$ in $\mathbf{P}(\mathcal{F})$ is equal to the image of $\mathcal{E}(-1)$ in $\mathcal{O}_{\mathbf{P}(\mathcal{F})}$, because the following sequence is well known to be exact [**5**, Ch. III, §6, no. 2, Prop. 4, p. 499]:

$$\mathcal{E}\,\mathcal{S}ym(\mathcal{F})[-1] \longrightarrow \mathcal{S}ym(\mathcal{F}) \longrightarrow \mathcal{S}ym(\mathcal{G}) \longrightarrow 0.$$

Since $\mathbf{P}(\mathcal{O}_U) \xrightarrow{\sim} U$ and $\mathcal{O}_{\mathbf{P}(\mathcal{O}_U)}(-1) = \mathcal{O}_{\mathbf{P}(\mathcal{O}_U)}$, therefore $p^{-1}U \to U$ is a closed embedding, with ideal $\mathcal{A}nn(\mathcal{I})|U$. A second look at (3.2.1) now reveals that $p_*\mathcal{O}_{X_2}|U$ is isomorphic to $\mathcal{I}|U$. Thus (2) holds. Hence, (d) implies (a).

Lemma (3.3) *Let R be a (commutative) ring, B an R-algebra, and I the kernel of the multiplication map $B \otimes_R B \to B$. View $B \otimes_R B$ as a B-algebra via the homomorphism u given by $u(b) := 1 \otimes b$. Then there is an isomorphism of B-modules,*

$$I \simeq (B/\operatorname{Im} R) \otimes_R B.$$

Proof. The homomorphism u splits the following exact sequence of B-modules:

$$0 \longrightarrow I \longrightarrow B \otimes_R B \longrightarrow B \longrightarrow 0.$$

So I is isomorphic to the cokernel of u. On the other hand, tensoring the exact sequence $R \to B \to B/\operatorname{Im} R \to 0$ with B yields the exact sequence,

$$B \longrightarrow B \otimes_R B \longrightarrow (B/\operatorname{Im} R) \otimes_R B \longrightarrow 0,$$

in which the first map is u. So the cokernel of u is also isomorphic to $(B/\operatorname{Im} R) \otimes_R B$. Thus the assertion holds.

Proposition (3.4) *Let $f: X \to Y$ be a finite map of locally Noetherian schemes.*

(1) *View the ideal $\mathcal{I}(\Delta)$ of the diagonal of $X \times_Y X$ as an $\mathcal{O}_X$-module via the second projection p_2. Then*

$$\mathcal{F}itt^0_X(\mathcal{I}(\Delta)) = \mathcal{F}itt^0_Y(f_*\mathcal{O}_X/\mathcal{I}m\,\mathcal{O}_Y)\,\mathcal{O}_X.$$

(2) *Off the image of $\overline{\Sigma}_2$ under the diagonal map, the structure map $p: X_2 \to X \times_Y X$ is a closed embedding, and its ideal is $\mathcal{A}nn(\mathcal{I})$. Off $\overline{\Sigma}_2$, the iteration map $f_1: X_2 \to X$ is finite, and*

$$\mathcal{F}itt^0_X(X_2) = \mathcal{F}itt^0_X(\mathcal{I}(\Delta)).$$

Proof. (1) The two Fitting ideals are defined because f is finite. The asserted equation holds locally because of (3.3), as the formation of a Fitting ideal commutes with base change. Therefore the equation holds globally.

(2) Off $\overline{\Sigma}_2$, the structure map p is a closed embedding by (3.2)(1). So $f_1 := p_2 p$ is finite, because p_2 is as f is.

It suffices to establish the asserted equality of ideals locally at each point $x \notin \overline{\Sigma}_2$. By (3.2)(1), $\mathcal{I}(\Delta)$ is generated by one element at each point w of $p_2^{-1}x$. Since p_2 is finite, there is, therefore, a neighborhood V of x such that, if $U := p_2^{-1}V$, then $\mathcal{I}(\Delta)|U$ is generated by a single element of $\Gamma(U, \mathcal{I}(\Delta))$. So (3.2)(2) yields the asserted equality on V.

Lemma (3.5) *Let R be a Noetherian local ring. Let $A := R/\delta$ where δ is a regular element (non-zero-divisor). Let F be a finitely generated A-module such that $A \subset F \subset K$, where K is the total fraction ring of A. Suppose that $F/A \neq 0$ and that the flat dimension $\mathrm{f.d}_R F$ is 1. Then*

$$Ann_R(F/A) = Fitt^0_R(F/A),$$

and $R/Ann_R(F/A)$ is an R-module of flat dimension 2, grade 2, and codimension 2.

Proof. Choose elements $x_1, \ldots, x_h$ of F whose images in $F/m_A F$ form a basis, and let E be the submodule of $R^{\oplus h}$ of relations among the x_i. Then E is free because $\mathrm{f.d}_R F = 1$, and E is of rank h because $F \subset K$. It is now easy to see that F/A is presented by a h by $h+1$ matrix.

The codimension of the R-module F/A is at least 2 because $F \subset K$. Furthermore, the grade of F/A is at least 2; that is, $Ann_R(F/A)$ contains a regular sequence of two elements. Indeed, if $Ann_R(F/A)/\delta R$ consisted of entirely of zero-divisors, then $Ann_R(F/A)$ would lie in the union of the associated primes of δ, so in one of them, say P. However, $(F/A)_P = 0$ because $A_P = K_P$.

Since $(h+1) - h + 1 = 2$, it follows from a general theorem of Eagon and Hochster that the grade of F/A is exactly 2, and from a general theorem of Buchsbaum and Eisenbud that $Ann_R(F/A)$ is equal to $Fitt^0_R(F/A)$; for both conclusions, see [4, top p. 232]. The general theorems may be avoided in the case at hand by the following argument, which also yields the remaining two assertions.

Since $\mathrm{f.d}_R F = 1$ and $\mathrm{f.d}_R A = 1$, obviously $\mathrm{f.d}_R F/A \le 2$. So, since F/A is presented by an h by $h+1$ matrix, there is an exact sequence

$$0 \longrightarrow R \longrightarrow R^{\oplus(h+1)} \longrightarrow R^{\oplus h} \longrightarrow F/A \longrightarrow 0.$$

Since the grade of F/A is at least 2, dualizing that exact sequence yields this one,

$$0 \longrightarrow R^{\oplus h} \longrightarrow R^{\oplus(h+1)} \longrightarrow R \longrightarrow R/I \longrightarrow 0,$$

where I is an ideal. In particular, $R/I = \mathrm{Ext}^2_R(F/A,\, R)$. Since any element of R that kills F/A also kills $\mathrm{Ext}^2_R(F/A,\, R)$, therefore

$$\mathit{Ann}_R(F/A) \subseteq I.$$

Now, the dual of the second exact sequence is obviously the first. Hence

$$\mathrm{Ext}^i_R(R/I,\, R) = \begin{cases} 0, & \text{if } i < 2; \\ F/A, & \text{if } i = 2. \end{cases}$$

Therefore, the grade of R/I is exactly 2. Moreover, I lies in $\mathit{Ann}_R(F/A)$, so the two ideals are equal. Finally, the Hilbert–Burch Theorem [8, Thm. 1, p. 122] yields that $I = \mathit{Fitt}^0_R(F/A)$.

The preceding argument also shows that $R/\mathit{Ann}_R(F/A)$ has flat dimension 2, as asserted, because $\mathit{Ann}_R(F/A) = I$. Hence its codimension is at most 2 by the Intersection Theorem of Peskine–Szpiro and Roberts (see the proof of (2.4)). Since, as was noted above, its codimension is at least 2, it is exactly 2, as asserted.

Proposition (3.6) *Under the conditions of* (2.1), *the target double-point scheme* N_2 *is of pure flat dimension* 2, *grade* 2, *and codimension* 2 *in* Y.

Proof. By (2.2), Z is a divisor in Y. So the assertion follows (3.5) applied locally.

Proposition (3.7) *If the conditions of* (2.1) *hold, then*

$$\begin{aligned} \mathcal{A}nn_Y(f_*\mathcal{O}_X/\mathcal{O}_Z) &= \mathcal{F}itt^0_Y(f_*\mathcal{O}_X/\mathcal{O}_Z) \\ \mathcal{C}_Z &= \mathcal{F}itt^0_Z(f_*\mathcal{O}_X/\mathcal{O}_Z) \\ \mathcal{C}_X &= \mathcal{F}itt^0_Z(f_*\mathcal{O}_X/\mathcal{O}_Z)\,\mathcal{O}_X. \end{aligned}$$

Proof. By (2.2), Z is a divisor in Y. So the first equation holds at each point of Y by (3.5). The second and third equations are easily derived from the first.

Proposition (3.8) *Under the conditions of* (2.1), *the formation of the double-point schemes,* M_2 *and* N_2, *commutes with base change.*

Proof. The assertion follows directly from (3.7) because the formation of a Fitting ideal commutes with base change.

Lemma (3.9) *Under the conditions of* (2.1), *suppose* f *is Gorenstein. Then*

$$p_{2*}[\mathcal{I}(\Delta)] = [M_2]$$

where $[\mathcal{I}(\Delta)]$ *is the fundamental cycle of the ideal of the diagonal, viewed simply as an* $\mathcal{O}_X$*-module.*

Proof. By (3.4)(1) and (3.7), $\mathcal{F}itt^0_X(\mathcal{I}(\Delta))$ is equal to $\mathcal{C}_X$. By (2.3), $\mathcal{C}_X$ is invertible. Hence, by standard algebra, $\mathcal{I}(\Delta)$ is of flat dimension 1 over X. Hence, at each generic point ξ of M_2, the length of $\mathcal{I}(\Delta)$ is equal to the colength of $\mathcal{C}_X$ [10, A.2.3, p. 411]. Therefore, $p_{2*}[\mathcal{I}(\Delta)]$ and M_2 are equal at ξ.

It remains to note that, if η is the generic point of a component of the support of $\mathcal{I}(\Delta)$, then $\dim \mathcal{O}_{X,p_2\eta} = 1$. However, the completion of the stalk $\mathcal{I}(\Delta)_\eta$ is a module of finite length and of flat dimension 1 over the completion of $\mathcal{O}_{X,p_2\eta}$. Hence its *0*th Fitting ideal is invertible, and is primary for the maximal ideal. Therefore, $\dim \mathcal{O}_{X,p_2\eta} = 1$.

Theorem (3.10) *Under the conditions of* (2.1), *suppose* f *is Gorenstein. Then*

$$[M_2] = f_{1*}[X_2] + D$$

where $D \geq 0$, *and the components of* D *are exactly the components of codimension* 1 *of* $\overline{\Sigma}_2$.

Proof. It follows from (3.2) that $[\mathcal{I}(\Delta)] = p_*[X_2] + C$ where C is a positive cycle whose components are exactly those components of the support of $\mathcal{I}(\Delta)$ that are images under the diagonal map of the components of $\overline{\Sigma}_2$. So, the assertion follows from (3.9) because $f_1 := p_2 p$ and because of (2.4).

Corollary (3.11) *Under the conditions of* (2.1), *suppose* f *is Gorenstein. Then*

$$[M_2] = f_{1*}[X_2]$$

off $\overline{\Sigma}_2$, *and that equation holds everywhere if and only if also* $\operatorname{cod}(\overline{\Sigma}_2, X) \geq 2$.

Proof. The assertion follows immediately from (3.10).

Corollary (3.12) *Under the conditions of* (2.1), *suppose that* X *and* Y *are of finite type over a base scheme* S, *and that* f *is a Gorenstein* S*-map. Then*

$$[M_2] = f_{1*}[X_2]$$

provided also one of the following two conditions is satisfied:

(1) *If* $\xi \in M_2$ *is the generic point of an arbitrary component, then* X/S *is smooth at* ξ, *and either* (a) $\dim_\xi(X/S) = 1$, *or* (b) $k(\xi)$ *is of characteristic 0, or simply,* (c) $k(\xi)/k(f\xi)$ *is separable.*

(2) *The map* f *is appropriately generic in the sense that, if* $\eta \in X_2$ *is the generic point of an arbitrary component, then*

$$\dim_\eta(X_2/S) = \dim_{f_1\eta}(X/S) - 1.$$

Proof. The assertion follows from (3.11) as $\operatorname{cod}(\overline{\Sigma}_2, X) \geq 2$. Indeed, by (2.4), M_2 is of pure codimension 1 in X. Let $\xi \in M_2$ be the generic point of a component. If (1) holds, then $\xi \notin \overline{\Sigma}_2$ by (2.5) applied to the geometric fiber of f over the image of ξ in S; compare with the proof of (2.6). If (2) holds, then $f_1^{-1}\xi$ is of dimension 0; so $\xi \notin \overline{\Sigma}_2$ by (3.2)(1).

(3.13) *Remark.* Ulrich [31] has proved a complement to (3.10), which suggests that the subscheme X_2' of $X \times_Y X$ defined by $\mathcal{A}nn(\mathcal{I}(\Delta))$ is a better external scheme of source double-points than X_2. Namely, under the conditions of (3.10),

$$[M_2] = f_{1*}[X_2']$$

provided that X is a complete intersection over Y at the generic point of every component of M_2.

Ulrich derives that assertion from the following one, in which the restriction to codimension 1 has been dropped: Let $f: X \to Y$ be a finite map of locally Noetherian schemes that is birational onto its image. Assume that X is locally a complete intersection of codimension s over Y, that Y satisfies S_{2s+1}, and that M_2 is of codimension at least s. Then X_2' is a perfect Y-scheme of grade $2s$, its ideal $\mathcal{A}nn(\mathcal{I}(\Delta))$ is equal to $\mathcal{F}itt^0_{X\times_Y X}(\mathcal{I}(\Delta))$, and its fundamental cycle $[X_2']$ is equal to $[\mathcal{I}(\Delta)]$. In particular, M_2 and X_2' are of pure codimension s, and X_2' is of flat dimension $2s$ over Y; moreover, X_2' has no embedded components, and it is Cohen–Macaulay if Y is. The assertion in the preceding paragraph follows because of (3.9) and (2.4).

4. References

[1] M. Artin and M. Nagata, *Residual intersections in Cohen–Macaulay rings,* J. Math. Kyoto Univ. **12–2** (1972), 307–323.

[2] L. Avramov and H.-B. Foxby, *Gorenstein local homomorphisms,* to appear in Bull. Amer. Math. Soc., April 1990.

[3] W. Brown, *Blow up sequences and the module of nth order differentials,* Can. J. Math. **28** (1976), 1289–1301.

[4] D. Buchsbaum and D. Eisenbud, *What annihilates a module?* J. Algebra **47** (1977), 231–243.

[5] N. Bourbaki, "Algebra," Hermann; Addison-Wesley, 1974.

[6] F. Catanese, *Commutative algebra methods and equations of regular surfaces,* in "Algebraic Geometry, Bucharest 1982.," L. Bădescu and D. Popescu (eds.), Lecture Notes in Math. **1056**, Springer-Verlag, 1984, pp. 68–111.

[7] F. Catanese, *Equations of pluriregular varieties of general type,* in "Geometry Today. Roma 1984," E. Arbarello, C. Procesi, E. Strickland (eds.) Prog. Math. **60**, Birkhäuser, 1985, pp. 47–67.

[8] D. Eisenbud, *Some directions of recent progress in commutative algebra,* in "Algebraic Geometry, Arcata 1974," R. Hartshorne (ed.), Proc. Symposia Pure Math., Vol. **29**, Amer. Math. Soc., 1975, pp. 111–128.

[9] W. Fulton, *A note on residual intersections and the double point formula,* Acta Math. **140** (1978), 93–101.

[10] W. Fulton, "Intersection theory," Ergebnisse der Mathematik und ihrer Grenzgebiete 3. Folge · Band 2, Springer-Verlag, 1984.

[11] K. Fischer, *The module decomposition of $I(\overline{A}/A)$,* Trans. Amer. Math. Soc. **186** (1973), 113–128.

[12] K. Fischer, *The decomposition of the module of n-th order differentials in arbitrary characteristic,* Can. J. Math. **30** (1978), 512–517.

[13] A. Grothendieck, with J. Dieudonné, "Eléments de Géométrie Algébrique I," Springer-Verlag, 1971.

[14] A. Grothendieck, with J. Dieudonné, "Éléments de Géometrie Algébrique IV_4," Publ. Math. I.H.E.S. **24**, 1965.

[15] L. Gruson and C. Peskine, *Courbes de l'espace projectif: variétés de sécantes,* in "Enumerative and Classical Algebraic Geometry," P. le Barz, Y. Hervier (eds.), Proc. Conf., Nice 1981. Progr. Math. **24**, Birkhäuser, 1982, pp. 1–31.

[16] S. Gusein-Zade, *Dynkin diagrams for singularities of functions of two variables,* Funktsional'nyi Analiz i Ego Prilozheniya **8** (1974), 23–30.

[17] R. Hartshorne, "Residues and duality," Springer Lecture Notes in Math. **20**, 1966.

[18] A. Hefez and S. Kleiman, *Notes on duality of projective varieties,* in "Geometry Today. Roma 1984," E. Arbarello, C. Procesi, E. Strickland (eds.) Prog. Math. **60**, Birkhäuser, 1985, pp. 143–184.

[19] S. Kleiman, *The enumerative theory of singularities,* in "Real and complex singularities," P. Holm (ed.), Proc. Conf., Oslo 1976, Sitjhoff & Noorhoof, 1977, pp. 297–396.

[20] S. Kleiman, *Multiple-point Formulas I: Iteration,* Acta Math. **147** (1981), 13–49.

[21] S. Kleiman, *Tangency and duality,* in "Proc. 1984 Vancouver Conf. in Algebraic Geometry," J. Carrell, A. V. Geramita, P. Russell (eds.), CMS Conf. Proc. **6**, Amer. Math. Soc., 1986, pp. 163–226.

[22] S. Kleiman,, *Multiple-point formulas II: the Hilbert scheme,* in "Enumerative Geometry," S. Xambó Descamps (ed.), Proc. Conf., Sitges 1987, Lecture Notes in Math. **1436**, Springer-Verlag, 1990, pp. 101–138.

[23] S. Kleiman, J. Lipman, and B. Ulrich, "The multiple-point cycles of a finite map of codimension one," in preparation.

[24] J. Lipman, "Notes on derived categories and derived functors," to appear.

[25] W. Marar and D. Mond, *Multiple point schemes for corank 1 maps,* J. London Math. Soc. (2) **39** (1989), 553–567.

[26] D. Mond, *Some remarks on the geometry and classification of germs of maps from surfaces to 3-space,* Topology **26** (1987), 361–383.

[27] D. Mond and R. Pellikaan, *Fitting ideals and multiple points of analytic mappings,* in "Algebraic geometry and complex analysis," E. Ramírez de Arellano (ed.), Proc. Conf., Pátzcuero 1987, Lecture Notes in Math. **1414**, Springer-Verlag, 1989, pp. 107–161.

[28] C. Peskine and L. Szpiro, *Modules de type fini et de dimension injective finie sur un anneau local noethérien,* C. R. Acad. Sci. Paris **266** (1968), 1117–1120.

[29] P. Roberts, *Le théorème d'intersection,* C. R. Acad. Sci. Paris **304** (1987), 177–180.

[30] B. Teissier, *Résolution simultanée - II. Résolution simultanée et cycles évanescents,* in "Séminaire sur les Singularités des Surfaces," M. Demazure, H. Pinkham, B. Teissier (eds.), Lecture Notes in Math. **777**, Springer-Verlag, 1980, pp. 82–146.

[31] B. Ulrich, "Algebraic properties of the double-point cycle of a finite map," in preparation.

[32] D. van Straten and T. de Jong, "Deformations of non-isolated singularities," Preprint, June 1988.

Department of Mathematics, 2–278 M.I.T., Cambridge, MA 02139, U.S.A.

Division of Mathematical Science, Purdue University, West Lafayette, IN 47907, U.S.A.

Department of Mathematics, Michigan State University, East Lansing, MI 48824-1027, U.S.A.

Fibré déterminant et courbes de saut sur les surfaces algébriques

J. Le Potier

Introduction

Considérons, sur le plan projectif $X = \mathbf{P}_2$ l'espace de modules $M_{\mathbf{P}_2}(2, 0, c_2)$ des classes d'équivalence de faisceaux semi-stables de rang 2, de classes de Chern $(0, c_2)$. Si F est un tel faisceau semi-stable, on dit qu'une droite $\ell \subset \mathbf{P}_2$ est de saut si $F|_\ell$ n'est pas trivial, ce qui revient à dire que $h^1(F(-1)|_\ell) \neq 0$. D'après le théorème de Grauert et Mülich, les droites de saut de F sont portées par une courbe de degré c_2 du plan projectif dual $\mathbf{P}_2^*$, courbe dont l'équation s'obtient comme déterminant du morphisme canonique sur $\mathbf{P}_2^*$

$$H^1(F(-2)) \otimes \mathcal{O}_{\mathbf{P}_2^*}(-1) \longrightarrow H^1(F(-1)) \otimes \mathcal{O}_{\mathbf{P}_2^*}$$

On obtient ainsi un morphisme

$$\gamma : M = M_{\mathbf{P}_2}(2, 0, c_2) \longrightarrow |\mathcal{O}_{\mathbf{P}_2^*}(c_2)|$$

Un faisceau qui n'est pas localement libre est dit singulier ; les classes de faisceaux singuliers constituent dans M une hypersurface ∂M. On sait (cf. Maruyama [13], Strømme [20], Hulek et Strømme [9]) que l'image de ∂M par γ est contenue dans le fermé des courbes réductibles et que les fibres de $\gamma|\partial M$ sont de dimension ≥ 1. D'autre part, l'image de M rencontre l'ouvert des courbes lisses, et la fibre de γ au-dessus d'une telle courbe lisse est obligatoirement finie : c'est un des résultats essentiels de l'article de Barth [1].

On se préoccupe ici de ce qui se passe au-dessus des courbes singulières, et même non réduites. Le point de départ de cet article est la démonstration du résultat suivant :

Théorème 0 . *Soit* U *l'ouvert de* $M_{\mathbf{P}_2}(2, 0, c_2)$ *des faisceaux localement libres. Alors le morphisme*

$$\gamma|_U : U \longrightarrow |\mathcal{O}_{\mathbf{P}_2^*}(c_2)|$$

est quasi-fini.

La méthode de Barth consistait à établir une correspondance entre fibrés stables et certaines thêta-caractéristiques sur la courbe de $\mathbf{P}_2^*$ des droites de saut. L'approche que nous proposons ici est totalement différente. Elle s'inspire de la construction de Donaldson du fibré déterminant sur la variété des fibrés μ-stables de rang 2 de classes de Chern $(0, c_2)$ sur une surface algébrique régulière X, et repose sur l'observation que le fibré déterminant s'étend, de manière naturelle, en un fibré $\mathscr{D}_1$ sur la variété projective des classes d'équivalence de faisceaux semi-stables de Gieseker et Maruyama. La méthode utilisée permet d'obtenir des variantes de l'énoncé ci-dessus sur $\mathbf{P}_1 \times \mathbf{P}_1$ polarisée par $\mathcal{O}(1, m)$ ou sur certaines surfaces de del Pezzo convenablement polarisées (théorème 6.10). La construction du fibré $\mathscr{D}_1$ est en fait possible dans le cas où la surface est quelconque, en rang et classes de Chern quelconques.

On considère donc une surface algébrique polarisée $(X, \mathcal{O}_X(1))$, propre et lisse, connexe, et on désigne par $M_X(r, c_1, c_2)$ l'espace de modules de classes d'équivalence de faisceaux semi-stables sur X de rang r, de classes de Chern $c_1 \in H^2(X, \mathbf{Z})$ et $c_2 \in \mathbf{Z}$, et par $\mathrm{Pic}_{c_1}(X)$ la composante du groupe de Picard des fibrés inversibles sur X de classes de Chern c_1. On dispose d'un morphisme

$$j : M_X(r, c_1, c_2) \longrightarrow \mathrm{Pic}_{c_1}(X)$$

qui associe à la classe du faisceau F le fibré inversible $\det F$. L'espace de modules $M_X(r, c_1, c_2)$ est un espace de modules grossier ; à toute famille F de faisceaux semi-stables de rang r de classes de Chern c_1 et c_2, paramétrée par une variété algébrique S est associé un morphisme $f_F : S \longrightarrow M_X(r, c_1, c_2)$, dit modulaire, qui dépend fonctoriellement de F. Sur $M_X(r, c_1, c_2)$ on peut alors construire des fibrés inversibles $\mathscr{L}_0$ et $\mathscr{L}_1$, bien déterminés à isomorphisme près par le choix d'un point $a \in X$ et d'une courbe lisse $Y \in |\mathcal{O}_X(1)|$ et caractérisés par la propriété universelle suivante : si $F = (F_s)_{s \in S}$ est une famille de faisceaux semi-stables comme ci-dessus, on a dans $\mathrm{Pic}(S)$:

$$\det(\mathrm{pr}_{1!}(F))^{\otimes -r} \otimes ((\det F)^{\otimes \chi}|_{S \times \{a\}}) = f_F^*(\mathscr{L}_0) \qquad (1)$$

où $\mathrm{pr}_1 : S \times X \longrightarrow S$ est la première projection et χ la caractéristique d'Euler-Poincaré de F_s pour $s \in S$. De même, dans $\mathrm{Pic}(S)$

$$\det(\mathrm{pr}_{1!}(F|_{S \times Y}))^{\otimes -r} \otimes ((\det F)^{\otimes \chi_1}|_{S \times \{a\}}) = f_F^*(\mathscr{L}_1) \qquad (2)$$

où χ_1 est la caractéristique d'Euler-Poincaré de $F_s|_Y$. Les fibrés $\mathscr{L}_0$ et $\mathscr{L}_1$ sont modifiés, en général, quand on change a et Y, mais leur classe dans le groupe de Picard relatif

$$\mathrm{Pic}_{\mathrm{rel}}(M_X(r, c_1, c_2)) = \mathrm{Pic}(M_X(r, c_1, c_2)) \big/ j^* \mathrm{Pic}(\mathrm{Pic}_{c_1}(X))$$

reste inchangée. On a alors (cf. corollaire (3.10)) :

Théorème 1 . *Pour* $m >> 0$, *le fibré* $\mathcal{L}_0 \otimes \mathcal{L}_1^{\otimes m}$ *est relativement ample par rapport à* $\mathrm{Pic}_{c_1}(X)$.

Ce fibré inversible est en effet celui qui permet à Gieseker d'obtenir une structure de variété projective sur $M_X(r, c_1, c_2)$.

Quand on le restreint à l'ouvert $U \subset M_X(r, c_1, c_2)$ des faisceaux localement libres μ-stables, le fibré $\mathcal{L}_1$ est une puissance $\mathcal{D}_1^{\otimes d_1}$ du fibré déterminant introduit dans le cas $r = 2, c_1 = 0$ (et X régulière) par Donaldson [11] . Le fibré $\mathcal{L}_1$ n'est pas en général relativement ample (cf. corollaire (5.7)). On a cependant :

Théorème 2 . *Pour* υ *entier convenable assez grand, il existe un faisceau algébrique cohérent* $\mathcal{E}_\upsilon$ *sur* $\mathrm{Pic}_{c_1}(X)$ *et un morphisme relatif au- dessus de* $\mathrm{Pic}_{c_1}(X)$

$$\varphi_\upsilon : M_X(r, c_1, c_2) \longrightarrow \mathbf{P}(\mathcal{E}_\upsilon)$$

tel que (1) $\varphi_\upsilon^* (\mathcal{O}(1)) = \mathcal{L}_1^{\otimes \upsilon}$; (2) *le morphisme* $\varphi_\upsilon |_U$ *est un plongement.*

Une version plus précise de cet énoncé sera donnée au chapitre 4 (théorème (4.1)). La construction de $\mathcal{L}_0$ et $\mathcal{L}_1$ peut se généraliser au cas où X est une variété projective de dimension quelconque ; nous avons ainsi obtenu une version du théorème 1 dans ce cas (corollaire (3.8)). Faute d'une variante sophistiquée du théorème de restriction de Mehta et Ramanathan [14], nous n'avons cependant pas pu étendre le théorème 2 en toute dimension.

Comme dans l'article de Drézet et Narasimhan [5] la construction de $\mathcal{L}_0$ et $\mathcal{L}_1$ repose sur la description de $M_X(r, c_1, c_2)$ comme quotient d'un ouvert Ω_{ss} d'un schéma de Grothendieck par l'action d'un groupe réductif (cf. §2) ; c'est un problème de descente qui se ramène à l'examen des représentations induites sur les stabilisateurs des points $q \in \Omega_{ss}$: le point clé est l'utilisation d'un lemme dû à Kempf (cf. §1). La méthode est calquée sur celle de Drézet et Narasimhan, une fois maîtrisée la difficulté liée au fait que l'ouvert Ω_{ss} n'a aucune raison d'être réduit, contrairement à ce qui se passe lorsque l'on travaille sur les courbes ou sur le plan projectif.

Le lien entre le fibré $\mathcal{L}_1$ (et le fibré inversible $\mathcal{D}_1$, qui étend dans le cas $r = 2$ et $c_1 = 0$ fibré déterminant de Donaldson) et les courbes de saut est donné au paragraphe 6 : le fibré $\mathcal{D}_1$ est essentiellement l'image réciproque de $\mathcal{O}(1)$ par le morphisme qui à un faisceau semi-stable F associe l'hypersurface $\gamma(F)$ des courbes $Y \in | \mathcal{O}_X(1) |$ de saut. Ce morphisme γ n'est défini que si le fibré anticanonique ω_X^* est ample et si le système linéaire ci-dessus contient des courbes lisses rationnelles ce qui revient pratiquement à limiter le champ d'application au plan projectif, à $\mathbf{P}_1 \times \mathbf{P}_1$ ou à certaines surfaces de del Pezzo, munies d'une polarisation adéquate.

Sommaire

1. Préliminaires

(1.1) Fibrés inversibles relativement amples Soient X et Y deux variétés algébriques, $f : X \longrightarrow Y$ un morphisme projectif et A un fibré inversible sur X. On pose $\mathrm{Pic}(X/Y) = \mathrm{Pic}\, X / f^*(\mathrm{Pic}(Y))$. On dit que A est relativement ample par rapport à Y si pour m > 0 convenable $A^{\otimes m}$ est, dans Pic(X/Y), l'image réciproque de $\mathcal{O}_{\mathbf{P}_N}(1)$ par un plongement $X \hookrightarrow Y \times \mathbf{P}_N$ au dessus de Y. C'est le cas si et seulement si pour tout $\mathcal{O}_X$-module cohérent F on a $R^q f_*(F \otimes A^{\otimes m}) = o$ pour m assez grand et $q > 0$. En particulier, cette notion est locale sur Y.

Dans le cas où $f : X \longrightarrow Y$ est un morphisme plat cette notion signifie simplement que A est ample sur les fibres de f :

Lemme (1.2). *Soient* $f : X \longrightarrow Y$ *un morphisme propre et plat de variété algébriques, A un fibré inversible sur X. Alors les assertions suivantes sont équivalentes :*

(1) *le fibré A est relativement ample par rapport à Y.*
(2) *le fibré A est ample sur les fibres de f.*
(3) *pour toute sous-variété intègre Z de dimension d contenue dans une fibre de f,* $\langle Z, c_1(A)^d \rangle > 0$.

Démonstration. On a évidemment (1) ⇒ (2). Montrons que (2) ⇒ (1) Soit $y \in Y$. Si A est ample sur la fibre X_y, on peut trouver un entier $m > 0$ tel que

(1) $H^q\left(A^{\otimes m}|_{X_y}\right) = o$ pour $q \geq 1$

(2) $A^{\otimes m}|_{X_y}$ est engendré par ses sections et le morphisme associé

$$X_y \longrightarrow \mathbf{P}^{\cdot}\left(H^0(A^{\otimes m}|_{X_y})\right)$$

est un plongement.

On a alors $R^q f_*(A^{\otimes m}) \otimes_{\mathcal{O}_y} \mathbf{C} = 0$. En effet, ceci se voit par récurrence descendante sur q ; c'est trivial pour $q > \dim X$. Supposons que ceci soit vrai en degré $> q$. Par changement de base, compte-tenu du fait que $A^{\otimes m}$ est plat sur Y, on a

$$R^q f_*(A^{\otimes m}) \underset{\mathcal{O}_y}{\otimes} \mathbf{C} \cong H^q(A^{\otimes m}|_{X_y})$$

d'où l'assertion. Mais $R^q f_* (A^m)$ est un faisceau cohérent sur Y, car f est propre. Alors, d'après le lemme de Nakayama, on a, pour $q > 0$, $R^q f_* (A^{\otimes m}) = 0$ au voisinage de y ; l'image directe $f_* (A^{\otimes m})$ est un fibré vectoriel au voisinage de y, et l'on a

$$f_* (A^{\otimes m}) \otimes_{O_y} C \cong \left(H^0 (A^{\otimes m}|_{X_y})\right)$$

(cf. Harsthorne [10], théorème 12.11 chapitre III). Par suite, le morphisme d'évaluation $f^* f_*(A^{\otimes m}) \longrightarrow A^{\otimes m}$ est surjectif au voisinage de la fibre X_y, et donc, puisque f est propre, sur un voisinage ouvert de la forme $f^{-1}(V)$, où V est un ouvert contenant y. Ainsi, on a un morphisme au dessus de V

$$f^{-1}(V) \longrightarrow \mathbf{P}\,(f_*(A^{\otimes m}))$$

qui par hypothèse est un plongement sur X_y, et par suite au voisinage de X_y. Ainsi, $A^{\otimes m}$ est relativement ample au voisinage de X_y.

(2) $\Leftrightarrow$ (3) : c'est le critère de Nakaï-Moishezon appliqué à la fibre X_y.

(1.3) Descente [5]. Soit G un groupe réductif opérant sur une variété algébrique X. On suppose qu'il existe un bon quotient $\pi : X \longrightarrow Y$ de X par l'action de G. Soit E un G-fibré vectoriel sur X ; on dit que E se quotiente par G s'il existe un fibré vectoriel F sur Y et un G-isomorphisme $E \cong \pi^* F$. Cette condition est équivalente à la suivante : pour toute orbite fermée O, il existe un voisinage ouvert affine G-invariant U de O sur lequel E est G-trivial.

Le lemme suivant, dû à Kempf, et démontré dans [5] dans le cas où X est intègre donne un critère pour que E se quotiente par G. Cet énoncé reste vrai même si X n'est pas intègre :

Lemme (1.4). *Le G-fibré vectoriel E se quotiente par G si et seulement si le stabilisateur de tout point d'orbite fermée opère trivialement sur E.*

La condition est évidemment nécessaire. Montrons qu'elle est suffisante. Soit O une orbite fermée. Sur O, le G-fibré E est évidemment trivial si le stabilisateur d'un point de O opère trivialement sur E. Une telle trivialisation s'étend à un voisinage ouvert affine invariant U de O ; pour voir qu'il existe une G-trivialisation, il suffit de vérifier qu'il existe un opérateur de Reynolds

$$R : \Gamma(U, E) \longrightarrow \Gamma(U, E)^G$$

Ceci résulte du lemme suivant (seul point dans lequel Drézet et Narasimhan utilisent l'hypothèse X intègre) :

Lemme (1.5). *L'orbite d'une section $s \in \Gamma(U, E)$ est contenue dans un sous-espace vectoriel de dimension finie.*

L'action de G sur E définit une famille algébrique de sections g.s de E sur U, c'est-à-dire une section σ sur $G \times U$ du fibré vectoriel $pr_2^*(E)$. Mais on a

$$\Gamma(G \times U, pr_2^*(E)) = \Gamma(G, \mathcal{O}_G) \otimes \Gamma(U, E)$$

donc $\sigma = \sum_{i=1}^{k} \varphi_i \otimes \psi_i$, avec $\varphi_i \in \Gamma(G, \mathcal{O}_G)$ et $\varphi_i \in \Gamma(U, E)$. On obtient $g.s = \sum_{i=1}^{k} \varphi_i(g)\, \psi_i$; ainsi g.s appartient au sous-espace vectoriel engendré par $\psi_1, ..., \psi_k$.

2. Le morphisme $H^{\perp\perp} \cap Z(c) \longrightarrow \mathrm{Pic}(M_X(c))$

Soit $(X, \mathcal{O}_X(1))$ une variété algébrique projective lisse polarisée de dimension n. On considère l'anneau de Grothendieck K(X) des classes de $\mathcal{O}_X$-modules cohérents ; il est équipé d'une forme quadratique entière : $u \to \chi(u^2)$ qui provient d'une forme quadratique sur $K_{top}(X)$. L'orthogonal d'un élément $u \in K(X)$ pour cette forme quadratique ne dépend que de l'image c de u dans l'anneau de Grothendieck des fibrés vectoriels topologiques $K_{top}(X)$. On le note Z(c). On désigne par H le sous-anneau de K(X) engendré par la classe $h = \mathcal{O}_Y$, où Y est une section hyperplane de X ; c'est aussi le sous-anneau engendré par $\mathcal{O}_X(1)$.

Exemple (2.1). Supposons que X soit une courbe lisse de genre g. Un élément $c \in K_{top}(X)$ est déterminé par son rang r, et son degré d, ou si on préfère, son rang r et sa caractéristique d'Euler-Poincaré χ. On a

$$\begin{aligned}\chi(c^2) &= 2r\,d + r^2(1-g)\\ &= 2r\,\chi - r^2(1-g)\end{aligned}$$

Pour $c = (r, \chi)$, l'orthogonal Z(c) est constitué des classes $w \in K(X)$ de rang r' et de degré d' tels que $rd' + \chi r' = 0$.

Exemple (2.2). Si X est une surface, $K_{top}(X) \cong \mathbf{Z} \times H^2(X, \mathbf{Z}) \times \mathbf{Z}$, l'isomorphisme étant donné par (r, c_1, χ). La forme quadratique sur $K_{top}(X)$ est donnée pour $c \in K_{top}(X)$ par

$$\chi(c^2) = 2r\,\chi + c_1^2 - r^2\,\chi(\mathcal{O}_X).$$

Soit $c \in K_{top}(X)$, de rang $r > 0$. On désigne par $M_X(c)$ l'espace de modules des classes d'équivalence de faisceaux semi-stables F de classe c (1). Par famille de $\mathcal{O}_X$-modules semi-stables de classe c paramétrée par la variété algébrique S, on entend un faisceau algébrique cohérent F sur $S \times X$, S-plat, tel que F(s) soit semi-stable de classe c pour tout $s \in S$. Une telle famille définit un morphisme

$$f_F : S \longrightarrow M_X(c)$$

dit modulaire, fonctoriel en S, qui fait de $M_X(c)$ un module grossier pour le foncteur

$S \to \{$classes d'isomorphisme de familles de $\mathcal{O}_X$-modules semi-stables de classe c paramétrées par S $\}$

(1) au sens de Gieseker et Maruyama [6, 12]

Soit F une telle famille, paramétrée par S ; considérons les projections

$$\begin{array}{ccc} S \times X & \xrightarrow{pr_2} & X \\ \big\downarrow{\scriptstyle pr_1} & & \\ S & & \end{array}$$

On pose, pour $u \in K(X)$,

$$\lambda_F(u) = \det\left(pr_{1!}(F.\, pr_2^*(u))\right)$$

où $F.\, pr_2^*(u)$ est le produit dans $K(S \times X)$. Pour vérifier que ceci a bien un sens même si S n'est pas lisse, on doit vérifier que si G est un $\mathcal{O}_X$-module cohérent, $pr_{1!}(F.\ pr_2^*(G))$ définit un élément du groupe de Grothendieck $K^0(S)$ construit à partir des faisceaux localement libres de type fini sur S . On peut bien sûr supposer que G est localement libre : alors $F.\, pr_2^*(G) = F \otimes pr_2^* G$; on est ramené à vérifier :

Lemme (2.3) . *Pour tout* $\mathcal{O}_{S\times X}$*-module cohérent* S*-plat* F *l'élément* $pr_{1!}(F)$ *de* $K(S)$ *provient d'un élément bien déterminé du groupe de Grothendieck* $K^0(S)$ *construit à partir des* $\mathcal{O}_S$*-modules localement libres de type fini.*

Démonstration. On peut trouver, en appliquant le théorème B de Serre, une résolution gauche

$$\mathcal{V}_{n-1} \otimes \mathcal{O}(-k_{n-1}) \longrightarrow \ldots \longrightarrow \mathcal{V}_0 \otimes \mathcal{O}(-k_0) \longrightarrow F \longrightarrow 0$$

où les k_i sont des entiers suffisamment grands, et les $\mathcal{V}_i$ des $\mathcal{O}_S$-modules localement libres de type fini. On pose, pour $0 \le i \le n-1$, $\mathcal{W}_i = \mathcal{V}_i \otimes \mathcal{O}(-k_i)$ et on prend pour $\mathcal{W}_n$ le noyau de la première flèche. On obtient ainsi une résolution gauche finie $\mathcal{W}. \longrightarrow F$ par des faisceaux localement libres $\mathcal{W}_i$ tels que $R^q pr_{1*}(\mathcal{W}_i) = 0$ pour $q < n$. D'après le théorème de changement de base [10] , $R^n pr_{1*}(\mathcal{W}_i)$ est localement libre de type fini. De plus $H_q(R^n pr_{1*}(\mathcal{W}.)) = R^{n-q} pr_{1*}(F)$. Ainsi, dans $K^0(S)$ l'élément $\sum_{i=0}^{n} (-1)^i\, pr_{1!}(\mathcal{W}_i)$ a un sens dans $K^0(S)$, et a pour image $pr_{1!}(F)$ dans $K(S)$.

Reste à vérifier que l'élément de $K^0(S)$ construit ne dépend pas de la résolution $\mathcal{W}.$ choisie. Si $\mathcal{W}.'$ est une autre résolution finie gauche construite suivant le même procédé, on peut supposer, quitte au besoin à coiffer les deux résolutions par une troisième, qu'il existe un morphisme de complexes $\mathcal{W}. \longrightarrow \mathcal{W}.'$ rendant commutatif le diagramme

$$\begin{array}{ccc} \mathcal{W}. & & \\ \downarrow & \searrow & \\ & & F \\ & \nearrow & \\ \mathcal{W}.' & & \end{array}$$

On obtient alors par image directe un quasi-isomorphisme

$$u : R^n pr_{1*}(\mathcal{W}.) \longrightarrow R^n pr_{1*}(\mathcal{W}.')$$

dont le mapping-cône est par conséquent acyclique. Alors, dans $K^0(S)$

$$\sum_{i=0}^{n} (-1)^i \mathrm{pr}_{1!}(\mathscr{W}_i) = \sum_{i=0}^{n} (-1)^i \mathrm{pr}_{1!}(\mathscr{W}_i')$$

ce qui achève la démonstration du lemme (2.3).

Lemme (2.4) . *Pour tout* $A \in \mathrm{Pic}(S)$ *et* $u \in Z(c)$, *on a* $\lambda_{A \otimes F}(u) = \lambda_F(u)$.

Ceci résulte du fait que dans $K(S)$, l'élément $\mathrm{pr}_{1!}(F \otimes \mathrm{pr}_2^*(u))$ est de rang 0 en raison du choix de u .

Soit $M_X^s(c)$ l'ouvert de $M_X(c)$ des classes de faisceaux stables.

Théorème (2.5) . *Soit* $c \in K_{top}(X)$.

1) *Soit* $u \in Z(c)$; *il existe une classe* $\lambda_X(u) \in \mathrm{Pic}(M_X^s(c))$ *et une seule satisfaisant à la propriété universelle suivante : pour toute famille* F de $\mathcal{O}_X$*-modules stables de classe* c , *paramétrée par* S , *on a dans* $\mathrm{Pic}(S)$

$$f_F^*(\lambda_X(u)) = \lambda_F(u)$$

2) *Supposons* $u \in Z(c) \cap H^{\perp\perp}$; *il existe une classe* $\lambda_X(u) \in \mathrm{Pic}(M_X(c))$ *et une seule satisfaisant à la propriété universelle suivante : pour toute famille* F *de faisceaux semi-stables de classe* c , *paramétrée par* S , *on a dans* $\mathrm{Pic}(S)$

$$f_F^*(\lambda_X(u)) = \lambda_F(u)$$

Démonstration . Comme c'est l'usage, on note $u(m)$ (resp. $c(m)$) l'élément obtenu dans $K(X)$ (resp. $K_{top}(X)$) par multiplication de u (resp. c) par $\mathcal{O}_X(m)$. Le sous-anneau $H \subset K(X)$ est évidemment invariant par l'opération $u \rightarrow u(m)$. On a pour m entier $Z(c(m)) \cong Z(c)$ et pour $u \in Z(c)$ et F famille de $\mathcal{O}_X$-modules de classe c , $\lambda_{F(m)}(u(-m)) = \lambda_F(u)$. L'espace de modules $M_X(c)$ s'identifie à $M_X(c(m))$, et compte tenu du diagramme commutatif

$$\begin{array}{ccc} & \overset{f_F}{\nearrow} & M_X(c) \\ S & & \downarrow \\ & \underset{f_{F(m)}}{\searrow} & M_X(c(m)) \end{array}$$

on peut remplacer F par $F(m)$, ce qui revient à changer c en $c(m)$. On peut alors supposer que les propriétés (2.6), (2.7) et (2.8) ci-dessous sont satisfaites.

(2.6) *Tout faisceau semi-stable* F *de classe* c , *est engendré par ses sections et* $H^q(F(m)) = 0$ *pour* $q > 0$ *et* $m \geq 0$.

On pose χ (F) = N . Ce nombre ne dépend que de c . Considérons le schéma de Grothendieck Groth ($\mathcal{O}_X{}^N$, c) des $\mathcal{O}_X$-modules cohérents quotients de $\mathcal{O}_X{}^N$ et de classe c [7]. C'est une variété projective sur laquelle on peut définir un fibré inversible naturel $\mathcal{A}$ de la manière suivante : on considère le morphisme "déterminant", à valeurs dans la composante $\mathrm{Pic}_{c_1}(X)$ des fibrés inversibles de classe de Chern c_1

$$\mathrm{Groth}\,(\mathcal{O}_X{}^N, c) \xrightarrow{\ j\ } \mathrm{Pic}_{c_1}(X)$$

Soit $\mathcal{P}$ un fibré de Poincaré sur $\mathrm{Pic}_{c_1}(X) \times X$. Le fibré universel quotient $\mathcal{F}$ sur Groth ($\mathcal{O}_X{}^N$, c) × X a une résolution localement libre finie, ce qui permet de définir le fibré inversible det $\mathcal{F}$, et on a d'après la propriété universelle de $\mathcal{P}$

$$\det \mathcal{F} \otimes ((j \times \mathrm{id}_X)^* \mathcal{P})^{-1} = \mathrm{pr}_1{}^* (\mathcal{A})$$

où $\mathcal{A} \in \mathrm{Pic}(\mathrm{Groth}\,(\mathcal{O}_X{}^N, c))$.

(2.7) *Le fibré* $\mathcal{A}$ *définit un morphisme* ψ *à valeurs dans un fibré en espaces projectif au-dessus de* $\mathrm{Pic}_{c_1}(X)$ *qui induit un plongement quand on le restreint à l'ouvert des quotients sans torsion.*

On a en effet, c étant convenable, $R^q\mathrm{pr}_{1*}(\mathcal{P}) = 0$ si $q > 0$ et $\mathrm{pr}_{1*}(\mathcal{P})$ est un fibré vectoriel sur $\mathrm{Pic}_{c_1}(X)$; d'autre part, le morphisme canonique sur Groth ($\mathcal{O}_X{}^N$, c) × X

$$\Lambda^r \mathcal{O}^N \longrightarrow \det \mathcal{F}$$

fournit un morphisme surjectif de fibrés vectoriels $j^* (\underline{\mathrm{Hom}}_{\mathrm{pr}_1} (\Lambda^r \mathcal{O}^N, \mathcal{P})) \longrightarrow \mathcal{A}$ et par suite un morphisme ψ, au-dessus de $\mathrm{Pic}_{c_1}(X)$:

$$\begin{array}{ccc} \mathrm{Groth}\,(\mathcal{O}_X{}^N, c) & \xrightarrow{\ \psi\ } & \mathbf{P}\,(\underline{\mathrm{Hom}}_{\mathrm{pr}_1} (\Lambda^r \mathcal{O}^N, \mathcal{P})) \\ & {\scriptstyle j}\searrow \quad \swarrow & \\ & \mathrm{Pic}_{c_1}(X) & \end{array}$$

tel que $\psi\,(\mathcal{O}(1)) = \mathcal{A}$. Sur l'ouvert des points sans torsion, ψ est un plongement : c'est l'analogue du plongement de Plucker [12].

Le groupe SL(N, C) opère sur Groth ($\mathcal{O}_X{}^N$, c) et j est SL(N, C) - équivariant. On peut alors considérer l'ouvert Ω_{ss} (resp. Ω_s) de Groth ($\mathcal{O}_X{}^N$, c) des points semi-stables (resp. stables) sous l'action de SL(N, C) .

Quitte à remplacer c par c(m) pour m convenable, on peut supposer que :

(2.8) *On a l'implication* (ii) ⇒ (i)

(i) $q \in \Omega_{ss}$

(ii) *le faisceau quotient* F_q *défini par* $q \in$ Groth $(\mathcal{O}_X{}^N, c)$ *est semi-stable, et le morphisme canonique* $C^N \longrightarrow H^0(F_q)$ *est un isomorphisme.*

Dans les cas $n = 1$, on a en fait l'équivalence (i) $\Leftrightarrow$ (ii) et ceci conduit à la construction de l'espace de modules $M_X(c)$ comme quotient de Mumford $\Omega_{ss}/SL(N, C)$. D. Gieseker dans le cas $\dim X = 2$ et M.Maruyama dans le cas $\dim X \geq 2$ montrent que l'ouvert R_{ss} des points $q \in$ Groth $(\mathcal{O}_X{}^N, c)$ satisfaisant à (ii) est encore l'image réciproque d'un sous-schéma à la fois ouvert et fermé de $\Omega_{ss}/SL(N, C)$, ce qui conduit encore à la construction de $M_X(c)$. ([6], théorème (0.7); [12], Proposition (4.10) et corollaire (5.9.1)).

Désignons par $\pi : R_{ss} \longrightarrow \Omega_{ss}/SL(N, C)$ la projection canonique.

(2.9) Construction de $\lambda_X(u)$. On considère le quotient universel $\mathcal{F}$ paramétré par Groth $(\mathcal{O}_X{}^N, c)$, et le fibré inversible $\lambda_{\mathcal{F}}(u)$ associé, auquel on veut appliquer le lemme de descente (1.4) : il est en effet muni d'une GL(N, C)-action, et compte-tenu du fait que $u \in Z(c)$, cette action se factorise à travers PGL(N, C). Sur l'ouvert R_s des points stables de R = Groth $(\mathcal{O}_X{}^N, c)$, le groupe PGL(N, C) opère librement ; par suite il existe un fibré inversible $\lambda_X(u) \in \mathrm{Pic}(M_X^s(c))$ et un seul tel que

$$\pi^*(\lambda_X(u)) = \lambda_{\mathcal{F}}(u)\,|_{R_s}$$

Sur l'ouvert R_{ss}, on doit vérifier que si $u \in H^{\perp\perp} \cap Z(c)$, le stabilisateur de $q \in R_{ss}$ opère trivialement sur la fibre de $\lambda_{\mathcal{F}}(u)$ en q. Les points $q \in R_{ss}$ d'orbite fermée correspondent aux faisceaux $F = F_q$ qui s'écrivent $F = F_1^{m_1} \oplus \ldots \oplus F_k^{m_k}$ où les faisceaux F_i sont stables, $m_i \geq 1$, $F_i \neq F_j$ pour $i \neq j$, et

$$\frac{\chi_{F_i}}{r_i} = \frac{\chi_F}{r} \qquad (\text{où } r_i = \mathrm{rg}(F_i)) \qquad (2)$$

L'isomorphisme $C^N \cong H^0(F)$ induit une décomposition $C^N \cong \bigoplus_{i=1}^{k} H^0(F_i) \otimes C^{m_i}$, et dans cette somme directe, le stabilisateur Stab(q) s'identifie aux automorphismes de F et correspond aux $g \in GL(N, C)$ de la forme $g = \bigoplus_{i=1}^{k} \mathrm{id}_{H^0(F_i)} \otimes g_i$ avec $g_i \in GL(m_i, C)$. La GL(N, C)-structure sur $\lambda_{\mathcal{F}}(u)$ définit un caractère de Stat (q) $\cong GL(m_1, C) \times \ldots \times GL(m_k, C)$ qu'on détermine facilement (cf. [4])

Lemme (2.10). *Pour tout* $u \in K(X)$ *le caractère* $\mathrm{Stab}(q) \longrightarrow C^*$ *associé à* $\lambda_{\mathcal{F}}(u)$ *au point* q *est donné par*

$$(g_1, \ldots, g_m) \longrightarrow \prod_{i=1}^{k} (\det g_i)^{\chi(F_i(u))}$$

---

(2) χ_F désigne le polynôme de Hilbert de F, r le rang.

Démonstration. On rappelle que $K(X)$ possède une filtration décroissante canonique $F^0 \supset F^1 \supset \ldots \supset F^n \supset \{0\}$ indexée par la codimension du support des faisceaux cohérents. Soit $u \in F^i K(X)$; on démontre la formule par récurrence descendante sur i ; pour $i > \dim X = n$, il n'y a rien à démontrer. Il suffit donc de vérifier la formule pour $G = \mathcal{O}_Y$, où Y est une sous-variété intègre de codimension i ; compte-tenu de la formule $\mathcal{O}_Y(m) = \mathcal{O}_Y \mod F^{i+1}$, il revient au même de vérifier la formule pour $G = \mathcal{O}_Y(m)$; on peut supposer que m est assez grand pour que

$$R^p pr_{1*} (\underline{Tor}_j (\mathcal{F}, pr_2^* (G)) = 0 \qquad \text{pour } p > 0 \text{ et } j \geq 0$$

de sorte que $pr_{1*} (\underline{Tor}_j (\mathcal{F}, pr_2^* (G))$ est localement libre et que $pr_{1!} (\mathcal{F}. pr_2^* (G))$ est donné, dans le groupe de Grothendieck engendré par les modules localement libres de type fini sur R_{ss} munis d'une action de $GL(N, C)$, par

$$\sum_{j=o}^{n} (-1)^j pr_{1*} (\underline{Tor}_j (\mathcal{F}, pr_2^* (G))$$

Tous les faisceaux $\underline{Tor}_j (\mathcal{F}, pr_2^* (G))$ introduits sont plats sur R_{ss} ; on peut leur appliquer le théorème du changement de base [10] qui donne, pour tout point $q \in R$, $F = F_q$ $H^p (\underline{Tor}_j (F, G)) = 0$ pour $p > 0$, $j \geq 0$ et $pr_{1*} \underline{Tor}_j (\mathcal{F}, pr_2^* (G))$ est localement libre, de fibre $H^0 (\underline{Tor}_j (F, G))$. Par suite, la fibre en q de $\lambda_{\mathcal{F}}(G)$ s'identifie en tant que représentation de Stab(q) à

$$\bigotimes_j \left(\Lambda^{max} H^0 (\underline{Tor}_j (F, G))\right)^{(-1)^j} = \bigotimes_{i,j} \left(\Lambda^{max} H^0 (\underline{Tor}_j (F_i, G)) \otimes C^{m_i}\right)^{(-1)^j}$$

Alors le caractère de $\text{Stab}(q) = \prod_{i=1}^{k} GL(m_i, C)$ associé est donné par

$$(g_1, \ldots, g_k) \longrightarrow \prod_{i=1}^{k} \det (g_i)^{\Sigma (-1)^j h^0 (\underline{Tor}_j (F_i, G))} = \prod_{i=1}^{k} \det (g_i)^{\chi (F_i, G)}$$

d'où le lemme.

On est ramené à démontrer que si $u \in Z(c) \cap H^{\perp\perp}$ on a $\chi(F_i(u)) = 0$, c'est-à-dire que u est orthogonal à $F_1, \ldots, F_k$. Or, dans $K(X) \otimes \mathbf{Q}$, l'hypothèse $\frac{\chi_{F_i}}{r_i} = \frac{\chi_F}{r}$ signifie que $\frac{F}{r} - \frac{F_i}{r_i}$ est orthogonal à H. D'autre part, l'hypothèse $u \in Z(c)$ signifie que u est orthogonal à F. Donc si $u \in Z(c) \cap H^{\perp\perp}$, u est orthogonal à F_i pour tout $i = 1, \ldots, k$.

Remarque (2.11). Considérons le sous-groupe $H'(c)$ de $K_{top}(X) \otimes \mathbf{Q}$ engendré par c et le sous-groupe de $H^{\perp}$ des classes qui s'écrivent

$$\frac{c'}{r'} - \frac{c''}{r''}$$

avec $M_X(c')$ et $M_X(c'')$ non vide, et $c' + c'' = c$. On pose $Z^{ss}(c) = H'(c)^{\perp}$. Alors la démonstration ci-dessus montre que $Z^{ss}(c)$ coïncide avec le sous-groupe de $Z(c)$ des éléments u tels que le fibré inversible $\lambda_{\mathcal{S}}(u)$ provienne de $M_X(c)$. On a

$$Z(c) \cap H^{\perp\perp} \subset Z^{ss}(c) \subset Z(c)$$

En général, le sous-groupe $Z^{ss}(c)$ est plus petit que $Z(c)$. C'est ce qui conduit Drézet à trouver sur certaines surfaces rationnelles des classes c telles que la variété $M_X(c)$ soit normale, mais non localement factorielle [4].

Remarque (2.12). Supposons que les entiers $\chi(c\, h^i)$ $(i = 0, \dots, n)$ soient premiers entr'eux. Alors

$$M_X^s(c) = M_X(c) .$$

En effet $\frac{c'}{r'} - \frac{c}{r} \in H^{\perp}$ équivaut à $\frac{\chi(c'.h^i)}{r'} = \frac{\chi(c.h^i)}{r}$ pour tout i. Ceci impose $r' = r$ et par suite $M_X(c)$ n'a que des points stables.

(2.13) La propriété universelle. Soit F une famille $\mathcal{O}_X$-modules semi-stables de classe c paramétrée par S. D'après le choix de c fait en (2.6) on a

$$R^q pr_{1*}(F) = 0 \qquad \text{pour } q > 0$$

et $G = pr_{1*}(F)$ est un faisceau localement libre de rang N. On désigne par $\mathcal{R}$ le fibré des repères de G, $\mathcal{S} = \mathcal{R}/\mathbb{C}^*$, $\pi' : \mathcal{S} \longrightarrow S$ la projection canonique. La variété $\mathcal{S}$ est un ouvert du fibré en espace projectif $\mathbf{P}(\underline{\mathrm{Hom}}(\mathcal{O}^N, G)$ et donc munie d'un faisceau inversible $\mathcal{O}(1)$. La variété $\mathcal{S}$ est munie d'une action libre de $PGL(N, \mathbb{C})$, de quotient S ; sur $\mathcal{S}$, l'inclusion $\mathcal{O}(-1) \hookrightarrow \pi'^* \underline{\mathrm{Hom}}(\mathcal{O}_{\mathcal{S}}^N, G)$ induit un isomorphisme canonique $\mathcal{O}_{\mathcal{S}}^N(-1) \cong \pi'^*(G)$, et en composant avec le morphisme d'évaluation on obtient un épimorphisme sur $\mathcal{S} \times X$

$$\mathcal{O}_{\mathcal{S}\times X}^N \to (\pi' \times id_X)^*(F) \otimes \mathcal{O}_{\mathcal{S}}(1) = \mathcal{F}$$

ce qui fournit d'après la propriété universelle du schéma de Grothendieck un morphisme

$$\mathcal{S} \xrightarrow{f} R = \mathrm{Groth}(\mathcal{O}^N, c)$$

équivariant pour l'action de $PGL(N, \mathbb{C})$, et un isomorphisme $F' \cong (f' \times id_X)^*(\mathcal{F})$. Pour $u \in K(X)$, on obtient par fonctorialité dans $\mathrm{Pic}(\mathcal{S})$, $\lambda_{F'}(u) = f'^*(\lambda_{\mathcal{S}}(u))$. D'autre part, si $u \in Z(c)$, $\lambda_{F'}(u) = \pi'^*(\lambda_F(u))$. Le morphisme f' prend ses valeurs dans l'ouvert R_{ss} et fournit par passage au quotient le morphisme modulaire f ; le diagramme commutatif

$$\begin{array}{ccc} \mathcal{S} & \xrightarrow{f'} & R_{ss} \\ {\scriptstyle \pi'}\downarrow & & \downarrow{\scriptstyle \pi} \\ S & \xrightarrow[f]{} & M_X(c) \end{array}$$

induit sur les groupes de Picard un diagramme commutatif

$$\begin{array}{ccc} \mathrm{Pic}(\mathcal{S}) & \longleftarrow & \mathrm{Pic}(R_{ss}) \\ {\scriptstyle \pi'^*}\uparrow & & \uparrow \\ \mathrm{Pic}(S) & \longleftarrow & \mathrm{Pic}\,(M_X(c)) \end{array}$$

Supposons que $u \in Z^{ss}(c)$. Alors $\lambda_{\mathcal{F}}(u)|_{R_{ss}}$ provient de $M_X(c)$ (cf. remarque 2.11) : $\lambda_{\mathcal{F}}(u) = \pi^*(\lambda_X(u))$. On obtient donc $\pi'^* f'^* (\lambda_X(u)) = \pi'^* (\lambda_X(u))$.

Pour obtenir $f^*(\lambda_X(u)) = \lambda_F(u)$, il suffit donc de constater que $\pi'^* : \mathrm{Pic}(S) \longrightarrow \mathrm{Pic}(\mathcal{S})$ est injectif. Ce morphisme se factorise à travers $\mathrm{Pic}^{PGL(N, C)}(\mathcal{S})$, et comme PGL(N, C) est un groupe réductif qui opère librement sur $\mathcal{S}$, le morphisme $\mathrm{Pic}(S) \longrightarrow \mathrm{Pic}^{PGL(N, C)}(\mathcal{S})$ est un isomorphisme. Ainsi, il suffit de vérifier le lemme suivant :

Lemme (2.14). ([5], prop. 3.2 ou [16] prop. 1.4 chapitre 1). *Soit* Y *une variété algébrique sur laquelle agit* PGL(N, C) *. Alors le morphisme d'oubli* $\mathrm{Pic}^{PGL(N, C)}(Y) \longrightarrow \mathrm{Pic}(Y)$ *est injectif.*

Démonstration . Il s'agit de vérifier que si $\chi : PGL(N, C) \times Y \longrightarrow C^*$ est un morphisme croisé, alors $\chi = 1$. Mais e désignant l'élément neutre de PGL(N, C), $\chi|_{\{e\} \times Y}$ est le morphisme constant égal à 1 . D'autre part, tous les morphismes $PGL(N, C) \longrightarrow C^*$ sont constants ; donc χ provient, par la projection pr_2, d'un élément de $H^0(\mathcal{O}_Y)^*$. Cet élément ne peut être que 1 .

Dans [5], cet énoncé est démontré pour une variété lisse, mais l'argument utilisé est valable pour une variété quelconque, même non réduite. En fait, on peut remplacer PGL(N, C) par n'importe quel groupe réductif tel que $\mathrm{Hom}\,(G, C^*) = 1$ (cf. [16]).

Ceci démontre la partie (2) du théorème. Si $u \in Z(c)$ le faisceau $\lambda_{\mathcal{F}}(u)|_{R_s}$ provient d'un élément $\lambda_X(u)$ de $\mathrm{Pic}\,(M_X^s(c))$, et le même argument donne, pour une famille de $\mathcal{O}_X$-modules stables F paramétrée par S l'égalité $\lambda_F(u) = f^*(\lambda_X(u))$ dans Pic(S) .

(2.15) Unicité de $\lambda_X(u)$. En considérant la famille universelle $\mathcal{F}$, on est ramené à vérifier que les morphismes images réciproques

$$\mathrm{Pic}\,(M_X(c)) \xrightarrow{\pi^*} \mathrm{Pic}\, R_{ss} \qquad\qquad \mathrm{Pic}\,(M_X^s(c)) \xrightarrow{\pi^*} \mathrm{Pic}\, R_s$$

sont injectifs. Mais ceci résulte du lemme (2.13), et du fait que $\mathrm{Pic}(M_X(c)) \longrightarrow \mathrm{Pic}^{PGL(N, C)}(R_{ss})$

est injectif ainsi que $\mathrm{Pic}(M_X^s(c)) \longrightarrow \mathrm{Pic}^{PGL(N,C)}(R_{ss})$.

3. Les fibrés inversibles $\mathcal{L}_i$

On pose pour $c \in K_{top}(X)$ de rang r, $\chi_i = \chi(c.h^i)$ pour $i = 0, \dots, n$. Ainsi $\chi_0 = \chi$ est la caractéristique d'Euler-Poincaré de c, $\chi(c\,h^n) = r \deg X$ où $\deg X$ est le degré de la variété projective polarisée X. On pose, dans $K(X)$

$$u_i = -r\,h^i + D_i$$

où D_i est un cycle de dimension 0 et de degré χ_i.

Lemme (3.1). *Pour tout* $i = 0, \dots, n$, *l'élément* u_i *appartient à* $H^{\perp\perp} \cap Z(c)$.

Le fait que $H^{\perp}$ est constitué de classes de rang $r = 0$ entraine $A^n(X) \subset H^{\perp\perp}$ et donc u_i appartient à $H^{\perp\perp}$. D'autre part, $\chi(c\,u_i) = -r\,\chi(c\,h^i) + \chi(c.\,D_i) = 0$; donc $u_i \in Z(c)$. D'où l'énoncé.

Il résulte du théorème (2.3) que l'on peut définir, à isomorphisme près, le fibré inversible sur $M_X(c)$

$$\mathcal{L}_i = \lambda_X(u_i)$$

On considère le morphisme canonique $M_X(c) \xrightarrow{j} \mathrm{Pic}_{c_1}(X)$ et on pose

$$\mathrm{Pic}_{rel}(M_X(c)) = \mathrm{Pic}\,M_X(c)/j^*\,\mathrm{Pic}(\mathrm{Pic}_{c_1}(X))$$

Proposition (3.2). *La classe de* $\mathcal{L}_i$ *dans* $\mathrm{Pic}_{rel}(M_X(c))$ *est indépendante du choix de* D_i.

Il suffit de vérifier le lemme suivant :

Lemme (3.3). *Soient* F *une famille plate de* $\mathcal{O}_X$*-modules de classe* c, *paramétrée par* S, *et* $D \in A^n(X)$. *On a* $\lambda_F(D) = \lambda_{\det F}(D)$ *dans* $\mathrm{Pic}(S)$

Démonstration. Les deux membres ont bien un sens pour toute famille plate F de $\mathcal{O}_X$-modules, et si l'on a une suite exacte $0 \longrightarrow F' \longrightarrow F \longrightarrow F'' \longrightarrow 0$ de tels modules, on a $\lambda_F(D) = \lambda_{F'}(D) \otimes \lambda_{F''}(D)$, de sorte que l'on peut supposer que F est une famille de fibrés vectoriels. D'autre part on peut par additivité supposer que D est un cycle de degré 1, défini par un point $a \in X$. On a alors $pr_{1!}(F \otimes pr_2^*(D)) = F|_{S\times\{a\}}$ et par suite

$$\lambda_F(D) = \det F|_{S\times\{a\}} = \lambda_{\det F}(D).$$

Corollaire (3.4). *Soient* $d_i = \mathrm{pgcd}(r, \chi_i)$, *et* $\mathcal{D}_i$ *le fibré sur* $M_X(c)$ *défini par l'élément de* K(X)

$$v_i = -\frac{r}{d_i} h^i + D'_i$$

où D'_i *est un 0-cycle de degré* $\frac{\chi_i}{d_i}$. *Alors, dans* $\mathrm{Pic}_{rel}(M_X(c))$, *on a* $\mathcal{D}_i^{\otimes d_i} = \mathcal{L}_i$.

Considérons la translation $M_X(c) \xrightarrow{\tau_m} M_X(c(m))$.

Proposition (3.5). *Soit* $\mathcal{L}_{m,i} \in \mathrm{Pic}(M_X(c(m)))$ *le fibré inversible construit sur* $M_X(c(m))$. *On a, dans* $\mathrm{Pic}_{rel}(M_X(c))$

$$\tau_m^* (\mathcal{L}_{m,i}) = \bigotimes_{j \geq 0} \mathcal{L}_{i+j}^{\otimes C^j_{m+j-1}}$$

Démonstration. Evidemment H et donc $H^{\perp}$ et $H^{\perp\perp}$ sont invariants par la translation $u \rightarrow u(m)$. De plus, $u \in Z(c(m))$ si et seulement si $u(m) \in Z(c)$ et $\lambda_F(u(m)) = \lambda_{F(m)}(u)$ pour $u \in Z(c(m))$. D'autre part, dans K(X)

$$\mathcal{O}_X(m) = (1-h)^{-m} = \sum_{j \geq 0} C^j_{m+j-1} h^j$$

Pour toute famille F de faisceaux semi-stables de classe c , paramétrée par S , on a dans Pic(S) , noté additivement,

$$\lambda_F(u(m)) = \sum_{j \geq 0} C^j_{m+j-1} \lambda_F(u h^j)$$

Prenons $u_i = -rh^i + \Delta_i$ où Δ_i est un cycle de degré $\chi(c(m)h^i)$. On a alors $u_i \in H^{\perp\perp} \cap Z(c)$ et

$$\lambda_{F(m)}(u_i) = \sum_{j \geq 0} C^j_{m+j-1} \lambda_F(-rh^{i+j} + D_{i+j}) + \lambda_F(\Delta_i - \sum_{j \geq 0} C^j_{m+j-1} (D_{i+j}))$$

où D_{i+j} est un cycle de degré $\chi(c\, h^{i+j})$. D'après la proposition (3.2) on obtient

$$\tau_m^* (\mathcal{L}_{m,i}) = \bigotimes_{j \geq 0} \mathcal{L}_{i+j}^{\otimes C^j_{m+j-1}}$$

Théorème (3.6). *Pour* m *assez grand* $\mathcal{L}_{m,0}$ *est relativement ample par rapport à* $\mathrm{Pic}_{c_1+mrh}(X)$.

Ceci signifie que pour m >> 0 , il existe, au-dessus de $\mathrm{Pic}_{c_1}(X)$ un plongement $M_X(c(m)) \xhookrightarrow{\varphi_m} \mathrm{Pic}_{c_1+mrh}(X) \times \mathbf{P}_k$ tel que $\varphi_m^*(\mathcal{O}_{\mathbf{P}_k}(1))$ soit une puissance positive de $\mathcal{L}_{m,0}$ dans $\mathrm{Pic}_{rel}(M_X(c))$.

Démonstration. On peut supposer, quitte à changer c en c(m), pour m assez grand, que c satisfait aux conditions (2.4), (2.5) et (2.6). On doit vérifier que $\mathscr{L}_0$ est relativement ample par rapport à $\mathrm{Pic}_{c_1}(X)$. On prend les notations du § 2.

Lemme (3.7). *Pour* $D \in A^n(X)$ *on a dans* $\mathrm{Pic}(R_{ss})$

$$\lambda_{\det \mathscr{F}}(D) = \mathscr{A}^{\otimes \deg D} \bmod j^*\left(\mathrm{Pic}\,(\mathrm{Pic}_{c_1}(X))\right).$$

Démonstration. Il suffit de le vérifier pour un cycle D de degré 1. Supposons D défini par un point a. On a par définition $\det \mathscr{F} = \mathscr{A} \otimes (j \times \mathrm{id})^* (\mathscr{P})$ et donc

$$\det \mathscr{F}\big|_{R_{ss} \times \{a\}} = \mathscr{A} \otimes j^* \left(\mathscr{P}\big|_{\mathrm{Pic}_{c_1}(X) \times \{a\}}\right)$$

D'où le résultat.

Démonstration du théorème (3.6). Pour $u = -r\mathscr{O}_X + D$, avec $D \in A^n(X)$ de degré $\chi = N$, on a d'après le lemme (3.3)

$$\lambda_{\mathscr{F}}(D) = (\det \mathrm{pr}_{1!}(\mathscr{F}))^{\otimes -r} \otimes \lambda_{\det \mathscr{F}}(D)$$

En raison du choix de c, $R^q\mathrm{pr}_{1*}(\mathscr{F})\big|_{R_{ss}} = 0$ pour $q > 0$ et $\mathrm{pr}_{1*}(\mathscr{F})\big|_{R_{ss}} \cong \mathscr{O}^N$. On obtient, après application du lemme (3.7)

$$\lambda_{\mathscr{F}}(u) = \det\,(H^0(\mathscr{O}_X)^N)^{\otimes -r} \otimes \mathscr{A}^{\otimes N} \bmod j^*\left(\mathrm{Pic}\,(\mathrm{Pic}_{c_1}(X))\right)$$

On rappelle que $\mathscr{A}$ est le fibré image réciproque de $\mathscr{O}(1)$ par le morphisme

$$\psi : R \hookrightarrow \mathbf{P}\,(\underline{\mathrm{Hom}}_{\mathrm{pr}_1}(\Lambda^n \mathscr{O}^N, \mathscr{P})$$

Les deux membres de l'égalité ci-dessus sont des PGL(N, C) - fibrés inversibles sur R, et l'égalité ci-dessus est donc vraie en tant que PGL(N, C) - fibrés vectoriels, d'après le lemme (2.14), donc en tant que SL(N, C) - fibrés vectoriels : $\lambda_{\mathscr{F}}(u) = \mathscr{O}(N)$. Elle implique que $\mathscr{O}(N)$ se quotiente sur $R_{ss}/SL(N, C) = M_X(c)$; le quotient est $\mathscr{L}_0 = \lambda_X(u) = \mathscr{O}_{M_X(c)}(1)$ obtenu est évidemment relativement ample par rapport à $\mathrm{Pic}_{c_1}(X)$ par construction de $M_X(c)$.

Corollaire (3.8). *Le fibré inversible sur* $M_X(c)$

$$A_m = \bigotimes_{i=0}^{n-1} \mathscr{L}_i \otimes C^i_{m+i-1}$$

est relativement ample par rapport à $\mathrm{Pic}_{c_1}(X)$ *pour* m *assez grand.*

Corollaire (3.9). *Si* X *est une courbe, le fibré inversible* $\mathscr{L}_0$ *sur* $M_X(c)$ *est relativement ample par rapport à* $\mathrm{Pic}_{c_1}(X)$.

Dans ce cas, $c \in K_{top}(X)$ est déterminé par (r, χ), où r est le rang et χ la caractéristique d'Euler-Poincaré. Si X est de genre > 1 et $r > 1$, le groupe $\mathrm{Pic}_{rel}(M_X(c))$ est en fait isomorphe à $\mathbf{Z}$, et a pour générateur la classe du fibré inversible $\mathscr{D} = \lambda_X\left(\frac{-r}{d}\mathcal{O}_X + D\right)$ où $d = \mathrm{pgcd}(r, \chi)$, où D est un diviseur de degré $\frac{\chi}{d}$ (Drézet et Narasimhan [5], Seshadri [17], Ramanan [17], Beauville [2]). On a ici $\mathscr{L}_0 = \mathscr{D}^{\otimes d}$ dans le groupe de Picard relatif. Le fibré $\mathscr{D}$ est le fibré associé au diviseur Θ généralisé de Drézet et Narasimhan.

Corollaire (3.10). *On suppose que* X *est une surface. Alors le fibré* $\mathscr{L}_0 \otimes \mathscr{L}_1^{\otimes m}$ *est relativement ample par rapport à* $\mathrm{Pic}_{c_1}(X)$ *pour* m *suffisamment grand.*

Exemple (3.11). On prend pour X le plan projectif $\mathbf{P}_2$, et on considère l'espace de modules $M_{\mathbf{P}_2}(2, 0, n)$ des faisceaux semi-stables de rang 2, de classe de Chern $c_1 = 0$, $c_2 = n$. Pour $n > 2$, on sait que $\mathrm{Pic}\, M(2, 0, n) \cong \mathbf{Z}^2$ [20, 3] ; les fibrés $\mathscr{D}_0$ et $\mathscr{D}_1$ introduits ci-dessus forment une base de ce groupe de Picard.

Lorsque c_2 est impair, on a les relations

$$\mathscr{D}_1^{\otimes 2} = \mathscr{L}_1$$
$$\mathscr{D}_1 = \varphi$$
$$\mathscr{L}_0 = \psi - 2(n-2)\varphi$$

φ et ψ désignant les générateurs introduits par Strømme [20]. Le cône ample est constitué des classes $a\varphi + b\psi$, avec a et b entiers strictement positifs. Ainsi, aucun des deux fibrés $\mathscr{L}_0$ et $\mathscr{L}_1$ n'est ample.

(3.12) L'ouvert de semi-stabilité. Soit F une famille plate quelconque de $\mathcal{O}_X$-modules de classe c paramétrée par S. On considère les fibrés inversibles $L_{F,i} = \lambda_F(u_i)$ sur S, où $u_i = -r\, h^i + D_i$, D_i cycle de degré $\chi(ch^i)$, et $\Lambda_{F,m} = \bigotimes_{i \geq 0} L_{F,i}^{\otimes C^i_{m+i-1}}$.

D'autre part, on considère le morphisme déterminant $S \xrightarrow{g} \mathrm{Pic}_{c_1}(X)$.

Proposition (3.13). *Soit* U_{ss} (resp. U_s) *l'ouvert des points* t *de* S *tels que* F(t) *soit semi-stable (resp. stable). Localement, au-dessus de* $\mathrm{Pic}_{c_1}(X)$, *l'ouvert* U_{ss} (resp. U_s) *est réunion finie d'ouverts principaux définis par des sections de* $\Lambda_{F,m}^{\otimes \ell}$ *pour* m *et* ℓ *assez grand.*

Démonstration. On choisit m assez grand pour que

(1) $R^q pr_{1*}(F(m)) = 0$ pour $q > 0$;
(2) $G = pr_{1*}(F(m))$ engendre $F(m)$.

Alors $G = pr_{1*}(F(m))$ est un fibré vectoriel de rang $N = \chi(c(m))$, dont on considère le fibré des repères $\mathscr{R} \longrightarrow S$; on pose $\mathscr{S} = \mathscr{R}/\mathbf{C}^*$, et on désigne par $\pi' : \mathscr{S} \longrightarrow S$ la projection canonique. Au § (2.13) nous avons obtenu un morphisme $SL(N, \mathbf{C})$ - équivariant associé à F(m)

$$\begin{array}{ccc} \mathcal{S} & \xrightarrow{f'} & R = \mathrm{Groth}(\mathcal{O}^N, c(m)) \\ & \searrow & \downarrow j \\ & & \mathrm{Pic}_{c_1 + mrh}(X) \end{array}$$

Au-dessus d'un ouvert affine V de $\mathrm{Pic}_{c_1}(X) \cong \mathrm{Pic}_{c_1 + mrh}(X)$ l'ouvert Ω_{ss} des points q $\in$ R semi-stables pour l'action de SL(N, C) est définie par la réunion de tous les ouverts principaux $D_s = \{s \neq o\}$ associés aux sections SL(N, C) invariantes de $\mathcal{O}_R(\ell)$ au-dessus de V, qui proviennent du plongement décrit au § (2.7). L'ouvert $R_{ss} \subset \Omega_{ss}$ des points q qui satisfont à la condition (ii) de (2.8) est la réunion de certains de ces ouverts principaux $D_{s_1}, \dots, D_{s_k}$, associés à des sections SL(N, C) -invariantes de $\mathcal{O}_R(\ell)$, avec ℓ convenable. L'ouvert $f'^{-1}(R_{ss})$ est donc le complémentaire du fermé de $\mathcal{S}$ défini par les équations $f'^*(s_i) = o$ $(i = 1, \dots, k)$. D'après la démonstration de (3.6), ces sections s'interprètent, au moins si V est assez petit, comme des sections SL(N, C) -invariantes du fibré inversible $\pi'^*(\lambda_{F(m)}(u)^{\otimes \ell})$, avec $u = -r\,\mathcal{O}_X + D$, D cycle de degré $\chi(c(m))$; ces sections proviennent donc de sections $\sigma_1, \dots, \sigma_k$ de $\lambda_{F(m)}(u)^{\otimes \ell}$ sur l'ouvert $g^{-1}(V)$. Comme on l'a vu dans la démonstration de la proposition (3.5), ce fibré n'est autre, modulo $g^*(\mathrm{Pic}_{c_1}(X))$, que $\Lambda_{F,m}^{\otimes \ell}$. Le même argument est valable pour U_s, d'où la proposition.

(3.14) Version équivariante. Soit G un groupe algébrique opérant sur S et F une famille équivariante de $\mathcal{O}_X$ -modules cohérents de classe c : on a donc, pour tout $g \in G$, un isomorphisme

$$\varphi_g : F \longrightarrow (g \times \mathrm{id})^* F$$

tel que $\varphi_{g'g} = g^*(\varphi_{g'})\,\varphi_g$ pour tout g et $g' \in G$. Alors le fibré $\Lambda_{F,m}$ se trouve équipé d'une action de G, et l'ouvert $U_{ss} \subset S$ est G -invariant. Plus précisément

Lemme (3.15). *Localement au dessus de* $\mathrm{Pic}_{c_1}(X)$, *le fermé* S_{inst} *des points* $t \in S$ *tels que* F(t) *soit instable (resp. non stable) est défini par des sections invariantes d'une puissance de* $\Lambda_{F,m}$.

Démonstration. On considère le diagramme introduit dans la démonstration de la proposition (3.9) :

$$\begin{array}{ccc} \mathcal{S} & \xrightarrow{f'} & R = \mathrm{Groth}(\mathcal{O}^N, c(m)) \\ {\scriptstyle \pi'}\downarrow & & \\ S & & \end{array}$$

Le fibré des repères $\mathscr{R}$ de $pr_{1*}(F(m))$ est équipé d'une action naturelle de $SL(N, C) \times G$ et le morphisme f' est équivariant pour cette action. Il en résulte que les sections $f'^*(s_i)$ sont des sections $SL(N, C) \times G$ invariantes de $\pi'^*(\lambda_{F(m)}(u)^{\otimes \ell})$ et par suite les sections σ_i de $\Lambda_{F,m}^{\otimes \ell}$ sont G -invariantes au dessus de l'ouvert V.

4. Le fibré inversible $\mathscr{L}_1$

Soit F un faisceau semi-stable sur X. On sait d'après un théorème de Mehta et Ramanathan que pour m assez grand, la restriction $F|_Y$ de F à une hypersurface $Y \in |\mathcal{O}_X(m)|$, lisse et de degré assez grand, est encore μ-semi-stable. Nous ignorons s'il est encore vrai que $F|_Y$ est semi-stable. Faute d'un tel résultat nous devrons nous limiter dans ce qui suit au cas où X est une surface.

Théorème (4.1). *Il existe une suite infinie d'entiers υ satisfaisant aux conditions suivantes :*

(1) *Le morphisme d'évaluation* $j^* j_* (\mathscr{L}_1^{\otimes \upsilon}) \xrightarrow{ev} \mathscr{L}_1^{\otimes \upsilon}$ *est surjectif.*

Soit $\varphi_\upsilon : M_X(c) \longrightarrow P(j_*(\mathscr{L}_1^{\otimes \upsilon}))$ *le morphisme associé.*

(2) *Soit* U *l'ouvert de* $M_X(c)$ *des classes de faisceaux localement libres* μ*-stables,* $\partial M_X(c)$ *son complémentaire. Alors* $\varphi_\upsilon|_U$ *est un plongement et* φ_υ *sépare les points de* U *et de* $\partial M_X(c)$.

Démonstration. On choisit m assez grand pour que pour tous $\mathcal{O}_X$-modules semi-stables F et F' de classe c, F étant localement libre, on ait $Ext^1(F', F(-m)) = 0$: c'est possible d'après le théorème de dualité et le théorème B de Serre, les familles étudiées étant limitées.

On considère pour $Y \in |\mathcal{O}_X(m)|$, lisse, l'ouvert $U_{Y,ss}$ des classes de faisceaux semi-stables F dont le gradué de Jordan-Holder satisfait à la condition suivante : $gr_i(F)|_Y$ est semi-stable pour tout i. D'après le théorème de restriction de Mehta et Ramanathan [14] ces ouverts recouvrent $M_X(c)$ si m est assez grand. D'autre part, d'après le lemme 3.13, chacun de ces ouverts est défini localement au-dessus de $Pic_{c_1}(X)$ par des sections de $\mathscr{L}_1^{\otimes m m'}$ pour m' assez grand : il suffit en effet d'appliquer ce lemme à la famille $G = F|_Y$ où F est la famille universelle quotient paramétrée par $R = Groth(\mathcal{O}^N, c(m''))$, muni de l'action de $SL(N, C)$, compte-tenu du lemme suivant :

Lemme (4.2). *Pour tout entier* $m'' \geq 0$ *le fibré* $\Lambda_{G,m''}$ *associé à la famille* $G = F|_Y$ *sur* $R \times Y$ *se quotiente sur* $M_X(c)$, *en un fibré inversible dont la classe dans* $Pic_{rel}(M_X(c))$, *est* $\mathscr{L}_1^{\otimes m}$.

Démonstration . Ce lemme résulte de la définition, du fait que dans $K(X)$, $\mathcal{O}_Y = mh$ mod $A^2(X)$, des lemmes (3.3) et (3.7).

Il en résulte que la classe de $\mathcal{L}_1^{\otimes \upsilon}$ dans $\mathrm{Pic}_{rel}(M_X(c))$ est celle d'un fibré engendré par ses sections pour υ assez grand et multiple de m . Autrement dit, pour un tel υ le morphisme d'évaluation $j^* j_* \mathcal{L}_1^{\otimes \upsilon} \longrightarrow \mathcal{L}_1^{\otimes \upsilon}$ est surjectif. On a donc un morphisme

$$\varphi_\upsilon : M_X(c) \longrightarrow \mathbf{P}(j_* (\mathcal{L}_1^{\otimes \upsilon}))$$

tel que $\varphi_\upsilon^* (\mathcal{O}(1)) = \mathcal{L}_1^{\otimes \upsilon}$

Pour vérifier l'assertion (2), on peut se placer au-dessus de la fibre, notée $M_X(c, L)$, de $M_X(c)$ au-dessus d'un point $L \in \mathrm{Pic}_{c_1}(X)$. En effet le diagramme

$$\begin{array}{ccc} M_X(c) & \longleftarrow & \mathcal{L}_1 \\ \Big\downarrow j & & \\ \mathrm{Pic}_{c_1}(X) & & \end{array}$$

est localement trivial dans la topologie étale de $\mathrm{Pic}_{c_1}(X)$ et on a par changement de base [10]

$$j_* (\mathcal{L}_1^{\otimes \upsilon}) \otimes_{\mathcal{O}_L} \mathbf{C} \cong H^0 (M_X(c, L) , \mathcal{L}_1^{\otimes \upsilon})$$

Considérons l'ouvert $U_{Y,s}$ des classes de faisceaux $F \in M_X(c)$ tels que $F|_Y$ soit stable. Un tel faisceau est obligatoirement μ-stable . D'après le théorème de restriction de Mehta et Ramanathan [15], on a si m est assez grand

$$U \subset \bigcup_Y U_{Y,s}$$

où U désigne l'ouvert des faisceaux localement libres μ-stable .

Considérons l'ouvert $V_{Y,L}$ des classes de $\mathcal{O}_X$-modules F de déterminant L , localement libres et tels que $F|_Y$ soit stable. Posons $c_Y = c|_Y$, $L_Y = L|_Y$, et désignons par $\mathcal{L}_{Y,0}$ le fibré relativement ample construit sur $M_Y(c_Y)$, et par $M_Y(c_Y, L_Y)$ la fibre au-dessus de L_Y . Considérons le diagramme

$$\begin{array}{ccc} M_X(c, L) \supset V_{Y,L} & \xrightarrow{\ r\ } & M_Y(c_Y, L_Y) \\ \Big\downarrow \varphi_{m\,m'} & & \Big\downarrow \varphi_{Y,m'} \\ \mathbf{P}^{\cdot} \left(H^0 (M_X(c, L) , \mathcal{L}_1^{\otimes m\,m'})\right) & \overset{r'}{\dashrightarrow} & \mathbf{P}^{\cdot} \left(H^0 (M_Y (c_Y, L_Y) , \mathcal{L}_{Y,0}^{\otimes m'})\right) \end{array}$$

où r est le morphisme de restriction. D'après le choix de m, le morphisme r est injectif et non ramifié [11]. On a d'autre part $r^*(\mathscr{L}_{Y,0})|_{V_{Y,L}} = \mathscr{L}_1^{\otimes m}$ d'après le lemme (4.2) ; les sections de $\mathscr{L}_{Y,0}^{\otimes m'}$ fournissent des sections de $\mathscr{L}_1^{\otimes mm'}$ d'après le paragraphe (3.14) . Ceci définit l'application rationnelle r'. Evidemment, le diagramme ci-dessus est commutatif. D'après le corollaire (3.9) appliqué à Y, $\varphi_{Y,m'}$ est un plongement si m' est assez grand. Il en résulte que $\varphi_{mm'}$ est injectif et non ramifié sur $V_{Y,L}$.

Or, deux points q et $q' \in U \cap M_X(c, L)$ qui n'appartiennent pas à un même ouvert $V_{Y,L}$ peuvent être séparés par des sections de $\mathscr{L}_1^{\otimes mm'}|_{M_X(c,L)}$, pourvu que m' soit assez grand : ceci résulte encore du lemme (3.15). Ainsi, le morphisme $\varphi_{mm'}$ est injectif et non ramifié si m' est assez grand.

Montrons que deux points $q \in U$ et $q' \in \partial M_X(c)$ peuvent être séparés par des sections de $\mathscr{L}_1^{\otimes mm'}|_{M_X(c,L)}$ pour $m' >> 0$. Choisissons en effet Y tel que q appartienne à $U_{Y,s}$. Si q' appartient à $U_{Y,ss}$, q' est la classe d'un faisceau semi-stable F' tel que $F'|_Y$ soit semi-stable. Si F est un faisceau μ-stable représentant q, on a $\mathrm{Hom}(F', F) = 0$. La suite exacte

$$\mathrm{Hom}(F', F) \longrightarrow \mathrm{Hom}(F'|_Y, F|_Y) \longrightarrow \mathrm{Ext}^1(F', F(-m))$$

et le choix de m montrent que $\mathrm{Hom}(F'|_Y, F|_Y) = 0$, et donc $F'|_Y$ n'est pas isomorphe à $F|_Y$. Puisque $F|_Y$ est stable, les points définis dans l'espace de modules $M_Y(c_Y, L_Y)$ sont distincts, donc séparés par $\varphi_{Y,m'}$, pourvu que m' soit assez grand. Alors F et F' sont séparés par les sections de $\mathscr{L}_1^{\otimes mm'}$. Si $q' \notin U_{Y,ss}$, on applique à nouveau le lemme (3.15) : il existe une section de $\mathscr{L}_1^{\otimes mm'}$ qui s'annule en q' et qui est non nulle au point q.

Il en résulte que si V désigne l'ouvert complémentaire de l'image de $\partial M_X(c)$ par φ_υ, on a $U = \varphi_\upsilon^{-1}(V)$. Alors $\varphi_\upsilon : U \longrightarrow V$ est injectif, non ramifié et propre. C'est donc un plongement fermé. D'où le théorème.

5. Le cas des surfaces rationnelles

Soit X une surface rationnelle. Le groupe de Picard de X est alors un groupe abélien libre de type fini, et les composantes $\mathrm{Pic}_{c_1}(X)$ introduites sont alors réduites à un point. On suppose de plus que le fibré canonique ω_X est de degré négatif.

Considérons d'abord l'espace de modules $M_X(r)$ des faisceaux semi-stables de rang r, de classes de Chern $c_1 = 0$, $c_2 = r$ c'est-à-dire de caractéristique d'Euler-Poincaré $\chi = 0$.

Proposition (5.1) . *Si* X *est une surface rationnelle dont le fibré canonique est de degré* <0 , *le fibré* $\mathcal{L}_0$ *construit sur* $M_X(r)$ *est trivial.*

En effet, le fibré canonique étant de degré négatif, pour tout faisceau semi-stable F sur X , on a

$$h^2(F) = \dim \operatorname{Hom}(F, \mathcal{O}_X)$$
$$= 0$$

D'autre part, on a aussi $h^0(F) = 0$, et donc $h^1(F) = 0$. Il en résulte que pour toute famille F de tels faisceaux semi-stables paramétrée par S , on a $R^q pr_{1*}(F) = 0$, et donc $pr_{1!}(F) = 0$. Par suite $\lambda_F(1) = 0$.

Corollaire (5.2) . *Considérons le fibré* $\mathcal{D}_1$ *construit en (3.4). Si* X *est une surface rationnelle dont le fibré canonique est de degré* <0 , *le fibré* $\mathcal{D}_1$ *(et donc* $\mathcal{L}_1$ *) construit sur* $M_X(r)$ *est ample.*

Exemple (5.3) . Prenons $X = \mathbf{P}_2$. Dans ce cas, Drézet a démontré [3] que si $r > 1$ $\operatorname{Pic}(M_X(r))$ est isomorphe à $\mathbf{Z}$; le fibré $\mathcal{D}_1$ ci-dessus est l'un des générateurs de ce groupe.

(5.4). L'extension de Maruyama. Considérons maintenant l'espace de modules $M_X(r,0,c_2)$ des $\mathcal{O}_X$-modules semi-stables de classe de Chern $c_1 = 0$, et c_2 . Cet espace de modules est non vide et non réduit à un point pourvu que $c_2 \geq r$ (cf. Maruyama, [13], proposition 5.2). Dans cet article, Maruyama construit un morphisme $M_X(r,0,c_2) \xrightarrow{u} M_X(c_2)$ en considérant pour tout faisceau semi-stable F de rang r de classes de Chern $c_1 = 0$ et c_2 le faisceau G défini par l'extension

$$0 \longrightarrow F \longrightarrow G \longrightarrow H^1(F) \otimes \mathcal{O} \longrightarrow 0$$

associé à l'élément de $\operatorname{Ext}^1(H^1(F) \otimes \mathcal{O}, F) \cong \operatorname{End}(H^1(F))$ correspondant à l'identité.

Proposition (5.5) . *Considérons les fibrés* $\mathcal{D}_1$ *construits respectivement sur* $M_X(r, 0, c_2)$ *et* $M_X(c_2)$. *On a alors dans* $\operatorname{Pic}(M_X(r, 0, c_2))$

$$u^*(\mathcal{D}_1) \cong \mathcal{D}_1$$

Démonstration . Considérons une famille F de $\mathcal{O}_X$-modules semi-stables de rang r , de classes de Chern $c_1 = 0$, $c_2 = r-\chi$, paramétrée par une variété algébrique S , et l'extension G qui en découle. On a $\chi_1 = r(1-g)$, où g est le genre d'une courbe $Y \in |\mathcal{O}_X(1)|$. Sur l'un ou l'autre des espaces de modules ci-dessus, le fibré $\mathcal{D}_1$ est construit à partir de l'élément $v_1 = -h + D$, où D est un cycle sur X de degré $1-g$, et on a dans le groupe de Grothendieck $K^0(S)$ associé aux fibrés vectoriels (cf. lemme (2.3))

$$pr_{1!}(G.\, v_1) = pr_{1!}(F.\, v_1)$$

par suite $\lambda_G(v_1) = \lambda_F(v_1)$. De la propriété universelle (2.12) découle l'égalité $u^*(\mathcal{D}_1) = \mathcal{D}_1$.

Cet énoncé permet de préciser le théorème (4.1). On sait déjà que $\mathscr{D}_1^{\otimes\upsilon}$ est engendré par ses sections pour υ assez grand convenable, et que le morphisme

$$\varphi_\upsilon : M_X(r, 0, c_2) \longrightarrow |\mathscr{D}_1^{\otimes\upsilon}|^*$$

induit un plongement quand on le restreint à l'ouvert U des faisceaux localement libres μ-stables.

Corollaire (5.6) . *On suppose* υ *assez grand convenable. Le morphisme de Maruyama se factorise suivant le diagramme*

$$\begin{array}{ccc} M_X(r, 0, c_2) & \xrightarrow{u} & M_X(c_2) \\ \downarrow{\scriptstyle\varphi_\upsilon} & \nearrow{\scriptstyle v} & \\ \operatorname{Im}\varphi_\upsilon & & \end{array}$$

où v *est un morphisme fini.*

En effet, on a un diagramme commutatif

$$\begin{array}{ccc} M_X(r, 0, c_2) & \xrightarrow{u} & M_X(c_2) \\ \downarrow{\scriptstyle\varphi_\upsilon} & & \downarrow{\scriptstyle\varphi_\upsilon} \\ \mathbf{P}^{\cdot}\big(\Gamma(M_X(r, 0, c_2), \mathscr{D}_1^{\otimes\upsilon})\big) & \overset{u'}{- - \rightarrow} & \mathbf{P}^{\cdot}\big(\Gamma(M_X(c_2), \mathscr{D}_1^{\otimes\upsilon})\big) \end{array}$$

où u' est l'application rationnelle induite par l'application linéaire $u^* : \Gamma(M_X(c_2), \mathscr{D}_1^{\otimes\upsilon}) \longrightarrow \Gamma(M_X(r, 0, c_2), \mathscr{D}_1^{\otimes\upsilon})$. Ce morphisme u' est évidemment fini sur $\operatorname{Im}\varphi_\upsilon$, et $u'|_{\operatorname{Im}\varphi_\upsilon}$ se factorise à travers $M_X(c_2)$. D'où le corollaire

Corollaire (5.7) . *Les fibres de* $\partial M_X(r, 0, c_2) \longrightarrow |\mathscr{D}_1^{\otimes\upsilon}|^*$ *sont de dimension* ≥ 1 .

C'est déjà le cas pour le morphisme u : ceci résulte de l'article de Maruyama [13] .

6. Applications aux courbes de saut

(6.1) Modules semi-stables de dimension 1. Soient $(X, \mathcal{O}_X(1))$ une surface algébrique projective polarisée, $h = c_1(\mathcal{O}_X(1))$. Considérons un $\mathcal{O}_X$-module cohérent θ de dimension 1 sur X . Pour un tel faisceau, on peut définir le degré $< c_1(\theta), h > = \deg\theta$ [(3)], qui est > 0 si $\theta \neq 0$ et la caractéristique d'Euler-Poincaré $\chi(\theta)$. Un tel module θ est dit semi-stable (resp. stable) si pour tout sous-module $0 \neq \theta' \neq \theta$ on a

$$\chi(\theta') \leq \frac{\deg\theta'}{\deg\theta}\chi(\theta) \qquad \text{(resp. <)}$$

(3) la forme d'intersection est notée < , >

Cette condition entraîne que θ est localement de Cohen-Macaulay. On appelle support schématique de θ la courbe Y définie par l'idéal de Fitting de θ. Les points singuliers de θ sont les points $x \in Y$ tels que θ_x n'est pas localement libre sur Y. Si Y est lisse, θ est un faisceau inversible sur Y.

On rassemble ici des résultats qui seront démontrés ailleurs.

Soit μ un nombre rationnel. Considérons la catégorie $\mathscr{C}(\mu)$ des $\mathcal{O}_X$-modules semi-stables θ de dimension 1 tels que $\chi(\theta) = \mu \deg \theta$. C'est une catégorie abélienne, artinienne et noethérienne ; donc tout module semi-stable $\theta \in \mathscr{C}(\mu)$ a une filtration de Jordan-Hölder. On dira que deux modules semi-stables θ et θ' de dimension 1 sont S-équivalents s'ils ont même gradué de Jordan-Hölder.

Soient NS(X) le groupe de Néron-Séveri de X, $c_1 \in NS(X)$, et $\chi \in \mathbf{Z}$. On considère le foncteur $\underline{\underline{P}}_X(c_1, \chi)$ qui associe à la variété algébrique S l'ensemble des classes d'isomorphisme de $\mathcal{O}_{S\times X}$-modules θ tels que pour tout $s \in S$, $\theta(s)$ soit un $\mathcal{O}_X$-module semi-stable de dimension 1 tel que $c_1(\theta(s)) = c_1$ et $\chi(\theta(s)) = \theta$.

Proposition (6.2) . (1) *La famille* $\underline{\underline{P}}_X(c_1, \chi)$ (.) *est limitée*

(2) *Il existe pour le foncteur* $\underline{\underline{P}}_X(c_1, \chi)$ *un espace de module grossier, noté* $P_X(c_1, \chi)$; *c'est une variété projective qui a pour ensemble sous-jacent l'ensemble des classes de* S-*équivalence de* $\mathcal{O}_X$*-modules semi-stables* θ *de dimension* 1, *tels que* $c_1(\theta) = c_1, \chi(\theta) = \chi$.

La construction est semblable à celle de Gieseker ou Maruyama (voir par exemple Simpson [19]) ; elle résulte du calcul des points semi-stables sous l'action de SL(N, C) sur le schéma de Grothendieck Groth $(\mathcal{O}_X^N(-i), c_1, \chi)$ des quotients θ de $\mathcal{O}_X^N$ de dimension 1, de classe de Chern $c_1(\theta) = c_1$ et de caractéristique d'Euler-Poincaré $\chi(\theta) = \chi$, avec i suffisamment grand. On a un morphisme canonique $P_X(c_1, \chi) \xrightarrow{\rho} \mathscr{C}_X(c_1)$ dans la variété $\mathscr{C}_X(c_1)$ des courbes de X de classe fondamentale c_1, qui associe à θ le support schématique de θ. Au-dessus d'une courbe lisse, la fibre est la composante du groupe de Picard des fibrés inversibles de caractéristique χ.

On suppose que $<\omega_X, c_1> < 0$. Dans ce cas, la variété $P_X(c_1, \chi)$ s'obtient en quotientant une variété lisse par l'action de SL(N, C). On obtient :

Proposition (6.3). *On suppose* $<\omega_X, c_1> < 0$. *Alors l'ouvert* P_s *des points stables de* $P_X(c_1, \chi)$ *est une variété lisse de dimension* $c_1^2 + 1$.

On suppose désormais que le fibré anticanonique $\omega_X{}^*$ est ample. Dans ce cas la surface X est rationnelle et la condition $<\omega_X, c_1> < 0$ est satisfaite. On peut montrer dans le cas où $c_1 = h = c_1(\mathcal{O}(1))$, que P_s est dense dans $P_X(h, \chi)$, ce qui permet de préciser l'énoncé (6.3) :

Proposition (6.4) . *Si le fibré anticanonique* $\omega_X{}^*$ *est ample, la variété* $P_X(h, \chi)$ *est irréductible, normale, de dimension* deg(X) + 1, *et l'ouvert des points stables est lisse.*

Supposons en outre que le système linéaire $|\mathcal{O}_X(1)|$ contienne des courbes lisses rationnelles, autrement dit que $\deg \omega_X + \deg X = -2$. L'espace projectif $|\mathcal{O}_X(1)|$ est alors de dimension $\deg X + 1$.

On s'intéresse aux modules semi-stables de caractéristique $\chi = 0$.

Proposition (6.5) . *Si le fibré anticanonique est ample et si* $\deg \omega_X + \deg X = -2$, *le morphisme* $\rho : P_X(h, 0) \longrightarrow |\mathcal{O}_X(1)|$ *est un isomorphisme.*

Ceci résulte essentiellement du fait que ρ induit un isomorphisme de l'ouvert $P' \subset P_X(h, \chi)$ des points dont les représentants ont au plus un point singulier, sur l'ouvert P'' de $|\mathcal{O}_X(1)|$ correspondant aux courbes Y ayant au plus un point singulier ordinaire. La fibre de ρ au-dessus de $\{Y\}$ est évidemment réduite à un point si Y est lisse. Si une telle courbe a un point singulier, elle est la réunion de deux courbes rationnelles lisses Y', Y'' qui se coupent transversalement, et les modules semi-stables de support Y, s'écrivent comme extension

$$0 \longrightarrow \theta' \longrightarrow \theta \longrightarrow \theta'' \longrightarrow 0$$

où θ' et θ'' sont des modules semi-stables de caractéristique nulle de support Y' et Y'' respectivement. Tous ces modules sont S-équivalents à $\theta' \oplus \theta''$. Il est d'autre part facile de constater que les complémentaires des ouverts P' et P'' sont de codimension ≥ 2, ce qui prouve que ρ est en fait un isomorphisme (Shafarevich [18], chapitre II).

(6.6) Courbes de saut . Dans ce qui suit on considère une surface X de fibré anticanonique ample et telle que $\deg \omega_X + \deg X = -2$. Soit F un faisceau semi-stable de rang $r > 0$, de classe de Chern $c_1 = 0$. Une courbe lisse $Y \in |\mathcal{O}_X(1)|$ est dite de saut pour F si et seulement si $F|_Y$ n'est pas trivial, c'est-à-dire si $h^1(F \otimes \theta) \neq 0$, θ désignant l'unique module semi-stable de caractéristique nulle de support Y. On étend cette définition aux courbes singulières:

Définition . *Une courbe* $Y \in |\mathcal{O}_X(1)|$ *est de saut pour* F *s'il existe un module semi-stable* θ *de caractéristique* 0 *et de support* Y *tels que* $h^1(F \otimes \theta) \neq 0$.

Proposition (6.7) . *Supposons que* F *soit un module semi-stable de rang* 2, *de classe de Chern* $c_1 = 0$, *et* c_2. *Alors les courbes de saut de* F *sont portées par une hypersurface* $\gamma(F)$ *de* $|\mathcal{O}_X(1)|$ *de degré* c_2.

Démonstration. Considérons, sur l'espace $P_X(h, 0)$ des modules semi-stables de dimension 1 le fibré inversible $L = \rho^*(\mathcal{O}(1))$: c'est aussi le fibré défini par l'hypersurface $\{\theta, h^0(\theta \otimes C_a) \neq 0\}$, où $a \in X$ est un point fixé de X. Comme dans le théorème 2.5, ce fibré est caractérisé par la propriété universelle suivante : si $\theta = (\theta_t)_{t \in S}$ est une famille de modules semi-stables de dimension 1, paramétrée par une variété algébrique S, $f_\theta : S \longrightarrow P_X(h, 0)$ le morphisme modulaire, on a

$$f_\theta^* (L) = pr_{1!} (\theta . C_a)$$

où $pr_1 : S \times X \longrightarrow S$ est la première projection, et $\theta . C_a$ est le produit, dans $K(S \times X)$ de θ et de $pr_2^*(C_a)$.

Lemme (6.8) . (a) *On a*

$$\det (pr_{1!}(F.\theta))^{-1} = f_\theta^*(L^{\otimes c_2})$$

(b) *Le fibré* $L^{\otimes c_2}$ *est muni d'une section dont le schéma des zéros a pour support l'ensemble des courbes* $Y \in |\mathcal{O}_X(1)|$ *de saut pour* F.

Démonstration. (a) Dans $K(X)$ on a $F = 2\mathcal{O} - c_2\, a$. En effet, $K(X)$ s'identifie à $K_{top}(X)$ et il suffit de constater que les deux membres ont même rang, même déterminant, même caractéristique. D'autre part, $pr_{1*}(\theta) = R^1\, pr_{1*}(\theta) = 0$, par suite $pr_{1!}(\theta) = 0$. Ainsi

$$pr_{1!}(F.\theta) = -c_2\, pr_{1!}(\theta . C_a)$$

et $(\det pr_{1!}(F.\theta))^{-1} = f_\theta^*(L^{\otimes c_2})$.

(b) Considérons, sur X, une présentation $0 \longrightarrow A \longrightarrow B \longrightarrow F \longrightarrow 0$ par des faisceaux localement libres, tels que $H^0(B \otimes \theta_t) = 0$ pour tout $t \in S$. Du fait que $\underline{Tor}_1^{S\times X}(F, \theta) = 0$, on tire la suite exacte $0 \longrightarrow A \otimes \theta \longrightarrow B \otimes \theta \longrightarrow F \otimes \theta \longrightarrow 0$ sur $S \times X$ et par image directe la suite exacte longue

$$0 \longrightarrow pr_{1*}(F \otimes \theta) \longrightarrow R^1\, pr_{1*}(A \otimes \theta) \xrightarrow{\ \upsilon\ } R^1\, pr_{1*}(B \otimes \theta) \longrightarrow R^1\, pr_{1*}(F \otimes \theta) \longrightarrow 0$$

Les $\mathcal{O}_S$-modules $R^1\, pr_{1*}(A \otimes \theta)$ et $R^1\, pr_{1*}(B \otimes \theta)$ sont localement libres, et le déterminant du morphisme υ fournit une section de $f_\theta^*(L^{\otimes c_2})$ dont le schéma des zéros a pour ensemble sous-jacent $\{t, h^1(F \otimes \theta_t) \neq 0\}$. Quand on représente $P_X(h, 0) \cong |\mathcal{O}_X(1)|$ comme quotient d'un schéma de Grothendieck $R = \mathrm{Groth}\,(\mathcal{O}_X^N(-i), h, 0)$ (avec N et i convenables) par l'action de $SL(N, C)$, on obtient une section $SL(N, C)$-invariante sur R, qui provient d'une section s_F de $L^{\otimes c_2}$ dont le schéma des zéros $\gamma(F)$ a pour ensemble sous-jacent l'ensemble des courbes $Y \in |\mathcal{O}_X(1)|$ de saut pour F.

Lemme (6.9) (Grauert-Mülich). *Il existe une courbe lisse* $Y \in |\mathcal{O}_X(1)|$ *telle que* $F|_Y$ *soit trivial.*

Démonstration. Soit $Z \subset |\mathcal{O}_X(1)| \times X$ la variété d'incidence ; considérons l'espace tangent relatif du morphisme $Z \xrightarrow{pr_2} X$. Pour $Y \in |\mathcal{O}_X(1)|$ on a la suite exacte

$$0 \longrightarrow T(Z/X)|_Y \longrightarrow H^0(\mathcal{O}_Y(1)) \otimes \mathcal{O}_Y \longrightarrow \mathcal{O}_Y(1) \longrightarrow 0$$

d'où il découle que $h^0(T(Z/X)|_Y) = 0$. Puisque Y est rationnelle, le fibré $T(Z/X)|_Y$ est somme directe de fibré inversible, de degré < 0 ; son rang est Y^2 et son degré $-Y^2$; par suite, chacun de ces fibrés est de degré -1 : ainsi $T(Z/X)|_Y$ est semi-stable. On peut alors utiliser par exemple l'énoncé de Hirschowitz [8] qui montre qu'alors $F|_Y \cong \mathcal{O}_Y{}^2$ pour $Y \in |\mathcal{O}_X(1)|$ suffisamment générale, ce qui prouve le lemme.

Il en résulte que le schéma $\gamma(F)$ est une hypersurface de $P_X(h, 0) \cong |\mathcal{O}_X(1)|$, de degré c_2, ce qui achève la démonstration de la proposition (6.7).

(6.10) Le morphisme $F \to \gamma(F)$. Considérons la variété produit $M_X(2, 0, c_2) \times P_X(h, 0)$. Elle est munie d'un fibré inversible $\mathcal{L}$ caractérisé par la propriété universelle suivante, analogue à celle de l'énoncé du théorème (2.5) : si $(F_s)_{s\in S}$ est une famille de faisceaux semi-stables de rang 2 de classes de Chern $(0, c_2)$ paramétrée par une variété algébrique S, et $(\theta_s)_{s\in S}$ une famille de modules de dimension 1, de classe h et de caractéristique 0, on a

$$\det\left(pr_{1!}(F \otimes \theta)\right) = (f_F \times f_\theta)^*(\mathcal{L})$$

où $f_F : S \longrightarrow M_X(2, 0, c_2)$ et $f_\theta : S \longrightarrow P_X(h, 0)$ sont les morphismes modulaires associés. De plus, $\mathcal{L}^{-1}$ est muni d'une section s : il suffit en effet de constater que, sur $S \times X$, F a une résolution localement libre $0 \longrightarrow A \longrightarrow B \longrightarrow F \longrightarrow 0$ telle que $pr_{1*}(B_t \otimes \theta_t) = 0$ pour $t \in S$, et de calquer la démonstration du lemme (6.8). La propriété universelle s'applique en particulier au cas où l'une ou l'autre des familles est constante. On obtient alors $\mathcal{L}^{-1} = \mathcal{D}_1 \otimes L^{\otimes c_2}$ et la section s définit un morphisme $\gamma : M(2, 0, c_2) \longrightarrow |L^{\otimes c_2}|$ dont l'application sous-jacente est $F \longrightarrow \gamma(F)$, et tel que $\gamma^*(\mathcal{O}(1)) = \mathcal{D}_1$. En particulier $\mathcal{D}_1$ est engendré par ses sections.

Théorème (6.11). *On suppose que le fibré anticanonique* $\omega_X{}^*$ *est ample, et que l'on a la relation* $\deg \omega_X + \deg X = -2$. *Soit* $U \subset M(2, 0, c_2)$ *l'ouvert des fibrés* μ*-stables. Alors le morphisme* $\gamma : U \longrightarrow |L^{\otimes c_2}|$ *est quasi-fini.*

Démonstration. On sait que $\gamma^*(\mathcal{O}(1)) = \mathcal{D}_1$. Supposons qu'il existe une courbe intègre C contenue dans une fibre de $\gamma|_U$; désignons par $\bar{C}$ l'adhérence de C dans $M_X(2, 0, c_2)$. On a alors $\gamma|_{\bar{C}} = \text{cte}$, donc $< c_1(\mathcal{D}_1), \bar{C} > = 0$. Or, d'après le théorème 4.2, l'image de $\bar{C}$ par le morphisme

$$M(2, 0, c_2) \longrightarrow |\mathcal{D}_1^{\otimes \upsilon}|$$

est une courbe pour υ convenable : donc $< c_1(\mathcal{D}_1^{\otimes \upsilon}), \bar{C} > > 0$. Ceci, compte-tenu de la relation $c_1(\mathcal{D}_1^{\otimes \upsilon}) = \upsilon\, c_1(\mathcal{D}_1)$, ceci conduit à une contradiction. D'où le théorème.

Le théorème s'applique en particulier au plan projectif $\mathbf{P}_2$, muni de la polarisation $\mathcal{O}(1)$ (théorème 0), ou $\mathcal{O}(2)$ (surface de Véronèse), à $\mathbf{P}_1 \times \mathbf{P}_1$ muni de la polarisation $\mathcal{O}(1,m)$.

Bibliographie

[1] W. Barth : *Moduli of vector bundles on the projective plane.* Invent. math. **42** (1977) p. 63-91.

[2] A. Beauville : *Fibrés de rang 2 sur une courbe, fibré déterminant et fonctions thêta.* Bull. Soc. Math. de France **116** (1988) p. 431-448.

[3] J.M. Drézet : *Groupe de Picard des variétés de modules de faisceaux semi-stables sur* $P_2(C)$. Ann. Inst. Fourier **38**,3 (1988) p. 105-168.

[4] J.M. Drézet : *Points non-factoriels des variétés de modules de faisceaux semi-stables sur une surface rationnelle.* Preprint Université Paris 7 (1989).

[5] J.M. Drézet et M.S. Narasimhan : *Groupe de Picard des variétés de modules de fibrés semi-stables sur les courbes algébriques.* Invent. Math. **97** (1989) p. 53-94.

[6] D. Gieseker : *On the moduli of vector bundles on an algebraic surface.* Ann. of Math. **106** (1977) p. 45-60.

[7] A. Grothendieck : *Techniques de construction et théorèmes d'existence en géométrie algébrique, IV : les schémas de Hilbert.* Séminaire Bourbaki, Exposé **221** (1960-61).

[8] A. Hirschowitz : *Sur la restriction des faisceaux semi-stables*, Ann. Sc. Ec. Norm. Sup. **14** (1980) p. 199-207.

[9] K. Hulek et S.A. Strømme : *Appendix to the paper "Complete families of stable vector bundles over* P_2*"*, Lecture Notes in Math., Springer, **1194** (1986) p. 34-40.

[10] R. Hartshorne : *Algebraic geometry*, Springer (1977).

[11] S.K. Donaldson: *Polynomial invariants for smooth four manifolds*, Preprint.

[12] M. Maruyama : *Moduli of stable sheaves, II*, J. Math. Kyoto Univ., **18** (1978) p. 557-614.

[13] M. Maruyama : *On a compactification of a moduli space of stable vector bundles on a rational surface.* Algebraic geometry and commutative algebra, in honor of M. Nagata (1987) p. 233-260.

[14] V.B. Mehta et A. Ramanathan: *Semistable sheaves on projective varieties and their restriction to curves.* Math. Annalen **258** (1982) p. 213-224.

[15] V.B. Mehta et A. Ramanathan : *Restriction of stable sheaves and representations of the fundamental group.* Invent. Math. **77** (1984) p. 163-172.

[16] D. Mumford et J. Fogarty : *Geometric invariant theory*, Springer (1982).

[17] S. Ramanan : *The moduli spaces of vector bundles over an algebraic curve.* Math. Ann. **200** (1973) p. 69-84.

[18] I.R. Shafarevich : *Basic algebraic geometry*, Springer (1977).

[19] C.T. Simpson : *Moduli of representations of the fundamental group of smooth projective variety*, Preprint.

[20] S.A. Strømme : *Ample divisors on fine moduli spaces on the projective plane.* Math. Z. **187** (1984) p. 405-423.

J. Le Potier
U.F.R. de Mathématiques et U.R.A. 212
Université Paris 7
2 place Jussieu
75251 Paris Cedex 05

COURBES MINIMALES DANS LES CLASSES DE BILIAISON

Mireille MARTIN-DESCHAMPS et Daniel PERRIN

On désigne par k un corps algébriquement clos, par R l'anneau de polynômes $k[X,Y,Z,T]$ et par $\mathbf{m}$ l'idéal (X,Y,Z,T).

Soit C une courbe (localement de Cohen-Macaulay) de $\mathbf{P}^3_k$ et $\mathcal{J}_C$ le faisceau d'idéaux qui définit C.

Un invariant essentiel attaché à C est le R-module gradué de longueur finie $M_C = \bigoplus_{n\in\mathbf{Z}} H^1\mathcal{J}_C(n)$ introduit par Hartshorne ([H]), et étudié par Rao ([R]). Nous le nommerons simplement module de Rao de la courbe C. La propriété fondamentale de M_C est de caractériser les classes de liaison des courbes de $\mathbf{P}^3_k$. Précisément, les modules M_C et $M_{C'}$ sont isomorphes à décalage de la graduation près (resp. à décalage et dualité près) si et seulement si il existe un entier n pair (resp. impair) et une suite de courbes $C = C_0, C_1, \ldots, C_n = C'$ telles que C_i soit algébriquement liée à C_{i+1} (cf. [R]). De plus, toujours d'après Rao, si M est un R-module gradué de longueur finie et si n est assez grand, le module décalé $M(n)$ est le module de Rao d'une courbe lisse C. L'ensemble des courbes de $\mathbf{P}^3_k$ est ainsi partagé en classes de biliaison (ou liaison paire) qui correspondent bijectivement aux modules. L'étude de cette partition, jointe à la stratification naturelle suivant les dimensions des espaces de cohomologie $H^i\mathcal{J}_C(n)$, nous paraît être une voie importante pour aborder la classification des courbes gauches (cf. [M-D,P 1,2]).

On note déjà (cf. [M]) que pour M non nul donné, il existe dans la classe de biliaison de M des courbes C **minimales** i.e., telles que si C' est une autre courbe de la classe, on ait $M_{C'} = M_C(h)$ avec $h \geq 0$, autrement dit, des courbes dont le module de Rao est le plus à gauche possible.

Posons, pour une courbe C :

$$s_0(C) = \inf\{n \in \mathbf{Z} | H^0\mathcal{J}_C(n) \neq 0\}, \quad e(C) = \sup\{n \in \mathbf{Z} | H^1\mathcal{O}_C(n) \neq 0\}.$$

Lazarsfeld et Rao ont montré qu'en particulier les courbes qui vérifient $e + 3 < s_0$ sont minimales et ils ont donné, dans ce cas, une description très simple de la classe de biliaison: si C est minimale et si C' est dans la classe, on passe de C à C' par une suite de biliaisons élémentaires (basic double linkage) suivies d'une déformation à cohomologie et module de Rao constants (cf. [LR]). L'intérêt de ces opérations de biliaison élémentaire est qu'on y contrôle parfaitement la variation du degré, du genre, et même de toute la cohomologie et des résolutions (cf. [M-D,P 1,2]). Cette propriété des classes de biliaison est connue sous le nom de propriété de Lazarsfeld-Rao.

On peut, dès lors, considérer qu'une classe de biliaison est parfaitement connue si on connait ses courbes minimales et si elle vérifie la propriété de Lazarsfeld-Rao. En particulier on connait alors les degrés et genres des courbes de la classe, ainsi que les valeurs possibles de la cohomologie (postulation, spécialité).

La propriété de Lazarsfeld-Rao a d'abord été prouvée pour certaines classes de biliaison particulières (cf. [BM 1,2,3]) puis en toute généralité par Ballico, Bolondi et Migliore (dans $\mathbf{P}^n_k$) ([BBM]) et par nous-mêmes ([M-D,P 1,2]).

Le but de cet article est de résoudre l'autre face du problème, i.e., de donner une construction explicite des courbes minimales associées à un module M et d'en montrer l'unicité (à déformation près). Pour plus de détails, le lecteur se référera à [M-D,P 1,2].

Nous utiliserons pour décrire une courbe C les deux résolutions canoniques de son faisceau d'idéaux $\mathcal{J}_C$:

a) une résolution dite de type E :

$$0 \to \mathcal{E} \to \mathcal{F} \to \mathcal{J}_C \to 0$$

où $\mathcal{F}$ est un fibré dissocié (i.e. une somme directe de $\mathcal{O}_{\mathbf{P}}$-modules inversibles), et $\mathcal{E}$ un fibré conoyau d'un morphisme injectif de fibrés dissociés ;

b) une résolution dite de type N :

$$0 \to \mathcal{P} \to \mathcal{N} \to \mathcal{J}_C \to 0$$

où $\mathcal{P}$ est un fibré dissocié et $\mathcal{N}$ un fibré noyau d'un morphisme surjectif de fibrés dissociés.

On notera que ces types de résolutions sont échangés par une liaison (ou plus généralement par une suite d'un nombre impair de liaisons), mais conservés par biliaison.

Pour construire les courbes minimales, la remarque de base, déjà faite par Rao, est que les fibrés $\mathcal{E}$ et $\mathcal{N}$ qui interviennent dans les deux résolutions de $\mathcal{J}_C$ sont liés au module de Rao. Précisément, si on a une résolution libre graduée $(*)$ de ce module :

$$(*) \qquad 0 \longrightarrow L_4 \xrightarrow{\sigma_4} L_3 \xrightarrow{\sigma_3} L_2 \xrightarrow{\sigma_2} L_1 \xrightarrow{\sigma_1} L_0 \xrightarrow{\sigma_0} M \longrightarrow 0,$$

si on pose $E_0 = \mathrm{Ker}\sigma_2$, $N_0 = \mathrm{Im}\sigma_2$ et si $\mathcal{E}_0$ et $\mathcal{N}_0$ sont les faisceaux associés aux modules E_0 et N_0, il existe des fibrés $\mathcal{L}$ et $\mathcal{L}'$ dissociés tels qu'on ait $\mathcal{E} = \mathcal{E}_0 \oplus \mathcal{L}$ et $\mathcal{N} = \mathcal{N}_0 \oplus \mathcal{L}'$.

On va donc chercher une courbe minimale comme une courbe pour laquelle $\mathcal{E}$ et $\mathcal{N}$ sont minimaux, i.e., $\mathcal{E} = \mathcal{E}_0$ et $\mathcal{N} = \mathcal{N}_0$ à un décalage près (attention, ces conditions ne suffisent pas à assurer que la courbe est minimale).

On construit la courbe minimale associée à un R-module M gradué de longueur finie, non nul, muni d'une résolution $(*)$ comme ci-dessus, en fabriquant un facteur direct $\mathcal{P}$ de $\mathcal{L}_2$ (faisceau associé au module L_2), de rang égal au rang de $\mathcal{N}_0$ diminué de 1 , tel que le conoyau de la restriction à $\mathcal{P}$ de la flèche $\tilde{\sigma}_2$ associée à σ_2 soit sans torsion, de rang 1 et de degré maximum pour ces conditions. Ce conoyau est alors, à décalage près, l'idéal de la courbe cherchée.

Plus précisément, si on pose :

$$\mathcal{L}_2 = \bigoplus_{k \in \mathbf{Z}} \mathcal{O}_{\mathbf{P}}(-k)^{l_2(k)},$$

on définit par récurrence, cf. 1.5 ci-dessous, en termes de certaines restrictions de la flèche $\tilde{\sigma}_2$, une fonction q de $\mathbf{Z}$ dans $\mathbf{N}$, avec $q(n) \leq l_2(n)$. On a alors le théorème :

Théorème. *Soit M un R-module gradué de longueur finie non nul muni d'une résolution de la forme $(*)$. Avec les notations précédentes, il existe une courbe C minimale dans la classe de biliaison associée à M, unique à déformation près (à cohomologie et module de Rao constants) dont le faisceau d'idéaux (décalé) admet les résolutions suivantes :*

$$0 \to \mathcal{E}_0 \to \mathcal{F} \to \mathcal{J}_C(h) \to 0$$

$$0 \to \mathcal{P} \to \mathcal{N}_0 \to \mathcal{J}_C(h) \to 0$$

avec

$$\mathcal{P} = \bigoplus_{k \in \mathbf{Z}} \mathcal{O}_{\mathbf{P}}(-k)^{q(k)}$$

$$\mathcal{F} = \bigoplus_{k \in \mathbf{Z}} \mathcal{O}_{\mathbf{P}}(-k)^{l_2(k)-q(k)}$$

et $h = \deg \mathcal{N}_0 - \deg \mathcal{P}$. On a donc $\mathcal{L}_2 = \mathcal{P} \oplus \mathcal{F}$. Ces résolutions déterminent entièrement la cohomologie de $\mathcal{J}_C$.

La fonction q est explicitement calculable en termes de rangs de certaines matrices extraites de σ_2. Ce calcul est illustré par plusieurs exemples (modules associés à une suite régulière, modules de Buchsbaum). On précise aussi le rapport entre la courbe minimale de la classe d'un module et celle de la classe du module dual.

On notera qu'en général la courbe minimale d'une classe n'est pas lisse, ni même réduite. Par exemple les courbes minimales qui ne vérifient pas la condition de Lazarsfeld-Rao ($e + 3 < s_0$) ne sont jamais lisses et connexes. Une question importante (ouverte) est de savoir à partir de quel décalage apparaissent les courbes lisses.

Pour voir comment le théorème ci-dessus s'insère dans les projets de classification et permet notamment de déterminer les cohomologies possibles dans les classes de biliaison, voir [M-D, P 1,2] Ch. V. On peut aussi sans doute l'utiliser pour construire des courbes jouissant de certaines bonnes propriétés (rang maximum, ...).

1. Deux fonctions de Z dans N associées à un $\mathcal{O}_\mathbf{P}$-module.

Définition 1.1. *Soient $\mathcal{B} \subset \mathcal{A}$ deux $\mathcal{O}_\mathbf{P}$-modules. On dit que $\mathcal{B}$ est* **maximal** *si pour tout $\mathcal{O}_\mathbf{P}$-module $\mathcal{B}'$ de même rang que $\mathcal{B}$ tel que $\mathcal{B} \subset \mathcal{B}' \subset \mathcal{A}$, on a $\mathcal{B} = \mathcal{B}'$.*

Dans le cas où $\mathcal{A}$ et $\mathcal{B}$ sont deux fibrés de rangs respectifs $r+1$ et r, $\mathcal{B}$ est maximal si et seulement si $\mathcal{A}/\mathcal{B}$ est l'idéal (tordu) d'une courbe, qui est le lieu de dégénérescence de l'injection de $\mathcal{A}$ dans $\mathcal{B}$. C'est par ce procédé que nous allons construire les courbes minimales.

Définition 1.2. *On appelle* **fibré dissocié** *une somme directe finie de faisceaux de la forme $\mathcal{O}_\mathbf{P}(n)$, $n \in \mathbf{Z}$.*

Dans tout ce qui suit, $\mathcal{N}$ est un $\mathcal{O}_\mathbf{P}$-module sans torsion, et on pose $N = H^0_*\mathcal{N}$. Une présentation graduée minimale (i.e. un début de résolution graduée minimale) de N est de la forme :

$$\sigma : \underset{n\in\mathbf{Z}}{\oplus} R(-n)^{l_\mathcal{N}(n)} \to N$$

où $l_\mathcal{N}$ est une fonction de $\mathbf{Z}$ dans $\mathbf{N}$ presque partout nulle définie de manière unique. C'est la **première fonction associée à** $\mathcal{N}$.

On note plus simplement l au lieu de $l_\mathcal{N}$ lorsqu'il n'y a pas d'ambiguïté.

Soit $a \in \mathbf{Z}$. On désigne par $\sigma_a : \bigoplus_{n\leq a} R(-n)^{l_\mathcal{N}(n)} \to N$ la restriction de σ, par $N_{\leq a}$ son image, qui est le sous-module de N engendré par les éléments de degré $\leq a$, et par $\mathcal{N}_{\leq a}$ le sous-$\mathcal{O}_\mathbf{P}$-module de $\mathcal{N}$, image de l'homomorphisme :

$$\underset{n\leq a}{\oplus} \mathcal{O}_\mathbf{P}(-n)^{h^0\mathcal{N}(n)} \to \mathcal{N}$$

défini par les sections globales de $\mathcal{N}(n)$, pour $n \leq a$ (qui est aussi l'image de $\widetilde{\sigma}_a$, homomorphisme de faisceaux associé à σ_a). On remarque que σ_a est une présentation graduée minimale de $N_{\leq a}$.

Puisque $\mathcal{N}$ est localement libre sur un ouvert de $\mathbf{P}^3$ qui contient les points génériques de tous les diviseurs intègres D de $\mathbf{P}^3$, on peut parler du rang de $\widetilde{\sigma}_a$ (resp. de $\widetilde{\sigma}_a|_D$) qui est le rang au point générique de $\mathbf{P}^3$ (resp. de D).

Définition 1.3. *On pose, pour tout $a \in \mathbf{Z}$:*

$$\alpha_a = \operatorname{rang} \widetilde{\sigma_a} = \operatorname{rang} \sigma_a$$
$$\beta_a = \inf\nolimits_D \operatorname{rang} \widetilde{\sigma_a}|_D$$

où D parcourt l'ensemble des diviseurs intègres de $\mathbf{P}^3$.
On a évidemment $0 \leq \beta_a \leq \alpha_a \leq \sum_{n\leq a} l(n)$. On pose aussi :

$$\begin{aligned} a_0 &= 1 + \sup\{a \in \mathbf{Z} \mid \alpha_a = \beta_a = \sum_{n\leq a} l(n)\} \text{ s'il existe} \\ &= +\infty \text{ sinon.} \\ a_1 &= \inf\{a \in \mathbf{Z} \mid \alpha_a = \beta_a = \operatorname{rang}\mathcal{N}\}. \end{aligned}$$

Proposition 1.4. *Lorsqu'il est fini, $a_0 - 1$ est aussi le plus grand entier a tel que l'une des propriétés équivalentes suivantes soit réalisée :*
i) $\mathcal{N}_{\leq a}$ est un sous-fibré maximal et dissocié de $\mathcal{N}$;
ii) $\widetilde{\sigma}_a$ est injectif et a un conoyau de rang constant sur un ouvert qui contient les points génériques de tous les diviseurs ;
iii) $\widetilde{\sigma}_a$ est injectif et a un conoyau sans torsion ;
iv) σ_a est injectif et a un conoyau sans torsion.
En particulier, le rang de $\mathcal{N}_{\leq a_0}$ est strictement supérieur à celui de $\mathcal{N}_{\leq a_0-1}$, donc on a $l(a_0) \neq 0$ et $\alpha_{a_0} > \alpha_{a_0-1}$.

Démonstration. $ii \Leftrightarrow \alpha_a = \beta_a = \sum_{n\leq a} l(n)$.

$i \Rightarrow iii$. Si $\mathcal{N}_{\leq a}$ est dissocié, de la forme $\bigoplus_{i=1}^{r} \mathcal{O}_{\mathbf{P}}(-n_i)$, on obtient une surjection $\widetilde{\sigma}_a : \bigoplus_{n\leq a} \mathcal{O}_{\mathbf{P}}(-n)^{l_{\mathcal{N}}(n)} \to \bigoplus_{i=1}^{r} \mathcal{O}_{\mathbf{P}}(-n_i)$, donc pour tout $i \in [1,r]$, on a $n_i \leq a$. Alors $H^0_*(\mathcal{N}_{\leq a}) = \bigoplus_{i=1}^{r} R(-n_i)$ est engendré par ses éléments de degré $\leq a$, et l'inclusion naturelle $N_{\leq a} \to H^0_*\mathcal{N}_{\leq a}$ est un isomorphisme. Puisque $N_{\leq a}$ est un R-module libre, σ_a qui en est une présentation minimale est un isomorphisme.

$iii \Rightarrow iv$. Supposons qu'on ait une suite exacte de faisceaux :

$$0 \longrightarrow \bigoplus_{n\leq a} \mathcal{O}_{\mathbf{P}}(-n)^{l(n)} \xrightarrow{\widetilde{\sigma}_a} \mathcal{N} \longrightarrow \operatorname{Coker}\widetilde{\sigma_a} \longrightarrow 0.$$

En prenant la cohomologie, on en déduit une suite exacte de R-modules :

$$0 \longrightarrow \bigoplus_{n\leq a} R(-n)^{l(n)} \xrightarrow{\sigma_a} N \longrightarrow H^0_*(\operatorname{Coker}\widetilde{\sigma_a}) \longrightarrow 0.$$

Donc $H^0_*(\operatorname{Coker}\widetilde{\sigma_a}) = \operatorname{Coker}\sigma_a$ n'a pas de torsion supportée par $\mathbf{m}$, et est sans torsion si et seulement si le faisceau associé $\operatorname{Coker}\widetilde{\sigma_a}$ l'est.

Les autres implications sont immédiates.

Définition 1.5. Deuxième fonction associée à $\mathcal{N}$. *On définit la fonction $q_{\mathcal{N}}$, ou plus simplement q lorsqu'il n'y a pas d'ambiguïté, par :*

$$\begin{aligned} q(n) &= l(n) \text{ si } n < a_0 \,; \\ \sum_{m\leq n} q(m) &= \inf(\alpha_n - 1, \beta_n) \text{ sinon.} \end{aligned}$$

Proposition 1.6. Propriétés de la fonction q. *Avec les notations précédentes :*
a) si $\mathcal{N}$ est dissocié, on a $l = q$ et $\sum_{n\in\mathbf{Z}} q(n) = \operatorname{rang}\mathcal{N}$;
b) si $\mathcal{N}$ n'est pas dissocié, on a

$$\sum_{m\leq a_1} q(m) = \sum_{n\in\mathbf{Z}} q(n) = \operatorname{rang}\mathcal{N} - 1$$

et pour tout $n \in \mathbf{Z}$, $0 \leq q(n) \leq l(n)$.
De plus a_0 *est fini, et on a* $a_0 \leq a_1$, $q(a_0) < l(a_0)$, *et* $q(n) = 0$ *pour tout* $n > a_1$.

Démonstration. b) Soit $n_0 = \sup\{n \in \mathbf{Z} \mid l(n) \neq 0\}$. On a $\mathcal{N} = \mathcal{N}_{\leq n_0}$, $\sigma_{n_0} = \sigma$, $\alpha_{n_0} = \beta_{n_0} = \text{rang}\ \sigma = \text{rang}\ \mathcal{N}$.

Si $n_0 < a_0$, on a $\mathcal{N} = \mathcal{N}_{\leq n_0} = \mathcal{N}_{\leq a_0-1}$ qui est dissocié, ce qui contredit l'hypothèse. Ceci prouve déjà que a_0 est fini et qu'on a $\sum_{m \leq n_0} q(m) = \sum_{n \in \mathbf{Z}} q(n) = \alpha_{n_0} - 1$.

L'encadrement $0 \leq q(n) \leq l(n)$, vrai pour $n < a_0$, se montre par récurrence sur n. Supposons qu'il soit vérifié pour un $n \geq a_0 - 1$.

On a une surjection $\mathcal{O}_{\mathbf{P}}(-n-1)^{l(n+1)} \to \mathcal{N}_{\leq n+1}/\mathcal{N}_{\leq n}$ et une suite exacte

$$\mathcal{O}_{\mathbf{P}}(-n-1)^{l(n+1)} \to \mathcal{N}/\mathcal{N}_{\leq n} \to \mathcal{N}/\mathcal{N}_{\leq n+1} \to 0$$

qui reste exacte sur tous les diviseurs. On en déduit les inégalités :

$$0 \leq \text{rang}\ \mathcal{N}_{\leq n+1}/\mathcal{N}_{\leq n} = \alpha_{n+1} - \alpha_n \leq l(n+1),$$

$$0 \leq \text{rang}\ (\mathcal{N}/\mathcal{N}_{\leq n})|_D - \text{rang}\ (\mathcal{N}/\mathcal{N}_{\leq n+1})|_D \leq l(n+1),$$

si D est un diviseur de $\mathbf{P}^3$, d'où

$$0 \leq \beta_{n+1} - \beta_n \leq l(n+1),$$

et d'après 1.4, on a également

$$1 \leq \alpha_{a_0+1} - \alpha_{a_0} \leq l(a_0+1).$$

On sait que $q(n+1) = \sum_{m \leq n+1} q(m) - \sum_{m \leq n} q(m)$. Il nous faut séparer les différents cas.

Premier cas : $n = a_0 - 1$. Alors $q(a_0) = \inf(\alpha_{a_0} - 1, \beta_{a_0}) - \alpha_{a_0-1} = \inf(\alpha_{a_0} - \alpha_{a_0-1} - 1, \beta_{a_0} - \beta_{a_0-1})$ et $q(a_0) \in [0, l(a_0)[$.

Deuxième cas : $n \geq a_0$. On a $q(n+1) = \inf(\alpha_{n+1} - 1, \beta_{n+1}) - \inf(\alpha_n - 1, \beta_n)$.

Si $\alpha_n = \beta_n$ et $\alpha_{n+1} = \beta_{n+1}$, alors $q(n+1) = \alpha_{n+1} - \alpha_n \in [0, l(n+1)]$.

Si $\alpha_n = \beta_n$ et $\alpha_{n+1} > \beta_{n+1}$, alors $q(n+1) = \beta_{n+1} - \alpha_n + 1 = \beta_{n+1} - \beta_n + 1 < \alpha_{n+1} - \alpha_n + 1$. Donc $q(n+1) \in [1, l(n+1)]$.

Si $\alpha_n > \beta_n$ et $\alpha_{n+1} = \beta_{n+1}$, alors $q(n+1) = \alpha_{n+1} - \beta_n - 1 = \beta_{n+1} - \beta_n - 1 \geq \alpha_{n+1} - \alpha_n$. Donc $q(n+1) \in [0, l(n+1) - 1]$.

Si $\alpha_n > \beta_n$ et $\alpha_{n+1} > \beta_{n+1}$, alors $q(n+1) = \beta_{n+1} - \beta_n \in [0, l(n+1)]$.

Si a_1 est inférieur à a_0, on a les égalités :

$$\sum_{m \leq a_1} q(m) = \sum_{m \leq a_1} l(m) = \alpha_{a_1} = \beta_{a_1} = \text{rang}\ \mathcal{N}$$

ce qui contredit le fait que q est une fonction positive et que $\sum_{n \in \mathbf{Z}} q(n) = \text{rang}\ \mathcal{N} - 1$. Donc on a $a_0 \leq a_1$, et

$$\sum_{m \leq a_1} q(m) = \inf(\alpha_{a_1} - 1, \beta_{a_1}) = \text{rang}\ \mathcal{N} - 1 = \sum_{n \in \mathbf{Z}} q(n),$$

et $q(n)$ est nul pour $n > a_1$.

Remarques 1.7.

a) S'il est fini, a_0 est le plus petit entier n tel que $l(n) \neq q(n)$.

b) Pour $n < a_0$, on a $\sum_{m \leq n} q(m) = \alpha_n = \beta_n$.

c) Soient $\mathcal{N}$ un $\mathcal{O}_{\mathbf{P}}$-module sans torsion et $\mathcal{L}$ un fibré dissocié, $\mathcal{N}' = \mathcal{N} \oplus \mathcal{L}$. On voit facilement qu'on a : $q_{\mathcal{N}'} = q_{\mathcal{N}} + l_{\mathcal{L}}$.

2. Existence d'un meilleur sous-fibré maximal d'un $\mathcal{O}_{\mathbf{P}}$-module sans torsion.

Soient $\mathcal{N}$ un $\mathcal{O}_{\mathbf{P}}$-module sans torsion, $N = H^0_*\mathcal{N}$, $\sigma : \oplus_{n \in \mathbf{Z}} R(-n)^{l(n)} \to N$ une présentation graduée minimale de N. Puisque pour tout $n \in \mathbf{Z}$, on a $0 \leq q(n) \leq l(n)$, il existe des facteurs directs de $\oplus_{n \in \mathbf{Z}} R(-n)^{l(n)}$ isomorphes à $\oplus_{n \in \mathbf{Z}} R(-n)^{q(n)}$, et leur rang est alors égal au rang de $\mathcal{N}$ diminué de 1 si $\mathcal{N}$ n'est pas dissocié. Nous allons montrer qu'un tel facteur direct (assez général) correspond à un sous-fibré maximal de $\mathcal{N}$, de corang 1. Le théorème suivant montre déjà que la fonction q est la "meilleure" possible.

Théorème 2.1. *Soit $r : \mathbf{Z} \to \mathbf{N}$ une fonction à support fini et désignons par $\widetilde{u} : \bigoplus_{n \in \mathbf{Z}} \mathcal{O}_{\mathbf{P}}(-n)^{r(n)} \to \mathcal{N}$ une injection dont l'image est un sous-fibré maximal et dissocié de $\mathcal{N}$.*
Avec les notations précédentes, on a, pour tout $n \in \mathbf{Z}$,

$$\sum_{m \leq n} r(m) \leq \sum_{m \leq n} q(m).$$

De plus, si pour $a \in \mathbf{Z}$ l'image de $\widetilde{u}$ est contenue dans $\mathcal{N}_{\leq a}$, on a, pour tout $n \in \mathbf{Z}$, $\sum_{m \leq n} r(m) \leq \sum_{m \leq a} q(m)$.

Démonstration. Soit $u : \bigoplus_{n \in \mathbf{Z}} R(-n)^{r(n)} \to N$ correspondant à $\widetilde{u}$. Il se relève en $v : \bigoplus_{n \in \mathbf{Z}} R(-n)^{r(n)} \to \bigoplus_{n \in \mathbf{Z}} R(-n)^{l(n)}$ tel que $u = \sigma v$. On a alors pour tout $b \in \mathbf{Z}$ un diagramme commutatif

$$\begin{array}{ccc} \bigoplus_{n \leq b} \mathcal{O}_{\mathbf{P}}(-n)^{r(n)} & \xrightarrow{\widetilde{u}_b} & \mathcal{N} \\ \downarrow \widetilde{v}_b & & \| \\ \bigoplus_{n \leq b} \mathcal{O}_{\mathbf{P}}(-n)^{l(n)} & \xrightarrow{\widetilde{\sigma}_b} & \mathcal{N} \end{array}$$

où $\widetilde{v}$ est associé à v, d'où

$$\alpha_b = \text{rang } \widetilde{\sigma}_b \geq \text{rang } \widetilde{u}_b = \sum_{n \leq b} r(n).$$

La restriction de $\widetilde{u}_b$ à tout diviseur de $\mathbf{P}^3$ est encore injective, donc on a aussi

$$\beta_b \geq \sum_{n \leq b} r(n).$$

Si de plus l'image de $\widetilde{u}$ est contenue dans $\mathcal{N}_{\leq a}$, on a

$$\inf(\alpha_a, \beta_a) \geq \sum_{n \leq b} r(n).$$

Distinguons deux cas :

1) $b < a_0$. Alors on a $\sum_{n\leq b} r(n) \leq \alpha_b = \sum_{n\leq b} q(n)$.

Si $b \leq a$, on a évidemment $\sum_{n\leq b} q(n) \leq \sum_{n\leq a} q(n)$, donc $\sum_{n\leq b} r(n) \leq \sum_{n\leq a} q(n)$.

Si $b > a$, on a alors $\sum_{n\leq a} q(n) = \alpha_a = \beta_a$, donc $\sum_{n\leq b} r(n) \leq \sum_{n\leq a} q(n)$.

2) $b \geq a_0$. On a :

$$\sum_{n\leq b} q(n) = \inf(\alpha_b - 1, \beta_b)$$

$$\sum_{n\leq a} q(n) = \inf(\alpha_a - 1, \beta_a) \text{ si } a \geq a_0$$

$$= \alpha_a = \beta_a \text{ si } a < a_0$$

On a donc

$$\sum_{n\leq b} r(n) \leq \inf(\sum_{n\leq a} q(n), \sum_{n\leq b} q(n))$$

sauf si on est dans un des cas suivants :

$$\alpha_b = \beta_b = \sum_{n\leq b} r(n) \text{ ou } \alpha_a = \beta_a = \sum_{n\leq b} r(n) \text{ et } a \geq a_0.$$

Dans le premier cas l'image de $\widetilde{u_b}$ est un sous-fibré maximal de $\mathcal{N}$, contenu dans $\mathcal{N}_{\leq b}$ qui a le même rang, donc il lui est égal, ce qui contredit le fait que, puisque $b \geq a_0$, $\mathcal{N}_{\leq b}$ ne peut être un sous-fibré maximal et dissocié de $\mathcal{N}$. Le deuxième cas se traite de la même manière.

On déduit de 2.1 la plus petite valeur non nulle de q :

Corollaire 2.2. *Soit $\mathcal{N}$ un $\mathcal{O}_{\mathbf{P}}$-module réflexif, l et q les fonctions associées. Alors on a*

$$\inf\{n \in \mathbf{Z} \mid l(n) \neq 0\} = \inf\{n \in \mathbf{Z} \mid q(n) \neq 0\}.$$

Démonstration. Soit $n_0 = \inf\{n \in \mathbf{Z} \mid l(n) \neq 0\} = \inf\{n \in \mathbf{Z} \mid h^0\mathcal{N}(n) \neq 0\}$. Pour $n \leq n_0$, $q(n)$ est nul. Puisque $\mathcal{N}$ est réflexif, une section non nulle de $H^0\mathcal{N}(n_0)$ correspond à une injection $\mathcal{O}_{\mathbf{P}}(-n_0) \to \mathcal{N}$ dont l'image est un sous-fibré maximal. D'après le théorème, on a $\sum_{n\leq n_0} r(n) = 1 \leq \sum_{n\leq n_0} q(n) = q(n_0)$. D'où le résultat.

Le théorème 2.1 exprimait la maximalité de q en termes de rangs. La proposition suivante porte sur les degrés :

Proposition 2.3. *Sous les hypothèses de 2.1, si $\mathcal{N}$ n'est pas dissocié et si $\sum_{n\in\mathbf{Z}} r(n) =$ rang $\mathcal{N} - 1$, on a aussi $\sum_{n\in\mathbf{Z}} nq(n) \leq \sum_{n\in\mathbf{Z}} nr(n)$ avec égalité si et seulement si $q = r$.*

Démonstration. On a, pour $m \gg 0$

$$\sum_{n\in\mathbf{Z}} nq(n) = (m+1)\sum_{n\in\mathbf{Z}} q(n) - \sum_{m'\leq m}\sum_{m''\leq m'} q(m'')$$

$$\sum_{n\in\mathbf{Z}} nr(n) = (m+1)\sum_{n\in\mathbf{Z}} r(n) - \sum_{m'\leq m}\sum_{m''\leq m'} r(m'')$$

et

$$\sum_{n\in\mathbf{Z}} q(n) = \sum_{n\in\mathbf{Z}} r(n) = \text{rang } \mathcal{N} - 1.$$

Il faut donc montrer $\sum_{m'\leq m}\sum_{m''\leq m'} r(m'') \leq \sum_{m'\leq m}\sum_{m''\leq m'} q(m'')$ pour $m \gg 0$ avec égalité si et seulement si $q = r$ ce qu'on obtient en ajoutant toutes les inégalités données par 2.1.

Le résultat suivant, pour la démonstration duquel on renvoie à [M-D, P 1,2], prouve l'existence et l'unicité de sous-fibrés maximaux $\mathcal{P}$ de $\mathcal{N}$ correspondant à la fonction q et montre qu'ils sont les "meilleurs" en ce sens que leur degré est le plus grand possible. Le conoyau $\mathcal{N}/\mathcal{P}$ est alors de la forme $\mathcal{J}_C(h)$, idéal (tordu) d'une courbe et à la maximalité du degré de $\mathcal{P}$ correspond la minimalité de h.

Théorème 2.4. *Soient $\mathcal{N}$ un $\mathcal{O}_{\mathbf{P}}$-module sans torsion de rang r, non dissocié et $\widetilde{\sigma} : \mathcal{L} = \bigoplus_{n\in\mathbf{Z}} \mathcal{O}_{\mathbf{P}}(-n)^{l(n)} \to \mathcal{N}$ l'homomorphisme correspondant à une présentation graduée minimale de $N = H^0_*\mathcal{N}$.*
1) Il existe un facteur direct $\mathcal{P} = \bigoplus_{n\in\mathbf{Z}} \mathcal{O}_{\mathbf{P}}(-n)^{q(n)}$ de $\mathcal{L}$, que $\widetilde{\sigma}$ identifie à un sous-fibré maximal et dissocié de $\mathcal{N}$.
2) L'ensemble des degrés des sous-fibrés maximaux et dissociés de $\mathcal{N}$ de rang $r-1$ a pour borne supérieure l'entier $d(\mathcal{N}) = -\sum_{n\in\mathbf{Z}} nq(n)$.
3) Tout sous-fibré maximal et dissocié de $\mathcal{N}$ de rang $r-1$ et de degré $d(\mathcal{N})$, est isomorphe à $\bigoplus_{n\in\mathbf{Z}} \mathcal{O}_{\mathbf{P}}(-n)^{q(n)}$ et s'obtient comme l'image par $\widetilde{\sigma}$ d'un facteur direct de $\bigoplus_{n\in\mathbf{Z}} \mathcal{O}_{\mathbf{P}}(-n)^{l(n)}$.

Remarque 2.5. Si $\mathcal{L}$ est un fibré dissocié, on a immédiatement, grâce à 1.7 c : $d(\mathcal{N} \oplus \mathcal{L}) = d(\mathcal{N}) + \deg \mathcal{L}$. De plus, si $\widetilde{u} : \bigoplus_{n\in\mathbf{Z}} \mathcal{O}_{\mathbf{P}}(-n)^{q(n)} \to \mathcal{N}$ correspond à un sous-fibré maximal et dissocié de $\mathcal{N}$, $\widetilde{u} \oplus id_{\mathcal{L}} : \bigoplus_{n\in\mathbf{Z}} \mathcal{O}_{\mathbf{P}}(-n)^{q(n)} \oplus \mathcal{L} \to \mathcal{N} \oplus \mathcal{L}$ correspond à un sous-fibré maximal et dissocié de $\mathcal{N} \oplus \mathcal{L}$.

Cette propriété admet une réciproque, utile dans l'étude des courbes minimales.

Proposition 2.6. *Soient $\mathcal{N}$ un $\mathcal{O}_{\mathbf{P}}$-module sans torsion, non dissocié, $\mathcal{L}$ un fibré dissocié et $\widetilde{v} : \bigoplus_{n\in\mathbf{Z}} \mathcal{O}_{\mathbf{P}}(-n)^{q(n)} \oplus \mathcal{L} \to \mathcal{N} \oplus \mathcal{L}$ un homomorphisme injectif dont l'image est un sous-fibré maximal de $\mathcal{N} \oplus \mathcal{L}$. Alors quitte à faire un automorphisme de $\bigoplus_{n\in\mathbf{Z}} \mathcal{O}_{\mathbf{P}}(-n)^{q(n)} \oplus \mathcal{L}$ et de $\mathcal{N} \oplus \mathcal{L}$, on peut supposer que le diagramme suivant :*

$$\begin{array}{ccc} \bigoplus_{n\in\mathbf{Z}} \mathcal{O}_{\mathbf{P}}(-n)^{q(n)} \oplus \mathcal{L} & \xrightarrow{\widetilde{v}} & \mathcal{N} \oplus \mathcal{L} \\ \downarrow p & & \downarrow p' \\ \mathcal{L} & = & \mathcal{L} \end{array}$$

où p et p' sont les projections canoniques, commute.

Remarque 2.7. La condition numérique du théorème 2.1, bien que peu suggestive, et l'existence de sous-fibrés de $\mathcal{N}$ qui la vérifient sont essentielles pour démontrer que la propriété de Lazarsfeld-Rao vaut pour toute classe de biliaison.

3. Application aux courbes. Existence de courbes minimales.

Théorème 3.1. *1) Dans toute classe de biliaison correspondant à un module de Rao non nul, il existe une courbe minimale C, unique à déformation à cohomologie et module de Rao constants près.*
2) Pour toute courbe C' de cette classe, il existe un entier $m \geq 1$, une suite de courbes $C = C_1, C_2, \ldots, C_m$ telle que C_{i+1} s'obtienne à partir de C_i par une biliaison élémentaire triviale, et C' à partir de C_m par une déformation à cohomologie et module de Rao constants.

Démonstration. Soient M un R-module gradué de longueur finie non nul, $\sigma_i : L_i \to L_{i-1}$ une résolution graduée libre minimale de M, $N_0 = \operatorname{Ker} \sigma_1$, $E_0 = \operatorname{Ker} \sigma_2$, $\mathcal{N}_0$ et $\mathcal{E}_0$ les faisceaux associés. On remarque que, puisque $H^1_*\mathcal{N}_0 = M$, $\mathcal{N}_0$ n'est pas dissocié.

La restriction de $\sigma_2 : L_2 = \bigoplus_{n\in\mathbf{Z}} R(-n)^{l_2(n)} \to N_0$ est une présentation graduée minimale, donc la fonction $l_{\mathcal{N}_0}$ n'est autre que la fonction l_2. Soit $q = q_{\mathcal{N}_0}$. D'après les résultats de 2., il existe une courbe C, un entier h , et une suite exacte :

$$0 \longrightarrow \mathcal{P}_0 = \bigoplus_{n\in\mathbf{Z}} \mathcal{O}_{\mathbf{P}}(-n)^{q(n)} \xrightarrow{\widetilde{\beta}_0} \mathcal{N}_0 \xrightarrow{\widetilde{\epsilon}_0} \mathcal{J}_C(h) \longrightarrow 0$$

où la flèche $\widetilde{\beta}_0$ s'obtient en composant une injection de $\mathcal{P}_0$ dans $\mathcal{L}_2$ (qui fait de $\mathcal{P}_0$ un facteur direct de $\mathcal{L}_2$) et la projection $\widetilde{\sigma}_2$. En particulier on a $H^1_*\mathcal{N}_0 = M = H^1_*\mathcal{J}_C(h)$ et C est dans la classe.

Soit C' une autre courbe de la classe de biliaison déterminée par M. En s'inspirant d'une remarque de Rao ([R]), on montre facilement qu'elle a une résolution (de type N, voir l'introduction) de la forme :

$$0 \longrightarrow \mathcal{P}' \xrightarrow{\widetilde{\beta}'} \mathcal{N}_0 \oplus \mathcal{L}' \xrightarrow{\widetilde{\epsilon}'} \mathcal{J}_{C'}(h') \longrightarrow 0$$

où $\mathcal{P}' = \bigoplus_{n\in\mathbf{Z}} \mathcal{O}_{\mathbf{P}}(-n)^{r'(n)}$ et $\mathcal{L}'$ sont dissociés, et $h' \in \mathbf{Z}$ est tel que $M = M_{C'}(h')$. On peut supposer de plus cette résolution minimale au sens qu'on ne peut simplifier aucun facteur direct inversible entre $\mathcal{P}'$ et $\mathcal{L}'$. La résolution de $\mathcal{J}_C$ peut encore s'écrire, en rajoutant le facteur $\mathcal{L}'$, sous la forme:

$$0 \longrightarrow \mathcal{P} \xrightarrow{\widetilde{\beta}} \mathcal{N}_0 \oplus \mathcal{L}' \xrightarrow{\widetilde{\epsilon}} \mathcal{J}_C \longrightarrow 0$$

où $\mathcal{P} = \mathcal{P}_0 \oplus \mathcal{L}' = \bigoplus_{n\in\mathbf{Z}} \mathcal{O}_{\mathbf{P}}(-n)^{r(n)}$ et où l'explicitation des flèches est laissée au lecteur.

D'après 1.7 c et 2.3 on a $\deg \mathcal{P}' \leq \deg \mathcal{P}_0 \oplus \mathcal{L}'$ et il y a égalité si et seulement si $\mathcal{P}' \simeq \mathcal{P}_0 \oplus \mathcal{L}'$. D'une part, on a $h \leq h'$ et C est minimale. D'autre part si C' est aussi minimale, (i.e., $h = h'$), alors $\mathcal{P}'$ et $\mathcal{P}_0 \oplus \mathcal{L}'$ sont isomorphes. La minimalité de la résolution, jointe à 2.6, entraîne $\mathcal{L}' = 0$.

Posons $\widetilde{\beta}_t = (1-t)\widetilde{\beta}_0 + t\widetilde{\beta}$ et $\mathcal{J}_{C_t}(h) = \operatorname{Coker} \widetilde{\beta}_t$. Il existe un ouvert de la droite affine contenant les points $t = 0$ et $t = 1$ tel que pour tout point t de cet ouvert, $\mathcal{J}_{C_t}$ soit l'idéal d'une courbe C_t. On passe donc de C à C' par une déformation à cohomologie et module de Rao constants.

Pour prouver le point 2) on note que l'on a, grâce à 2.1 et 1.7 c : pour tout $n \in \mathbf{Z}$, $\sum_{m\leq n} r'(m) \leq \sum_{m\leq n} r(m)$. Le résultat est alors une conséquence du lemme suivant, qui est montré, avec une autre formulation, dans [LR] (proposition 1.4) :

Lemme 3.2. *Soient C et C' deux courbes ayant des résolutions*

$$0 \longrightarrow \bigoplus_{n\in\mathbf{Z}} \mathcal{O}_{\mathbf{P}}(-n)^{r(n)} \xrightarrow{\widetilde{\beta}} \mathcal{N} \xrightarrow{\widetilde{\epsilon}} \mathcal{J}_C \longrightarrow 0$$

$$0 \longrightarrow \bigoplus_{n\in\mathbf{Z}} \mathcal{O}_{\mathbf{P}}(-n)^{r'(n)} \xrightarrow{\widetilde{\beta}'} \mathcal{N} \xrightarrow{\widetilde{\epsilon}'} \mathcal{J}_{C'}(h) \longrightarrow 0$$

où $\mathcal{N}$ est un fibré qui vérifie $H^2_*\mathcal{N} = 0$ et où l'on a $\sum_{m\leq n} r'(m) \leq \sum_{m\leq n} r(m)$ pour tout $n \in \mathbf{Z}$. Alors il existe une suite de courbes comme dans l'énoncé du théorème.

On peut préciser les résolutions de types E et N des courbes minimales :

Proposition 3.3. *Notons encore $\sigma_2 : \bigoplus_{n\in\mathbf{Z}} R(-n)^{l_2(n)} \to N_0$ la présentation graduée minimale de N_0. Une courbe minimale C a des résolutions de type E et de type N de la forme suivante :*

$$0 \longrightarrow \mathcal{E}_0 \xrightarrow{\widetilde{\alpha}_0} \bigoplus_{n\in\mathbf{Z}} \mathcal{O}_{\mathbf{P}}(-n)^{l_2(n)-q(n)} \xrightarrow{\widetilde{\pi}_0} \mathcal{J}_C(h) \longrightarrow 0,$$

$$0\longrightarrow \bigoplus_{n\in\mathbf{Z}} \mathcal{O}_{\mathbf{P}}(-n)^{q(n)} \xrightarrow{\widetilde{\beta}_0} \mathcal{N}_0 \xrightarrow{\widetilde{\epsilon}_0} \mathcal{J}_C(h)\longrightarrow 0.$$

En particulier, on a : $s_0(C) = a_0 + h$.

Démonstration. Vu la construction de $\widetilde{\beta}_0$, il suffit de remarquer qu'on a un diagramme commutatif de suites exactes :

$$\begin{array}{ccccccccc}
 & & & & 0 & & 0 & & \\
 & & & & \downarrow & & \downarrow & & \\
 & & & & \mathcal{P}_0 & = & \mathcal{P}_0 & & \\
 & & & & \downarrow & & \downarrow{\scriptstyle \widetilde{\beta}_0} & & \\
0 & \longrightarrow & \mathcal{E}_0 & \xrightarrow{\widetilde{\lambda}} & \mathcal{L}_2 & \xrightarrow{\widetilde{\sigma}_2} & \mathcal{N}_0 & \longrightarrow & 0 \\
 & & \| & & \downarrow & & \downarrow{\scriptstyle \widetilde{\epsilon}_0} & & \\
0 & \longrightarrow & \mathcal{E}_0 & \xrightarrow{\widetilde{\alpha}_0} & \mathcal{L}_2/\mathcal{P}_0 & \xrightarrow{\widetilde{\pi}_0} & \mathcal{J}_C(h) & \longrightarrow & 0 \\
 & & & & \downarrow & & \downarrow & & \\
 & & & & 0 & & 0 & &
\end{array}$$

où $\widetilde{\lambda}$ est l'injection canonique, et ce diagramme donne la deuxième suite exacte de l'énoncé.

Puisque σ_2 est une présentation graduée minimale de N_0, on voit que l'homomorphisme $\pi_0 : \bigoplus_{n\in\mathbf{Z}} R(-n)^{l_2(n)-q(n)} \to H^0_*\mathcal{J}_C(h)$ est aussi une présentation graduée minimale, ce qui, vu 1.7 a, entraîne l'égalité annoncée.

Puisqu'une biliaison élémentaire triviale augmente le genre et le degré, et qu'une déformation à cohomologie constante les conserve, on a le théorème suivant :

Théorème 3.4. *Une courbe minimale a un genre et un degré minimaux parmi les courbes de sa classe de biliaison.*

Remarque 3.5. Soient M un R-module gradué de longueur finie, M^* son dual, $\sigma_i : L_i \to L_{i-1}$ une résolution graduée libre minimale de M, $N_0 = \mathrm{Ker}\ \sigma_1$, $E_0 = \mathrm{Ker}\ \sigma_2$, $\mathcal{N}_0$ et $\mathcal{E}_0$ les faisceaux associés. On montre (cf. [M-D,P 1,2]) que la duale de la résolution graduée de M est une résolution graduée (libre minimale) de $M^*(4)$ (où $M^* = \mathrm{Ext}^4_R(M,R)$ est le module dual de M), et que les rôles de N_0 et $E_0^\vee$ sont échangés par dualité. En comparant les fonctions $q = q_{\mathcal{N}_0}$ et $q' = q_{\mathcal{E}_0^\vee}$, on peut établir le résultat suivant, qui caractérise les courbes de la classe duale : soient a_0 et a_1 les entiers définis en 1.3 ; si h est le décalage d'une courbe minimale de la classe de biliaison définie par M, toute courbe minimale de la classe de biliaison définie par M (resp. M^*) est liée par $(a_0+h)\times(a_1+h)$ à une courbe minimale de la classe de biliaison définie par M^* (resp. M).

4. Exemples.

Dans ce qui suit, nous appliquons notre méthode à deux types d'exemples. Dans le premier, le module est un quotient de R par une suite régulière, on dispose d'une courbe de la classe de liaison (la réunion de deux intersections complètes), mais cette courbe n'est pas minimale en général. Dans le second, le module est de Buchsbaum et on n'a pas, a priori, de courbe de la classe.

Soit M un module gradué muni d'une résolution $(*)$ comme dans l'introduction. On pose $N_0 = \mathrm{Im}\sigma_2$, on note φ_a la restriction de σ_2 à $L_{2,\leq a}$ et t_a le rang de $L_{2,\leq a}$. La proposition suivante va permettre de faire les calculs :

Proposition 4.1. *Avec les notations de 1.3, on a :*
1) L'entier $a_0 - 1$ est le plus grand entier $a \in \mathbf{Z}$ tel que les t_a-mineurs de φ_a n'aient pas de facteur commun non trivial (en particulier ils sont alors non tous nuls).
2) L'entier α_a $(= \operatorname{rang} \varphi_a)$ est la dimension d'un plus grand mineur non nul de φ_a.
3) β_a est le plus grand entier β tel que les β-mineurs de φ_a n'aient pas de facteur commun non trivial.

Exemple 1 : module associé à une suite régulière.

Soit f_1, f_2, f_3, f_4 une suite régulière d'éléments homogènes de $\mathbf{m}$, (en particulier, les f_i sont deux à deux sans facteur commun). Posons $n_i = \deg f_i$ et supposons les f_i ordonnés de sorte que l'on ait $n_1 \leq n_2 \leq n_3 \leq n_4$. Soit $M = R/(f_1, f_2, f_3, f_4)$. C'est un R-module de longueur finie qui est le module de Rao de la courbe $C = C_1 \cup C_2$ où C_1 (resp. C_2) est l'intersection complète d'équations (f_1, f_2) (resp. (f_3, f_4)), ou aussi des courbes analogues obtenues en permutant les indices. Il admet pour résolution minimale le complexe de Koszul associé à la suite f_1, f_2, f_3, f_4 :

$$0 \to R(-\nu) \to \bigoplus_{i=1}^{4} R(n_i - \nu) \to \bigoplus_{i<j} R(-n_i - n_j) \xrightarrow{\sigma_2} \bigoplus_{i=1}^{4} R(-n_i) \to R \to M \to 0$$

(avec $\nu = n_1 + n_2 + n_3 + n_4$). On note qu'on a les inégalités :

$$n_1 + n_2 \leq n_1 + n_3 \leq \begin{matrix} n_1 + n_4 \\ \\ n_2 + n_3 \end{matrix} \leq n_2 + n_4 \leq n_3 + n_4$$

mais qu'il n'y a pas d'inégalité automatique entre $n_1 + n_4$ et $n_2 + n_3$; on pose :

$$\mu = \sup(n_1 + n_4, n_2 + n_3).$$

En utilisant 4.1 on obtient les valeurs suivantes de α_a et β_a :

a	n_1+n_2	n_1+n_3	n_1+n_4	n_2+n_3	μ	n_2+n_4	n_3+n_4
α_a	1	2	3	2	3	3	3
β_a	1	1	1	2	2	3	3

Proposition 4.2. *1) On a $a_0 = n_1 + n_3$, $a_1 = n_2 + n_4$.*
2) Calcul de q :
a) Si $n_1 + n_2 < \mu$, on a $q(n) = 0$, sauf $q(n_1 + n_2) = q(\mu) = 1$.
b) Si $n_1 + n_2 = \mu$, on a $q(n) = 0$, sauf $q(n_1 + n_2) = 2$.

Démonstration. L'assertion 1), résulte de la définition de a_0 (cf. 1.3). Comme le fibré $\mathcal{N}_0$ est de rang 3 et non dissocié, on a $\sum_{n \in \mathbf{Z}} q(n) = 2$ et $q(n) \geq 0$ (cf. 1.6). D'autre part, on a $q(n_1 + n_2) \geq 1$ (cf. 2.2). Le plus petit entier n tel que $\sum_{m \leq n} q(m) = 2$ est alors le plus petit entier n tel que $\inf(\alpha_n - 1, \beta_n) = 2$. C'est donc μ.

Les résultats 2.4, 3.1 et 3.3 nous permettent alors de préciser la courbe minimale C_0 dans la classe de liaison associée à M.

Corollaire 4.3. *Soit C_0 une courbe minimale associée à M (unique à déformation à cohomologie et module de Rao constants près).*
1) On a $M_{C_0} = M(\mu - n_3 - n_4)$.
2) On a les résolutions minimales suivantes de $I_{C_0}(\mu - n_3 - n_4)$:
a) Type N :

$$0 \to P \to N_0 \to I_{C_0}(\mu - n_3 - n_4) \to 0,$$

avec $N_0 = \operatorname{Im}\sigma_2$ et $P = R(-n_1 - n_2) \oplus R(-\mu)$.

b) Type E :

$$0 \to E_0 \to F \to I_{C_0}(\mu - n_3 - n_4) \to 0,$$

avec $E_0 = \mathrm{Ker}\sigma_2$ *et* $F = R(-n_1 - n_3) \oplus R(-\mu') \oplus R(-n_2 - n_4) \oplus R(-n_3 - n_4)$ *où l'on a posé* $\mu' = \inf(n_2 + n_3, n_1 + n_4)$.
3) Le degré de C_0 *est* $d = \mu(n_1 + n_2) - n_1 n_3 - n_2 n_4$.

Remarques 4.4.

1) Sauf si trois des n_i sont égaux, on a $\mu - n_3 - n_4 < 0$. Ceci signifie que les courbes C réunions d'intersections complètes qui vérifient $M_C = M$ ne sont pas minimales. C'est d'ailleurs évident a priori puisque les trois courbes obtenues à partir de C par permutation des indices sont de degrés $n_1n_2 + n_3n_4$, $n_1n_3 + n_2n_4$, $n_1n_4 + n_2n_3$ distincts, et qu'il y a unicité du degré d'une courbe minimale. On notera que l'on a $h^1\mathcal{J}_C(\mu - n_3 - n_4) \neq 0$ et que donc, si $\mu - n_3 - n_4 < 0$, la courbe minimale n'est pas réduite.

2) On a $s_0(C_0) = \mu + n_1 - n_4$; $e(C_0) = 2\mu - n_3 - n_4 - 4$. On note que la relation $e + 3 < s_0$ n'est vérifiée que si $n_1 = n_2$ et $n_3 = n_4$.

3) Les résultats de 2.4 permettent d'affirmer qu'il existe un facteur direct $\mathcal{L}$ de $\mathcal{L}_2$, isomorphe à $\mathcal{O}_{\mathbf{P}}(-n_1 - n_2) \oplus \mathcal{O}_{\mathbf{P}}(-\mu)$ tel que $\sigma_2(\mathcal{L})$ soit un sous faisceau maximal de $\mathcal{N}_0$. Dans le cas présent on peut décrire explicitement un plongement convenable de $\mathcal{L}$ dans $\mathcal{L}_2$: on choisit des polynômes homogènes f, g de degrés respectifs $\mu - n_1 - n_4$ et $\mu - n_2 - n_3$, non nuls et tels que f, g, et les f_i soient deux à deux sans facteur commun. On désigne par e_{ij}un vecteur d base de $\mathcal{O}_{\mathbf{P}}(-n_i - n_j)$. On envoie alors les vecteurs de base de $\mathcal{O}_{\mathbf{P}}(-n_1 - n_2) \oplus \mathcal{O}_{\mathbf{P}}(-\mu)$ respectivement sur e_{12} et $fe_{14} + ge_{23}$. La composée φ de σ_2 avec ce plongement admet pour matrice :

$$\begin{pmatrix} f_2 & ff_4 \\ -f_1 & gf_3 \\ 0 & -gf_2 \\ 0 & ff_1 \end{pmatrix}$$

Les 2-mineurs de cette matrice n'ont pas de facteurs communs et définissent l'idéal de C_0 qui est donc engendré par les polynômes f_1f_2, $ff_1f_4 + gf_2f_3$, gf_2^2, ff_1^2. On vérifie aussitôt que C_0 est liée par les surfaces d'équations f_1f_2 et $ff_1f_4 + gf_2f_3$ à la réunion C' des intersections complètes $(f_1, f_3) \cup (f_2, f_4)$. On peut encore lier C' par les surfaces f_1f_2 et f_3f_4 à $C'' = (f_1, f_4) \cup (f_2, f_3)$. En définitive, C'' est donc obtenue à partir de C_0 par une biliaison élémentaire de type $(n_1 + n_2, n_3 + n_4 - \mu)$ ce qui confirme 4.3.1. Cette remarque permet de calculer plus aisément les invariants de C_0 et notamment le degré et le genre.

Exemple 4.5. Prenons $n_1 = n_2 = 1$, $n_3 = n_4 = a$ avec $a > 1$. Dans ce cas, on a $M = M_C$ avec C réunion disjointe de deux courbes planes de degré a. Cette courbe n'est pas minimale. Une courbe minimale C_0 est de degré 2 et de genre $-a$, il s'agit d'une structure double sur une droite qu'on peut par exemple décrire par les équations XY, X^2, Y^2, $XA + YB$ avec A, B de degré a .

Exemple 2 : module de Buchsbaum.

On suppose cette fois que le module M est de Buchsbaum : $M = M_0 \oplus M_1 \oplus \cdots M_r$ avec $r \geq 0$ et $\dim_k M_i = \rho(i)$. On suppose $\rho(0) > 0$ et $\rho(r) > 0$. La structure de R-module de M est triviale et on a la résolution minimale obtenue en additionnant des complexes de Koszul :

$$0 \to \bigoplus_{n=0}^{r} R(-4-n)^{\rho(n)} \to \bigoplus_{n=0}^{r} R(-3-n)^{4\rho(n)} \to \bigoplus_{n=0}^{r} R(-2-n)^{6\rho(n)}$$
$$\xrightarrow{\sigma_2} \bigoplus_{n=0}^{r} R(-1-n)^{4\rho(n)} \to \bigoplus_{n=0}^{r} R(-n)^{\rho(n)} \to M \to 0$$

Le noyau N_0 de σ_2 est naturellement décomposé en somme directe $N_0 = \bigoplus_{n=0}^{r} N_{0,n}$ des modules correspondants pour chaque complexe, et la matrice de σ_2 dans une base convenablement choisie est une matrice formée de blocs diagonaux 4×6 tous égaux à la matrice :

$$\begin{pmatrix} Y & Z & T & 0 & 0 & 0 \\ -X & 0 & 0 & Z & T & 0 \\ 0 & -X & 0 & -Y & 0 & T \\ 0 & 0 & -X & 0 & -Y & -Z \end{pmatrix}$$

Le rang (resp. le rang sur les diviseurs) d'un tel bloc est égal à 3. On en déduit :

Proposition 4.6.
1) On a $a_0 = 2$, $a_1 = r + 2$.
2) On a, pour tout $n \geq 0$, $\alpha_{n+2} = \beta_{n+2} = 3\sum_{i=0}^{n} \rho(i)$.
3) On a $q(2) = 3\rho(0) - 1$; $q(n+2) = 3\rho(n)$ *pour* $n = 1, \ldots, r$ *et* $q(n) = 0$ *sinon.*

Corollaire 4.7. *On pose* $\alpha = \sum_{i=0}^{r} \rho(i)$ *et* $h = 2\alpha - 2$. *Soit* C_0 *une courbe minimale associée à* M.
1) On a $M_{C_0} = M(h)$.
2) Les résolutions minimales de $I_{C_0}(h)$ *sont les suivantes :*
a) Type N : $0 \to P \to N_0 \to I_{C_0}(h) \to 0$ *avec*

$$P = \bigoplus_{i=1}^{r} R(-2-i)^{3\rho(i)} \oplus R(-2)^{3\rho(0)-1}.$$

b) Type E : $0 \to E_0 \to F \to I_{C_0}(h) \to 0$, *avec* $E_0 = \mathrm{Ker}\sigma_0$ *et*

$$F = \bigoplus_{i=1}^{r} R(-2-i)^{3\rho(i)} \oplus R(-2)^{3\rho(0)+1}.$$

3) Les invariants de C_0 *sont les suivants :* $s_0 = 2\alpha$; $e = 2\alpha + r - 4$.
4) Si $r = 0$, *et si* C_0 *est générale, elle est lisse, et irréductible si* $\rho(0) > 1$.
5) Si $r > 0$, *et si* C_0 *est générale, elle est réunion (non disjointe) d'une courbe lisse (minimale pour le module de Buchsbaum* M_0*) et d'une courbe réduite.*

Démonstration. Le décalage $h = 2\alpha - 2$ est toujours > 0 sauf si $r = 0$ et $\rho(0) = 1$, c'est-à-dire dans le cas de la classe de liaison de deux droites. On en déduit que, sauf dans ce cas, on a $h^1\mathcal{J}_{C_0} = 0$ de sorte que C_0 est connexe.

Si $r = 0$, $\mathcal{N}_0(2)$ est engendré par ses sections, donc une courbe C_0 générale est lisse.

Si $r > 0$, on a une décomposition $\mathcal{N}_0 = \bigoplus_{n=0}^{r} \mathcal{N}_{0,n}$. On déduit des résultats de [C] qu'une courbe générale C_0 est réduite. De plus, pour tout $n > 0$, $\mathcal{N}_{0,n}(2)$ n'a pas de sections. Soit $\widetilde{\beta}_0$ la flèche de $\mathcal{P} = \bigoplus_{i=1}^{r} \mathcal{O}_{\mathbf{P}}(-2-i)^{3\rho(i)} \oplus \mathcal{O}_{\mathbf{P}}(-2)^{3\rho(0)-1}$ dans $\mathcal{N}_0$. La restriction $\widetilde{\beta}_0'$ de $\widetilde{\beta}_0$ au sous-fibré $\mathcal{O}_{\mathbf{P}}(-2)^{3\rho(0)-1}$ de $\mathcal{P}$ envoie donc $\mathcal{O}_{\mathbf{P}}(-2)^{3\rho(0)-1}$ dans $\mathcal{N}_{0,0}$. Le lieu de dégénérescence de $\widetilde{\beta}_0'$ est contenu dans celui de $\widetilde{\beta}_0$, c'est-à-dire C_0, et c'est une courbe C_0', minimale pour le module de Buchsbaum M_0, qui est donc une composante de C_0.

Remarques 4.8.

1) Par liaison on voit que toute courbe de Buchsbaum, sauf la réunion de deux droites, est connexe.

2) On note que la condition de Lazarsfeld-Rao ($e + 3 < s_0$) est satisfaite pour C_0 si et seulement si $r = 0$.

3) La courbe C_0 est de rang maximum si et seulement si on a $r + 2\alpha - 2 < s_0 = 2\alpha$, i.e., si $r = 0$ ou $r = 1$. Vu 3.1 on en déduit qu'il n'existe pas de courbes de Buchsbaum de rang maximum si $r > 2$ et que pour $r \leq 2$ il en existe pour tout décalage $h \geq 0$ de M_{C_0}.

4) Le cas $r = 0$ (cf. [BM1]).

On suppose $r = 0$, $\rho(0) = a > 0$. On peut calculer le degré et le genre à l'aide des résolutions : $d = 2a^2$, $g = (2a-3)(2a-1)(2a+1)/3$. Si C est une courbe de la classe on a alors par biliaison des minorations des invariants de C. Par exemple, si C n'est pas minimale, on a $d_C \geq 2a^2 + 2a$ et il existe des courbes de tout degré d vérifiant cette inégalité. Comme C_0 est de rang maximum on peut même supposer C de rang maximum.

5) Le cas $r = 1$ (cf. [BM2]).

Cette fois on a $r = 1$, $\rho(0) = a$, $\rho(1) = b$, $a, b > 0$. Les calculs sont analogues. On a par exemple $d = 2(a+b)^2 + 2b$ pour la courbe minimale et $d \geq 2(a+b)^2 + 2b + a + b$ sinon. Si l'inégalité est stricte il existe une courbe de la classe qui a le degré d et est de rang maximum.

Références bibliographiques

[BBM] E. Ballico, G. Bolondi et J. Migliore, The Lazarsfeld-Rao problem for liaison classes of two-codimensional subschemes of $\mathbf{P}^n$, à paraître in Amer. J. of Maths.

[BM1] G. Bolondi et J. Migliore, Classification of maximal rank curves in the liaison class L_n, Math. Ann., 277, 1987, 585–603.

[BM2] G. Bolondi et J. Migliore, Buchsbaum Liaison classes, J. of Algebra, 123, 1989, 2, 426–456.

[BM3] G. Bolondi et J. Migliore, The Lazarsfeld-Rao and Zeuthen problems for Buchsbaum curves, preprint.

[C] M.-C. Chang, A filtered Bertini-type theorem, J.reine angew. Math. 397, 1989, 214–219.

[H] R. Hartshorne, On the classification of algebraic space curves, in Vector bundles and differential equations, Proceedings, Nice, France, 1979, Progress in Math.7, Birkhäuser.

[LR] R. Lazarsfeld et A. P. Rao, Linkage of general curves of large degree, Lecture Notes 997, Springer Verlag, 1983, 267–289.

[MD-P1] M. Martin-Deschamps et D. Perrin, Sur la classification des courbes gauches I, Rapport de recherche du LMENS, Ecole Normale Supérieure, 45 rue d'Ulm, 75230 PARIS Cedex 05, mai 1989.

[MD-P2] M. Martin-Deschamps et D. Perrin, Sur la classification des courbes gauches, à paraître dans Astérisque, décembre 1990.

[M] J. Migliore, Geometric Invariants of Liaison, J. of Algebra, 99, (1986), 548–572.

[R] A. P. Rao, Liaison among curves in $\mathbf{P}^3$, Invent. Math., 50, 1979, 205–217.

Labo. Math., Ecole Normale Supérieure, Unité associée au C.N.R.S., 45 rue d'Ulm, 75230 Paris Cedex 05.

Fano 3-folds

Shigeru MUKAI

Abstract: In the beginning of this century, G. Fano initiated the study of 3-dimensional projective varieties $X_{2g-2} \subset P^{g+1}$ with canonical curve sections in connection with the Lüroth problem.[1] After a quick review of a modern treatment of Fano's approach (§1), we discuss a new approach to Fano 3-folds via vector bundles, which has revealed their relation to certain homogeneous spaces (§§2 and 3) and varieties of sums of powers (§§5 and 6). We also give a new proof of the genus bound of prime Fano 3-folds (§4). In the maximum genus ($g = 12$) case, Fano 3-folds $X_{22} \subset P^{13}$ yield a 4-dimensional family of compactifications of C^3 (§8).

A compact complex manifold X is *Fano* if its first Chern class $c_1(X)$ is positive, or equivalently, its anticanonical line bundle $\mathcal{O}_X(-K_X)$ is ample. If $\mathcal{O}_X(-K_X)$ is generated by global sections and $\Phi_{|-K_X|}$ is birational, then its image is called the anticanonical model of X. In the case $\dim X = 3$, every smooth curve section $C = X \cap H_1 \cap H_2 \subset P^{g-1}$ of the anticanonical model $X \subset P^{g+1}$ is canonical, that is, embedded by the canonical linear system $|K_C|$. Conversely, every projective 3-fold $X_{2g-2} \subset P^{g+1}$ with a canonical curve section is obtained in this way. The integer $\frac{1}{2}(-K_X)^3 + 1$ is called the *genus* of a Fano 3-fold X since it is equal to the genus of a curve section of the anticanonical model.

A projective 3-fold $X_{2g-2} \subset P^{g+1}$ with a canonical curve section is a complete intersection of hypersurfaces if $g \leq 5$. In particular, the Picard group of X is generated by $\mathcal{O}_X(-K_X)$. We call such a Fano 3-fold *prime.* If a Fano 3-fold X is not prime, then either $-K_X$ is divisible by an integer ≥ 2 or the Picard number ρ of X is greater than one. See [15], [7] and [9] for the classification in the former case, and [24] and [26] in the latter case.

§1 Double projection The anticanonical line bundle $\mathcal{O}_X(-K_X)$ is very ample if X is a prime Fano 3-fold of genus ≥ 5 (*cf.* [15] and [41]). To classify prime Fano 3-folds $X_{2g-2} \subset P^{g+1}$ of genus $g \geq 6$, Fano investigated the *double projection* from a line[2] ℓ on X_{2g-2}, that is, the rational map associated to the linear system $|H - 2\ell|$ of hyperplane sections singular along ℓ.

Example 1 *Let $X_{16} \subset P^{10}$ be a prime Fano 3-fold of genus 9. Then the double projection $\pi_{2\ell}$ from a line $\ell \subset X_{16}$ is a birational map onto P^3. The union D of conics which intersect ℓ is a divisor of X and contracted to a space curve $C \subset P^3$ of genus 3 and degree 7. The inverse rational map $P^3 - \to X_{16} \subset P^{10}$ is given by the linear system $|7H - 2C|$ of surfaces of degree 7 which are singular along C.*

The key for the analysis of $\pi_{2\ell}$ is the notion of flop. Let X^- be the blow-up of X along ℓ. Since other lines intersect ℓ, X^- is not Fano. But X^- is *almost Fano* in the sense that $|-K_{X^-}|$ is free and gives a birational morphism contracting no divisors. The anticanonical model $\bar{X}$ of X^- is the image of the projection $X - \to P^8$ from ℓ. The strict transform $D^- \subset X^-$ of D is relatively negative over $\bar{X}$. By the theory of flops ([33], [19]), there exists another almost Fano 3-fold X^+ which has the same anticanonical model as X^- and such that the strict transform $D^+ \subset X^+$ of D^- is relatively ample over $\bar{X}$. X^+ is called the D^--*flop*[3] of X^-.

---

[1] A surface dominated by a rational variety is rational by Castelnuovo's criterion. But this does not hold any more for 3-folds. See [5], [44] and [18].

[2] The existence of a line is proved by Shokurov [42].

[3] The smoothness of X^+ follows from [19, 2.4] or from the classification [6, Theorem 15] of the singularity of $\bar{X}$.

Theorem ([23], [17])*Let X, ℓ and D be as in Example 1. Then the D^--flop X^+ of the blow-up X^- of X along ℓ is isomorphic to the blow-up of P^3 along a space curve of genus 3 and degree 7.*

For the proof, the theory of extremal rays ([22]) is applied to the almost Fano 3-fold X^+. If X is a prime Fano 3-fold of genus 10, then X^+ is isomorphic to the blow-up of a smooth 3-dimensional hyperquadric $Q^3 \subset P^4$ along a curve of genus 2 and degree 7. In the case genus 12, X^+ is the blow-up of a quintic del Pezzo 3-fold[4] $V_5 \subset P^6$ along a quintic normal rational curve.

§2 Bundle method A line on $X_{2g-2} \subset P^{g+1}$ can move in a 1-dimensional family. Hence the double projection method does not give a canonical biregular description of $X_{2g-2} \subset P^{g+1}$. In the case $g = 9$, *e.g.*, there are infinitely many different space curves[5] $C \subset P^3$ which give the same Fano 3-fold $X_{16} \subset P^{10}$. By the same reason, the double projection method does not classify $X_{2g-2} \subset P^{g+1}$ over fields which are not algebraically closed. Even when a Fano 3-fold X is defined over $k \subset C$, it may not have a line defined over k. Our new classification makes up these defects. It is originated to solve the following:

Problem:[6] Classify all projective varieties $X^n_{2g-2} \subset P^{g+n-2}$ of dimension $n \geq 3$ with a canonical curve section.[7]

We restrict ourselves to the case where every divisor on X is cut out by a hypersurface. In contrast with the case $g \leq 5$, the dimension n of X cannot be arbitrarily large in the case $g \geq 6$. In each case $7 \leq g \leq 10$, the maximum dimension $n(g)$ is attained by a homogeneous space Σ_{2g-2}.

Table

g	$n(g)$	$\Sigma_{2g-2} \subset P^{g+n(g)-2}$	$r(E)$	$\chi(E)$	$c_1(E)c_2(E)$
6	6	Hyperquadric section of the cone of the Grassmann variety[8] $G(2,5) \subset P^9$	2	5	4
7	10	10-dimensional spinor variety $SO(10,C)/P \subset P^{15}$	5	10	48
8	8	Grassmann variety $G(2,6) \subset P^{14}$	2	6	5
9	6	$Sp(6,C)/P \subset P^{13}$	3	6	8
10	5	$G_2/P \subset P^{13}$	5	7	12
12	3	$G(V,3,N) \subset P^{13}$ (See Theorem 3.)	3	7	10

We claim that every variety $X \subset P$ with canonical curve section of genus $g \geq 6$ is a linear section of the above $\Sigma_{2g-2} \subset P^{g+n(g)-2}$. Since each Σ_{2g-2} has a natural morphism to a Grassmann variety, vector bundles play a crucial role in our classification. Instead of a line, we show the existence of a good vector bundle E on X. Instead of the double projection, we embed X into a Grassmann variety by the linear system $|E|$ and describe its image. The vector bundle is first constructed over a general (K3) surface section S of X and then

[4]A smooth projective variety $V^n_d \subset P^{d+n-2}$ with a normal elliptic curve section is called *del Pezzo*. The anticanonical class $-K_V$ is linearly equivalent to $(n-1)$ times hyperplane section. All quintic del Pezzo 3-folds are isomorphic to each other (see [15] and [9]).

[5]The isomorphism classes of curves C are uniquely determined by the Torelli theorem since the intermediate Jacobian variety of X is isomorphic to the Jacobian variety of C.

[6]Roth [36] [37] studied this problem by generalizing the double projection method.

[7]The anticanonical class of X^n_{2g-2} is $(n-2)$-times hyperplane section. In the case $n = 2$, X^2_{2g-2} is a (polarized) K3 surface. The integer g is called the genus of X.

[8]$G(s,n)$ denotes the Grassmann variety of s-dimensional subspaces of a fixed n-dimensional vector space.

extended to X applying a Lefschetz type theorem (*cf.* [8]).[9] The numerical invariants of E are as in the above table.[10] All higher cohomology groups of E vanish and E is generated by its global sections. The morphism[11] $\Phi_{|E|} : X \longrightarrow G(H^0(E), r(E))$ is an embedding if $g \geq 7$. The first Chern class $c_1(E)$ is equal to $2c_1(X)$ if $g = 7$ and equal to $c_1(X)$ otherwise. E is characterized by the following two properties:

1) $r(E)$, $c_1(E)$ and $c_2(E)$ are as above, and
2) the restriction[12] of E to a general surface section is stable.

In the case $g = 9$, $|E|$ embeds X into the 9-dimensional Grassmann variety $G(V, 3)$, where $V = H^0(X, E)$. Consider the natural map

$$\lambda_2 : \bigwedge^2 H^0(X, E) \longrightarrow H^0(X, \bigwedge^2 E).$$

The kernel is generated by a nondegenerate bivector σ on V. Hence the image of X is contained in the zero locus $G(V, 3, \sigma)$ of the global section of $\bigwedge^2 \mathcal{E}$ corresponding to σ, where $\mathcal{E}$ is the universal quotient bundle on $G(V, 3)$. $G(V, 3, \sigma)$ is a 6-dimensional homogeneous space of $Sp(V, \sigma)$ and a projective variety $\Sigma_{16} \subset P^{13}$ with a canonical curve section of genus 9. In the case $\dim X = 3$, we have

Theorem 2 *A prime Fano 3-fold $X_{16} \subset P^{10}$ of genus 9 is isomorphic to the intersection of Σ_{16} and a linear subspace P^{10} in P^{13}.*

By the above characterization, E is defined over $k \subset C$ if X is so. Hence the theorem holds true for every Fano 3-fold $X_{16} \subset P^{10}_k$ over $k \subset C$ such that $X \otimes C$ is prime.

The results are similar for $g = 7, 8$ and 10. In the case $g = 7$ and 10, the natural mappings $\sigma_2 : S^2 H^0(X, E) \longrightarrow H^0(X, S^2 E)$ and $\lambda_4 : \bigwedge^4 H^0(X, E) \longrightarrow H^0(X, \bigwedge^4 E)$ are considered instead of λ_2. In the case $g = 6$, X is a double cover of a linear section of $G(2, 5) \subset P^9$ if the linear subspace P passes through the vertex of the Grassmannian cone. Otherwise, X is isomorphic to the complete intersection of a 6-dimensional hyperquadric $Q \subset P$ and $G(2, 5) \subset P^9$.

§3 Fano 3-fold of genus 12 A prime Fano 3-fold[13] X of genus 12 cannot be an ample divisor of a 4-fold. But the vector bundle E gives a canonical description of X in the 12-dimensional Grassmann variety $G(V, 3), V = H^0(X, E)$. Consider the natural map $\lambda_2 : \bigwedge^2 H^0(X, E) \longrightarrow H^0(X, \bigwedge^2 E)$ as in the case $g = 9$. Its kernel N is of dimension 3. Let $\{\sigma_1, \sigma_2, \sigma_3\}$ be a basis of N.

Theorem 3 *A prime Fano 3-fold $X_{22} \subset P^{12}$ of genus 12 is isomorphic to the common zero locus $G(V, 3, N)$ of the three global sections of $\bigwedge^2 \mathcal{E}$ corresponding to σ_1, σ_2 and σ_3, where $\mathcal{E}$ is the universal quotient bundle on $G(V, 3)$.*

The third Chern number $\deg c_3(E)$ is equal to 2. Hence every general global section of E vanishes at two points. Conversely, since V is of dimension 7, there exists a nonzero global section $s_{x,y}$ vanishing at x and y for every pair of distinct points x and y. If x and y are general, then $s_{x,y}$ is unique up to constant multiplications. The correspondence $(x, y) \mapsto [s_{x,y}]$ gives the birational mappings $\Pi : S^2 X - \rightarrow P_*(V) \simeq P^6$ and $\Pi_x : X - \rightarrow P_*(V_x) \simeq P^3$ for

[9] By our assumption on X and by [21], there exists a surface section with Picard number one. Hence every member of $|\mathcal{O}_S(-K_X)|$ is irreducible. We use this property to analyze $\Phi_{|E|}$.

[10] The bundle method works for other values of g, *e.g.*, 18 and 20, and gives a description of generic polarized K3 surfaces (see [30]).

[11] For a vector space V, $G(V, r)$ denotes the Grassmann variety of r-dimensional quotient space of V.

[12] The restriction of E is rigid and characterized by its numerical invariants and stability ([27, §3]).

[13] Prime Fano 3-folds of genus 12 were omitted in [38, Chap. V, §7] and first constructed by Iskovskih [16].

general x, where $V_x \subset V$ is the space of global sections of E which vanish at x. In particular, X is rational. The birational mapping Π_x is the same as the triple projection of $X_{22} \subset P^{13}$ from x.

The bundle method gives another canonical description of prime Fano 3-folds of genus 12 in the variety of twisted cubics ([29, §3]). This description is useful to analyze the double projection of $X_{22} \subset P^3$ from a line.

Remark 4 *The third Betti number of a prime Fano 3-fold of genus $g \geq 7$ is equal $2(n(g)-3)$. In particular, prime Fano 3-folds of genus 12 have the same homology group as P^3.*

§4 Genus bound The descriptions given in §§2 and 3 complete the classification of prime Fano 3-folds by virtue of Iskovskih's genus bound:

Theorem 5 *The genus g of a prime Fano 3-fold satisfies $g \leq 10$ or $g = 12$.*

This is proved in the course of the classification by the double projection method. Here we sketch a simple proof using a correspondence between the moduli spaces of K3 surfaces and curves. Let $\mathcal{F}_g$ be the moduli space of polarized K3 surfaces (S, h) of degree $2g-2$. A smooth member of $|h|$ is a curve of genus g. Hence we obtain the rational map ϕ_g from the P^g-bundle $\mathcal{P}_g := \coprod_{(S,h)\in\mathcal{F}_g} |h|$ over $\mathcal{F}_g$ to the moduli space $\mathcal{M}_g$ of stable curves of genus g. The key observation is this.

Proposition 6 *If a prime Fano 3-fold of genus g exists, then the rational map $\phi_g : \mathcal{P}_g - \rightarrow \mathcal{M}_g$ is not generically finite.*

By a simple deformation argument, we have that the generic hyperplane section (S, h) of the generic prime Fano 3-fold is generic in $\mathcal{F}_g$. Take a generic pencil P of hyperplane sections of $X_{2g-2} \subset P^{g+1}$. The isomorphism classes of the members of P vary since the pencil P contains a singular member. But every member of P contains the base locus of P, which is a curve of genus g. This shows the proposition.

Since $\dim \mathcal{P}_g = g + 19$ and $\dim \mathcal{M}_g = 3g - 3$, $\dim \mathcal{P}_g \leq \dim \mathcal{M}_g$ holds if and only if $g \geq 11$. We recall the proof of the generic finiteness of ϕ_{11} in [25]. Let $C \subset P^5$ be a sextic normal elliptic curve and S a smooth complete intersection of three hyperquadrics containing C. Let H be a general hyperplane section of S and put $\Gamma = H \cup C$. Then S is a K3 surface and Γ is a stable curve of genus 11.

Theorem ([25, (1.2)]) *For every embedding $i : \Gamma \longrightarrow S'$ of Γ into a K3 surface S', there exists an isomorphism $I : S \longrightarrow S'$ whose restriction to Γ coincides with i.*

This implies that the point $\xi \in \mathcal{P}_{11}$ corresponding to (S, Γ) is isolated in $\phi_{11}^{-1}(\phi_{11}(\xi))$. Hence ϕ_{11} is generically finite and a prime Fano 3-fold of genus 11 does not exist. The non-existence of prime Fano 3-folds of genus $g \geq 13$ is proved in a similar way. Note that the elliptic curve C induces an elliptic fibration of S, which we denote by $\pi : S \longrightarrow P^1$. We consider the case in which π has two singular fibres of the following types:

i) $E_1 \cup E_2 \cup E_3$ with $(E_2.E_3) = (E_3.E_1) = (E_1.E_2) = 1$, and

ii) $E_2' \cup E_4$ with $(E_2'.E_4) = 2$,

where E_ν is isomorphic to P^1 and satisfies $(E_\nu.H) = \nu$ for every $1 \leq \nu \leq 4$. It is easy to construct a stable curve Γ_g of genus $g \geq 13$ on S from Γ by adding fibres of π. For example, $\Gamma \cup E_3, \Gamma \cup E_4$ and $\Gamma \cup E_2 \cup E_3$ are of genus 13, 14 and 15, respectively. Note that to add one general fibre of π increases the genus by 6. By the above theorem, it is easy to show that every embedding of Γ_g into a K3 surface S' is extended to an isomorphism from S onto S'. Hence we have

Theorem 7 *The rational map $\phi_g : \mathcal{P}_g - \rightarrow \mathcal{M}_g$ is generically finite if and only if $g = 11$ or $g \geq 13$.*

This completes the proof of Theorem 5.

Remark 8 *The map* ϕ_g *is generically of maximal rank except for* $g = 10, 12$. *In the case* $g = 10$, *the image of* ϕ_{10} *is a divisor of* $\mathcal{M}_{10}$, *though* $\dim \mathcal{P}_{10} > \dim \mathcal{M}_{10}$ *(see [28]).*

§5 Theory of polars Prime Fano 3-folds of genus 12 are related to the classical problem on sums of powers, which is a polynomial version of the Waring problem. Let F_d be a homogeneous polynomial of degree d in n variables.

1) Are there N linear forms $f_1, \cdots, f_N$ such that $F_d = \sum_1^N f_i^d$?

2) If so, then how many?

In the following cases, every general F_d is a sum of d-th powers of N linear forms and the expression is unique:

(1) $n = 2$ and $d = 2N$ (Sylvester [43]),

(2) $n = 4$, $d = 3$ and $N = 5$(Sylvester's pentahedral theorem [34] [39]), and

(3) $n = 3$, $d = 5$ and $N = 7$(Hilbert [14, p. 153], Richmond [34] and Palatini [32]).

We consider the case $n = 3$. Let C and Γ be the plane curves defined by F_d and $\prod_1^N f_i$, respectively. Γ is called a *polar N-side* of C if $F_d = \sum_1^N f_i^d$. The name comes from the following:

Example 9 *Let C be a smooth conic and ℓ_1, ℓ_2 and ℓ_3 three distinct lines. Then the following are equivalent:*

(1) $\Delta = \ell_1 + \ell_2 + \ell_3$ is a polar 3-side of C in the above sense, and

(2) the triangle Δ is self polar with respect to C, that is, each side is the polar of its opposite vertex.

§6 Variety of sums of powers We regard the set of polar N-sides of $C : F_d(x, y, z) = 0$ as a subvariety of the projective space of plane curves of degree N. We denote its closure[14] by $VSP(C, N)$ or $VSP(F_d, N)$. The homogeneous forms of degree N form a vector space of dimension $\frac{1}{2}(d+1)(d+2)$. The N-ples of linear forms form a vector space of dimension $3N$. Hence the dimension of $VSP(C, N)$ is expected to be $3N - \frac{1}{2}(d+1)(d+2)$ for general C. In the case $(d, N) = (2, 3)$, this is true.

Proposition 10 *If C is a smooth conic, then $VSP(C, 3)$ is a smooth quintic del Pezzo 3-fold.*

Let V_2 be the vector space of quadratic forms. If $\Delta : f_1 f_2 f_3 = 0$ is a polar 3-side of C, then the defining equation F_2 of C is contained in the subspace $< f_1^2, f_2^2, f_3^2 >$ of V_2. Therefore, Δ determines a 2-dimensional subspace W of $V^* := V_2/CF_2$. Hence we have the morphism from $VSP(C, 3)$ to the 6-dimensional Grassmann variety $G(2, V^*) \subset P^9$. Let $q : V_2 \longrightarrow C$ be the linear map associated to the dual conic of C. For a pair of quadratic forms f and g, consider the three minors $J_i(f, g)$, $i = 1, 2, 3$, of the Jacobian matrix

$$\begin{pmatrix} f_x & f_y & f_z \\ g_x & g_y & g_z \end{pmatrix}$$

and put $\sigma_i(f, g) = q(J_i(f, g))$. Then σ_i are skew-symmetric forms on V_2 and F_2 is their common radical. Therefore, each σ_i determines a hyperplane H_i of $P^9 = P_*(\bigwedge^2 V^*)$. $VSP(C, 3)$ is isomorphic to the quintic del Pezzo 3-fold $G(2, V^*) \cap H_1 \cap H_2 \cap H_3 \subset P^6$.

Now we consider plane quartic curves $C : F_4(x, y, z) = 0$. The dimension count

$$3N - 15 \stackrel{?}{=} \dim VSP(C, N)$$

does not hold for $N = 5$.

[14] The closure is taken in the symmetric product $Sym^N P^2$. But this is a temporary definition. In practice, we choose a suitable model of $Sym^N P^2$ to define $VSP(C, N)$.

Let $\{\partial_1 = \partial^2/\partial x^2, \cdots, \partial_6 = \partial^2/\partial z^2\}$ be a basis of the space of homogeneous second order partial differential operators.

Theorem (Clebsch [4]) *If a plane quartic curve $C : F_4(x,y,z) = 0$ has a polar 5-side, then*

$$\Omega(F) := \det(\partial_i \partial_j F)_{1 \le i,j \le 6} = 0.$$

In particular, general plane quartic curves have no polar 5-sides.

In other words, polar 5-sides are not equally distributed to quartic curves. Once a quartic curve has a polar 5-side, it has a 1-dimensional family of polar 5-sides. (The same happens for polar 2-sides of conics.)

Polar 6-sides of plane quartics were studied by Rosanes [35] and Scorza [40]. The dimension count is correct for $N = 6$ and we obtain 3-folds.

Theorem 11 *(1) If a quartic curve C has no polar 5-sides or no complete quadrangles as its polar 6-sides, then the variety $VSP(C,6)$ of polar 6-sides of C is a prime Fano 3-fold of genus 12.*

(2) Conversely every prime Fano 3-fold X of genus 12 is obtained in this way. The isomorphism class of C is uniquely determined by that of X.

By virtue of Theorem 3, it suffice to show that $G(V,3,N)$ is isomorphic to $VSP(C,6)$. Let V_3 be the vector space of cubic forms. If $\Gamma : f_1 f_2 \cdots f_6 = 0$ is a polar 6-side of C, then the partial derivatives F_x, F_y and F_z of the defining equation F_4 are contained in $< f_1^3, f_2^3, \cdots, f_6^3 >$. Hence Γ determines a 3-dimensional subspace of $V^* := V_3/ < F_x, F_y, F_z >$ and we obtain a morphism ϕ from $VSP(C,6)$ to $G(3,V^*)$. Three skew-symmetric forms σ_1, σ_2 and σ_3 on V^* are defined as in the case of $VSP(F_2,3)$ and the image of ϕ is contained in $G(3,V^*,\sigma_1,\sigma_2,\sigma_3)$.

Conversely, let V and $N \subset \bigwedge^2 V$ be as in Theorem 3. The multiplication in the exterior algebra $\bigwedge^\bullet V$ induces the map $\sigma_3 : S^3 N \longrightarrow \bigwedge^6 V$. This is surjective and its kernel is of dimension 3.

Lemma 12 *There exists a quartic polynomial $F(x,y,z) \in S^4 N$ whose partial derivatives F_X, F_Y and F_Z form a basis of the kernel of σ_3, where $\{x,y,z\}$ is a basis of N and $\{X,Y,Z\}$ is its dual.*

The conics on (the anticanonical model of) $G(V,3,N)$ is parametrized by the projective plane[15] $P_*(N)$. For every point x of $G(V,3,N)$, there exist exactly six conics $\{Z_{\lambda_i}\}_{1 \le i \le 6}$, $\lambda_i \in P_*(N)$, passing through x, counted with their multiplicities. Let Λ_i, $1 \le i \le 6$, be the lines on $P(N)$ with coordinates λ_i. Then $\Gamma = \sum_1^6 \Lambda_i$ is a polar 6-side of the plane curve C on $P(N)$ defined by the quartic form $F(x,y,z)$ in the lemma. This correspondence $x \mapsto \Gamma$ gives the inverse of the above morphism ϕ.

Remark 13 *(1) Assume that a plane quartic C' has a polar 5-side and that the 5 lines are in general position.*

When C in Theorem 11 deforms to C', the variety $VSP(C,6)$ deforms to a Fano 3-fold X' with an ordinary double point. X' is isomorphic to the anticanonical model of $P(E)$, where E is a stable vector bundle on P^2 with $c_1 = 0$ and $c_2 = 4$ (cf. [2]).

(2) If C is a general plane sextic curve, then the variety $VSP(C,10)$ is a polarized K3 surface of genus 20.

§7 Almost homogeneous Fano 3-fold Varieties of sums of powers give two examples of almost homogeneous spaces of $SO(3,C)$ and their compactifications. We apply Theorem 11 to a double conic, say $2C_0 : (XZ + Y^2)^2 = 0$.

[15] For a vector space V, $P_*(V)$ is the projective space of 1-dimensional subspaces of V. $P(V)$, or $P^*(V)$, is its dual.

The variety $VSP(2C_0, 6)$ is a Fano 3-fold and has an action of $SO(3, C)$. It is easy to check

$$30(XZ+Y^2)^2 = 25Y^4 + \sum_{i=0}^{4}(\zeta^i X + Y + \zeta^{-i}Z)^4,$$

where ζ is a fifth root of unity. The polar 6-side

$$\Gamma : Y\prod_{i=0}^{4}(\zeta^i X + Y + \zeta^{-i}Z) = 0$$

intersects the 2-sphere C_0 at the 12 vertices of a regular icosahedron. The stabilizer group at Γ of $SO(3, C)$ is the icosahedral group $\simeq A_5$. Hence we have

Theorem 14 *The variety $VSP(2C_0, 6)$ is a smooth equivariant compactification of $SO(3, C)/Icosa$.*

Similarly the quintic del Pezzo 3-fold $VSP(C_0, 3)$ is a smooth equivariant compactification of the quotient of $SO(3, C)$ by an octahedral group $\simeq S_4$ by Proposition 10. These two compactifications are described in [31, §§3 and 6] by another method. We remark that Q^3 and P^3 also are almost homogeneous spaces of $SO(3, C)$. The stabilizer groups are a tetrahedral group $\simeq A_4$ and a dihedral group of order 6 $\simeq S_3$, respectively.

§8 Compactification of C^3 There are four types of Fano 3-folds with the same homology group as P^3: P^3 itself, $Q^3 \subset P^4$, $V_5 \subset P^6$ and the 6-dimensional family of prime Fano 3-folds $X_{22} \subset P^{13}$ of genus 12 (see Remark 4). These Fano 3-folds are related to not only $SO(3, C)$ but also C^3, the affine 3-space. It is well-known that P^3 and Q^3 are smooth compactifications of C^3 with irreducible boundary divisors. The quintic del Pezzo 3-fold $V_5 \subset P^6$ is a compactification of C^3 in two ways (see [10] and [13]).

Furushima has found that the almost homogeneous Fano 3-fold $U_{22} := VSP(2C_0, 6)$ also is a compactification of C^3. This fact is proved in three ways using

i) the defining equation ([31] p. 506) of $U_{22} \subset P^{12}$ (see [11]),

ii) the double projection of $U_{22} \subset P^{13}$ from a line (see [12]), and

iii) the action of a torus $C^* \subset SO(3, C)$ on U_{22} (see [1] and [20]).

In the last case, U_{22} is decomposed into a disjoint union of affine spaces by virtue of [3]. The four compactifications by P^3, Q^3 and V_5 are rigid but that by U_{22} is not. In fact, by a careful analysis of the double projection of $VSP(C, 6) \subset P^{13}$ from a line, we have

Theorem 15 *The variety $VSP(C, 6)$ in Theorem 11 is a compactification of C^3 if C has a non-ordinary singular point.*

The variety $VSP(C, 6)$ has a line ℓ (on its anticanonical model) with normal bundle $\mathcal{O}(1) \oplus \mathcal{O}(-2)$ corresponding to a non-ordinary singular point of C. Let D the union of conics which intersect ℓ as in Example 1. Then the complement of D is isomorphic to C^3.

References

[1] Akyildiz, E. and J.B. Carrell: A generalization of the Kostant-Macdonald identity, Proc. Natl. Acad. Sci. USA, **86** (1989), 3934–3937.

[2] Barth, W.: Moduli of vector bundles on the projective plane, Inv. Math. **42** (1977), 63–91.

[3] Bialynicki-Birula, A.: Some theorems on actions of algebraic groups, Ann. of Math., **98** (1973), 480–497.

[4] Clebsch, A.: Über Curven vierter Ordnung, J. für Math. **59**(1861), 125–145.

[5] Clemens, C.H. and P.A. Griffiths: The intermediate Jacobian of the cubic threefold, Ann. of Math. **95** (1972), 281–356.

[6] Cutkosky, S.D.: On Fano 3-folds, Manuscripta Math. **64** (1989), 189–204.

[7] Fujita, T.: On the structure of polarized manifolds with total deficiency one, I, J. Math. Soc. Japan **32** (1980), 709–725.

[8] Fujita, T.: Vector bundles on ample divisors, J. Math. Soc. Japan **33** (1981), 405–414.

[9] Fujita, T.: On the structure of polarized manifolds with total deficiency one, II, J. Math. Soc. Japan **33** (1980), 415–434.

[10] Furushima, M.: Singular del Pezzo surfaces and analytic compactifications of 3-dimensional complex analytic space $\mathbf{C}^3$, Nagoya Math. J. **104** (1986), 1–28.

[11] Furushima, M.: Complex analytic compactifications of $\mathbf{C}^3$, to appear in Compositio Math.

[12] Furushima, M.: A note on an example of a compactification of $\mathbf{C}^3$, 'Algebraic geometry and Hodge theory', Hokkaido Univ. Tech. Rep. Ser. in Math., **16** (1990), 103–115.

[13] Furushima, M. and N. Nakayama: A new construction of a compactification of $\mathbf{C}^3$, Tôhoku Math. J. **41** (1989), 543–560.

[14] Hilbert, D.: Letter adressée à M. Hermite, Gesam. Abh. vol. II, pp.148–153.

[15] Iskovskih, V.A.: Fano 3-folds I, Math. USSR Izv. **11** (1977), 485–527.

[16] Iskovskih, V.A.: Fano 3-folds II, Math. USSR Izv. **12** (1978), 469–506.

[17] Iskovskih, V.A.: Lectures on 3-dimensional algebraic manifolds : Fano manifolds (in Russian), Moscow University, 1988.

[18] Iskovskih, V.A. and Ju. I. Manin: Three-dimensional quartics and counterexamples to the Lüroth problem, Math. USSR Sbornik **15** (1971), 141–166.

[19] Kollár, J.: Flops, Nagoya Math. J. **113** (1989), 15–36.

[20] Konarski, J.: Some examples of cohomological projective spaces via $\mathbf{C}^+$-actions, 'Group actions and invariant theory (Montreal, PQ, 1988)', 73–84, CMS Conf. Proc. **10**, Amer. Math. Soc., Providence, RI, 1989.

[21] Moishezon, B.G.: Algebraic homology classes on algebraic varieties, Math. USSR Izv., **1** (1967) , 209–251.

[22] Mori, S.: Threefolds whose canonical bundles are not numerically effective, Ann. of Math. **110** (1979), 593–606.

[23] Mori, S.: Lectures on 3-dimensional algebraic varieties, Nagoya University, Fall, 1987.

[24] Mori, S. and S. Mukai: Classification of Fano 3-folds with $B_2 \geq 2$, Manuscripta Math. **36** (1982), 147–162.

[25] Mori, S. and S. Mukai: Uniruledness of the moduli space of curves of genus 11, Algebraic Geometry (Proceedings, Tokyo/Kyoto 1982), Lecture Notes in Math. n^o1016, Springer Verlag, 1983, pp. 334–353.

[26] Mori, S. and S. Mukai: On Fano threefolds with $B_2 \geq 2$, Adv. Stud. Pure Math. **1** (1983), 101–129, Kinokuniya and North-Holland.

[27] Mukai, S.: On the moduli space of bundles on K3 surfaces: I, 'Vector Bundles on Algebraic Varieties (Proceedings of the Bombay Conference 1984)', Tata Institute of Fundamental Research Studies **11**, pp. 341–413, Oxford University Press, 1987.

[28] Mukai, S.: Curves, K3 surfaces and Fano 3–folds of genus ≤ 10, in 'Algebraic Geometry and Commutative Algebra in Honor of Masayoshi Nagata', pp. 357–377, 1988, Kinokuniya, Tokyo.

[29] Mukai, S.: Biregular classification of Fano threefolds and Fano manifolds of coindex 3, Proc. Natl. Acad. Sci. USA **86** (1989), 3000–3002.

[30] Mukai, S.: Polarized K3 surfaces of genus 18 and 20, to appear in 'Vector Bundles and Special Projective Embeddings (Proceedings, Bergen 1989)'.

[31] Mukai, S. and H. Umemura: Minimal rational 3-folds, in 'Algebraic Geometry (Proceedings, Tokyo/Kyoto 1982)', Lecture Notes in Math. n^o1016, Springer, 1983, pp. 490–518.

[32] Palatini, F.:Sulla rappresentazione delle forme ternarie mediante la somma di potenze di forme lineari, Rom. Acc. L. Rend. (5) **12** (1903), 378–384.

[33] Reid, M.: Minimal models of canonical 3-folds, Adv. Stud. Pure Math. **1** (1983), 131–180, Kinokuniya and North-Holland.

[34] Richmond, H.W.: On canonical forms, Quart. J. Math. **33** (1902), 331–340.

[35] Rosanes, J.:Über ein Prinzip der Zuordnung algebraischer Formen, J. für r. u. angew. Math. **76**(1873), 312–330.

[36] Roth, L.: Algebraic varieties with canonical curve sections, Ann. di mat. (4) **29** (1949), 91–97.

[37] Roth, L.: On fourfolds with canonical curve sections, Proc. Cambridge Phil. Soc. **46** (1950), 419–428.

[38] Roth, L.: Algebraic threefold with special regard to problems of rationality, Springer Verlag, 1955.

[39] Segre, B.: The non-singular cubic surfaces, Oxford Clarendon Press, 1942.

[40] Scorza, G.: Sopra la teoria delle figure polari delle curve piane del 4. ordine, Ann. di mat. (3) **2**(1899), 155–202.

[41] Shokurov, V.V.: Smoothness of the general anticanonical divisor of a Fano 3-fold, Math. USSR Izv. **14** (1980), 395–405.

[42] Shokurov, V.V.: The existence of lines on Fano 3-folds, Math. USSR Izv. **15** (1980), 173–209.

[43] Sylvester, J.J.: An essay on canonical forms, supplement to a sketch of a memoir on elimination, transformation and canonical forms, Collected Works Vol. I, pp. 203–216, Cambridge University Press, 1904.

[44] Tyurin, A.N.: Five lectures on three-dimensional varieties, Russ. Math. Survey, **27** (1972), 1–53.

Department of Mathematics
School of Sciences
Nagoya University
Furō-chō, Chikusa-ku
Nagoya, 464-01 Japan

Polarized K3 surfaces of genus 18 and 20

Dedicated to Professor Hisasi Morikawa on his 60th Birthday

Shigeru MUKAI

A surface, i.e., 2-dimensional compact complex manifold, S is of type K3 if its canonical line bundle $\mathcal{O}_S(K_S)$ is trivial and if $H^1(S, \mathcal{O}_S) = 0$. An ample line bundle L on a K3 surface S is a polarization of genus g if its self intersection number (L^2) is equal to $2g - 2$, and called *primitive* if $L \simeq M^k$ implies $k = \pm 1$. The moduli space $\mathcal{F}_g$ of primitively polarized K3 surfaces (S, L) of genus g is a quasi-projective variety of dimension 19 for every $g \geq 2$ ([15]). In [12], we have studied the generic primitively polarized K3 surfaces (S, L) of genus $6 \leq g \leq 10$. In each case, the K3 surface S is a complete intersection of divisors in a homogeneous space X and the polarization L is the restriction of the ample generator of the Picard group $Pic\, X \simeq Z$ of X.

In this article, we shall study the generic (polarized) K3 surfaces (S, L) of genus 18 and 20. (Polarization of genus 18 and 20 are always primitive.) The K3 surface S has a *canonical* embedding into a homogeneous space X such that L is the restriction of the ample generator of $Pic\, X \simeq Z$. S is not a complete intersection of divisors any more but a complete intersection in X with respect to a homogeneous vector bundle $\mathcal{V}$ (Definition 1.1): S is the zero locus of a global section s of $\mathcal{V}$. Moreover, the global section s is uniquely determined by the isomorphism class of (S, L) up to the automorphisms of the pair $(X, \mathcal{V})$. As a corollary, we obtain a description of birational types of $\mathcal{F}_{18}$ and $\mathcal{F}_{20}$ as orbit spaces (Theorem 0.3 and Corollary 5.10).

In the case of genus 18, the ambient space X is the 12-dimensional variety of 2-planes in the smooth 7-dimensional hyperquadric $Q^7 \subset P^8$. The complex special orthogonal group $G = SO(9, C)$ acts on X transitively. Let $\mathcal{F}$ be the homogeneous vector bundle corresponding to the fourth fundamental weight $w_4 = (\alpha_1 + 2\alpha_2 + 3\alpha_3 + 4\alpha_4)/2$ of the root system

$$B_4 : \overset{\alpha_1}{\circ} - \overset{\alpha_2}{\circ} - \overset{\alpha_3}{\circ} \Longrightarrow \overset{\alpha_4}{\circ} \tag{0.1}$$

of G. $\mathcal{F}$ is of rank 2 and the determinant line bundle $\mathcal{O}_X(1)$ of $\mathcal{F}$ generates the Picard group of X. The vector bundle $\mathcal{V}$ is the direct sum of five copies of $\mathcal{F}$.

Theorem 0.2 *Let $S \subset X$ be the common zero locus of five global sections of the homogeneous vector bundle $\mathcal{F}$. If S is smooth and of dimension 2, then $(S, \mathcal{O}_S(1))$ is a (polarized) K3 surface of genus 18, where $\mathcal{O}_S(1)$ is the restriction of $\mathcal{O}_X(1)$ to S.*

Remark Consider the variety X of lines in the 5-dimensional hyperquadric $Q^5 \subset P^6$ instead. This is a 7-dimensional homogeneous space of $SO(7, C)$ and has a homogeneous vector bundle of rank 2 on it. The zero locus Z of its general global section is a Fano 5-fold of index 3 and a homogeneous space of the exceptional group of type G_2. See [12] and [13] for other descriptions of Z and its relation to K3 surfaces of genus 10.

The space $H^0(X, \mathcal{F})$ of global sections of $\mathcal{F}$ is the (16-dimensional) spin representation U^{16} of the universal covering group $\tilde{G} = Spin(9, C)$ (see [5]). Let $G(5, U^{16})$ be the

Grassmann variety of 5-dimensional subspaces N of U^{16} and $G(5,U^{16})^s$ be its open subset consisting of stable points with respect to the action of $\tilde{G}$. The orthogonal group G acts on $G(5,U^{16})$ effectively and the geometric quotient $G(5,U^{16})^s/G$ exists as a normal quasi-projective variety ([14]).

Theorem 0.3 *The generic K3 surface of genus 18 is the common zero locus $S_N \subset X$ of a 5-dimensional space N of global sections of the rank 2 homogeneous vector bundle $\mathcal{F}$. Moreover, the classification (rational) map $G(5,U^{16})^s/SO(9,C) - \to \mathcal{F}_{18}$ is birational.*

Remark The spin representation U^{16} is the restriction of the half spin representation H^{16} of $Spin(10,C)$ to $Spin(9,C)$. The quotient $SO(10,C)/SO(9,C)$ is isomorphic to the complement of the 8-dimensional hyperquadric $Q^8 \subset P^9$. Hence $\mathcal{F}_{18}$ is birationally equivalent to a P^9-bundle over the 10-dimensional orbit space $G(5,H^{16})^s/SO(10,C)$.

Let $\mathcal{E}$ be the homogeneous vector bundle corresponding to the first fundamental weight $w_1 = \alpha_1+\alpha_2+\alpha_3+\alpha_4$ and E_N its restriction to S_N. The uniqueness of the expression of S_N is essentially a consequence of the rigidity of E_N and the following:

Proposition 0.4 *Let E be a stable vector bundle on a K3 surface S and assume that E is rigid, i.e., $\chi(sl(E)) = 0$. If a semi-stable vector bundle F has the same rank and Chern classes as E, then F is isomorphic to E.*

This follows from the Riemann-Roch theorem

$$\begin{aligned} \dim Hom(E,F) + \dim Hom(F,E) &\geq \chi(E^\vee \otimes F) \\ &= \chi(E^\vee \otimes E) = \chi(sl(E)) + \chi(\mathcal{O}_S) = 2 \end{aligned}$$

on a K3 surface (*cf.* [11], Corollary 3.5).

In the case of genus 20, the ambient homogeneous space X is the (20-dimensional) Grassmann variety $G(V,4)$ of 4-dimensional quotient spaces of a 9-dimensional vector space V and the homogeneous vector bundle $\mathcal{V}$ is the direct sum of three copies of $\bigwedge^2 \mathcal{E}$, where $\mathcal{E}$ is the (rank 4) universal quotient bundle on X. The generic K3 surface of genus 20 is a complete intersection in $G(V,4)$ with respect to $(\bigwedge^2 \mathcal{E})^{\oplus 3}$ in a unique way (Theorem 5.1 and Theorem 5.9). This description of K3 surfaces is a generalization of one of the three descriptions of Fano threefolds of genus 12 given in [13].

This work was done during the author's stay at the University of California Los Angeles in 1988-89, whose hospitality he gratefully acknowledges.

Table of Contents

Notation and Conventions. All varieties are considered over the complex number field C. A vector bundle E on a variety X is a locally free $\mathcal{O}_X$-module. Its rank is denoted by $r(E)$. The determinant line bundle $\bigwedge^{r(E)} E$ is denoted by $\det E$. The dual vector bundle $\mathcal{H}om(E,\mathcal{O}_X)$ of E is denoted by $E^\vee$. The subbundle of $\mathcal{E}nd(E) \simeq E \otimes E^\vee$ consisting of trace zero endomorphisms of E is denoted by $sl(E)$.

1 Complete intersections with respect to vector bundles

We generalize Bertini's theorem for vector bundles. Let $s \in H^0(U, E)$ be a global section of a vector bundle E on a scheme U. Let $\mathcal{O}_U \longrightarrow E$ be the multiplication by s and $\eta : E^\vee \longrightarrow \mathcal{O}_U$ its dual homomorphism. The subscheme $(s)_0$ of U defined by the ideal $I = Im\,\eta \subset \mathcal{O}_U$ is called the scheme of zeroes of s.

Definition 1.1 *(1) Let $\{e_1, \cdots, e_r\}$ be a local frame of E at $x \in U$. A global section $s = \sum_{i=1}^r f_i e_i$, $f_i \in \mathcal{O}_X$, of E is nondegenerate at x if $s(x) = 0$ and if $(f_1, \cdots, f_r)$ is a regular sequence. s is nondegenerate if it is so at every point x of $(s)_0$.*

(2) A subscheme Y of U is a complete intersection with respect to E if Y is the scheme of zeroes of a nondegenerate global section of E.

In the case U is Cohen-Macaulay, a global section s of E is nondegenerate if and only if the codimension of $Y = (s)_0$ is equal to the rank of E.

The wedge product by $s \in H^0(U, E)$ gives rise to a complex

$$\Lambda^\bullet : \quad \mathcal{O}_U \longrightarrow E \longrightarrow \bigwedge^2 E \longrightarrow \cdots \longrightarrow \bigwedge^{r-1} E \longrightarrow \bigwedge^r E \tag{1.2}$$

called the *complex* of s. The dual complex

$$K^\bullet : \quad \bigwedge^r E^\vee \longrightarrow \bigwedge^{r-1} E^\vee \longrightarrow \cdots \longrightarrow \bigwedge^2 E^\vee \longrightarrow E^\vee \longrightarrow \mathcal{O}_U \tag{1.3}$$

is called the Koszul complex of Y.

Proposition 1.4 *If Y is a complete intersection with respect to E, then the Koszul complex $K^\bullet$ is a resolution of the structure sheaf $\mathcal{O}_Y$ of Y, that is, the sequence*

$$0 \longrightarrow K^\bullet \longrightarrow \mathcal{O}_Y \longrightarrow 0$$

is exact.

In particular, the conormal bundle I/I^2 of Y in U is isomorphic to $E^\vee$ and we have the adjunction formula

$$K_Y \equiv (K_U + \det E)|_Y. \tag{1.5}$$

Since the pairing

$$\bigwedge^i E \times \bigwedge^{r-i} E \longrightarrow \det E$$

is nondegenerate for every i, we have

Lemma 1.6 *The complex $K^\bullet$ is isomorphic to $\Lambda^\bullet \otimes (\det E)^{-1}$.*

Let $\pi : P(E) \longrightarrow X$ be the P^{r-1}-bundle associated to E in the sense of Grothendieck and $\mathcal{O}_P(1)$ the tautological line bundle on it. By the construction of $P(E)$, we have the canonical isomorphisms

$$\pi_* \mathcal{O}_P(1) \simeq E \quad \textit{and} \quad H^0(P(E), \mathcal{O}(1)) \simeq H^0(U, E). \tag{1.7}$$

The two linear systems associated to E and $\mathcal{O}_P(1)$ have several common properties.

Proposition 1.8 *A vector bundle E is generated by its global sections if and only if the tautological line bundle $\mathcal{O}_P(1)$ is so.*

Proposition 1.9 *Let σ be the global section of the tautological line bundle $\mathcal{O}_P(1)$ corresponding to $s \in H^0(U, E)$ via (1.7). If U is smooth, then the following are equivalent:*
i) s is nondegenerate and $(s)_0 \subset U$ is smooth, and
ii) the divisor $(\sigma)_0 \subset P(E)$ is smooth.

Proof. Since the assertion is local, we may assume E is trivial, i.e., $E \simeq \mathcal{O}_U^{\oplus r}$. Let $f_1, \cdots, f_r$ be a set of generators of the ideal I defining $(s)_0$. A point $x \in (s)_0$ is singular if and only if $df_1, \cdots, df_r$ are linearly dependent at x, that is, there is a set of constants $(a_1, \cdots, a_r) \neq (0, \cdots, 0)$ such that $a_1 df_1 + \cdots + a_r df_r = 0$ at x. This condition is equivalent to the condition that the divisor

$$(\sigma)_0 : f_1 X_1 + \cdots + f_r X_r = 0$$

in $P(E) \simeq U \times P^{r-1}$ is singular at $x \times (a_1 : \cdots : a_r)$. Therefore, ii) implies i). Since $(\sigma)_0$ is smooth off $(s)_0 \times P^{r-1}$, i) implies ii). *q.e.d.*

By these two propositions, Bertini's theorem (see [7, p. 137]) is generalized for vector bundles:

Theorem 1.10 *Let E be a vector bundle on a smooth variety. If E is generated by its global sections, then every general global section is nondegenerate and its scheme of zeroes is smooth.*

2 A homogeneous space of $SO(9, C)$

Let X be the subvariety of $Grass(P^2 \subset P^8)$ consisting of 2-planes in the smooth 7-dimensional hyperquadric

$$Q^7 : q(X) = X_1X_4 + X_2X_5 + X_3X_6 + X_7X_8 + X_9^2 = 0 \tag{2.1}$$

in P^8. The special orthogonal group $G = SO(9, C)$ acts transitively on X. Let P be the stabilizer group at the 2-plane $X_4 = X_5 = \cdots = X_9 = 0$. X is isomorphic to G/P and the reductive part L of P consists of matrices

$$\begin{pmatrix} A & 0 & 0 \\ 0 & {}^tA^{-1} & 0 \\ 0 & 0 & B \end{pmatrix}$$

with $A \in GL(3, C)$ and $B \in SO(3, C)$. We denote the diagonal matrix

$$[x_1, x_2, x_3, x_1^{-1}, x_2^{-1}, x_3^{-1}, x_4, x_4^{-1}, 1]$$

by $< x_1, x_2, x_3, x_4 >$. All the invertible diagonal matrices $< x_1, x_2, x_3, x_4 >$ form a maximal torus H of G contained in L. Let $X(H) \simeq Z^{\oplus 4}$ be the character group and $\{e_1, e_2, e_3, e_4\}$ its standard basis. For a character $\alpha = a_1e_1 + a_2e_2 + a_3e_3 + a_4e_4$, let $\underline{g}^\alpha$ be the α-eigenspace $\{Z : Ad < x_1, x_2, x_3, x_4 > \cdot Z = x_1^{a_1}x_2^{a_2}x_3^{a_3}x_4^{a_4} Z\}$ of the adjoint action Ad on the Lie algebra $\underline{g}$ of G. Then we have the well-known decomposition

$$\underline{g} = \underline{h} \oplus \bigoplus_{0 \neq \alpha \in X(H)} \underline{g}^\alpha.$$

A character α is a *root* if $\underline{g}^\alpha \neq 0$. In our case, there are 16 positive roots

$$e_1, e_2, e_3, e_4 \text{ and } e_i \pm e_j (1 \leq i \leq j \leq 4)$$

ant their negatives. The basis of roots are

$$\alpha_1 = e_1 - e_2,\ \alpha_2 = e_2 - e_3,\ \alpha_3 = e_3 - e_4 \text{ and } \alpha_4 = e_4.$$

Since e_1, e_2, e_3 and e_4 are orthonormal with respect to the Killing form, the Dynkin diagram of G is of type B_4 (see (0.1)). The fundamental weights are

$$\begin{aligned} w_1 &= e_1, w_2 = e_1 + e_2, w_3 = e_1 + e_2 + e_3 \text{ and} \\ w_4 &= \frac{1}{2}(e_1 + e_2 + e_3 + e_4) \end{aligned} \tag{2.2}$$

(Cf. [4]). The positive roots of $L \simeq GL(3, C) \times SO(3, C)$ with respect to H are

$$\alpha_1, \alpha_2, \alpha_1 + \alpha_2 \text{ and } \alpha_4. \tag{2.3}$$

The root basis of L is of type $A_2 \coprod A_1$ and the Weyl group W' is a dihedral group of order 12. There are 12 positive roots of G other than (2.3) and their sum is equal to $5(e_1 + e_2 + e_3) = 5w_3$. Hence by [2, §16], we have

Proposition 2.4 *X is a 12-dimensional Fano manifold of index 5.*

Let $\tilde{G}$ be the universal covering group of G, and $\tilde{H}$ and $\tilde{L}$ the pull-back of H and L, respectively. The character group $X(\tilde{H})$ of $\tilde{H}$ is canonically isomorphic to the weight lattice. Let $\rho_i, 1 \leq i \leq 4$, be the irreducible representation of $\tilde{L}$ with the highest weight w_i. Since the W'-orbit of w_4 consists of two weights $p_+ = w_4$ and $p_- = w_4 - e_4$ and since p_- is the reflection of p_+ by e_4, the representation ρ_4 is of dimension 2. ρ_1 is induced from the vector representation of the $GL(3, C)$-factor of L. From the equality

$$e_1 + e_2 + e_3 = p_+ + p_- = w_3,$$

we obtain the isomorphism

$$\bigwedge^3 \rho_1 \simeq \bigwedge^2 \rho_4 \simeq \rho_3. \tag{2.5}$$

Let $\mathcal{E}$ (resp. $\mathcal{O}_X(1), \mathcal{F}$) be the homogeneous vector bundle over $X = G/P$ induced from the representation ρ_1 (resp. ρ_3, ρ_4). $\mathcal{O}_X(1)$ is the positive generator of $Pic X$. $\mathcal{E}$ and $\mathcal{F}$ are of rank 2 and 3, respectively. By the above isomorphism, we have

$$\bigwedge^3 \mathcal{E} \simeq \bigwedge^2 \mathcal{F} \simeq \mathcal{O}_X(1). \tag{2.6}$$

We shall study the property of the zero locus of a global section of $\mathcal{F}^{\oplus 5}$ in the next section. For its study we need vanishing of cohomology groups of homogeneous vector bundles on X and apply the theorem of Bott[3]. The sum δ of the four fundamental weights w_i in (2.2) is equal to $(7e_1 + 5e_2 + 3e_3 + e_4)/2$. The sum of all positive roots is equal to 2δ.

Theorem 2.7 *Let $\mathcal{E}(w)$ be the homogeneous vector bundles on $X = \tilde{G}/\tilde{P}$ induced from the representation of $\tilde{L}$ with the highest weight $w \in X(\tilde{H})$. Then we have*

1) $H^i(X, \mathcal{E}(w))$ vanishes for every i if there is a root α with $(\alpha.\delta + w) = 0$, and

2) Let i_0 be the number of positive roots α with $(\alpha.\delta + w) < 0$. Then $H^i(X, \mathcal{E}(w))$ vanishes for every i except i_0.

We apply the theorem to the following four cases:
1) $w = jw_3 + nw_4$ and $\mathcal{E}(w) \simeq S^n\mathcal{F}(j)$ for $n \leq 6$,
2) $w = w_1 + jw_3 + nw_4$ and $\mathcal{E}(w) \simeq \mathcal{E} \otimes S^n\mathcal{F}(j)$ for $n \leq 5$,
3) $w = 2w_1 + jw_3 + nw_4$ and $\mathcal{E}(w) \simeq S^2\mathcal{E} \otimes S^n\mathcal{F}(j)$ for $n \leq 5$, and
4) $w = w_1 + w_2 + (j-1)w_3 + nw_4$ and $\mathcal{E}(w) \simeq sl(\mathcal{E}) \otimes S^n\mathcal{F}(j)$ for $n \leq 5$.

Proposition 2.8 *(1) The cohomology group $H^i(X, S^n\mathcal{F}(j))$ vanishes for every (i,n,j) with $0 \leq n \leq 6$ except the following:*

i	0	3	9	12
n	n	6	6	n
j	≥ 0	-4	-7	$\leq -n-5$

(2) The cohomology group $H^i(X, \mathcal{E} \otimes S^n\mathcal{F}(j))$ vanishes for every (i,n,j) with $0 \leq n \leq 5$ except the following:

i	0	2	2	11	11	11	11	12
n	n	4	5	2	3	4	5	n
j	≥ 0	-3	-3	-6	-7	-9	-9	$\leq -n-6$

(3) The cohomology group $H^i(X, S^2\mathcal{E} \otimes S^n\mathcal{F}(j))$ vanishes for every (i,n,j) with $0 \leq n \leq 5$ except the following:

i	0	2	2	10	11	11	11	11	11	12
n	n	4	5	2	3	4	4	5	5	n
j	≥ 1	-3	-3	-6	-8	-8	-9	-9	-10	$\leq -n-7$

(4) The cohomology group $H^i(X, sl(\mathcal{E}) \otimes S^n\mathcal{F}(j))$ vanishes for every (i,n,j) with $0 \leq n \leq 5$ except the following:

i	0	1	1	1	1	3	9	11	11	11	11	12
n	n	2	3	4	5	4	4	2	3	4	5	n
j	≥ 1	-1	-1	-1	-1	-3	-6	-6	-7	-8	-9	$\leq -n-6$

3 K3 surfaces of genus 18

In this section, we prove Theorem 0.2 and prepare the proof of Theorem 0.3. Let X, $\mathcal{E}$ and $\mathcal{F}$ be as in Section 2. Let N be a 5-dimensional subspace of $H^0(X, \mathcal{F})$ and $\{s_1, \cdots, s_5\}$ a basis of N. The common zero locus $S_N \subset X$ of N coincides with the zero locus of the global section $s = (s_1, \cdots, s_5)$ of $\mathcal{F}^{\oplus 5}$. Let Ξ_{SCI} be the subset of $G(5, H^0(X, \mathcal{F}))$ consisting of $[N]$ such that S_N is smooth and of dimension 2. $\mathcal{F}$ is generated by its global section and $\dim X - r(\mathcal{F}^{\oplus 5}) = 2$. Hence by Theorem 1.10, we have

Proposition 3.1 *Ξ_{SCI} is a non-empty (Zariski) open subset of $G(5, H^0(X, \mathcal{F}))$.*

We compute the cohomology groups of vector bundles on S_N with $[N] \in \Xi_{SCI}$, using the Koszul complex

$$\text{(3.2)} \qquad \Lambda^\bullet : \ \mathcal{O}_X \longrightarrow \mathcal{F}^{\oplus 5} \longrightarrow \bigwedge^2(\mathcal{F}^{\oplus 5}) \longrightarrow \cdots \longrightarrow \bigwedge^9(\mathcal{F}^{\oplus 5}) \longrightarrow \bigwedge^{10}(\mathcal{F}^{\oplus 5}).$$

The terms $\Lambda^i = \bigwedge^i(\mathcal{F}^{\oplus 5})$ of this complex have the following symmetry:

Lemma 3.3 $\Lambda^i \simeq \Lambda^{10-i} \otimes \mathcal{O}_X(i-5)$.

Proof. By (1.6), Λ^i is isomorphic to $(\Lambda^{10-i})^\vee \otimes \mathcal{O}_X(5)$. Since $\mathcal{F}$ is of rank 2, $\mathcal{F}^\vee$ is isomorphic to $\mathcal{F} \otimes \mathcal{O}_X(-1)$, which shows the lemma. *q.e.d.*

We need the decomposition of Λ^i into irreducible homogeneous vector bundles. $\Lambda^i = \bigwedge^i(\mathcal{F}^{\oplus 5})$, $i \leq 5$, has the following vector bundles as its irreducible factors:

(3.4)

Λ^0	$\mathcal{O}$
Λ^1	$\mathcal{F}$

Λ^2	$S^2\mathcal{F}, \mathcal{O}(1)$
Λ^3	$S^3\mathcal{F}, \mathcal{F}(1)$

Λ^4	$S^4\mathcal{F}, S^2\mathcal{F}(1), \mathcal{O}(2)$
Λ^5	$S^5\mathcal{F}, S^3\mathcal{F}(1), \mathcal{F}(2)$

Proposition 3.5 *If $[N] \in \Xi_{SCI}$, then S_N is a K3 surface.*

Proof. By Proposition 2.4 and (2.6), the vector bundle $\mathcal{F}^{\oplus 5}$ and the tangent bundle T_X have the same determinant bundle. Hence S_N has a trivial canonical bundle by the adjunction formula (1.5). By Proposition 1.4 and Lemma 1.6, we have the exact sequence $0 \longrightarrow K^\bullet \longrightarrow \mathcal{O}_{S_N} \longrightarrow 0$ and the isomorphism $K^\bullet \simeq \Lambda^\bullet \otimes \mathcal{O}(-5)$. By (1) of Proposition 2.8 and the Serre duality $H^i(K^{10}) \simeq H^{12-i}(\mathcal{O}_X)^\vee$, we have

$$H^1(K^1) = H^2(K^2) = \cdots = H^{10}(K^{10}) = 0$$

and

$$H^1(K^0) = H^2(K^1) = \cdots = H^{10}(K^9) = H^{11}(K^{10}) = 0.$$

Therefore, the restriction map $H^0(\mathcal{O}_X) \longrightarrow H^0(\mathcal{O}_{S_N})$ is surjective and $H^1(\mathcal{O}_{S_N})$ vanishes. Hence S_N is connected and regular. *q.e.d.*

Let F_N be the restriction of $\mathcal{F}$ to S_N. The complex $K^\bullet \otimes \mathcal{F}$ gives a resolution of F_N. Since $S^n\mathcal{F} \otimes \mathcal{F} \simeq S^{n+1}\mathcal{F} \oplus S^{n-1}\mathcal{F}(1)$, we have the following four series of vanishings by (1) of Proposition 2.8:

(a) $H^1(\mathcal{F} \otimes K^2) = H^2(\mathcal{F} \otimes K^3) = \cdots = H^9(\mathcal{F} \otimes K^{10}) = 0$,

(b) $H^1(\mathcal{F} \otimes K^1) = H^2(\mathcal{F} \otimes K^2) = \cdots = H^9(\mathcal{F} \otimes K^9) = H^{10}(\mathcal{F} \otimes K^{10}) = 0$,

(c) $H^1(\mathcal{F}) = H^2(\mathcal{F} \otimes K^1) = \cdots = H^{10}(\mathcal{F} \otimes K^9) = H^{11}(\mathcal{F} \otimes K^{10}) = 0$

and

(d) $H^2(\mathcal{F}) = H^3(\mathcal{F} \otimes K^1) = \cdots = H^{11}(\mathcal{F} \otimes K^9) = H^{12}(\mathcal{F} \otimes K^{10}) = 0$.

By (c) and (d), both $H^1(F_N)$ and $H^2(F_N)$ vanish. By (a) and (b), the sequence

$$0 \longrightarrow H^0(\mathcal{F} \otimes K^1) \longrightarrow H^0(\mathcal{F}) \longrightarrow H^0(F_N) \longrightarrow 0$$

is exact. So we have proved

Proposition 3.6 *If* $\dim S_N = 2$, *then we have*

(1) $H^1(S_N, F_N) = H^2(S_N, F_N) = 0$, and

(2) the restriction map $H^0(X, \mathcal{F}) \longrightarrow H^0(S_N, F_N)$ is surjective and its kernel coincides with N.

Let E_N be the restriction of $\mathcal{E}$ to S_N. Arguing similarly for the complexes $K^\bullet \otimes \mathcal{E}$, $K^\bullet \otimes S^2\mathcal{E}$ and $K^\bullet \otimes sl(\mathcal{E})$, we have the following by Proposition 2.8:

Proposition 3.7 *If* $\dim S_N = 2$, *then we have*

(1) all the higher cohomology groups of E_N and S^2E_N vanish,

(2) the restriction maps $H^0(X, \mathcal{E}) \longrightarrow H^0(S_N, E_N)$ and $H^0(X, S^2\mathcal{E}) \longrightarrow H^0(S_N, S^2E_N)$ are bijective, and

(3) all the cohomology groups of $sl(E_N)$ vanish.

Corollary 3.8 *The natural map* $S^2H^0(S_N, E_N) \longrightarrow H^0(S_N, S^2E_N)$ *is surjective and its kernel is generated by a nondegenerate symmetric tensor.*

Proof. The assertion holds for the pair of X and $\mathcal{E}$ since $H^0(X, S^2\mathcal{E})$ is an irreducible representation of G. Hence it also holds for the pair S_N and E_N by (2) of the proposition. *q.e.d.*

Corollary 3.9 $\chi(E_N) = \dim H^0(X, \mathcal{E}) = 9$ *and* $\chi(sl(E_N)) = 0$.

Theorem 0.2 is a consequence of Proposition 3.5 and the following:

Proposition 3.10 *The self intersection number of* $c_1(E_N)$ *is equal to 34.*

Proof. By (3.9), and the Riemann-Roch theorem, we have

$$9 = \chi(E_N) = (c_1(E_N)^2)/2 - c_2(E_N) + 3 \cdot 2$$

and

$$0 = \chi(sl(E_N)) = -c_2(sl(E_N)) + 8 \cdot 2 = 2(c_1(E_N)^2) - 6c_2(E_N) + 16,$$

which imply $(c_1(E_N)^2) = 34$ and $c_2(E_N) = 14$. *q.e.d.*

4 Proof of Theorem 0.3

We need the following general fact on deformations of vector bundles on K3 surfaces, which is implicit in [10].

Proposition 4.1 *Let E be a simple vector bundle on a K3 surface S and (S', L') be a small deformation of $(S, \det E)$. Then there is a deformation (S', E') of the pair (S, E) such that* $\det E' \simeq L'$.

Proof. The obstruction $ob(E)$ for E to deform a vector bundle on S' is contained in $H^2(S, \mathcal{E}nd(E))$. Its trace is the obstruction for $\det E$ to deform a line bundle on S', which is zero by assumption. Since the trace map $H^2(S, \mathcal{E}nd(E)) \longrightarrow H^2(S, \mathcal{O}_S)$ is injective by the Serre duality, the obstruction $ob(E)$ itself is zero. q.e.d.

We fix a 5-dimensional subspace N of $H^0(X, \mathcal{F}) \simeq U^{16}$ belonging to Ξ_{SCI} and consider deformations of the polarized K3 surface $(S_N, \mathcal{O}_S(1))$, where $\mathcal{O}_S(1)$ is the restriction of $\mathcal{O}_X(1)$ to S_N.

Proposition 4.2 *Let (S, L) be a small deformation of $(S_N, \mathcal{O}_S(1))$. Then there exists a vector bundle E on S which satisfies the following:*

i) $\det E \simeq L$,

ii) The pair (S, E) is a deformation of (S_N, E_N),

iii) E is generated by its global sections and $H^1(S, E) = H^2(S, E) = 0$,

iv) the natural linear map

$$S^2H^0(S, E) \longrightarrow H^0(S, S^2E)$$

is surjective and its kernel is generated by a nondegenerate symmetric tensor, and

v) the morphism $\Phi_{|E|} : S \longrightarrow G(H^0(E), 3)$ associated to E is an embedding.

Proof. The existence of E which satisfies i) and ii) follows from Proposition 4.1. The pair (S_N, E_N) satisfies iii) and iv) by Proposition 3.7and Corollary 3.8. Since (S, E) is a small deformation of (S_N, E_N), E satisfies iii) and v). Since $H^1(S_N, S^2 E_N)$ vanishes, E satisfies iv), too. *q.e.d.*

By iii) of the proposition, we identify S with its image in $G(H^0(E), 3)$. By iv) of the proposition, (the image of) S lies in the 12-dimensional homogeneous space X of $SO(9, C)$. By Proposition 3.6, S is contained in the common zero locus of a 5-dimensional subspace N' of $H^0(X, \mathcal{F})$. Since S is a small deformation of S_N, S is also a complete intersection with respect to $\mathcal{F}^{\oplus 5}$. Therefore, we have shown

Proposition 4.3 *The image of the classification morphism* $\Xi_{SCI} \longrightarrow \mathcal{F}_{18}$, $[N] \mapsto (S_N, \mathcal{O}_S(1))$, *is open.*

The Picard group of a K3 surface S is isomorphic to $H^{1,1}(S, Z) = H^2(S, Z) \cap H^0(\Omega^2)^{\perp}$. Since the local Torelli type theorem holds for the period map of K3 surfaces ([1, p. 254]), the subset $\{(S, L) | Pic\, S \neq Z \cdot [L]\}$ of $\mathcal{F}_g$ is a countable union of subvarieties. Hence by the proposition and Baire's property, we have

Proposition 4.4 *There exists* $[N] \in \Xi_{SCI}$ *such that* $(S_N, \mathcal{O}_S(1))$ *is Picard general, i.e.,* $Pic\, S_N$ *is generated by* $\mathcal{O}_S(1)$.

The stability of vector bundles is easy to check over a Picard general variety.

Proposition 4.5 *If* $(S_N, \mathcal{O}_S(1))$ *is Picard general, then* E_N *is* μ*-stable (with respect to* $\mathcal{O}_S(1)$*).*

Proof. Let B be a locally free subsheaf of E_N. By our assumption, $\det B$ is isomorphic to $\mathcal{O}_S(b)$ for an integer b. In the case B is a line bundle, we have $b \leq 0$ since $\dim H^0(B) \leq \dim H^0(E_N) = 9$. In the case B is of rank 2, $\bigwedge^2 B \simeq \mathcal{O}_S(b)$ is a subsheaf of $\bigwedge^2 E_N \simeq E_N^{\vee} \otimes \mathcal{O}_S(1)$. We have $H^0(S_N, E_N^{\vee}) = H^2(S_N, E_N)^{\vee} = 0$ by (1) of Proposition 3.7 and the Serre duality. Hence b is nonpositive. Therefore we have $b/r(B) < 1/3$ for every B with $r(B) < r(E_N) = 3$. If F is a subsheaf of E_N, then its double dual $F^{\vee\vee}$ is a locally free subsheaf of E_N. Hence $c_1(F)/r(F) < c_1(E_N)/r(E_N)$ for every subsheaf F of E_N with $0 < r(F) < r(E_N)$. *q.e.d.*

Let Ξ be the subset of Ξ_{SCI} consisting of $[N]$ such that E_N is stable with respect to $\mathcal{O}_S(1)$ in the sense of Gieseker [6]. Ξ is non-empty by the above two propositions.

Theorem 4.6 *Let* M *and* N *be 5-dimensional subspaces of* $H^0(X, \mathcal{F})$ *with* $[M], [N] \in \Xi$.

(1) If $(S_M, \mathcal{O}_S(1))$ *and* $(S_N, \mathcal{O}_S(1))$ *are isomorphic to each other, then* $[M]$ *and* $[N]$ *belong to the same* $SO(9, C)$*-orbit.*

(2) The automorphism group of $(S_N, \mathcal{O}_S(1))$ *is isomorphic to the stabilizer group of* $SO(9, C)$ *at* $[N] \in G(5, U^{16})$.

Proof. Let $\phi : S_M \longrightarrow S_N$ be an isomorphism such that $\phi^* \mathcal{O}_S(1) \simeq \mathcal{O}_S(1)$. Two vector bundles E_M and $\phi^* E_N$ have the same rank and Chern classes. Since E_N is rigid by (3) of Proposition 3.7 and since both are stable, there exists an isomorphism $f : E_M \longrightarrow \phi^* E_N$ by Proposition 0.4. By (2) of Proposition 3.7,

$$H^0(f) : H^0(S_M, E_M) \longrightarrow H^0(S_M, \phi^* E_N) \simeq H^0(S_N, E_N)$$

is an automorphism of $V = H^0(X, \mathcal{E})$. By Corollary 3.8, $H^0(f)$ preserves the 1-dimensional subspace Cq of S^2V. Hence replacing f by cf for suitable constant c, we may assume that $H^0(f)$ belongs to the special orthogonal group $G = SO(V, q)$. Let $L \in \tilde{G} = Spin(V, q)$ be a lifting of $H^0(f)$. Since $\mathcal{F}$ is homogeneous, there exists an isomorphism $\ell : F_M \longrightarrow \phi^* F_N$ such that $H^0(\ell) = L$. By (2) of Proposition 3.6, L maps $M \subset H^0(X, \mathcal{F})$ onto $N \subset H^0(X, \mathcal{F})$, which shows (1). Putting $M = N$ in this argument, we have (2). *q.e.d.*

Let $\Phi : \Xi \longrightarrow \mathcal{F}_{18}$, $[N] \mapsto (S_N, \mathcal{O}_S(1))$ be the classification morphism. By the theorem, every fibre of Φ is an orbit of $SO(9, C)$. By the openness of stability condition ([8]) and Proposition 4.3, the image of Φ is Zariski open in $\mathcal{F}_{18}$. Hence we have completed the proof of Theorem 0.3.

5 K3 surfaces of genus 20

Let V be a vector space of dimension 9 and $\mathcal{E}$ the (rank 4) universal quotient bundle on the Grassmann variety $X = G(V, 4)$. The determinant bundle of $\mathcal{E}$ is the ample generator $\mathcal{O}_X(1)$ of $Pic X \simeq Z$. We denote the restrictions of $\mathcal{E}$ and $\mathcal{O}_X(1)$ to S_N by E_N and $\mathcal{O}_S(1)$, respectively.

Theorem 5.1 *Let N be a 3-dimensional subspace of $H^0(G(V,4), \bigwedge^2 \mathcal{E}) \simeq \bigwedge^2 V$ and $S_N \subset G(V,4)$ the common zero locus of N. If S_N is smooth and of dimension 2, then the pair $(S_N, \mathcal{O}_S(1))$ is a K3 surface of genus 20.*

The tangent bundle T_X of $G(V, 4)$ is isomorphic to $\mathcal{E} \otimes \mathcal{F}$, where $\mathcal{F}$ is the dual of the universal subbundle. Hence $G(V, 4)$ is a 20-dimensional Fano manifold of index 9. S_N is the scheme of zeroes of the section

$$s : \mathcal{O}_X \longrightarrow (\bigwedge^2 \mathcal{E}) \otimes_C N^\vee \simeq (\bigwedge^2 \mathcal{E})^{\oplus 3}$$

induced by $\mathcal{O}_X \otimes_C N \longrightarrow \bigwedge^2 \mathcal{E}$. The vector bundle $(\bigwedge^2 \mathcal{E})^{\oplus 3}$ is of rank $6 \cdot 3 = 18$, and has the same determinant as T_X. Hence, by Theorem 1.10 and the adjunction formula (1.5), we have

Proposition 5.2 *Let Ξ_{SCI} be the subset of $G(3, \bigwedge^2 V)$ consisting of $[N]$ such that S_N is smooth and of dimension 2. Then Ξ_{SCI} is non-empty and S_N has a trivial canonical bundle for every $[N] \in \Xi_{SCI}$.*

We show vanishing of cohomology groups of vector bundles on S_N using the Koszul complex

(5.3)

$$\Lambda^\bullet : \ \mathcal{O}_X \longrightarrow (\bigwedge^2 \mathcal{E})^{\oplus 3} \longrightarrow \bigwedge^2 (\bigwedge^2 \mathcal{E})^{\oplus 3} \longrightarrow \cdots \longrightarrow \bigwedge^{17} (\bigwedge^2 \mathcal{E})^{\oplus 3} \longrightarrow \bigwedge^{18} (\bigwedge^2 \mathcal{E})^{\oplus 3}$$

of s and the Bott vanishing. $G(V, 4)$ is a homogeneous space of $GL(9, C)$. The stabilizer group P consists of the matrices of the form $\begin{pmatrix} A & B \\ 0 & D \end{pmatrix}$ with $A \in GL(4, C), B \in M_{4,5}(C)$ and $D \in GL(5, C)$. The set H of invertible diagonal matrices is a maximal torus. The roots of $GL(9, C)$ are $e_i - e_j, i \neq j$, for the standard basis of the character group $X(H)$ of H. We take $\Delta = \{e_i - e_{i+1}\}_{1 \leq i \leq 8}$ as a root basis. The reductive part L of P is isomorphic to $GL(4, C) \times GL(5, C)$ and its root basis is $\Delta \setminus \{e_4 - e_5\}$. Let $\rho(a_1, a_2, a_3, a_4)$ be the irreducible representation of $GL(4, C)$ (or L) with the highest weight

$w = (a_1, a_2, a_3, a_4) = a_1e_1 + a_2e_2 + a_3e_3 + a_4e_4, a_1 \geq a_2 \geq a_3 \geq a_4$. We denote by $\mathcal{E}(a_1, a_2, a_3, a_4)$ the homogeneous vector bundle on X induced from the representation $\rho(a_1, a_2, a_3,$ The universal quotient bundle $\mathcal{E}$ is $\mathcal{E}(1,0,0,0)$ and its exterior products $\bigwedge^2 \mathcal{E}$, $\bigwedge^3 \mathcal{E}$ and $\bigwedge^4 \mathcal{E}$ are $\mathcal{E}(1,1,0,0)$, $\mathcal{E}(1,1,1,0)$ and $\mathcal{E}(1,1,1,1)$, respectively.

We apply the Bott vanishing theorem ([3]). One half δ of the sum of all the positive roots is equal to $4e_1 + 3e_2 + 2e_3 + e_4 - e_6 - 2e_7 - 3e_8 - 4e_4$ and we have

$$\delta + w = (4 + a_1)e_1 + (3 + a_2)e_2 + (2 + a_3)e_3 + (1 + a_4)e_4 - e_6 - 2e_7 - 3e_8 - 4e_9.$$

All the cohomology groups of $\mathcal{E}(a_1, a_2, a_3, a_4)$ vanish if a number appears more than once among the coefficients. For the convenience of later use we state the vanishing theorem for $\mathcal{E}(a_1, a_2, a_3, a_4) \otimes \mathcal{O}_X(-9)$:

Proposition 5.4 *The cohomology group $H^i(X, \mathcal{E}(a_1, a_2, a_3, a_4) \otimes \mathcal{O}_X(-9))$ vanishes for every i if one of the following holds:*
i) $\lambda \leq a_\lambda \leq \lambda + 4$ for some $1 \leq \lambda \leq 4$, or
ii) $a_\mu - a_\nu = \mu - \nu$ for some pair $\mu \neq \nu$.

To apply this to the Koszul complex (5.4), we need the decomposition of $\Lambda^i = \bigwedge^i(\bigwedge^2 \mathcal{E})^{\oplus 3}$ into the sum of irreducible homogeneous vector bundles. Since $(\bigwedge^2 \mathcal{E})^\vee \simeq (\bigwedge^2 \mathcal{E})(-1)$, we have the following in the same manner as Lemma 3.3:

Lemma 5.5 $\Lambda^i \simeq \Lambda^{18-i} \otimes \mathcal{O}_X(i - 9)$.

Put $\rho_2 = \rho(1,1,0,0)$. It is easy to check the following:

i	2	3	4	5	6
$\bigwedge^i \rho_2$	$\rho(2,1,1,0)$	$\rho(3,1,1,1)$ $\oplus\rho(2,2,2,0)$	$\rho(3,2,2,1)$	$\rho(3,3,2,2)$	$\rho(3,3,3,3)$

The representation $\bigwedge^i(\rho_2^{\oplus 3})$ is isomorphic to

$$\bigoplus_{p+q+r=i} (\bigwedge^p \rho_2) \otimes (\bigwedge^q \rho_2) \otimes (\bigwedge^r \rho_2).$$

By the computation using the Littlewood-Richardson rule ([9, Chap. I, §9]), we have

Proposition 5.6 *The set W_i of the highest weights of the irreducible components of the representation $\bigwedge^i \rho_2^{\oplus 3}$ is as follows:*

i	W_i
1	$\{(1,1,0,0)\}$
2	$\{(2,2,0,0),(2,1,1,0),(1,1,1,1)\}$
3	$\{(3,3,0,0),(3,2,1,0),(3,1,1,1),(2,2,2,0),(2,2,1,1)\}$
4	$\{(4,3,1,0),(4,2,2,0),(4,2,1,1),(3,3,2,0),(3,3,1,1),(3,2,2,1),(2,2,2,2)\}$
5	$\{(5,3,2,0),(5,3,1,1),(5,2,2,1),(4,4,2,0),(4,3,3,0)\} \cup W_3 + (1,1,1,1)$
6	$\{(6,3,3,0),(6,3,2,1),(6,2,2,2),(5,4,3,0),(4,4,4,0)\} \cup W_4 + (1,1,1,1)$
7	$\{(7,3,3,1),(7,3,2,2),(6,4,4,0),(5,5,4,0)\} \cup W_5 + (1,1,1,1)$
8	$\{(8,3,3,2),(6,5,5,0)\} \cup W_6 + (1,1,1,1)$
9	$\{(9,3,3,3),(6,6,6,0)\} \cup W_7 + (1,1,1,1)$

The sets of the highest weights appearing in the decompositions of $\mathcal{E} \otimes \Lambda^\bullet$, $\bigwedge^2 \mathcal{E} \otimes \Lambda^\bullet$ and $\mathcal{E} \otimes \mathcal{E}^\vee \otimes \Lambda^\bullet$ are easily computed from Lemma 5.5 and the proposition using the following formula:

$$\rho(1,0,0,0)\otimes\rho(a_1,a_2,a_3,a_4)=\bigoplus_{\substack{\sum b_i=1+\sum a_i\\ a_i\le b_i\le a_i+1}}\rho(b_1,b_2,b_3,b_4),$$

$$\rho(1,1,0,0)\otimes\rho(a_1,a_2,a_3,a_4)=\bigoplus_{\substack{\sum b_i=2+\sum a_i\\ a_i\le b_i\le a_i+1}}\rho(b_1,b_2,b_3,b_4)$$

and

$$\rho(0,0,0,-1)\otimes\rho(a_1,a_2,a_3,a_4)=\bigoplus_{\substack{\sum b_i=-1+\sum a_i\\ a_i-1\le b_i\le a_i}}\rho(b_1,b_2,b_3,b_4).$$

Applying Proposition 5.5 to the exact sequence

$$0\longrightarrow\Lambda^\bullet\otimes\mathcal{O}_X(-9)\longrightarrow\mathcal{O}_{S_N}\longrightarrow 0,$$

$$0\longrightarrow\Lambda^\bullet\otimes\mathcal{E}\otimes\mathcal{O}_X(-9)\longrightarrow E_N\longrightarrow 0,$$

$$0\longrightarrow\Lambda^\bullet\otimes\bigwedge^2\mathcal{E}\otimes\mathcal{O}_X(-9)\longrightarrow\bigwedge^2 E_N\longrightarrow 0$$

and

$$0\longrightarrow\Lambda^\bullet\otimes sl(\mathcal{E})\otimes\mathcal{O}_X(-9)\longrightarrow sl(E_N)\longrightarrow 0,$$

we have the following in a similar way to Propositions 3.6 and 3.7:

Proposition 5.7 *If* $\dim S_N=2$, *then we have*

(1) the restriction map $H^0(\mathcal{O}_X)\longrightarrow H^0(\mathcal{O}_{S_N})$ *is surjective and* $H^1(\mathcal{O}_{S_N})$ *vanishes,*

(2) all higher cohomology groups of E_N *and* $\bigwedge^2 E_N$ *vanish,*

(4) the restriction map $H^0(X,\mathcal{E})\longrightarrow H^0(S_N,E_N)$ *is bijective,*

(4) the restriction map $H^0(X,\bigwedge^2\mathcal{E})\longrightarrow H^0(S_N,\bigwedge^2 E_N)$ *is surjective and its kernel coincides with* N*, and*

(5) all cohomology groups of $sl(E_N)$ *vanish.*

Corollary 5.8 $\chi(E_N)=\dim H^0(X,\mathcal{E})=9$ *and* $\chi(sl(E_N))=0$.

If $[N]$ belongs to Ξ_{SCI}, then S_N is a K3 surface by Proposition 5.2 and (1) of Proposition 5.7. We have $(c_1(E_N)^2)=38$ by the corollary in a similar manner to Proposition 3.10. This proves Theorem 5.1.

Theorem 5.9 *Let* Ξ *be the subset of* $G(3,\bigwedge^2 V)$ *with* $\dim V=9$ *consisting of* $[N]$ *such that* S_N *is a K3 surface and* E_N *is stable with respect to* $\mathcal{O}_S(1)$*. Then we have*

(1) Ξ *is a non-empty Zariski open subset,*

(2) the image of the classification morphism $\Phi:\Xi\longrightarrow\mathcal{F}_{20}$ *is open,*

(3) every fibre of Φ *is an orbit of* $PGL(V)$*, and*

(4) the automorphism group of (S_N,L_N) *is isomorphic to the stabilizer group of* $PGL(V)$ *at* $[N]\in G(3,\bigwedge^2 V)$.

There exists a 3-dimensional subspace N of $\bigwedge^2 V$ such that the polarized K3 surface $(S_N,\mathcal{O}_S(1))$ is Picard general. Let B be a locally free subsheaf of E_N. In the cases $r(B)=1$ and 3, we have $b\le 0$ in the same way as Proposition 4.4. In the case B is of rank 2, $\bigwedge^2 B\simeq\mathcal{O}_S(b)$ is a subsheaf of $\bigwedge^2 E_N$. Since $\bigwedge^2 E_N\simeq\bigwedge^2 E_N^\vee\otimes\mathcal{O}_S(1)$, we have

$$Hom\,(\mathcal{O}(1),\bigwedge^2 E_N)\simeq H^0(S_N,(\bigwedge^2 E_N)(-1))\simeq H^2(S_N,\bigwedge^2 E_N)^\vee=0$$

by the Serre duality and (2) of Proposition 5.7. Hence b is nonpositive. Therefore, E_N is μ-stable if $(S_N, \mathcal{O}_S(1))$ is Picard general. This shows (1) of the theorem. The rest of the proof of Theorem 5.9 is the same as that of Theorem 0.3 in Section 4.

Let $G(3, \bigwedge^2 V)^s$ be the stable part of $G(3, \bigwedge^2 V)$ with respect to the action of $SL(V)$.

Corollary 5.10 *The moduli space $\mathcal{F}_{20}$ of K3 surfaces of genus 20 is birationally equivalent to the moduli space $G(3, \bigwedge^2 V)^s / PGL(V)$ of nets of bivectors on V.*

References

[1] Barth, W., C. Peters and A. Van de Ven: Compact complex surfaces, Springer, 1984.

[2] Borel A. and F. Hirzebruch: Characteristic classes and homogeneous spaces I, Amer. J. Math. **80** (1958), 458-538.

[3] Bott, R.: Homogeneous vector bundles, Ann. of Math. **66** (1957), 203-248.

[4] Bourbaki, N.: Éléments de mathematique, Groupes et algébre de Lie, Chapitres 4, 5 et 6, Hermann, Paris, 1968.

[5] Chevalley, C.: The algebraic theory of spinors, Columbia University Press, 1955.

[6] Gieseker, D.: On the moduli of vector bundles on an algebraic surface, Ann. of Math. **106** (1977), 45-60.

[7] Griffiths, P.A. and J. Harris: Principles of Algebraic Geometry, John Wiley, New-York, 1978.

[8] Maruyama, M.: Openness of a family of torsion free sheaves, J. Math. Kyoto Univ. **16** (1976), 627-637.

[9] Macdonald, I.G.: Symmetric functions and Hall polynomials, Clarendon Press, Oxford, 1979.

[10] Mukai, S.: Symplectic structure of the moduli space of sheaves on an abelian or K3 surface, Invent. Math. **77** (1984), 101-116.

[11] Mukai, S.: On the moduli space of bundles on K3 surfaces: I, in 'Vector Bundles on Algebraic Varieties (Proceedings of the Bombay Colloquium 1984)', Tata Institute of Fundamental Research Studies, **11**, pp. 341-413, Oxford University Press, 1987.

[12] Mukai, S.: Curves, K3 surfaces and Fano 3-folds of genus ≤ 10, in 'Algebraic Geometry and Commutative Algebra in Honor of Masayoshi Nagata', pp. 357-377, 1988, Kinokuniya, Tokyo.

[13] Mukai, S.: Biregular classification of Fano threefolds and Fano manifolds of coindex 3, Proc. Nat. Acad. Sci. **86** (1989), 3000-3002.

[14] Mumford, D.: Geometric invariant theory, Springer, 1965.

[15] Piatetskij-Shapiro, I.I. and I.R. Shafarevich: A Torelli theorem for algebraic surfaces of type K3, Izv. Akad. Nauk. SSSR Ser. Mat. **35** (1971), 503-572.

Department of Mathematics
Nagoya University
School of Sciences
464 *Furō-chō, Chikusa-ku*
Nagoya, Japan

PROJECTIVE COMPACTIFICATIONS OF COMPLEX AFFINE VARIETIES

STEFAN MÜLLER-STACH

§1.Introduction

In this paper we look at the following two problems:

Compactification problem: *Let X be a smooth complex projective variety , A a divisor on X such that $X\backslash A$ is biregular to an affine homology cell , i.e. a smooth affine scheme with the homology groups of a point. Classify all pairs (X, A) !*

Characterisation problem: *Let $V = \mathbf{Spec}(R)$ be a smooth complex affine scheme. Characterise V by topological or algebraic properties of V itself and at infinity (i.e. of suitable smooth compactifications).*

We will present known results about these two problems as well as their relation to other topics like Fano varieties and homogenous spaces.
Our problems are obviously related by the process of compactification which is due to Nagata in this category. Therefore results about one of these problems will give information about the solution of the other.
The compactification problem was first stated by Hirzebruch (see [H]) in the category of complex manifolds: Let X be a complex manifold with an analytic subset A such that $X\backslash A$ is biholomorphic to $\mathbf{C}^n$. Then (X, A) is called a *complex analytic compactification* of $\mathbf{C}^n$. Problems 26 and 27 in [H] ask for the classification of all pairs (X, A). By the theorem of Hartogs it follows that A has to be a divisor in X, hence this is a special case of our compactification problem in the complex analytic case.
The characterisation problem was attacked by several people, including Ramanujam [R] and [Mi].
In the sequel we give an overview of some of the results, in particular concerning the case of 3-dimensional varieties.

§2.Basic properties of compactifications and Fano varieties

In this paragraph we state the elementary properties of (complex analytic) compactifications. First let X be a smooth projective variety of dimension n and A a divisor on X, such that $X\backslash A$ is a homology cell.

Proposition 1: *On the cohomology level we have*

$$H^k(X, \mathbf{Z}) = H^k(A, \mathbf{Z}) \qquad for \qquad 0 \leq k \leq 2n-1.$$

Corollary 1: *(a) The second Betti number* $b_2(X) = dim H^2(X, \mathbf{R})$ *is the number of irreducible components of A .*
(b) $b_1(X) = b_{2n-1}(X) = b_{2n-1}(A) = 0$ *(hence* $H^2(X, \mathbf{Z})$ *is torsion free)*
(c) A is connected.

The Proposition is proved using the long exact cohomology sequence for the pair (X, A), Poincaré duality and the homology cell property.

Proposition 2: *Let* (X, A) *be as above and assume that* $X \backslash A$ *is biholomorphic to* $\mathbf{C}^n$. *Then*

$$H^q(X, \mathcal{O}_X) = 0 \quad for \quad q = 1, 2$$

If $b_2(X) = 1$, then the Picard group of X is $\mathbf{Z}$ and X is a *Fano variety* (i.e. $-K_X$ is ample).
Proof. The statement about $H^q(X, \mathcal{O}_X)$ follows from Hodge decomposition and Prop.1, since $H^2(X, \mathbf{Z})$ is generated by the (1,1)-classes of the irreducible components of A. If $b_2(X) = 1$, then $H^2(X, \mathbf{Z}) = \mathbf{Z}$ and the exponential sequence gives $Pic(X) = \mathbf{Z}$. If K_X denotes the canonical divisor of X, then we have to show that mK_X has no sections for large m. This is provided by the following theorem.

Theorem 1: (Kodaira [K]) *Let* (X, A) *be an arbitrary smooth analytic compactification of* $\mathbf{C}^n$. *Then X has Kodaira dimension* $-\infty$, *i.e. all plurigenera* $P_m(X) = h^0(X, mK_X)$ *vanish.*

§3. Basic examples for compactifications

There are 3 different obvious ways of constructing compactifications of affine n-space $\mathbf{A}^n$:

(1) Take a homogenous space where the algebraic cell decomposition contains one n-cell isomorphic to $\mathbf{A}^n$ in the top dimension (for example projective space, quadrics or more general grassmannians)
(2) Use a (Zariski) locally trivial fibration with compactifications as base and fibre spaces
(3) Blow up a known compactfication in centers contained in the boundary.

We will give basic examples:
(a) $(X, A) = (\mathbf{P}^n, \mathbf{P}^{n-1})$ is an obvious example for a compactification of $\mathbf{A}^n$. This yields the following well known conjecture:

Problem 1: *Is* $(\mathbf{P}^n, \mathbf{P}^{n-1})$ *the only compactification of* $\mathbf{C}^n$ *with smooth boundary?*

This is proved for $n \leq 6$ by Van de Ven and Fujita [VdV,F], if X is a Kähler manifold.

(b) $(\mathbf{Q}^n, \mathbf{Q}_0^{n-1})$ is a compactification of $\mathbf{A}^n$, where $\mathbf{Q}^n$ is a smooth quadric and $\mathbf{Q}_0^{n-1}$ is a tangent hyperplane section with one node.

One may see this by blowing up one point P on the smooth quadric $\mathbf{Q}^n$ contained in $\mathbf{P}^{n+1}$ and considering the map ϕ given by projection from P to $\mathbf{P}^n$ which is well defined after blow up. The image of $\mathbf{Q}_0^{n-1}$ is a hyperplane H in $\mathbf{P}^n$ and ϕ maps $\mathbf{Q}^n \backslash \mathbf{Q}_0^{n-1}$ isomorphically onto $\mathbf{P}^n \backslash H$.

(c) Any $\mathbf{P}^m$–bundle over $\mathbf{P}^k$ with $k + m = n$ provides an obvious example with second Betti number two.

(d) There is one example due to Furushima [Fu1], which was not known before 1985: (see also [E] for more examples in higher dimensions)
Take the compactification $(\mathbf{Q}^3, \mathbf{Q}_0^2)$ as above and choose a twisted cubic C which is contained in the boundary. Then blow up $\mathbf{Q}^3$ in C. Let M be the resulting manifold. The strict transform of $\mathbf{Q}_0^2$ is a rational scroll which can be contracted inside M onto a rational curve L, such that the resulting 3-dimensional variety X is smooth. Since by construction X is a Fano variety the classification of Fano varieties [I,M] shows that X is isomorphic to $\mathbf{V}_5$ the projectively unique determined Fano threefold of index two with the invariants $b_2(X) = 1$ and $b_3(X) = 0$ (the index is the largest integer that divides the canonical divisor of X in $Pic(X)$). $\mathbf{V}_5$ can also be constructed as a 3-fold hyperplane section of the 6-dimensional Grassmannian $Gr(2, 5)$ contained in $\mathbf{P}^9$.

(e) We should also mention another example due to Furushima [Fu2], which has a quite interesting history. By [M] there is a family of Fano threefolds of index one with $b_2(X) = 1$ and $b_3(X) = 0$. Due to other useful invariants these are denoted by $\mathbf{A}_{22}$. In the classification proofs for Fano threefold these varieties were overlooked by Fano. Then one example was found by Iskovskikh, another one by Mukai and Umemura [MU] which has a $SL(2, \mathbf{C})$–action. Finally Mukai gave a complete classification using vector bundles on $K3$–surfaces which occur naturally as hyperplane sections in the study of Fano threefolds of index one. It is exactly the example of Mukai-Umemura which is known to be a compactification of $\mathbf{C}^3$ amongst all the varieties in this family. Furushima showed this by using a description of the *Mukai-Umemura variety* as an equivariant completion, which gives explicit equations for $\mathbf{A}_{22}$. Using these equations he was able to perform a coordinate change, which makes the $\mathbf{C}^3$ actually visible! We remark that $\mathbf{V}_5$ may also be described in these terms [E]. It is remarkable that only the "most" homogenous variety of type $\mathbf{A}_{22}$ is known to be a compactification. This leads to

Problem 2: *Which varieties of type* $\mathbf{A}_{22}$ *are compactifications?*

§4. The compactification problem

Let (X, A) be a projective analytic compactification of $\mathbf{C}^n$ or less general of $\mathbf{A}^n$ in the algebraic category. The projectivity is already a consequence of the other assumptions if $n \leq 2$ or $n = 3$ and $b_2(X) = 1$ by [RV],[Mo] and [PS,P].

For $n = 1$ we clearly have $(X, A) = (\mathbf{P}^1, point)$ by Corollary 1. For $n = 2$ and $b_2 = 1$ we have

Theorem 2:(Remmert-Van de Ven)
If (X, A) is an irreducible compactification of $\mathbf{C}^2$ (i.e. $b_2(X) = 1$) then X is isomorphic to $\mathbf{P}^2$ and A is embedded as a line.
Proof:[RV], see also [VdV] for the case of homology cells! With the help of Enriques-Kodaira classification this also follows from theorem 1.

In the case $n = 2$ and $b_2(X) \geq 2$ the following is true:

Theorem 3:(Morrow) *X is a rational surface, A a tree of rational curves and (X, A) is obtained from $(\mathbf{P}^2, \mathbf{P}^1)$ by blow ups and blow downs on the boundary. A complete list of minimal boundary graphs exists.*
Proof:[Mo]

For $n = 3$ and $b_2(X) = 1$ we have the results of Brenton and Morrow [BM], Furushima [Fu1,2], Peternell and Schneider [PS,P] which turn into the following

Theorem 4:
(X, A) is an irreducible analytic compactification of $\mathbf{C}^3$ if and only if it is one of the following pairs:

$\mathbf{P}^3$ $\mathbf{P}^2$ *(embedded as a plane)*
$\mathbf{Q}^3$ $\mathbf{Q}^2_0$ *(quadric cone)*
$\mathbf{V}_5$ *normal del Pezzo surface with A_4 double point or non normal*
$\mathbf{A}_{22}$ *non normal K3 surface with double line*

Corollary 4: *X is rational with $b_3(X) = 0$ hence with trivial intermediate Jacobian in the sense of Griffiths.-*

Remark: The classification of the varieties A_{22} can be found in [M] and [MU].

Now consider the case $n = 3$ and $b_2(X) \geq 2$. In this situation it is fairly natural to blow X up along the boundary such that we may assume A has normal crossings. By theorem 1 we know $P_m(X) = 0$ for all $m \geq 1$. Hence - by a theorem of Miyaoka [Miy] - the canonical divisor K_X cannot be numerically effective. From the theory of Mori [KMM] we conclude that there exists an extremal ray on X. Using the classification theory of threefolds and induction on the number of components of A it should be possible to classify all pairs (X, A) completely. But the extremal rays create some singularities and destroy the normal crossing property. Hence we are forced to assume more general singularities like terminal singularities for X and arbitrary rational singularities for the boundary surfaces. This program is not worked out completely but we have one result in this direction:

Theorem 5:([MS])
If (X, A_1, A_2) is a smooth projective analytic compactification of $\mathbf{C}^3$ such that A_1 and A_2 have isolated singularities and intersect transversally, then X is rational, the boundary surfaces birationally ruled with rational singularities and one of the following cases occur:

(1) X is a $\mathbf{P}^2$*-bundle over* $\mathbf{P}^1$
(2) X is a $\mathbf{P}^1$*-bundle over* $\mathbf{P}^2$
(3) X is the blow up of one of the four irreducible compactifications

Corollary 5: $b_3(X) = 0$ *unless X is the blow up of a curve. Hence the intermediate Jacobian of X is the Jacobian of a curve.*

Remarks: Isolated singularities on the boundary is equivalent to assuming that A_1 and A_2 are normal surfaces since they are hypersurfaces in a smooth variety. In [MS] you can find more details and a detailed description of the boundary surfaces. The transversality assumption may be weakened to the case where $A_1 \cap A_2$ is a tree of smooth curves, but the author can show this only for a small number of components. In the proof of theorem 5 in [MS] Fano threefolds play a very important role and actually the methods used in the classification of Fano threefolds with $b_2 \geq 2$ [MM] are quite similar. We finish this chapter by a remark about the algebraicity of those compactifications: Up to now any example of a complex analytic compactification of $\mathbf{C}^n$ is algebraic in the sense that the complement of the boundary divisor is biregular to $\mathbf{A}^n$. It is natural to ask:

Problem 3:*Is every complex analytic compactification already algebraic in this sense?*

Note that this is not at all true if one considers compactifications of $\mathbf{C}^* \times \mathbf{C}^*$ or $\mathbf{C} \times \mathbf{C}^*$ instead of $\mathbf{C}^2$ because there exist compactifications of these Stein surfaces such that the embedding cannot be realized in a way that the complement is biregular to the corresponding affine variety $\mathbf{A}^* \times \mathbf{A}^*$ or $\mathbf{A} \times \mathbf{A}^*$ (see [U] and [YVT] for more literature on this topic).

§.The characterisation problem

The most striking result in this direction is given by:

Theorem 6:(Ramanujam [R])
Let $V = \mathbf{Spec}(R)$ *be a smooth complex affine surface which is topological contractible and simply connected at infinity. Then V is biregular to the affine plane* $\mathbf{A}^2 = \mathbf{Spec}\mathbf{C}[T_1, T_2]$.

Here *simply connected at infinity* means that $\pi_1^\infty(V) = 1$, where $\pi_1^\infty = \pi_1(\partial N)$ and N is the tubular neighborhood of any smooth compactification (X, A) of V. In this way our two problems are strongly related!
In this context it is also interesting to look at the work of Iitaka, Fujita, Kawamata, Miyanishi, Tsunoda, Gurjar, Shastri and Sugie on compactifications of affine varieties, log del Pezzo surfaces and the affine cancellation problem. See [Mi] for more explanations. The most important invariant in this direction is Iitaka's *logarithmic Kodaira dimension:*
Let V be a normal affine variety and (X, A) a smooth compactification such that A has normal crossings. Then we define:

$$log\kappa(V) = trdeg\{\bigoplus_{m\geq 1} H^0(X, m(K_X + A))\} - 1$$

(or $-\infty$ if $H^0(X, m(K_X + A)) = 0 \forall m \geq 1$).

It can be shown that this definition does not depend on the particular choice of (X, A).

For example it is known that a contractible normal affine surface V with $log\kappa(V) = -\infty$ is biregular to the affine plane $\mathbf{A}^2$. If we could prove theorem 5 in the general case, then the proof would tell us probably a characterisation criterion for $\mathbf{A}^3$. The proof of theorem 5 in [MS] does not use too much properties of $\mathbf{C}^3$.
Miyanishi has obtained a characterisation of $\mathbf{A}^3$ by a mixture of topological and algebraic properties [Mi].
Acknowledgements: This work was supported by Deutsche Forschungsgemeinschaft. I am also grateful to Mark Green for his encouragement and like to thank him and Rob Lazarsfeld for their hospitality during my time at UCLA.
Note added in proof: Meanwhile Mukai has solved problem 2 by showing that there is a 4-dimensional family of compactifications in the index 1 case.

References
[BM] Brenton,L., Morrow,J.: *Compactifications of* $\mathbf{C}^n$, Trans. Amer.Math.Soc.**246**,139-158(1979)
[E] Esser, J.: *Kompaktifizierungen des* $\mathbf{C}^n$, Diplomarbeit Bonn (1989)
[F] Fujita, T.: *On topological characterisations of complex projective spaces and affine linear spaces*, Proc.Japan Acad.**56**(1980)
[Fu1] Furushima, M.: *Singular del Pezzo surfaces and analytic compactifications of* $\mathbf{C}^3$, Nagoya Math. J.**104**,1–28 (1986)
[Fu2] Furushima, M.: *Complex analytic compactifications of* $\mathbf{C}^n$, Preprint Nagoya (1988)
[H] Hirzebruch, F.: *Some problems on differentiable and complex manifolds*,Ann.Math.**60**, 212-236 (1954); see also Collected works Springer(1988)
[I] Iskovskikh,V.A.: *Lectures on threedimensional algebraic varieties: Fanovarieties*, (in Russian) Moskow university (1988)
[KMM] Kawamata, Y., Matsuda, K. and Matsuki, K.: *Introduction to the minimal model problem*, Adv.Stud.in Pure Math.**10**,283–360 (1987)
[K] Kodaira, K.: *Holomorphic mappings of polydisks into compact complex manifolds*, J.Diff.Geom.**6**,33–46 (1971)
[Miy] Miyaoka, Y.: *On the Kodaira dimension of minimal threefolds*, Math.Ann.**281**,325-332 (1988)
[Mi] Miyanishi, M.: *Characterisations of affine 3-space*,Proc.of the Alg.Geom.Seminar Singapore (1987)
[MM] Mori, S. and Mukai, S.: *On Fano 3-folds with* $b_2 \geq 2$, Adv.Stud.in Pure Math.**1**,101–129 (1983)
[Mo] Morrow, J.: *Compactifications of* $\mathbf{C}^2$, Bull.Amer.Math. Soc.**78**,813–816 (1972)
[MS] Müller-Stach, S.: *Compactifications of* $\mathbf{C}^3$ *with reducible boundary divisor*, Math.Ann. **286** (1990)
[M] Mukai, S.: *Biregular classification of Fano 3-folds and Fano manifolds of coindex 3*,Proc.Natl.Acad.Sci. USA **86**,3000–3002 (1989)

[MU] Mukai, S., Umemura, H.:*Minimal rational threefolds*, Springer Lecture Notes **1016**, 490–518 (1983)
[PS] Peternell,Th., Schneider,M.:*Compactifications of* $\mathbf{C}^3$**I**, Math.Ann.**280**,129–146 (1988)
[P] Peternell,Th.:*Compactifications of* $\mathbf{C}^3$ **II/III**, Math.Ann.**283**,121–137 (1989) and Math. Zeitschrift **205**, 213-222 (1990)
[R] Ramanujam, C.P.:*A characterisation of the affine plane as an algebraic variety*, Ann.of Math.**94**,69–88 (1971)
[RV] Remmert, R., Van de Ven, A.: *Zwei Sätze über die komplex projektive Ebene*, Niew.Arch. Wisk. **8**,147–157 (1960)
[U] Ueda, T.: *Compactifications of* $\mathbf{C} \times \mathbf{C}^*$ *and* $(\mathbf{C}^*)^2$,Tohoku Math.J.**31**,81–90 (1979)
[VdV] Van de Ven, A.: *Analytic compactifications of complex homology cells*, Math.Ann. **147**, 189–204 (1962)
[YVT] Yo Van Tan: *On the compactification problem for Stein surfaces*,Comp.Math.**71**,1–12 (1989)

Stefan J. Müller-Stach
Department of Mathematics
University of California
Los Angeles, California 90024-1555

ON GENERALIZED LAUDAL'S LEMMA

Rosario Strano (*)

Introduction. Let C be an integral curve in $\mathbf{P}^3$ of degree d and let $\Gamma = C \cap H$ be the generic plane section. The idea of using Γ in order to deduce properties of C goes back to Castelnuovo and has been successfully used by many authors especially in the following problem: determine the maximum genus G(d, s) of a smooth curve of degree d not lying on a surface of degree $< s$ (see [HH], [E] for more informations). For this reason this method is called Castelnuovo method.

In this paper we discuss the following problem: *given the Hilbert function of* Γ, *deduce some information about the least degree of a surface containing* C.

Section 1 is devoted to recall some known results in this direction; the main result is Laudal's lemma which says that, if C is integral of degree $d > \sigma^2+1$ and Γ is contained in a curve of H of degree σ, then C is contained in a surface of degree σ (see [L], Corollary p.147 and [GP], Lemme).

On the other hand very little is known about curves of degree d, $\binom{\sigma+2}{2} \le d \le \sigma^2+1$, such that Γ is contained in a curve of H of degree σ and C is not contained in a surface of degree σ. In the remaining sections we give some partial results in the previous situation. In particular the following two theorems are proved for an integral curve C .

a) *If* Γ *is contained in a curve of* H *of degree* σ *and* $d > \sigma^2-\sigma+4$, *then* C *lies on a surface of degree* $\sigma+1$.

b) *If* Γ *is contained in two curves of* H *of degree* σ *and* $d > \sigma^2-2\sigma+6$, *then* C *lies on a surface of degree* $\sigma+1$.

Moreover some application of these theorems is given.

1. Let k be a field of characteristic 0 and let $\mathbf{P}^3 = \mathbf{P}^3_k$. Let $C \subset \mathbf{P}^3$ be a curve (i.e. a 1-dimensional closed subscheme which is locally C.M. and equidimensional). Let $\Gamma = C \cap H$ be the generic plane section of C. Denote by $\mathcal{J}_C$ and by $\mathcal{J}_\Gamma$ the ideal sheaves of C and Γ in $\mathbf{P}^3$ and H respectively. Finally let $R = k[x_1, x_2, x_3]$ be the polynomial ring and J the homogeneous ideal of Γ in R .

First we recall the following "lifting lemma" ([S], Teorema 4).

(*) Lecture given at the meeting on «Projective Varieties». Trieste, June 1989.
Work done with financial support of M.P.I.

THEOREM 1. *Suppose that* $\mathrm{Tor}_1^R(J,k)_n = 0$ *for every* $0 \le n \le \sigma+2$. *Then every curve of* H *of degree* σ *containing* Γ *can be lifted to a surface of degree* σ *containing* C.

Observe that $\dim_k \mathrm{Tor}_1^R(J,k)_n$ is the number of syzygies in degree n in a minimal free resolution of J.

From this result we can easily deduce the following corollaries ([S], Corollari 1,2; [E], Lemme II,7).

1) *Suppose that* C *is not contained in a quadric,* $d>4$, *and that* Γ *is a complete intersection; then* C *is a complete intersection.*

2) (Laudal's lemma) *If* C *is integral of degree* $d>\sigma^2+1$ *and* Γ *is contained in a curve of* H *of degree* σ, *then* C *is contained in a surface of degree* σ.

3) *If* C *is integral of degree* $d>\sigma^2-\sigma+2$ *and* Γ *is contained in two curves of* H *of degree* σ, *then* C *is contained in a surface of degree* σ.

In [E] Ellia uses 3) in order to determine the maximum genus $G(d,s)$ of a smooth curve of degree d not lying on a surface of degree $<s$, in the range:

$$s^2-3s+5 \le d \le s^2-2s+1.$$

Before going on we notice that in [R] Re gives a generalization of Theorem 1 and of 1) to every variety X in $\mathbf{P}^r$. We observe that, from this result, it is easy to deduce the following generalization of Laudal's lemma in higher dimension.

4) *Let* X *be an integral, non degenerate variety of codimension* 2 *in* $\mathbf{P}^r$ *and let* $\Gamma = X \cap H$ *be the generic hyperplane section of* X. *If* $\deg(X) > \sigma(\sigma+1)$ *and* Γ *is contained in a hypersurface of* H *of degree* σ, *then* X *is contained in a hypersurface of degree* σ.

Coming back to our curve C in $\mathbf{P}^3$, it is natural to ask the following questions :

PROBLEM 1. Let C be smooth and assume $\binom{\sigma+2}{2} \le d \le \sigma^2+1$, $H^0(\mathcal{I}_\Gamma(\sigma)) \ne 0$, $H^0(\mathcal{I}_C(\sigma)) = 0$. What can we say about C ? Is there any lower bound for the genus g(C) ? In particular, can C be rational ?

PROBLEM 2. (Generalized Laudal's Lemma) Let C be integral. Determine a function $d(\sigma,t)$ such that, if $H^0(\mathcal{I}_\Gamma(\sigma)) \ne 0$ and $d > d(\sigma,t)$, then $H^0(\mathcal{I}_C(\sigma+t)) \ne 0$.

In the following sections we will give some partial results concerning the above problems.

2. In this section we give a general result we will use later. Let, as above, $R = k[x_1,x_2,x_3]$, $\mathbf{m} = (x_1,x_2,x_3)$, and define, for $i>0$

$$\omega_i = \mathrm{Ext}_R^3(R/\mathbf{m}^i, R)(-i-2).$$

LEMMA 1. ω_i *is a graded R-module of finite length. In every degree* j *the homogeneous component* $(\omega_i)_j$ *has dimension* $\binom{i-j+1}{2}$ *and the multiplication is as follows*

$$e_{i_1 i_2 \dots i_r} x_s = \begin{cases} 0 & \text{for } s \neq i_1, i_2, \dots, i_r \\ e_{i_1 i_2 \dots i_{t-1} i_{t+1} \dots i_r} & \text{for } s = i_t \end{cases}$$

where $\{e_{i_1 i_2 \dots i_r}\}$, $1 \leq i_1 \leq i_2 \leq \dots \leq i_r \leq 3$, *is a suitable basis of the component of* ω_i *in degree* i-r-1 .

Proof. From the exact sequence

$$0 \to m^{i-1}/m^i \to R/m^i \to R/m^{i-1} \to 0$$

we obtain the exact sequence

$$0 \to \omega_{i-1}(-1) \to \omega_i \to k^{i(i+1)/2} \to 0$$

and this gives us the structure of ω_i as k-vector space. As for the R-module structure it is enough to consider the map $(\omega_i)_0 \to (\omega_{i-1})_0$ induced by the multiplication by x_s. To this end remember (See [BE]) that R/m^i has a resolution of the form

$$0 \to \overset{3}{L_i} \otimes R(-i-2) \to \overset{2}{L_i} \otimes R(-i-1) \to \overset{1}{L_i} \otimes R(-i) \to R \to R/m^i \to 0$$

where $\overset{q}{L_p}$ is the image of the map

$$\partial : S_{p-1}(V) \otimes \overset{q}{\wedge} V \to S_p(V) \otimes \overset{q-1}{\wedge} V$$

and V is the vector space generated by x_1, x_2, x_3. In particular $\overset{3}{L_p}$ has a basis given by the images of the elements of the form $x_{i_1} x_{i_2} \dots x_{i_{p-1}} \otimes x_1 \wedge x_2 \wedge x_3$ and with respect to this basis the multiplication by $x_s : \overset{3}{L_{p-1}} \to \overset{3}{L_p}$ sends $\partial(x_{i_1} x_{i_2} \dots x_{i_{p-2}} \otimes x_1 \wedge x_2 \wedge x_3)$ to $\partial(x_s x_{i_1} x_{i_2} \dots x_{i_{p-2}} \otimes x_1 \wedge x_2 \wedge x_3)$. By dualizing the resolution of R/m^i we get the result. □

Now, for a general plane $H \subset \mathbf{P}^3$, consider the map

$$\varphi_{H^i}(t) : H^1(\mathbf{P}^3, \mathcal{I}_C(t-i)) \to H^1(\mathbf{P}^3, \mathcal{I}_C(t))$$

induced by the multiplication by H^i. We have the following theorem.

THEOREM 2. *With the above notations we have an inclusion of* k*-vector spaces*

$$\operatorname{Ker} \varphi_{H^i}(t) / \operatorname{Ker} \varphi_{H^i}(t) \cap \operatorname{Im} \varphi_H(t-i) \to \operatorname{Tor}_1^R(J, \omega_i)_{t+2}$$

Proof. Let $\alpha \in \operatorname{Ker}\varphi_{H^i}(t)$. As in [S] and [S1] we have $\overline{\alpha}\, x_{r_1}x_{r_2}\ldots x_{r_i}=0$ for every $1\leq r_1, r_2, \ldots, r_i \leq 3$, where $\overline{\alpha}$ is the restriction of α to H under the map $H^1(\mathcal{I}_C(t-i)) \to H^1(\mathcal{I}_\Gamma(t-i))$. This means that, if $\beta \in H^0(\mathcal{O}_\Gamma(t-i))$ induces $\overline{\alpha}$, there are polynomials $Q_{r_1r_2\ldots r_i} \in R_t$ such that $\beta x_{r_1}x_{r_2}\ldots x_{r_i} = Q_{r_1r_2\ldots r_i}$ in $H^0(\mathcal{O}_\Gamma(t))$. The polynomials $Q_{r_1r_2\ldots r_i}$ satisfy the following conditions

$$Q_{r_1\, r_2\ldots r_i}\, x_{r_j} \equiv Q_{r_1\, r_2\ldots r_{i-1}\, r_j}\, x_{r_i} \quad (\mathrm{mod}\ J_{t+1})$$

for every $1\leq r_1, r_2, \ldots, r_i, r_j \leq 3$.

Now we consider the above resolution of R/m^i and dualize it; we obtain the following resolution of ω_i:

$$0 \to R(-2-i) \to (\overset{1}{L_i})^*\otimes R(-2) \to (\overset{2}{L_i})^*\otimes R(-1) \to (\overset{3}{L_i})^*\otimes R \to \omega_i \to 0$$

If we tensor this by $\otimes_R R/J$, we see that the polynomials $Q_{r_1r_2\cdots r_i}$ define a cocycle in $(\overset{1}{L_i})^*\otimes(R/J)(-2)$ of degree t+2, i.e. an element of $\operatorname{Tor}_2^R(R/J,\omega_i)_{t+2} = \operatorname{Tor}_1^R(J,\omega_i)_{t+2}$.

It is easy to see that this map is well defined: in fact if β' also induces $\overline{\alpha}$, the associated cocycle $Q'_{r_1\, r_2\ldots r_i}$ satisfies $Q_{r_1\, r_2\ldots r_i} - Q'_{r_1\, r_2\ldots r_i} \equiv Q\, x_{r_1}x_{\rho_2}\ldots x_{r_i}$ $(\mathrm{mod} J_t)$ with $Q\in R_{t-i}$, hence defines the same element in $\operatorname{Tor}_2^R(R/J,\omega_i)_{t+2}$. Moreover if the cocycle defined by $Q_{r_1\, r_2\ldots r_i}$ is a coboundary, then $Q_{r_1\, r_2\ldots r_i} \equiv Q x_{r_1}x_{r_2}\ldots x_{r_i}$ (mod J_t), hence $\beta\, x_{r_1}x_{r_2}\ldots x_{r_i} = Q x_{r_1}x_{r_2}\ldots x_{r_i}$ in $H^0(\mathcal{O}_\Gamma(t))$ for every $1\leq r_1, r_2, \ldots, r_i \leq 3$, and then $\beta = Q$ in $H^0(\mathcal{O}_\Gamma(t-i))$, i.e. $\overline{\alpha}=0$. □

3. From now on we suppose that C is integral. We know that $\Gamma = C\cap H$ is a set of d points with the Uniform Position Property (see [H]). Let J, as before, be the homogeneous ideal of Γ in R. In this section we study three particular cases for a minimal resolution of J. More precisely, let

$$0 \to \sum_{i=1}^{m-1} R(-b_i) \to \sum_{i=1}^{m} R(-a_i) \to J \to 0$$

be a minimal resolution of J and let α_j (resp.β_j) be the number of a_i (resp.b_i) equal to j. Let $\sigma = \min\{a_i \mid a_i \neq 0\}$.

Case i) $\alpha_\sigma = 1$, $\alpha_{\sigma+1} = 2$, $\beta_{\sigma+2} = 1$, $\beta_{\sigma+3} = 0$.

Case ii) $\alpha_\sigma = 2$, $\alpha_{\sigma+1} = 1$, $\beta_{\sigma+2} = 1$, $\beta_{\sigma+3} = 0$.

Case iii) $\alpha_\sigma = 3$, $\beta_{\sigma+1} = 1$, $\beta_{\sigma+2} = 0$.

We conclude in all cases that C is contained in a surface of degree $\sigma+1$.

Case i). In this case we prove that $h^0(\mathcal{I}_C(\sigma+1)) \geq 3$.

Consider the exact sequence

$$0 \to H^0(\mathcal{I}_C(\sigma)) \to H^0(\mathcal{I}_C(\sigma+1)) \to H^0(\mathcal{I}_\Gamma(\sigma+1)) \to H^1(\mathcal{I}_C(\sigma)) \to H^1(\mathcal{I}_C(\sigma+1))$$

and suppose $H^0(\mathcal{J}_C(\sigma))=0$. Since $h^0(\mathcal{J}_\Gamma(\sigma+1))=5$ it is enough to show that $\dim \operatorname{Ker} \varphi_H(\sigma+1) \leq 2$. In fact suppose that $\dim \operatorname{Ker} \varphi_H(\sigma+1) \geq 3$ and let $\gamma_1, \gamma_2, \gamma_3$ be linearly independent elements in $\operatorname{Ker} \varphi_H(\sigma+1)$. Tensoring the resolution of J by $\omega_h(\sigma+3)$, we have, by the hypotheses on J, a map

$$f_h: \omega_h(1) \longrightarrow \omega_h(2)^2 \oplus \omega_h(3)$$

and we have $\operatorname{Tor}_1^R(J,\omega_h)_{\sigma+3} = K_h = (\operatorname{Ker} f_h)_0$.

First we prove that we may assume $f_h = (x_1, x_2, x_3^2)$: the only non trivial thing is to show that in the syzygy $L_1F_1+L_2F_2+AG$ in degree $\sigma+2$, with $G \in H^0(\mathcal{J}_\Gamma(\sigma))$ and $F_1, F_2 \in H^0(\mathcal{J}_\Gamma(\sigma+1))$, we have $A \neq 0$. In fact if $A=0$ then $F_1=L_2F$ and $F_2=-L_1F$ where F is a form of degree σ passing through d-1 points of Γ: if $d \geq \binom{\sigma+2}{2}$ then by the UP Property we have $F \in H^0(\mathcal{J}_\Gamma(\sigma))$ and this is impossible; if $d = \binom{\sigma+2}{2}-1$ it follows that $\alpha_{\sigma+1} = \beta_{\sigma+2} = \sigma$, which is not the case.

Now we examine K_h for $h=1, 2, \ldots$.

If $h=1$, then $K_1=0$ which implies, by Theorem 2, that $\gamma_i = \delta_i H$ $(i=1,2,3)$, $\delta_i \in \operatorname{Ker} \varphi_{H^2}(\sigma+1)$.
If $h=2$, then $\dim K_2 = 1$ which implies that $\gamma_i = \zeta_i H^2$ $(i=1,2)$, with $\zeta_i \in \operatorname{Ker} \varphi_{H^3}(\sigma+1)$.
If $h=3$,then $\dim K_3=1$ which implies that $\gamma_1 = \eta H^3$, $\eta \in \operatorname{Ker} \varphi_{H^4}(\sigma+1)$.
If $h=4$, then $K_4=0$: in fact the map $f_4: k^6 \longrightarrow k^3 \oplus k^3 \oplus k$ is given, in the basis of Lemma 1, by $f_4(\ldots, c_{ij}, \ldots) = (c_{11}, c_{12}, c_{13}; c_{21}, c_{22}, c_{23}; c_{33})$.

In the same way we prove that $K_h=0$ for $h \geq 4$; hence, for every t there exists λ_t, $\lambda_t \in H^1(\mathcal{J}_C(\sigma-t))$ s.t. $\gamma_1 = \lambda_t H^t$. But this is absurd since $H^1(\mathcal{J}_C(n))=0$ for $n \ll 0$.

Case ii). In this case we prove that $h^0(\mathcal{J}_C(\sigma+1)) \geq 3$. Since $h^0(\mathcal{J}_\Gamma(\sigma+1))=7$ let assume that $\dim \operatorname{Ker} \varphi_H(\sigma+1) \geq 5$. As before we have a map

$$f_h: \omega_h(1) \longrightarrow \omega_h(2) \oplus \omega_h(3)^2.$$

Since $d \geq \binom{\sigma+2}{2}-2$ it is easy to see, by the UP Property of Γ, that every $F \in H^0(\mathcal{J}_\Gamma(\sigma))$ is irreducible for $\sigma>2$ (for $\sigma \leq 2$ this case does not occur). From this it follows that we can assume $f_h=(x_1, x_2x_3, x_2^2+ax_3^2)$ with $a \neq 0$. Hence we have: $K_1=0$, $\dim K_2=1$, $\dim K_3=2$.

If $h=4$, then $\dim K_4 = 1$: in fact the map $f_4: k^6 \longrightarrow k^3 \oplus k \oplus k$ is given, in the basis of Lemma 1, by $f_4(\ldots, c_{ij}, \ldots) = (c_{11}, c_{12}, c_{13}; c_{23}; c_{22}+ac_{33})$.
If $h=5$ then $K_5=0$: in fact the map $f_5: k^{10} \longrightarrow k^6 \oplus k^3 \oplus k^3$ is given, in the basis of Lemma 1, by $f_5(\ldots, c_{ijk}, \ldots) = (c_{111}, c_{112}, c_{113}, c_{122}, c_{123}, c_{133}; c_{123}, c_{223}, c_{233}; c_{122}+ac_{133}, c_{222}+ac_{233}, c_{223}+ac_{333})$.
In the same way we prove that $K_h=0$ for $h \geq 5$, and then, as before, we have the absurd.

Case iii). In this case we prove that $h^0(\mathcal{I}_C(\sigma)) \neq 0$; more precisely we have $h^0(\mathcal{I}_C(\sigma)) \geq 2$. In fact, tensoring the resolution of J by $\omega_h(\sigma+2)$, we have a map

$$f_h: \omega_h(1) \longrightarrow \omega_h(2)^3$$

which we can assume to be given by (x_1, x_2, x_3). We see, as before, that $K_1 = 0$, $\dim K_2 = 1$, $K_h = 0$ for $h > 2$.

4. In this section we prove the main theorems of the paper.

THEOREM 3. *Let* C *be an integral curve in* $\mathbf{P}^3$ *of degree* d *and let* $\Gamma = C \cap H$ *be the generic plane section. Assume* $H^0(\mathcal{I}_\Gamma(\sigma)) \neq 0$ *and* $d > \sigma^2 - \sigma + 4$. *Then* $h^0(\mathcal{I}_C(\sigma+1)) \geq 3$.

Proof. Assume $h^0(\mathcal{I}_C(\sigma)) = 0$ and let consider the Hilbert function HF of Γ. Let $\Delta HF(n) = HF(n) - HF(n-1)$. We know that, since C is integral, ΔHF is of decreasing type (see [H]). Moreover, by corollary 3) of Theorem 1, we have that $h^0(\mathcal{I}_\Gamma(\sigma)) = 1$. Hence ΔHF has the form

$$\Delta HF = \{ 1, 2, ..., \sigma, \sigma, h_{\sigma+1}, h_{\sigma+2}, ... \}.$$

We have various possibilities. We use repeatedly a theorem by Maggioni-Ragusa (see [MR1], Theorem 1.5) which gives bounds for the numbers α_j and β_j defined above.

1) $h_{\sigma+1} = \sigma$. In this case we have $\beta_j = 0$ for $j \leq \sigma+2$; hence, by Theorem 1, we have $h^0(\mathcal{I}_C(\sigma)) \neq 0$.

2) $h_{\sigma+1} = \sigma-1$. Same as in 1).

3) $h_{\sigma+1} = \sigma-2$. In this case we consider $h_{\sigma+2}$; by the assumption on the degree d it follows that $h_{\sigma+2} = \sigma-3$ and $h_{\sigma+3} \geq \sigma-5$. Hence ΔHF has one of the following two forms:

$$\{1, 2, ..., \sigma, \sigma, \sigma\text{-}2, \sigma\text{-}3, \sigma\text{-}4, h_{\sigma+4}, ...\} \quad \text{or} \quad \{1, 2, ..., \sigma, \sigma, \sigma\text{-}2, \sigma\text{-}3, \sigma\text{-}5, h_{\sigma+4}, ...\}$$

and in both situations we are in case i) of Section 3.

4) $h_{\sigma+1} < \sigma-2$. This case is impossible by the assumption on the degree d. $\quad\boxed{.}$

THEOREM 4. *Let* C *be an integral curve in* $\mathbf{P}^3$ *of degree* d *and let* $\Gamma = C \cap H$ *be the generic plane section. Assume* $h^0(\mathcal{I}_\Gamma(\sigma)) > 1$ *and* $d > \sigma^2 - 2\sigma + 6$. *Then* $h^0(\mathcal{I}_C(\sigma+1)) \geq 3$.

Proof. We assume $H^0(\mathcal{I}_C(\sigma)) = 0$, and let

$$\Delta HF = \{ 1, 2, ..., \sigma, h_\sigma, h_{\sigma+1}, h_{\sigma+2}, ... \}.$$

By the assumption on the degree d, we can have $h_\sigma = \sigma-1$ or $h_\sigma = \sigma-2$.

1) $h_\sigma = \sigma-1$. Again by the assumption on the degree d we have $h_{\sigma+1} \geq \sigma-3$. If $h_{\sigma+1} = \sigma-2$, then we have $H^0(\mathcal{I}_C(\sigma)) \neq 0$. It remains the case $h_{\sigma+1} = \sigma-3$; it is necessarily $h_{\sigma+2} = \sigma-4$, hence ΔHF has the form

$$\Delta HF = \{1, 2, ..., \sigma, \sigma\text{-}1, \sigma\text{-}3, \sigma\text{-}4, h_{\sigma+3}, ...\}$$

and we are in case ii) of Section 3.
2) $h_\sigma=\sigma-2$. We must have $h_{\sigma+1}=\sigma-3$, and we are in case iii) of Section 3. □

REMARK 1. In a forthcoming paper, in collaboration with Ph.Ellia, we will use Theorems 3 and 4 in order to determine the maximum genus G(d, s) of a smooth curve of degree d not lying on a surface of degree <s, in the range $s^2-5s+11\leq d\leq s^2-3s+4$.The value of G(d,s) confirms a conjecture of Hartshorne-Hirschowitz (see [HH] Conjecture 5.6).

5.In this section we will apply Theorem 3 in order to describe the integral curves in the range $\sigma^2-\sigma+5\leq d\leq\sigma^2+1$, such that $H^0(\mathcal{I}_\Gamma(\sigma))\neq 0$ and $H^0(\mathcal{I}_C(\sigma))=0$.
First we prove a result which has been stated without proof by Gruson-Peskine (see [GP], Remarque).

PROPOSITION 1. *Let* C *be an integral curve of degree* $d=\sigma^2+1$, $\sigma>3$, *such that* $H^0(\mathcal{I}_\Gamma(\sigma))\neq 0$ *and* $H^0(\mathcal{I}_C(\sigma))=0$. *Then* C *is the zero set of a section of* E(σ), *where* E *is a null correlation bundle.*

Proof. By Theorem 3 we have that ΔHF of Γ has the form

$$\Delta HF=\{1, 2, ..., \sigma, \sigma, \sigma-2, \sigma-3, \sigma-4, ..., 1\}$$

and C lies on two irreducible surfaces S, S' of degree σ+1. Let X be the curve linked to C by means of these two surfaces. X has degree 2σ and, by liaison, the ΔHF of the generic plane section of X is

$$\Delta HF=\{1, 2, 2, ..., 2, 1\}.$$

From this, by Theorem 1, it follows that X lies on a quadric Q. Let Y be the curve linked to X by means of Q and S. Y is a curve of degree 2.
Now we prove that Y has genus -1. If Y had genus 0, then $H^0(\mathcal{I}_Y(1))\neq 0$, which would imply $H^0(\mathcal{I}_C(\sigma))\neq 0$. Assume that the genus of Y is $-\alpha<-1$. In this case we prove that all surfaces of degree σ+1 passing through X will contain the line L=SuppY, hence C is not integral.
In fact from the exact sequence

$$0\rightarrow\mathcal{I}_U\rightarrow\mathcal{I}_X\rightarrow\omega_Y(1-\sigma)\rightarrow 0$$

where $U=Q\cap S$, we obtain

$$0\rightarrow\mathcal{I}_U(\sigma+1)\rightarrow\mathcal{I}_X(\sigma+1)\rightarrow\omega_Y(2)\rightarrow 0.$$

But, using the restriction $\omega_Y(2)\rightarrow\omega_L(3-\alpha)$, we see that every global section of $\omega_Y(2)$ vanishes on L.

We obtain that the genus of C is $\sigma^3-2\sigma^2+\sigma-1$; since $H^0(\mathcal{I}_X(2))\neq 0$, we get $H^0(\omega_C(4-2\sigma))\neq 0$, hence $\omega_C(4-2\sigma)\cong\mathcal{O}_C$ because $\deg(\omega_C(4-2\sigma))=0$.
From this the result follows easily, using the fact that every stable rank 2 bundle with $c_1=0$ and $c_2=1$ is a null correlation bundle. □

Before going on, we prove a lemma which will be useful in the sequel.

Let C be a curve and let $\Gamma = C \cap H$ be the generic plane section. Suppose that the Hilbert function of Γ is not of decreasing type. Then we know ([MR], Theorem 2.9) that all the curves of H of a certain degree n passing through Γ contain a fixed curve F; let $\Gamma' = F \cap \Gamma$ and let Γ'' be the residual subscheme which is defined in $\mathcal{O}_H$ by the sheaf of ideals $\mathcal{I}_\Gamma : \widetilde{F}$, where $\widetilde{F}$ is a local equation of F.

LEMMA 2. *In the above situation there exists a curve* $C' \subset C$ *whose generic plane section* $C' \cap H$ *is* Γ'.

Proof. Let us fix two generic planes H and H_1 and let $H_\lambda = H + \lambda H_1$; if $U \subset \mathbf{A}^1 = \mathrm{Spec}\,k[t]$ denotes the open set that parametrizes the planes H_λ which are generic with respect to C, for $\lambda \in U$ we define $\Gamma_\lambda, F_\lambda, \Gamma'_\lambda, \Gamma''_\lambda$ as above, with H_λ instead of H.

Let $\overline{k}$ be an algebraic closure of k(t); first we observe that the scheme $\Gamma_t = \Gamma \cap H_t$, $H_t = H + tH_1$, is rational over k(t), and hence the linear system of the curves of H_t of degree n passing through Γ_t has a basis consisting of curves which are rational over k(t). Let F_t be the GCD of the elements of $H^0(H_t, \mathcal{I}_{\Gamma_t}(n))$. It follows from this that also the curve $F_t \subset H_t$ is defined over k(t) and its homogeneous ideal in $\overline{k}[x_0, x_1, x_2, x_3]$ has the form $(H+tH_1, F(t))$, with $F(t) \in k[t, x_0, x_1, x_2, x_3]$. Therefore all the subschemes $\Gamma_t, F_t, \Gamma'_t, \Gamma''_t$, are rational over k(t).

Let us consider $\mathbf{P}^3 \times U \to U$ and the closed subschemes $\mathbf{\Gamma}, \mathbf{F}, \mathbf{\Gamma}', \mathbf{\Gamma}''$, whose fibres over k(t) are $\Gamma_t, F_t, \Gamma'_t, \Gamma''_t$ respectively; we note that the fibres, for $\lambda \in U$, are $\Gamma_\lambda, F_\lambda, \Gamma'_\lambda, \Gamma''_\lambda$ respectively. Let S, C' be the closures of the images of $\mathbf{F}, \mathbf{\Gamma}'$ in $\mathbf{P}^3$. S is a surface and $S \cap H_\lambda \subset F_\lambda \cup L$, where $L = H \cap H_1$, for every $\lambda \in U$. $C' \subset S \cap C$ and since we can assume $L \cap C = \emptyset$, we see that $C' \cap H_\lambda = \Gamma'_\lambda$ for every $\lambda \in U$. Hence C' is the smallest subscheme of $\mathbf{P}^3$ containing all the Γ_λ's, for $\lambda \in U$. From this it follows that C' is a curve, i.e. C' has no 0-dimensional components, isolated or embedded.

Now we have to show that the section of C' with a generic plane $\overline{H}$ of $\mathbf{P}^3$ is $\overline{\Gamma}'$. Let $H_\mu = H + \mu\overline{H}$ and let $\overline{U} \subset \mathbf{A}^1$ be defined as above. Assume that $C' \cap H_\mu$ is not contained in Γ_μ for generic $\mu \in \overline{U}$. Then in H_μ the subscheme defined by the ideal $\mathcal{I}_{C' \cap H_\mu} : \widetilde{F}_\mu$ is not empty for generic $\mu \in \overline{U}$. Since $\mathbf{P}^3 \times \overline{U} \to \overline{U}$ is a closed map, it follows that the set of μ's for which subscheme defined by the ideal $\mathcal{I}_{C' \cap H_\mu} : \widetilde{F}_\mu$ is not empty, is closed in $\overline{U}$; hence it coincides with $\overline{U}$. But this is absurd since for $\mu = 0$, $C' \cap H = \Gamma'$. □

REMARK 2. In the proof of lemma 2 we did not use the fact that the Hilbert function of Γ is not of decreasing type, but only the hypothesis that all the elements in $H^0(H, \mathcal{I}_\Gamma(n))$, for some n, have a GCD of positive degree. Hence lemma 2 holds under this weaker hypothesis.

PROPOSITION 2. *Let* C *be an integral curve of degree* $d = \sigma^2 + 1 - r$, $\sigma > r+3$, $r > 0$, *such that* $H^0(\mathcal{I}_\Gamma(\sigma)) \neq 0$ *and* $H^0(\mathcal{I}_C(\sigma)) = 0$. *Then* C *is linked, in a complete intersection* $S \cap S'$ *of two surfaces of degree* $\sigma+1$, *to a curve* D *of degree* $2\sigma + r$ *of the following kind:* D *contains a curve* X *of degree* 2σ *which lies on a quadric* Q *and the curve linked to* X *in the complete intersection*

$S\cap Q$ is a curve of degree 2 and genus -1; moreover $X=D\cap Q$. If Z is the residual curve of D with respect to Q we have an exact sequence

$$0 \to \mathcal{I}_D \to \mathcal{I}_X \to \mathcal{O}_Z(-2) \to 0.$$

Proof. As in the Proposition 2, C lies on two irreducible surfaces S, S' of degree $\sigma+1$, and ΔHF of Γ has the form

$$\Delta HF=\{1, 2, ..., \sigma, \sigma, \sigma-2, \sigma-3, ...\}.$$

Let D be the curve linked to C by means of these two surfaces; if we consider the Hilbert function of the generic plane section E of D, it has the form

$$\Delta HF=\{1, 2, 3, ..., 2, 2, 1\}.$$

Then there is a conic $F\subset H$ such that $E'=E\cap F$ has degree 2σ and the residual scheme E" has degree r. It follows from Lemma 2 that D contains a curve X whose generic plane section E' has the Hilbert function of the form

$$\Delta HF=\{1, 2, 2, ..., 2, 2, 1\}.$$

By Theorem 1, X lies on a quadric Q and $H\cap Q=F$. Let Y be the curve linked to X by means of Q and S; Y is a curve of degree 2. As in Proposition 1 we prove that Y has genus -1. In fact if the genus of Y is 0, we have $H^0(\mathcal{I}_Y(1))\neq 0$ and by liaison $H^2(\mathcal{I}_X(\sigma-2))\neq 0$. From the exact sequence

$$0 \to \mathcal{I}_D \to \mathcal{I}_X \to \mathcal{I}_X/\mathcal{I}_D \to 0$$

we have a surjection $H^2(\mathcal{I}_D(\sigma-2)) \to H^2(\mathcal{I}_X(\sigma-2))$ and hence, again by liaison, $H^0(\mathcal{I}_C(\sigma))\neq 0$. By the same argument as in Proposition 1, we see that the genus of Y is ≥ -1.

Now we prove that $X=D\cap Q$. In fact assume that $X\neq D\cap Q$; then every surface S' of degree $\sigma+1$ containing D is such that $S'\cap Q$ will meet Y outside of X.

From the exact sequence

$$0 \to \mathcal{I}_U \to \mathcal{I}_X \to \omega_Y(1-\sigma) \to 0$$

where $U=Q\cap S$, we obtain

$$0 \to \mathcal{I}_U(\sigma+1) \to \mathcal{I}_X(\sigma+1) \to \omega_Y(2) \to 0$$

and, since $H^0(\omega_Y(2))\cong H^0(\mathcal{O}_Y)$, we see that every section in $H^0(\omega_Y(2))$ which vanishes on a point of Y will vanish on a component L of Y. Hence every surface S' of degree $\sigma+1$ containing D contains the line L, and this is absurd since C is integral.

Now let Z be the residual scheme of D with respect to Q, i.e. the scheme defined in $\mathbf{P}^3$ by the sheaf of ideals $\mathcal{I}_D:\tilde{Q}$, where $\tilde{Q}$ is a local equation of Q. It is easy to see that Z has no 0-dimensional components, isolated or embedded, i.e. $Z\subset D$ is a curve.

Moreover the generic plane section $Z\cap H$ is E": in fact in $\mathcal{O}_\mathbf{P}=\mathcal{O}_{\mathbf{P}^3}$ we have

$$(\mathcal{I}_D:\tilde{Q})+\tilde{H}\mathcal{O}_\mathbf{P}=(\mathcal{I}_D+\tilde{H}\mathcal{O}_\mathbf{P}):\tilde{Q}.$$

In order to see this, we restrict to an affine open in $\mathbf{P}^3$ and let $a\in\mathcal{O}_\mathbf{P}$ such that $a\tilde{Q}\in\mathcal{I}_D+\tilde{H}\mathcal{O}_\mathbf{P}$; then $a\tilde{Q}=b\tilde{H} \mod \mathcal{I}_D$, hence $b\tilde{H}=0 \mod \mathcal{I}_X$; but H is generic, so $\tilde{H}$ is not a zero divisor in $\mathcal{O}_X$, hence $b=c\tilde{Q} \mod \mathcal{I}_D$, and consequently $a\in\mathcal{I}_D:\tilde{Q}+\tilde{H}\mathcal{O}_\mathbf{P}$.

Finally consider the exact sequence

$$0 \to \mathcal{I}_D \to \mathcal{I}_X \to \mathcal{I}_X/\mathcal{I}_D \to 0.$$

Since $\mathcal{I}_X/\mathcal{I}_D \cong \mathcal{I}_X\mathcal{O}_D \cong \widetilde{Q}\mathcal{O}_D$, from the exact sequence

$$0 \to \mathcal{I}_Z(-2) \to \mathcal{O}_P(-2) \to \widetilde{Q}\mathcal{O}_D \to 0$$

it follows that $\mathcal{I}_X/\mathcal{I}_D \cong \mathcal{O}_Z(-2)$. □

REMARK 3. We note that, when Z is a plane curve (in particular when $r=1$), the curves described in Proposition 2 are exactly the curves of maximum genus $G(d,\sigma+1)$ (see [E] Theoreme). For $r>1$, in the same way as in [E] (Lemme III.2) it is possible to give examples of smooth curves C with $H^0(\mathcal{I}_\Gamma(\sigma))\neq 0$, $H^0(\mathcal{I}_C(\sigma))=0$ and genus $<G(d,\sigma+1)$. In any event the genus of C is bounded from below: in fact it is easy to see that $e(C)\geq\sigma$, where $e(C)=\max\{n\,|\,H^1(C,\mathcal{O}(n))\neq 0\}$.

REFERENCES

[BE] D.A.Buchsbaum-D.Eisenbud, *Generic free resolutions and a family of Generically Perfect Ideals*. Adv. in Math. **18** (1975), 245-301.

[E] P.Ellia, *Sur le genre maximal des courbes gauches de degree* d *non sur une surface de degree* s-1. Crelle J. (to appear)

[GP] L.Gruson-C.Peskine, *Section plane d'une courbe gauche: postulation.* Progress in Math. n. 24, Birkhauser, Boston, 1982, 33-35.

[H] J.Harris, *The genus of space curves*. Math. Ann. **249** (1980), 191-204.

[HH] R.Hartshorne-A.Hirschowitz, *Nouvelles courbes de bon genre dans l'espace projectif.* Math. Ann. **280** (1988), 353-367.

[L] O.A.Laudal, *A generalized trisecant lemma.* Algebraic Geometry, L.N.M.n.687, Springer-Verlag, Berlin, 1978, 112-149.

[MR] R.Maggioni-A.Ragusa, *Connections between Hilbert function and geometric properties for a finite set of points in* $\mathbf{P}^2$. Le Matematiche **39** (1984), 153-170.

[MR1] R.Maggioni-A.Ragusa, *Construction of smooth curves of* $\mathbf{P}^3$ *with assigned Hilbert function and generators' degrees.* Le Matematiche **42** (1987), 195-210.

[R] R.Re, *Sulle sezioni iperpiane di una varietà proiettiva.* Le Matematiche **42** (1987), 211-218.

[S] R.Strano, *Sulle sezioni iperpiane delle curve.* Rend.Sem.Mat.e Fis.Milano **57** (1987), 125-134.

[S1] R.Strano, *A characterization of complete intersection curves in* $\mathbf{P}^3$. Proc.A.M.S. **104** (1988), 711-715.

Author's Address: Dipartimento di Matematica Università di Catania
Viale A.Doria, 6. 95125 Catania Italia

SUR LA STABILITÉ DES SOUS-VARIÉTÉS LAGRANGIENNES DES VARIÉTÉS SYMPLECTIQUES HOLOMORPHES

Claire VOISIN (URA D0752)

0. Introduction. — On considère dans cet article des variétés kählériennes symplectiques irréductibles (cf. [2]) ; une telle variété Y est donc une variété analytique compacte kählérienne de dimension complexe $2n$ munie d'une deux-forme holomorphe ω unique à coefficient près et partout non dégénérée. Les sous-variétés lagrangiennes X de Y sont les sous-variétés analytiques connexes de Y de dimension n satisfaisant : $\omega_{|X} = 0$. On se propose de montrer le théorème 0.1 suivant concernant les déformations de Y "préservant X". Soit $\mathcal{M}$ un ouvert simplement connexe de la famille complète des déformations de Y ; on note $\mathcal{M}_X$ la sous-variété analytique de $\mathcal{M}$ définie par la condition : $t \in \mathcal{M}_X \Leftrightarrow$ il existe une déformation X_t de X contenue dans Y_t. Soit j l'inclusion $X \hookrightarrow Y$; comme $\mathcal{M}$ est simplement connexe, on a un isomorphisme naturel $\alpha_t : H^2(Y_t, \mathbb{C}) \simeq H^2(Y, \mathbb{C})$ pour $t \in \mathcal{M}$, et on note j_t^* le composé $j^* \circ \alpha_t$; soit enfin ω_t la deux-forme holomorphe de Y_t (définie à un coefficient près) ; ω_t appartient naturellement à $H^2(Y_t, \mathbb{C})$, et bien sûr, si $O \in \mathcal{M}$ est tel que $Y_0 = Y$, on a $j_0^*(\omega_0) = 0$. Le résultat est alors le suivant.

0.1 Théorème. — $\mathcal{M}_X = \{t \in \mathcal{M} \mid j_t^*(\omega_t) = 0, \ \textit{dans} \ H^2(X, \mathbb{C})\}$.

Notant enfin que $\operatorname{Ker} j^* \subset H^2(Y, \mathbb{C})$ est défini sur $\mathbb{Z}$, et est une sous-structure de Hodge de $H^2(Y, \mathbb{C})$, contenant $H^{2,0}(Y)$ on voit facilement que l'orthogonal de $\operatorname{Ker} j^*$ relativement à la forme d'intersection naturelle sur $H^2(Y, \mathbb{C})$ (cf. [2]) est défini sur $\mathbb{Z}$ et de type (1,1), de sorte qu'il a même rang que son intersection L_X avec $H^2(Y, \mathbb{Q})$ et que L_X est contenu dans $NS(Y) \otimes \mathbb{Q}$ (où $NS(Y)$ dénote le groupe de Néron-Séveri de Y) ; le corollaire suivant est une conséquence facile du théorème.

0.2 Corollaire. — *$\mathcal{M}_X$ est la famille des déformations de Y préservant le sous-groupe $L_X \subset NS(Y)$.*

En d'autres termes l'existence de sous-variétés lagrangiennes de Y ne dépend, au moins localement, que de la structure du groupe de Picard de Y. On donne en § 3 quelques exemples illustrant ce phénomène. Les autres paragraphes sont organisés de la façon suivante : en § I, on étudie la variété $\mathcal{M}_{[X]}$ constituée des déformations de Y telles que la classe de X (dans $H^{2n}(Y_t, \mathbb{Z})$) reste de type (n, n) relativement à la décomposition de Hodge de Y_t ; $\mathcal{M}_{[X]}$ contient évidemment $\mathcal{M}_X$, et on montre aisément que $\mathcal{M}_{[X]}$ satisfait l'égalité du théorème 0.1. Au paragraphe 2, on montre que $\mathcal{M}_X = \mathcal{M}_{[X]}$, c'est-à-dire que si $[X]$ reste de type (n, n) dans $H^{2n}(Y_t, \mathbb{C})$, X se déforme effectivement en $X_t \subset Y_t$. L'égalité $\mathcal{M}_X = \mathcal{M}_{[X]}$ n'est nullement évidente dans ce cas car X ne satisfait pas en général la condition de semi-régularité de Bloch (cf. [5]), comme le montrent certains des exemples du paragraphe 3.

§ 1 1.1. — Soient Y une variété kählerienne symplectique, et X une sous-variété lagrangienne. La dimension complexe de Y est paire égale à $2n$, et la classe de cohomologie de X est un élément $[X]$ de $H^{2n}(Y, \mathbb{Z})$, qui est de type (n, n) dans la décomposition de Hodge : $H^{2n}(Y, \mathbb{C}) = \oplus_{p+q=2n} H^{p,q}(Y)$. Comme X est lagrangienne on a, avec les notations du paragraphe 0, $j^*\omega = 0$, et donc, notant $\mu_{[X]}$ le cup-produit par la classe $[X] : H^k(Y) \to H^{k+2n}(Y)$, $\mu_{[X]}(\omega) = 0$ dans $H^{2n+2}(Y, \mathbb{C})$. Soit $\mathcal{M}$ un ouvert connexe et simplement connexe de la famille universelle des déformations de Y, tel qu'il existe une section holomorphe $(\omega_t)_{t \in \mathcal{M}}$ du sous-fibré holomorphe $\mathcal{H}^{2,0} \subset \mathcal{H}^2$ prolongeant $\omega = \omega_0$, où $\mathcal{H}^2$ est le faisceau holomorphe plat sur $\mathcal{M}$, de fibre en t $H^2(Y_t, \mathbb{C})$. Notant $H^{2n}_{\mathbb{Z}}$ le faisceau de $\mathbb{Z}$-modules trivial sur $\mathcal{M}$, de fibre en t naturellement isomorphe à $H^{2n}(Y_t,\mathbb{Z})$, et $\mathcal{H}^{2n+2}$ le fibré vectoriel holomorphe plat, de fibre $H^{2n+2}(Y_t,\mathbb{C})$, on peut étendre $[X]$ en une section de $H^{2n}_{\mathbb{Z}}$ localement constante sur $\mathcal{M}$, qu'on notera de la même façon, et l'on obtient alors une section holomorphe $t \mapsto \mu_{[X]}(\omega_t)$ du fibré $\mathcal{H}^{2n+2}$. Définissant de même $\mathcal{H}^{2n}$ et notant $F^n\mathcal{H}^{2n} \subset \mathcal{H}^{2n}$ le sous-fibré vectoriel de fibre en t le sous-espace $\bigoplus_{\substack{p\geq n\\ p+q=2n}} H^{p,q}(Y_t) \subset H^{2n}(Y_t,\mathbb{C})$, la projection de $[X]$ dans le quotient $\mathcal{H}^{2n}/F^n\mathcal{H}^{2n}$ est holomorphe et s'annule en $t \in \mathcal{M}$ si et seulement si $[X]$ est de type (n,n) dans $H^{2n}(Y_t,\mathbb{C})$. Soit $\mathcal{M}^0_{[X]}$ la composante connexe passant par 0 (où $Y_0 = Y$) de la sous-variété de $\mathcal{M}$ définie par l'annulation de cette projection. De même, notons $\mathcal{M}'^0_{[X]}$ la composante connexe passant par 0 de la variété $\mathcal{M}'_{[X]} := \{t \in \mathcal{M}/\mu_{[X]}(\omega_t) = 0, \text{dans} H^{2n+2}(Y_t,\mathbb{C})\}$; on a alors la proposition suivante :

1.2 PROPOSITION. — *Les variétés $\mathcal{M}^0_{[X]}$ et $\mathcal{M}'^0_{[X]}$ sont égales, lisses, de codimension dans $\mathcal{M}$ égale au rang $r_{[X]}$ de l'application $\mu_{[X]} : H^2(Y,\mathbb{C}) \to H^{2n+2}(Y,\mathbb{C})$.*

Démonstration : Montrons d'abord l'égalité ensembliste $\mathcal{M}^0_{[X]} = \mathcal{M}'^0_{[X]}$. Soit $t \in \mathcal{M}'_{[X]}$; dans $H^{2n}(Y_t,\mathbb{C})$ la classe $[X]$ se décompose en composantes de type (p,q), soit $[X] = \sum_{p+q=2n} [X]^{p,q}_t$. Comme X est réelle, on a $[X]^{p,q}_t = \overline{[X]^{q,p}_t}$, et pour montrer que $t \in \mathcal{M}_{[X]}$, il suffit de montrer que $[X]^{p,q}_t = 0$ pour $p < n$. Mais par hypothèse, on a : $\mu_{[X]}(\omega_t) = 0$ dans $H^{2n+2}(Y_t,\mathbb{C})$, ce qui entraîne, puisque ω_t est de type $(2,0)$, que $[X]^{p,q}_t.\omega_t = 0$, dans $H^q(\Omega^{p+2}_{Y_t})$ pour tout couple (p,q). Or ω_t est une deux forme holomorphe non dégénérée sur Y_t, et donc ω_t^k fournit un isomorphisme : $\Omega^{n-k}_{Y_t} \simeq \Omega^{n+k}_{Y_t}$, pour tout k. On en déduit que le cup-produit par ω_t^k fournit un isomorphisme : $H^q(\Omega^{n-k}_{Y_t}) \simeq H^q(\Omega^{n+k}_{Y_t})$ pour tout couple (k,q), puis que le cup-produit par $\omega_t : H^q(\Omega^p_{Y_t}) \to H^q(\Omega^{p+2}_{Y_t})$ est injectif pour $p < n$; la condition $[X]^{p,q}_t.\omega_t = 0$ dans $H^{p+2,q}(Y_t)$ entraîne donc $[X]^{p,q}_t = 0$, pour $p < n$, et donc on a bien $t \in \mathcal{M}_{[X]}$; de $\mathcal{M}'_{[X]} \subset \mathcal{M}_{[X]}$, on déduit évidemment $\mathcal{M}'^0_{[X]} \subset \mathcal{M}^0_{[X]}$. Pour montrer l'inclusion inverse, il suffit de montrer que l'intersection $\mathcal{M}'^0_{[X]} \cap \mathcal{M}^0_{[X]}$ est ouverte dans $\mathcal{M}^0_{[X]}$, puisque par hypothèse, elle contient 0, donc est non vide, et que d'autre part elle est aussi fermée dans $\mathcal{M}^0_{[X]}$ qui est connexe. Soit $t \in \mathcal{M}'^0_{[X]} \cap \mathcal{M}^0_{[X]}$. Au point t, la classe $[X]$ est de type (n,n) et satisfait $\mu_{[X]}(\omega_t) = 0$ dans $H^{2n+2}(Y_t,\mathbb{C})$; sur $\mathcal{M}^0_{[X]}$ la classe $[X]$ reste de type (n,n) et fournit donc un morphisme de structures de Hodge : $H^2(Y_t,\mathbb{C}) \to H^{2n+2}(Y_t,\mathbb{C})$ se décomposant (de façon holomorphe), en ses différentes composantes :
$\mu^{2,0}_{[X]} : \mathcal{H}^{2,0} \to \mathcal{H}^{n+2,n} := F^{n+2}\mathcal{H}^{2n+2}/F^{n+3}\mathcal{H}^{2n+2}$
$\mu^{1,1}_{[X]} : \mathcal{H}^{1,1} := F^1\mathcal{H}^2/F^2\mathcal{H}^2 \to \mathcal{H}^{n+1,n+1} := F^{n+1}\mathcal{H}^{2n+2}/F^{n+2}\mathcal{H}^{2n+2}$, et $\mu^{0,2}_{[X]} : \mathcal{H}^{0,2} \to \mathcal{H}^{n,n+2}$ (définis de manière analogue).
Les rangs de $\mu^{2,0}_{[X]}$, $\mu^{1,1}_{[X]}$, $\mu^{0,2}_{[X]}$ étant semi-continus on peut trouver un voisinage U de t dans $\mathcal{M}^0_{[X]}$, tel que pour $t' \in U$, on ait : $\text{rang}(\mu^{2,0}_{[X]}(t')) \geq \text{rang}(\mu^{2,0}_{[X]}(t))$, et de même pour les autres composantes. Mais par ailleurs $\mu_{[X]}$ s'identifie, de façon $\mathcal{C}^\infty$, à la somme directe de ses trois composantes, et est de rang constant, puisque $[X]$ est une section localement constante de $H^{2n}_{\mathbb{Z}}$. On en déduit évidemment que chacun des rangs des composantes $\mu^{i,j}_{[X]}$ reste constant sur U, et comme $\mu^{2,0}_{[X]}$ est nul au point t, $\mu^{2,0}_{[X]}$ reste nul sur U, ce qui montre bien que U est contenu dans $\mathcal{M}'^0_{[X]}$. L'égalité ensembliste $\mathcal{M}'^0_{[X]} = \mathcal{M}^0_{[X]}$ est donc montrée. La lissité de $\mathcal{M}^0_{[X]}$ et $\mathcal{M}'^0_{[X]}$ (et donc leur égalité schématique) est alors très facile. En effet, $\mathcal{M}'^0_{[X]}$ est définie par l'annulation de $\mu_{[X]}(\omega_t)$ qui varie dans le sous-fibré plat

$\mu_{[X]}(\mathcal{H}^2) \subset \mathcal{H}^{2n+2}$, de rang $r_{[X]}$. On a donc certainement codim$\mathcal{M}'^0_{[X]} \leq r_{[X]}$. Pour montrer que $\mathcal{M}'^0_{[X]}$ est lisse de codimension $r_{[X]}$, il suffit donc de montrer que le rang du système d'équations $\mu_{[X]}(\omega_t) = 0$ est exactement $r_{[X]}$; or ceci est clair, car notant ∇ la connexion de Gauss-Manin sur $\mathcal{H}^2$ et $\mathcal{H}^{2n+2}$, on a pour $\chi \in H^1(T_Y) = T\mathcal{M}_{(0)}$, $\nabla_\chi \mu_{[X]}(\omega_t)) = \mu_{[X]}(\nabla_\chi(\omega_t))$; les éléments $\nabla_\chi(\omega_t)$ engendrent un sous-espace vectoriel de $F^1H^2(Y,\mathbb{C})$ se projetant isomorphiquement sur $H^{1,1}(Y)$, et comme la multiplication par $[X]$ est réelle, et nulle sur $H^{2,0}(Y)$, elle a même image que sa restriction à F^1H^2, qui se factorise par $H^{1,1}(Y)$; cela entraîne bien que la différentielle en 0 du système d'équations $\mu_{[X]}(\omega_t) = 0$ est de rang égal à $r_{[X]}$. Le raisonnement est identique pour $\mathcal{M}^0_{[X]}$, ce qui achève la preuve de la proposition 1.2.

1.3 Avant d'énoncer le corollaire 1.4, on rappelle l'existence d'une forme d'intersection naturelle q rationnelle et non dégénérée sur $H^2(Y)$, satisfaisant : $H^{2,0}(Y)$ est l'orthogonal de $F^1H^2(Y)$ pour $q_\mathbb{C}$ (cf. [2]). De la rationalité de q et de celle du sous-espace $\operatorname{Ker}\mu_{[X]} \subset H^2(Y)$, on déduit que l'orthogonal $L_{[X]} \subset H^2(Y,\mathbb{C})$ de $\operatorname{Ker}\mu_{[X]}$ est défini sur $\mathbb{Q}$ et de type $(1,1)$, puisque $\operatorname{Ker}\mu_{[X]}$ contient $H^{2,0}$; de plus, comme q est non dégénérée, on a rang $L_{[X]} = r_{[X]}$. L'intersection $L^\mathbb{Q}_{[X]}$ de $L_{[X]}$ avec $H^2(Y,\mathbb{Q})$ est donc un sous-espace de $NS(Y)\otimes\mathbb{Q}$, de rang $r_{[X]}$ et l'on a :

1.4 COROLLAIRE. — $\mathcal{M}^0_{[X]}$ *est aussi la composante connexe passant par* 0 *de la famille des déformations de* Y *préservant le sous-espace* $L^\mathbb{Q}_{[X]} \subset NS(Y)\otimes\mathbb{Q}$.

Démonstration : D'après la proposition 1.2, on a localement $\mathcal{M}_{[X]} = \{t \in \mathcal{M}/\omega_t \in \operatorname{Ker}\mu_{[X]}\}$; comme ω_t engendre $H^{2,0}(Y_t)$, et que l'orthogonal pour q de $H^{2,0}(Y_t)$ est égal à $F^1H^2(Y_t)$, on a aussi $\mathcal{M}_{[X]} = \{t \in \mathcal{M}/L_{[X]} \subset F^1H^2(Y)\}$ au moins localement ce qui équivaut bien sûr à l'énoncé du corollaire; l'égalité schématique résulte du fait que les deux sous-variétés considérées sont lisses de codimension $r_{[X]}$ (prop. 1.2 et [2]).

Pour conclure cette section, et faire le lien avec le théorème 0.1, on établit le lemme suivant :

1.5 LEMME. — *Soit* $X \subset Y$ *une sous-variété lagrangienne connexe; alors le rang* $r_{[X]}$ *du cup-produit* $\mu_{[X]}$: $H^2(Y,\mathbb{C}) \to H^{2n+2}(Y,\mathbb{C})$ *est égal au rang de la restriction* $j^* : H^2(Y,\mathbb{C}) \to H^2(X,\mathbb{C})$.

Démonstration : Il suffit de le montrer pour la cohomologie à coefficients réels. Comme $\mu_{[X]} = j_* \circ j^*$, on a évidemment $\operatorname{Ker} j^* \subset \operatorname{Ker}\mu_{[X]} \subset H^2(Y,\mathbb{R})$; pour montrer l'inclusion inverse, on choisit une classe de Kähler $\lambda \in H^2(Y,\mathbb{R})$; le résultat étant évident si $n = \dim X = 1$, on peut supposer $n \geq 2$, et définir la forme d'intersection q_λ suivante sur $H^2(X,\mathbb{R})$: $q_\lambda(\alpha,\beta) = \int_X \lambda^{n-2}.\alpha.\beta$. Si maintenant α et β sont des éléments de $H^2(Y,\mathbb{R})$, on trouve :$q_\lambda(j^*\alpha, j^*\beta) = \int_Y \mu_{[X]}(\alpha).\beta.\lambda^{n-2}$; si $\mu_{[X]}(\alpha) = 0$, on a donc : $\forall\beta \in H^2(Y,\mathbb{R})$, $q_\lambda(j^*\alpha, j^*\beta) = 0$, c'est-à-dire $j^*\alpha$ est dans le noyau de la restriction de q_λ à $\operatorname{Im} j^*$. Pour montrer que $j^*\alpha = 0$, il suffit donc de voir que la restriction de q_λ à $\operatorname{Im} j^*$ est non dégénérée; or cela résulte du théorème de l'index. Par hypothèse l'image de j^* est contenue dans $H^2(X,\mathbb{R}) \cap H^{1,1}(X) := H^{1,1}_\mathbb{R}(X)$; sur $H^{1,1}_\mathbb{R}(X)$ la forme q_λ est non dégénérée, de signature $(1, h^{1,1}-1)$, et plus précisément on a $q_\lambda(j^*\lambda) > 0$, q_λ est négative définie sur l'orthogonal de $j^*\lambda$. Comme $\operatorname{Im} j^*$ contient un élément de self-intersection > 0, on voit facilement que q_λ est non dégénérée sur $\operatorname{Im} j^*$.

1.6 Remarque : L'hypothèse "X lagrangienne" n'est pas nécessaire; il suffit d'utiliser le fait que j^* est un morphisme de structures de Hodge, et la positivité de la forme q_λ sur l'espace $(H^{2,0}(X) \oplus H^{0,2}(X)) \cap H^2(X,\mathbb{R})$ pour obtenir le même énoncé en toute généralité.

Du lemme 1.5 et de la proposition 1.2, on déduit finalement l'égalité suivante : (cf. Théorème 0.1).

1.7 PROPOSITION. — *$\mathcal{M}^0_{[X]}$ est égale à la composante connexe passant par 0 de la sous-variété $\mathcal{M}''_{[X]}$ de $\mathcal{M}$ définie par la condition : $t \in \mathcal{M}''_{[X]} \Leftrightarrow j_t^*(\omega_t) = 0$ dans $H^2(X, \mathbb{C})$.*

1.8 Remarque : D'après le lemme 1.5, on voit que l'espace $L^{\mathbb{Q}}_{[X]}$ défini en 1.3 a le même rang que la restriction $j^* : H^2(Y) \to H^2(X)$; cependant, l'exemple d'une surface K3 dont le groupe de Picard est engendré par la classe d'une courbe elliptique montre que la restriction de j^* à $L^{\mathbb{Q}}_{[X]}$ n'est pas nécessairement injective, ou encore que $L^{\mathbb{Q}}_{[X]}$ n'est pas nécessairement un supplémentaire de $\operatorname{Ker} j^*$.

§2 2.1. — On se propose dans cette section de montrer l'égalité $\mathcal{M}_X = \mathcal{M}^0_{[X]}$ (notons que $\mathcal{M}_X$ est connexe par définition) qui signifie que localement les déformations Y_t de Y pour lesquelles la classe de X reste de type (n,n) dans $H^{2n}(Y_t, \mathbb{C})$ sont aussi celles pour lesquelles il existe une déformation $X_t \subset Y_t$ de $X \subset Y$. Notons que ce résultat est bien connu dans le cas des surfaces : Y est alors en effet une surface K3 ou un tore complexe, et $X \subset Y$ est une courbe lisse de fibré inversible associé $\mathcal{O}_Y(X) =: L$ sur Y . Si Y est une surface K3, l'égalité $\mathcal{M}_X = \mathcal{M}_{[X]}$ résulte de la nullité du groupe $H^1(Y, L)$, elle-même due à la connexité de X , si Y est un tore complexe, on a L ample et $H^1(Y, L) = 0$, dès que $L^2 > 0$, la seule possibilité restante étant $L^2 = 0$, et X est une courbe elliptique, fibre de l'application $Y \to \operatorname{Pic}^0(\mathrm{Y})$ associée à L , et donc préservée par les déformations de Y qui préservent L.

2.2. — D'après 1.7, on a l'égalité locale : $\mathcal{M}^0_{[X]} = \{t \in \mathcal{M} / j_t^*(\omega_t) = 0 \text{ dans } H^2(X, \mathbb{C})\}$, et cela entraîne immédiatement la description suivante de l'espace tangent de $\mathcal{M}^0_{[X]}$ en 0 : $T\mathcal{M}^0_{[X](0)} = \{u \in H^1(T_Y) / j^*(\operatorname{int} u(\omega)) = 0, \text{ dans } H^1(\Omega_X)\}$. Notons enfin que si $\mathcal{Y} \xrightarrow{p} \mathcal{M}$ est une famille complète (locale) de déformations de Y, $\mathcal{M}_X \subset \mathcal{M}$ est l'image par p de la composante connexe passant par X du schéma de Hilbert relatif $H_{\mathcal{Y}/\mathcal{M}}$, lequel est propre au-dessus de $\mathcal{M}$ puisque les fibres Y_t sont kählériennes. $\mathcal{M}_X$ est donc un sous-espace analytique de $\mathcal{M}$. La démonstration procède alors de la façon suivante. On montre d'abord.

2.3 LEMME. — *Les espaces tangents de Zariski $T\mathcal{M}_{X(0)}$ et $T\mathcal{M}_{[X](0)}$ en 0 à $\mathcal{M}_X$ et $\mathcal{M}_{[X]}$ respectivement, sont égaux.*

Ce lemme montre l'égalité désirée au premier ordre. La proposition 2.4 étudie alors les obstructions d'ordre supérieur à déformer X avec Y, lorsque Y reste dans $\mathcal{M}_{[X]}$; on montre précisément :

2.4 PROPOSITION. — *Soit $u \in H^1(T_Y)$ un élément de $T\mathcal{M}_X$; alors il existe une série formelle $\theta(t) = \sum_{i>0} \theta_i t^i$ où les θ_i sont des sections de $T_Y \otimes A^{0,1}(Y)$, satisfaisant la condition d'intégrabilité :$\bar\partial\theta + 1/2[\theta, \theta] = 0$, telle que l'image de θ dans $N_Y \otimes A^{0,1}(X)$ (par l'application évidente que l'on notera α_X) soit nulle, et telle que la classe de θ_1 dans $H^1(T_Y)$ soit égale à u.*

Si $\theta(t)$ est convergente, cela signifie que $\theta(t)$ définit une structure complexe Y_t sur Y, pour laquelle $X \subset Y$ est une sous-variété analytique, ou encore que la courbe $t \to Y_t$ est contenue dans $\mathcal{M}_X$. En général, l'existence d'une telle courbe formelle contenue dans $\mathcal{M}_X$, et ayant un vecteur tangent arbitraire en 0, élément de $T\mathcal{M}_{X(0)}$, suffit à assurer que $\mathcal{M}_X$ est lisse; comme $\mathcal{M}_X$ est évidemment contenu dans $\mathcal{M}_{[X]}$, et que par le lemme 2.3, elles ont même espace tangent en 0, elles sont égales. Le théorème 0.1 résulte alors de 1.7 et le corollaire 0.2 résulte de 1.4 et 1.5. Il n'y a donc plus qu'à montrer le lemme 2.3 et la proposition 2.4.

2.5 Preuve du lemme 2.3 : Considérons les deux flèches naturelles $\alpha : T_Y \to T_{Y|X}$ et $\beta : T_{Y|X} \to N_X$; elles induisent des applications (notées de la même façon) $\alpha : H^1(T_Y) \to H^1(T_{Y|X})$ et $\beta : H^1(T_{Y|X}) \to H^1(N_X)$. Il est alors bien connu que l'espace tangent à $\mathcal{M}_X$ en 0 est le sous-espace $\alpha^{-1}(\operatorname{Ker}\beta) = \operatorname{Ker}\beta \circ \alpha$ de $H^1(T_Y)$.

Reprenant la notation ω pour la deux-forme holomorphe de Y, on a par hypothèse $\omega_{|X} = 0$, et l'espace tangent à X est en tout point de X totalement isotrope maximal pour ω ; ω fournit donc une application naturelle $\tilde{\omega} : T_Y \to \Omega_Y$, qui est un isomorphisme, et dont la restriction à X passe en un isomorphisme :$\tilde{\omega}_X : N_X \simeq \Omega_X$.

Ces applications induisent des applications, notées de la même façon, au niveau des groupes de cohomologie H^1 des faisceaux considérés, et l'on a évidemment : $\tilde{\omega}_X \circ \beta \circ \alpha = j^* \circ \tilde{\omega} : H^1(T_Y) \to H^1(\Omega_X)$. Comme $\tilde{\omega}_X$ est un isomorphisme, on a $\operatorname{Ker}\beta \circ \alpha = \operatorname{Ker}\tilde{\omega}_X \circ \beta \circ \alpha = \operatorname{Ker} j^* \circ \tilde{\omega} \subset H^1(T_Y)$. Or, d'après 2.2, $\operatorname{Ker} j^* \circ \tilde{\omega}$ est aussi l'espace tangent à $\mathcal{M}_{[X]}$ en 0, ce qui montre le lemme.

2.6 Remarque : En général, si $X \subset Y$ est une sous-variété quelconque de dimension n, on a une application naturelle $\gamma : H^1(N_X) \to H^{n+1}(\Omega_Y^{n-1})$ telle que le diagramme suivant commute :

$$(*) \qquad \begin{array}{ccc} H^1(T_Y) & \xrightarrow{\beta\circ\alpha} & H^1(N_X) \\ & {\scriptstyle \gamma_{[X]}} \searrow \quad \swarrow {\scriptstyle \gamma} & \\ & H^{n+1}(\Omega_Y^{n-1}) & \end{array}$$

(cf. [5]).

La théorie des variations de structure de Hodge dit que $\operatorname{Ker}\gamma_{[X]}$ est l'espace tangent à $\mathcal{M}_{[X]}$, tandis que comme mentionné plus haut $\operatorname{Ker}\beta \circ \alpha$ est l'espace tangent à $\mathcal{M}_X$. La condition de semi-régularité de Bloch est l'injectivité de γ et assure (cf. [5]) l'égalité $\mathcal{M}_X = \mathcal{M}_{[X]}$. Dans le cas considéré ici, cette condition n'est pas réalisée en général (cf. § 3), et au premier ordre elle est simplement remplacée par la condition $\operatorname{Ker}\beta\circ\alpha = \operatorname{Ker}\gamma\circ\beta\circ\alpha$, qui s'identifie à l'égalité $\operatorname{Ker} j^* = \operatorname{Ker}\mu_{[X]}$ du lemme 1.5, si l'on note qu'en faisant agir ω sur chacun des espaces ci-dessus, le triangle $(*)$ s'identifie à :

$$(**) \qquad \begin{array}{ccc} H^1(\Omega_Y) & \xrightarrow{j^*} & H^1(\Omega_X) \\ & {\scriptstyle \mu_{[X]}} \searrow \quad \swarrow {\scriptstyle j_*} & \\ & H^{n+1}(\Omega_Y^{n-1}) & \end{array}$$

à condition bien sûr que $\mu_{[X]}(\omega) = 0$.

2.7 Preuve de la proposition : On utilise le résultat de [4], [13], [15], qui dit que $\mathcal{M}$ est lisse, et la preuve est d'ailleurs directement inspirée de la démonstration de Tian, telle que présentée par Friedman dans [7].

Pour θ une section de $T_Y \otimes A^{0,k}(Y)$, on note $\tilde{\omega}(\theta)$ le produit intérieur de ω par θ (comme on l'a fait au lemme 2.5). Partant de $u \in T\mathcal{M}_{X(0)}$, on va construire par récurrence sur n

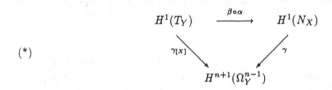

une section $\theta_n \in T_Y \otimes A^{0,k}(Y)$, telle que $\bar{\partial}\theta_1 = 0$ et $\theta_1 = u$ dans $H^1(T_Y)$, et satisfaisant les trois conditions :

a) $\bar{\partial}\theta_n + 1/2(\sum_{i=1}^{n-1}[\theta_i, \theta_{n-i}]) = 0$

b) $\partial(\tilde{\omega}(\theta_n)) = 0$, dans $\Omega_Y^2 \otimes A^{0,1}(Y)$

c) $\alpha_X(\theta_n) = 0$, dans $N_X \otimes A^{0,1}(X)$.

1) Si $n = 1$: l'application $\tilde{\omega}$ fournit un isomorphisme entre $T_Y \otimes A^{0,k}(Y)$ et $\Omega_Y \otimes A^{0,k}(Y)$, compatible avec les applications $\bar{\partial}$, et induit en particulier un isomorphisme entre $H^k(T_Y)$ et $H^k(\Omega_Y)$; la théorie de Hodge dit qu'on peut choisir une forme η de type $(1,1)$ ∂- et $\bar{\partial}$- fermée représentant $\tilde{\omega}(u) \in H^1(\Omega_Y)$. Posant $u' = \tilde{\omega}^{-1}(\eta) \in T_Y \otimes A^{0,1}(Y)$, on a donc $u' = u$ dans $H^1(T_Y)$, et $\partial(\tilde{\omega}(u')) = 0$. D'autre part, par hypothèse, et par le lemme 2.3, u' satisfait : $j^*(\tilde{\omega}(u')) = 0$ dans $H^1(\Omega_X)$: comme $\tilde{\omega}(u')$ est ∂- et $\bar{\partial}$- fermée, on en déduit par le "lemme $\partial\bar{\partial}$" appliqué à X qu'il existe une fonction φ sur X telle que $j^*(\tilde{\omega}(u')) = \partial\bar{\partial}\varphi$. Etendant φ en une fonction $\tilde{\varphi}$ sur Y, et posant $\theta_1 = u' - \tilde{\omega}^{-1}(\partial\bar{\partial}\tilde{\varphi})$, on voit que θ_1 satisfait les conditions a), b) et c), le dernier point résultant du fait déjà noté que pour $\theta \in T_Y \otimes A^{0,1}(Y)$, $j^*(\tilde{\omega}(\theta)) = 0$ dans $\Omega_X \otimes A^{0,1}(X)$ est équivalent à $\alpha_X(\theta) = 0$, dans $N_X \otimes A^{0,1}(X)$.

2) Supposons maintenant a), b), c) satisfaits pour $i < n$, avec $n \geq 2$; on va d'abord trouver θ'_n satisfaisant a) et b). Il est bien connu que $\sum_{i=1}^{n-1}[\theta_i, \theta_{n-i}]$ est un élément $\bar{\partial}$-fermé de $T_Y \otimes A^{0,2}(Y)$, dont la classe dans $H^2(T_Y)$ mesure l'obstruction à étendre la déformation à l'ordre $n-1$ de Y donnée par $\sum_{i=1}^{n-1} t^i\theta_i$ (les θ_i satisfaisant la condition a)) en une déformation à l'ordre n de Y. Cette obstruction est nulle du fait de la lissité de $\mathcal{M}$, et donc $\sum_{i=1}^{n-1}[\theta_i, \theta_{n-i}]$ est $\bar{\partial}$-exact, ainsi que $\tilde{\omega}(\sum_{i=1}^{n-1}[\theta_i, \theta_{n-i}])$. On a par ailleurs le lemme suivant :

2.8 Lemme. — *Si θ et θ' sont des éléments de $T_Y \otimes A^{0,1}(Y)$ satisfaisant : $\partial\tilde{\omega}(\theta) = 0 = \partial\tilde{\omega}(\theta')$, alors $\partial\tilde{\omega}([\theta, \theta']) = 0$.*

Démonstration : Ecrivons localement $\theta = \sum \chi_i \otimes d\bar{z}_i$ et $\theta' = \sum \chi'_i \otimes d\bar{z}_i$, où les z_i sont des coordonnées holomorphes et les χ_i, χ'_i sont des champs C^∞ de type $(1,0)$; alors $[\theta, \theta'] = \sum_{i,j}[\chi_i, \chi'_j] \otimes d\bar{z}_i \wedge d\bar{z}_j$ et $\tilde{\omega}(\theta) = \sum_i \mathrm{int}\chi_i(\omega) \otimes d\bar{z}_i$, $\tilde{\omega}(\theta') = \sum_j \mathrm{int}\chi'_j(\omega) \otimes d\bar{z}_j$; l'hypothèse entraîne donc : $\forall i$, $\mathrm{int}\chi_i(\omega)$ est ∂-fermée, ainsi que $\mathrm{int}\chi'_i(\omega)$, et l'égalité $\tilde{\omega}([\theta, \theta']) = \sum_{i,j} \mathrm{int}[\chi_i, \chi'_j](\omega) \otimes d\bar{z}_i \wedge d\bar{z}_j$ montre qu'il suffit de prouver : si χ, χ' sont des champs de type $(1,0)$, tels que int $\chi(\omega)$ et int $\chi'(\omega)$ soient ∂-fermées, alors $\mathrm{int}[\chi, \chi'](\omega)$ est ∂-fermée. Or cela résulte immédiatement du fait que ω est ∂-fermée et du résultat analogue très classique de géométrie différentielle.

2.9 D'après le lemme 2.8, et l'hypothèse de récurrence, $\tilde{\omega}(-1/2(\sum_{i=1}^{n-1}[\theta_i, \theta_{n-i}]))$ est une forme ∂ et $\bar{\partial}$-fermée, et aussi $\bar{\partial}$-exacte. Par le "lemme $\partial\bar{\partial}$", on en déduit qu'elle peut s'écrire $\bar{\partial}\eta$, où η est une forme de type $(1,1)$ ∂-fermée. Posant $\theta'_n = \tilde{\omega}^{-1}(\eta)$, on voit que θ'_n satisfait les conditions a) et b). On peut bien sûr ajouter à θ'_n un élément φ de $T_Y \otimes A^{0,1}(Y)$, satisfaisant les conditions $\partial\tilde{\omega}(\varphi) = 0 = \bar{\partial}\tilde{\omega}(\varphi)$, et c'est ce que l'on va faire pour obtenir c). La condition c) étant satisfaite pour $i < n$ il est très facile de voir que $\alpha_X(-1/2(\sum_{i=1}^{n-1}[\theta_i,\theta_{n-i}])) = 0$ dans $N_X \otimes A^{0,2}(X)$, ce qui se réécrit encore : $\alpha_X(\bar\partial\theta'_n) = 0$ dans $N_X \otimes A^{0,2}(X)$; $\alpha_X(\theta'_n)$ fournit un élément de $H^1(N_X)$ qui mesure l'obstruction à étendre la déformation $X_{n-1} \subset Y_{n-1}$ donnée par le développement $\sum_{i=1}^{n-1} t^i\theta_i$, où les θ_i satisfont a) et c), en une déformation $X_n \subset Y_n$, où la déformation à l'ordre n Y_n de Y est donnée par le développement $\sum_{i=1}^{n-1} t^i\theta_i + t^n\theta'_n$. Bien sûr, en appliquant l'isomorphisme $\tilde\omega_X : N_X \simeq \Omega_X$ défini en 2.5, à $\alpha_X(\theta'_n)$, on obtient la forme $j^*(\tilde\omega(\theta'_n))$, de type $(1,1)$ sur X; pour construire $\theta_n = \theta'_n + \varphi$, satisfaisant a), b), c), où φ doit donc satisfaire :$\partial\tilde\omega(\varphi) = 0 = \bar\partial\tilde\omega(\varphi)$, et $j^*(\tilde\omega(\varphi)) = -j^*(\tilde\omega(\theta'_n))$, il suffit clairement de montrer : la forme $j^*(\tilde\omega(\theta'_n))$ sur X est la restriction d'une forme de type $(1,1)$ $\partial-$ et $\bar\partial$-fermée sur Y. Comme par construction $\tilde\omega(\theta'_n)$ est de type $(1,1)$ ∂-fermée et que sa restriction à X est $\partial-$ et $\bar\partial$-fermée, il faut montrer :

2.10 Lemme. — *Soit η une forme sur X de type $(1,1)$ ∂ et $\bar\partial$-fermée ; si η est la restriction d'une forme de type $(1,1)$ ∂-fermée sur Y, alors η est aussi la restriction d'une forme de type $(1,1)$ ∂ et $\bar\partial$-fermée sur Y.*

Démonstration : Soit $\eta = j^*(\tilde\eta)$ avec $\partial\tilde\eta = 0$; alors $\bar\eta = j^*(\tilde{\bar\eta})$, avec $\bar\partial\ \tilde{\bar\eta} = 0$. Donc la classe de $\bar\eta$ dans $H^1(\Omega_X)$ est dans l'image de la restriction $j^* : H^1(\Omega_Y) \to H^1(\Omega_X)$; il en va donc de même pour η, et la décomposition de Hodge de $H^2(Y,\mathbb{C})$ fournit une forme η' sur Y, $\partial-$ et $\bar\partial$-fermée, de type $(1,1)$ telle que $j^*(\eta') = \eta$ dans $H^1(\Omega_X))$. Donc $\eta - j^*(\eta')$ est ∂- et $\bar\partial$-fermée, et $\bar\partial$-exacte; le lemme $\partial\bar\partial$ entraîne alors que $\eta - j^*(\eta') = \partial\bar\partial\psi$, où ψ est une fonction sur X, et donc si $\tilde\psi$ est une extension de ψ à Y, η est la restriction de $\eta' + \partial\bar\partial\tilde\psi$, qui est de type $(1,1)$, $\partial-$ et $\bar\partial$-fermée sur Y.

Ceci achève la preuve de la proposition 2.4. Comme noté plus haut, celle-ci entraîne l'égalité $\mathcal{M}_X = \mathcal{M}^0_{[X]}$, et aussi, par les résultats de § 1, le théorème 0.1 et son corollaire 0.2.

2.11 Remarque : Il est facile de voir que le théorème 0.1 reste vrai dans le cas où X n'est pas connexe. Par contre, l'égalité $\mathcal{M}_X = \mathcal{M}^0_{[X]}$ n'a plus nécessairement lieu si X n'est pas connexe (comme le montre l'exemple de deux courbes rationnelles disjointes sur une surface K3); cela reflète le fait que le lemme 1.5 n'est plus vrai dans ce cas.

§ 3 **Exemples.** —

1) Dans [2] A. Beauville montre que si S est une surface K3, le schéma de Hilbert $S^{[r]}$ des zéro-cycles de degré r de S est une variété symplectique irréductible (kählérienne en vertu d'un théorème de Varouchas). Si S contient une courbe C lisse, on a une immersion du produit symétrique $C^{(r)}$ de C dans $S^{[r]}$, qui fournit une sous-variété lagrangienne de $S^{[r]}$. Il est facile de voir que l'image de la restriction $j^* : H^2(S^{[r]}) \to H^2(C^{(r)})$ est engendrée par les classes des diviseurs $D_c :=$ image de $c \times C^{(r-1)}$ dans $C^{(r)}$, et de la diagonale de $C^{(r)}$. (Notons que bien que l'image de j^* soit engendrée par des classes de diviseurs la restriction de j^* à $\mathrm{NS}(S^{[r]})$ n'a pas nécessairement la même image, cf. 1.8).

Deux cas se produisent donc : a) si C est rationnelle, on a rang $j^* = 1$ et donc le théorème principal entraîne que la sous-variété lagrangienne $C^{(r)}$ est préservée sur une hypersurface de la famille complète des déformations de $S^{[r]}$. Toujours à cause du théorème, on voit facilement que cette hypersurface est transverse à la famille des déformations de $S^{[r]}$ qui restent du type $T^{[r]}$, où T est une déformation de S; dans ce cas, $S^{[r]}$ se déforme donc sur une variété symplectique irréductible qui n'est plus du type "produit symétrique", et qui contient un espace projectif $\mathbb{P}^r$.

b) Par contre, si C n'est pas rationnelle, le rang de j^* est égal à 2, et les déformations de $S^{[r]}$ préservant la sous-variété lagrangienne $C^{(r)}$ forment une sous-variété de codimension deux, qui est donc nécessairement égale à la famille des variétés $T^{[r]}$, où T varie dans l'hypersurface de la famille complète des déformations de S, sur laquelle le diviseur C est préservé.

2) Si l'on considère maintenant les variétés $S^{[2]}$, éclatement du produit symétrique $S^{(2)}$ le long de sa diagonale Δ, et si l'on note $\tau : E \to \Delta$ le diviseur exceptionnel, fibré en coniques au-dessus de $\Delta \simeq S$, pour une courbe $C \subset S$ lisse, on obtient une sous-variété lagrangienne de $S^{[2]}$ en considérant la surface réglée $\tau^{-1}(C) \subset E \subset S^{[2]}$. Il est facile de voir dans ce cas que le rang de la restriction j^* est égal à deux. Les déformations de $S^{[2]}$ préservant la sous-variété lagrangienne $\tau^{-1}(C)$ sont donc les variétés $T^{[2]}$ où T contient une déformation de C.

3) Si maintenant S est une surface K3 munie d'un diviseur ample H tel que $H^2 = 2g-2$, et si S ne contient pas de courbe C telle que $h^0(H(-C)) \geq 2$, on a une immersion naturelle de la grassmannienne G des droites de $\mathbf{P}(H^0(H)) = \mathbf{P}^g$ dans $S^{[2g-2]}$. On obtient ainsi une sous-variété lagrangienne de $S^{[2g-2]}$. Comme $H^2(G)$ est de rang 1 et que j^* n'est pas nul, cette sous-variété se déforme avec $S^{[2g-2]}$ en codimension un dans la famille complète des déformations de $S^{[2g-2]}$. En appliquant le théorème, on voit facilement que l'intersection de cette hypersurface avec la famille des déformations de $S^{[2g-2]}$ qui restent du type produit symétrique est constituée des variétés $T^{[2g-2]}$, où H est une classe algébrique sur T, de sorte que l'on peut construire des déformations qui ne sont plus du type produit symétrique, et qui contiennent une grassmannienne.

4) Si S est une surface K3 dont le groupe de Picard est engendré par un diviseur H très ample avec $H^2 = 2g-2$ et g pair, $g = 2s$, il existe un unique fibré vectoriel stable E de rang 2 sur S de déterminant H, satisfaisant $c_2(E) = s+1$ et $h^0(E) = s+2$. On voit alors facilement qu'on a une immersion de $\mathbf{P}(H^0(E)) = \mathbf{P}^{s+1}$ dans $S^{[s+1]}$, qui fournit une sous-variété lagrangienne de $S^{[s+1]}$. Encore une fois, le théorème montre que localement, sur la famille des déformations de $S^{[s+1]}$, l'existence de ce sous-espace $\mathbf{P}^{[s+1]}$ caractérise les déformations de S pour lesquelles la classe de H reste algébrique, et aussi qu'on peut déformer $S^{[s+1]}$ sur une variété qui n'est plus du type produit symétrique, en préservant cet espace projectif.

5) Si S est une surface quartique générique dans $\mathbf{P}^3$, la surface T des bitangentes de S est naturellement immergée dans $S^{[2]}$. C'est une sous-variété lagrangienne puisque les zéro-cycles $Z_\ell = x_1 + x_2$ de S découpés par une bitangente ℓ de S satisfont $2Z_\ell = \ell.S$ qui est constant dans $CH_0(S)$, ce qui entraîne le résultat par le théorème de Mumford [12]. Notons que $H^1(\Omega_T) = H^1(N_T)$ est de rang bien plus grand que celui de $H^3(\Omega_{S^{[2]}})$, ce qui entraîne que la condition de semi-régularité 2.6 ne peut être réalisée dans ce cas. Le rang de la restriction j^* est égal à 1 dans ce cas, comme il résulte du fait que le rang générique de NS(T) est égal à 1. Encore une fois, on voit qu'on peut déformer $S^{[2]}$ sur une variété qui n'est plus de ce type et qui contient une déformation de T. Cet exemple diffère des précédents par le fait que T a un fibré canonique très ample.

6) Surfaces K3 admettant une involution. Soit S une surface K3 obtenue comme un revêtement double d'une surface T, qui peut être un plan, une quadrique, ou une surface d'Enriques. La surface T satisfait $h^{2,0} = 0$ et son image naturelle dans $S^{[2]}$ est donc une sous-variété lagrangienne. L'application de restriction j^* est surjective dans tous les cas, et il est facile de voir qu'on peut déformer $S^{[2]}$ en une variété qui n'est plus du type produit symétrique, en préservant la sous-variété lagrangienne T. (En effet par le théorème et son corollaire 0.2, il suffit de voir que le diviseur exceptionnel E (cf. Ex. 2)) n'est pas dans l'orthogonal de $\operatorname{Ker} j^*$; or le contraire entraînerait que $\operatorname{Ker} j^*$ est contenu dans l'orthogonal de E, qui est encore égal à l'image "diagonale" de $H^2(S)$ dans $H^2(S^{[2]})$ (cf. [2]). Il est facile de voir que cela n'est pas le cas. Par contraste avec les exemples précédents, la surface d'Enriques fournit un exemple où la codimension de la famille sur laquelle la sous-variété

lagrangienne est préservée est assez grande (égale à 10).

7) La variété F des droites d'une cubique X de dimension 4 (cf. [2], [3]) fournit un exemple explicite de variété symplectique qui n'est pas du type produit symétrique d'une surface K3 et qui contient des sous-variétés lagrangiennes. La surface des droites contenues dans une section hyperplane lisse de X est une telle sous-variété de F, se déformant avec F aussi longtemps que F reste la variété des droites d'une cubique, c'est-à-dire dans une hypersurface de la famille des déformations de F (cf. [14]).

Si maintenant X contient un plan P, F contient le plan dual P^v, qui constitue évidemment une sous-variété lagrangienne de F. Comme $H^2(P^v)$ est de rang 1, on a rang $j^* = 1$ dans ce cas, et le plan P^v est préservé dans une hypersurface de la famille des déformations de F. Il est facile de voir que cette hypersurface est transverse à la famille des déformations de F comme variété des droites d'une cubique.

Tous ces exemples illustrent le contenu du théorème, au sens où ils exhibent des sous-variétés lagrangiennes d'une variété générique dans une famille de déformations d'une variété symplectique avec groupe de Picard fixé. La question suivante est assez naturelle. Soit S une surface K3, ayant un groupe de Picard fixé. Peut-on construire toutes les sous-variétés lagrangiennes des produits symétriques $S^{[k]}$ par l'un des procédés décrits ci-dessus, ou par des procédés similaires ?

BIBLIOGRAPHIE

[1] M. Artin. — On the solutions of analytic equations, *Invent. Math* **5**, (1968), 277-291.

[2] A. Beauville. — Variétés kählériennes dont la première classe de Chern est nulle, *J. Differential Geometry* **18**, (1983), 755-782.

[3] A. Beauville, R. Donagi. — The variety of lines of a cubic fourfold, *C. R. Acad. Sci. Paris* **301 série I**, (1985), 703-706.

[4] F. Bogomolov. — Hamiltonian kähler manifolds, *Soviet. Math. Doklady* **19**, (1978), 1462-1465.

[5] S. Bloch. — Semi-regularity and De Rham cohomology, *Invent Math* **17**, (1972), 51-66.

[6] O. Debarre. — Un contre-exemple au théorème de Torelli pour les variétés symplectiques irréductibles, C. R. Acad. Sci. Paris, t. 299 série I, (1984).

[7] R. Friedman. — On threefolds with trivial canonical bundle (preprint).

[8] P. Griffiths, J. Harris. — Infinitesimal variation of Hodge structure II : an infinitesimal invariant of Hodge class, Comp. Math., vol. 50, 207-265.

[9] K. Kodaïra. — On stability of compact submanifolds of complex manifolds, Amer. J. of Math., vol. 85, (1963), 79-94.

[10] S. Mukai. — Biregular classification of Fano threefolds and Fano manifolds of coindex three, Proc. Natl. Acad. Sci. USA, vol. 86, (1989), 3000-3002.

[11] S. Mukai. — Symplectic structure on the moduli space of sheaves on an abelian or K3 surface, *Invent. Math.* **77**, (1984), 101-116.

[12] D. Mumford. — Rational equivalence of zéro cycles on surfaces, J. Math. Kyoto Univ., vol. 9, (1968), 195-204.

[13] G. Tian. — Smoothness of the universal deformation space of compact Calabi-Yau manifolds and its Petersson-Weil metric, in "Mathematical aspects of string theory", S.T.Yau ed., World Scientific, Singapore, (1987), pp. 629-646.

[14] C. Voisin. — Théorème de Torelli pour les cubiques de P^5, *Invent. Math.* **86**, (1986), 577-601.

[15] A.N. Todorov. — The Weil-Petersson geometry of the moduli space of $SU(N \geq)$ (Calabi-Yau) manifolds I, (preprint de l'I.H.E.S.), (1988).

Claire VOISIN
URA D0752 du CNRS
Université de Paris-Sud
Mathématique, bât. 425, ORSAY

INTRODUCTION TO GAUSSIAN MAPS ON AN ALGEBRAIC CURVE

Jonathan Wahl

Abstract: For two line bundles L and M on a smooth projective curve, the natural multiplication map $\mu_{L,M}: \Gamma(L)\otimes\Gamma(M)\to\Gamma(L\otimes M)$ was classically studied by Castelnuovo, M. Noether, and Petri. We study the "Gaussian" map $\Phi_{L,M}$ from the kernel of $\mu_{L,M}$ to $\Gamma(K\otimes L\otimes M)$; when M=L, it is really $\Phi_L:\wedge^2\Gamma(L)\to\Gamma(K\otimes L^{\otimes 2})$, and is the classical Gauss map. We find criteria for the surjectivity of $\Phi_{L,M}$, and explain the relationship of the cokernel of $\Phi_{K,L}$ to deformation theory. Special note is paid to the map Φ_K, which is surjective for a general curve of genus ≥12, but not for a curve on a K-3 surface. Most of the paper reviews known results, but there is a new discussion about the kernel of the Gaussian.

INTRODUCTION

Let L be a line bundle on a projective variety X. The original Gaussian map is the homomorphism

$$\Phi_L:\wedge^2\Gamma(X,L)\to\Gamma(X,\Omega^1_X\otimes L^{\otimes 2}),$$

defined essentially by $\Phi(s\wedge t)=sdt-tds$. In other words, if on an open affine U, L|U is free with generator T, then one defines

$$\Phi(fT\wedge gT)=(fdg-gdf)T\otimes T.$$

It is remarkable though easily verified that this definition makes sense, and the construction is natural. Thus, if $f:Y\to X$ is a morphism,then using pull-back one has a commutative diagram

$$\begin{array}{ccc}\wedge^2\Gamma(X,L) & \to & \Gamma(X,\Omega^1_X\otimes L^{\otimes 2})\\ \downarrow & & \downarrow\\ \wedge^2\Gamma(Y,f^*L) & \to & \Gamma(Y,\Omega^1_Y\otimes f^*L^{\otimes 2}).\end{array}$$

If L is very ample, providing an embedding $X\subset\mathbb{P}^n$, then Φ_L is a Gauss map in the usual sense (hence the name Gaussian, suggested by R. Lazarsfeld). To see this, suppose for convenience that X is a smooth curve. The Gauss map sends X to the Grassmannian G(1,n) of lines in $\mathbb{P}^n$ (associating to each point of X its tangent line), followed by the Plücker embedding into $\mathbb{P}^N$, $N=n(n+1)/2-1$. At a point $P\in X$, with local coordinate t, we assume a basis for $\Gamma(L)$ is written $1,f_1(t),\ldots,f_n(t)$, where the $f_i(t)$ are regular functions vanishing at P. The Plücker embedding of the point of G(1,n) corresponding to the tangent line is found by considering the 2≠2 minors of the matrix

$$\begin{bmatrix} 1 & f_1(t) & f_2(t) & \ldots & f_n(t)\\ 0 & f_1'(t) & f_2'(t) & \ldots & f_n'(t)\end{bmatrix},$$

which is the same as the Gaussian map Φ_L. The map $X \to \mathbb{P}^N$ is also called the "first associated curve."

If K is the canonical line bundle of a non-hyperelliptic curve, the map Φ_K (sometimes referred to as the "Wahl map") was introduced in [W1] because of the relevance of its corank to the deformation theory of the cone over the canonical curve. An appealing consequence was

Theorem 1: If X is a smooth curve which lies on a K-3 surface, then the Gaussian Φ_K is not surjective.

In fact, if Φ_K is surjective and $X \subset \mathbb{P}^{g-1}$ is the hyperplane section of $Y \subset \mathbb{P}^g$, then Y is the cone over the curve X. We describe our proof in Theorem 4.2 below, and also outline the key ideas of a proof of [BM] which avoids the use of deformation theory. (There is another recent proof due to F. L. Zak.)

There are many curves with surjective Gaussian. For instance, complete intersection curves in $\mathbb{P}^n$ $(n \geq 3)$, with the exception of a few multidegrees, have Φ_K surjective [W1]. Mori-Mukai [MM] and Mukai [Mk] have proved that the general curve of genus g lies on a K-3 surface iff $g \leq 9$ or $g = 11$; this motivates the basic

Theorem 2 ([CHM]): The map Φ_K is surjective for the general curve of genus 10 or ≥12.

Note that Theorems 1 and 2 together imply Mukai's result that a general curve of genus 10 does not lie on a K-3 surface. We mention in §4 a new proof of Theorem 2 by Claire Voisin [V], which relates the non-surjectivity of Φ_K to the existence of line bundles which are not normally generated. We also discuss some of the numerous recent results on the stratification of the moduli space $\mathcal{M}_g$ of curves by the corank of Φ_K. While one knows that the corank is $\leq 3g-2$, with equality if and only if X is hyperelliptic, one still does not know the correct upper bound for non-hyperelliptic curves.

More general Gaussian maps are naturally defined for 2 line bundles L and M on a projective variety X. On $X \times X$, consider the external tensor product $L \boxtimes M = p_1^* L \otimes p_2^* M$ and the powers I^i of the ideal sheaf of the diagonal $\Delta \subset X \times X$. The Künneth formula gives the identification $\Gamma(X \times X, L \boxtimes M) = \Gamma(X,L) \otimes \Gamma(X,M)$; one therefore has a filtration $\{\mathcal{R}_i(L,M)\}$ of the tensor product by

$$\mathcal{R}_i(L,M) = \Gamma(X \times X, I^i \otimes (L \boxtimes M)),$$

which we call the **diagonal filtration of the tensor product**. Since $I/I^2 \cong \Omega^1_X$, we have $I^i/I^{i+1} \cong S^i \Omega^1_X$. Thus, the cohomology sequence gives

$$0 \to \mathcal{R}_{i+i}(L,M) \to \mathcal{R}_i(L,M) \to \Gamma(X, L \otimes M \otimes S^i \Omega^1_X);$$

the last map is the **ith generalized Gaussian** Φ_i. It is clear that any particular Φ_i is surjective if

L and M are sufficiently ample, e.g. if $H^1(X \times X, I^{i+1}(L \boxtimes M)) = 0$. Now, $\mathcal{R}_0(L,M) = \Gamma(L) \otimes \Gamma(M)$, and Φ_0 is nothing but the multiplication map $\mu_{L,M}$. Next, $\mathcal{R}_1(L,M) = \mathrm{Ker}\ \mu_{L,M}$ (usually written $\mathcal{R}(L,M)$, and $\Phi_1 = \Phi_{L,M}$ is a Gaussian map, defined (in (1.1) below) similarly to Φ_L. In fact, when $L = M$, we have $\mathcal{R}(L,L) \supset \Lambda^2\Gamma(L)$, and $\Phi_{L,L}$ is essentially Φ_L.

When X is a curve, surjectivity results for $\Phi_{L,M}$ should be seen in the context of similar results for $\mu_{L,M}$. These include:

(1) (Noether) If X is non-hyperelliptic, then μ_{K,K^i} is surjective, all $i \geq 1$.

(2) (Petri) If L is very ample (g>0), then $\mu_{K,L}$ is surjective.

(3) (Castelnuovo) If $\deg L \geq 2g$, $\deg M \geq 2g+1$, then $\mu_{L,M}$ is surjective.

We have already seen that the surjectivity of Φ_K is quite delicate, though with a few exceptions Φ_{K,K^i} is surjective for $i \geq 2$. As for the Gaussians $\Phi_{L,M}$, we quoted at the conference a result from [W3], that surjectivity follows for $\deg L \geq 2g+2$ and $\deg M \geq 5g+2$. In conversation thereafter, R. Lazarsfeld described a simple argument that allows one to sharpen the bounds (Lemma 2.3 below). The point is to use L to embed the curve so that it is ideal-theoretically generated by quadrics. We incorporate his argument in our Theorem 2.2:

Theorem 3: Suppose $\deg L$, $\deg M \geq 2g+2$ and $\deg L + \deg M \geq 6g+3$. Then the Gaussian $\Phi_{L,M}$ is surjective.

Similar results on $\Phi_{K,L}$ follow by the same argument, once the canonical curve is defined by quadrics. Missing cases and some improved bounds (using the Clifford index Cliff(X), a measure of the specialness of X) are handled in the thesis of S. Tendian [T] (see 2.6). Most recently, nearly complete results in this direction have been found by Bertram, Ein, and Lazarsfeld [BEL]. However, it is still an open question (2.5) to determined on a general curve the smallest degree for L past which $\Phi_{K,L}$ is surjective.

Up to this point (and in fact in our Conference talk), the only question we have considered is the surjectivity of certain $\Phi_{L,M}$. But it is also natural to look at $\mathrm{Ker}\ \Phi_{L,M}$, as well as at higher order Gaussians. For instance, if L is a g^r_d on a curve X, the injectivity of Petri's map

$$\mu_0 = \mu_{L,K-L} : \Gamma(L) \otimes \Gamma(K-L) \to \Gamma(K)$$

implies the smoothness at L of W^r_d, the space of g^r_d's in $\mathrm{Pic}^d X$. But in [G], Griffiths studies the higher Gaussians $\mu_i = \Phi_i(L, K-L)$, giving for instance a smoothness interpretation of the injectivity of $\Phi_{L,K-L}$ (see (5.2) below). As a second example, for very ample L on a smooth projective X, the Gaussian $\Phi_2(L,L)$: $I_2 \to \Gamma(K^{\otimes 2} \otimes L^{\otimes 2})$ (where $I_2 = \mathrm{Ker} S^2\Gamma(L) \to \Gamma(L^{\otimes 2})$) is essentially the "second fundamental form" of complex differential geometry [GH]. We discuss in § 5 three questions concerning the kernel of $\Phi_L : \Lambda^2\Gamma(L) \to \Gamma(K \otimes L^{\otimes 2})$ on a curve:

(a) For which L, with $h^0(L)=4$, is Φ_L **not** injective?

(b) For general X, what is the smallest degree of an L with Φ_L not injective?

(c) When is Ker Φ_L generated by elements of rank 4?

(A rank 4 element is one expressible as $\alpha\wedge\beta + \gamma\wedge\delta$, where α, β, γ, and δ are linearly independent; i.e., it has rank 4 as an alternating form on $\Gamma(L)^*$.) As for (a), the obvious examples come from three times a pencil; are there any others? This question is related (Prop. 5.4) to the existence of linearly normal contact curves in $\mathbb{P}^3$ other than the twisted cubic, and suggests an alternate proof of a theorem of Robert Bryant [Bry]: every curve embeds into $\mathbb{P}^3$ as a contact curve. A bound for (b) is given in (5.8), using Brill-Noether theory. Question (c) is motivated by analogy with the theorem (e.g., [ACGH], p. 255) that the space of quadrics vanishing on a non-hyperelliptic canonical curve is generated by quadrics of rank ≤4. In that case, one knows that the tangent cone to the double points on the theta-divisor give such rank 4 quadrics. There is also a method to produce quadratic equations "determinantly" when the degree of L is large [EKS]. We do not know any geometric way to produce rank 4 elements of Ker Φ_L, although one might expect a role to be played by contact curves in $\mathbb{P}^3$.

We mention briefly another situation in which Gaussians are of interest--the case when X is a complex homogeneous space G/P (e.g., Grassmannians, flag manifolds, etc.); here, G is a simple and simply-connected algebraic group, P a parabolic subgroup. If L on X is ample, it is automatically very ample, and $\Gamma(L)$ is an irreducible G-module (we always assume characteristic 0). If L and M are ample line bundles, the multiplication map $\mu_{L,M}: \Gamma(L)\otimes\Gamma(M)\to \Gamma(L\otimes M)$ is surjective, being a non-0 G-equivariant map to an irreducible G-module. Recently, S. Kumar has proved [K] our conjecture from [W4]: if L and M are ample on G/P, the Gaussian map $\Phi_{L,M}$ is surjective. In this situation, the diagonal filtration provides a G-filtration on the tensor product of two irreducible G-modules, with subquotients contained in $\Gamma(G/P, S^i\Omega^1\otimes L\otimes M)$; surjectivity results on higher order Gaussians therefore can provide new information on the decomposition of the tensor product. However, we shall not discuss these aspects in this paper.

The first section of the paper reviews the basics, while §2 considers surjectivity of $\Phi_{L,M}$ on a curve. §3 explains the relationship between the cokernel of the Gaussian $\Phi_{K,L}$ and deformation theory of the cone over a curve X embedded via L. The Gaussian Φ_K is the subject of §4, while the fifth section examines other questions on Gaussians, including higher-order ones. We will sometimes say a line bundle L is a g^r_d if L has degree d, and $h^0(L)=r+1$.

It is a pleasure to thank the organizers of this Trieste conference for the quality of the scientific program and the good time which they provided. We also thank Mark Green and Rob Lazarsfeld for a number of helpful ideas. This research was partially supported by NSF Grant DMS 8801855.

§1. Gaussian maps

(1.1) As in the introduction, fix two line bundles L and M on a smooth projective variety X. The map

$$\Phi=\Phi_{L,M} : \mathcal{R}(L,M) \rightarrow \Gamma(\Omega^1 \otimes L \otimes M)$$

may be defined as follows: Let $\alpha=\Sigma \ell_i \otimes m_i \in \mathcal{R}(L,M)$, where $\ell_i \in \Gamma(L)$, $m_i \in \Gamma(M)$. On an affine open $U \subset X$ such that $L|U$, $M|U$ are free, with generators S and T, write $\ell_i = f_i S$, $m_i = g_i T$ (f_i, g_i functions on U). By hypothesis, $\Sigma f_i g_i = 0$. Define locally

$$\Phi(\alpha)=\Sigma(f_i dg_i - g_i df_i) S \otimes T \in \Gamma(U, \Omega^1 \otimes L \otimes M).$$

Of course, one needs to check that this expression is independent of the choice of local generators S and T, and that $\Sigma \ell_i \otimes m_i = 0$ implies that $\Phi(\alpha)=0$ (see, e.g., [W2], p. 123). One should also check that this definition agrees with the one given in the Introduction, which uses the diagonal on $X \times X$.

(1.2) It is important to note the **naturality** of the Gaussian maps for $f:Y \rightarrow X$, f^*L, and f^*M. That is, there is a commutative diagram

$$\begin{array}{ccc} \mathcal{R}(L,M) & \rightarrow & \Gamma(\Omega^1_X \otimes L \otimes M) \\ \downarrow & & \downarrow \\ \mathcal{R}(f^*L, f^*M) & \rightarrow & \Gamma(\Omega^1_Y \otimes f^*L \otimes f^*M). \end{array}$$

(1.3) If $L=M$, then $\mathcal{R}(L,L) = \mathrm{Ker}\ (S^2\Gamma(L) \rightarrow \Gamma(L^2)) \oplus \wedge^2 \Gamma(L)$, where $\wedge^2\Gamma(L)$ is a subspace of $\mathcal{R}(L,L)$ via $\ell_1 \wedge \ell_2 \mapsto \frac{1}{2}(\ell_1 \otimes \ell_2 - \ell_2 \otimes \ell_1)$. Since $\Phi_{L,L}$ clearly vanishes on the symmetric tensors, it is essentially defined by the map

$$\begin{array}{rcl} \Phi_L : \wedge^2\Gamma(L) & \rightarrow & \Gamma(\Omega^1 \otimes L^2) \\ fS \wedge gS & \mapsto & (fdg - gdf) S \otimes S. \end{array}$$

The Gaussian was originally formulated in this way in [W1]. One may use this definition to compute if a particularly simple basis of $\Gamma(L)$ is available (e.g., when $L = kM$ is a multiple of a pencil M, and $h^0(L) = k+1$).

(1.4) It is clear from the definition that Φ_L is injective on decomposable vectors of the form $\ell_1 \wedge \ell_2$, since $fdg - gdf = f^2 d(g/f) = 0$ implies $f \wedge g = 0$. Dually, if $X \subset \mathbb{P}^r$ is a nondegenerate curve, there is no linear H^{r-2} which intersects every tangent line of X. Since the indecomposables form a subvariety of $\wedge^2\Gamma(L)$ (the cone over the Grassmannian) of dimension $2h^0(L)-3$, one concludes as in [W3]:

Proposition 1.5: The Gaussian $\Phi_L : \wedge^2\Gamma(L) \rightarrow \Gamma(\Omega^1 \otimes L^2)$ has rank $\geq 2h^0(L)-3$. If $L = kM$, where $h^0(M)=2$ and $h^0(L)=k+1$, then rank $\Phi_L = 2h^0(L)-3$.

(1.6) The most important example is of course $\mathcal{O}(1)$ on $\mathbb{P}^r$:

$$\Phi_{\mathcal{O}(1)} : \wedge^2\Gamma(\mathcal{O}_{\mathbb{P}}(1)) \rightarrow \Gamma(\Omega^1_{\mathbb{P}}(2))$$

$$X_i \wedge X_j \quad \mapsto X_i dX_j - X_j dX_i.$$

It is easy to check by hand that this map is an isomorphism. One clever way to do this is to note that both sides are irreducible $SL(r+1)$-modules of the same dimension, and the map is non-zero and $SL(r+1)$-equivariant. The same two approaches give surjectivity of the Gaussian for $\mathcal{O}(n)$, $n \geq 0$, and also for $\Phi_{\mathcal{O}(m),\mathcal{O}(n)}$, $m,n \geq 0$.

Lemma 1.7: If L is very ample, giving an embedding $X \subset \mathbb{P}^r$, then the Gaussian Φ_L is equivalent to the restriction map

$$\Gamma(\mathbb{P}^r, \Omega^1_{\mathbb{P}}(2)) \to \Gamma(X, \Omega^1_X(2)).$$

Proof: Look at the natural diagram (1.2) for $X \subset \mathbb{P}^r$; then use (1.6).

(1.8) Another description of the Gaussian will allow a generalization (Proposition 1.10) of the last lemma. Consider the Atiyah class extension, given by the class of L in $H^1(\Omega^1_X)$:

$$0 \to \Omega^1(L) \to P(L) \to L \to 0.$$

If $L|U$ has generator T, one writes a local section of $P(L)$ in the form

$$\Sigma f_i dg_i \cdot T + hT \cdot (dT/T), \quad f_i, g_i, h \text{ local functions;}$$

then $P(L) \to L$ is given by the "residue map" to $hT \in \Gamma(U, L|U)$. There is a $\mathbb{C}$-linear section $d: L \to P(L)$, sending a local section hT to $dh \cdot T + hT \cdot (dT/T)$. The induced $d: \Gamma(L) \to \Gamma(P(L))$ gives a sheaf map

$$d: \Gamma(L) \otimes \mathcal{O}_X \to P(L),$$

for which composing with $P(L) \to L$ gives the usual map

$$\Gamma(L) \otimes \mathcal{O}_X \to L.$$

Following [L2], write the kernel of the last map as $\mathcal{M}_L$:

$$0 \to \quad \mathcal{M}_L \to \Gamma(L) \otimes \mathcal{O}_X \to L.$$

Tensoring everything in sight with M yields the diagram

$$\begin{array}{ccccccccc} & 0 \to & \mathcal{M}_L \otimes M & \to & \Gamma(L) \otimes M & \to & L \otimes M & & \\ (1.8.1) & & \downarrow & & d \downarrow & & \downarrow & & \\ & 0 \to & \Omega^1(L) \otimes M & \to & P(L) \otimes M & \to & L \otimes M & \to & 0. \end{array}$$

By definition, $\mathcal{R}(L,M) = \Gamma(\mathcal{M}_L \otimes M)$, and the Gaussian map is easily checked to be obtained by taking global sections of the first vertical map (cf. [W2]).

(1.9) In the last construction, suppose L is very ample, giving $X \subset \mathbb{P}^r$. The Euler sequence on $\mathbb{P}^r$

$$0 \to \Omega^1_{\mathbb{P}}(1) \to \Gamma(L) \otimes \mathcal{O}_{\mathbb{P}} \to L \to 0,$$

restricted to X, shows that

$$\mathcal{M}_L = \Omega^1_{\mathbb{P}}(1) \otimes \mathcal{O}_X.$$

The first vertical map in (1.8.1) is then deduced from the exact sequence

$$0 \to N^*(1) \to \Omega^1_{\mathbb{P}} \otimes \mathcal{O}_X(1) \to \Omega^1_X(1) \to 0,$$

where N^* is the conormal bundle of X in $\mathbb{P}^r$. We summarize in

Proposition 1.10: Let L be a very ample line bundle, giving an embedding $X \subset \mathbb{P}^r$, with conormal bundle N^*. Then

(a) The Gaussian map $\Phi_{L,M}$ is the restriction

$$H^0(X, \Omega^1_{\mathbb{P}} \otimes \mathcal{O}_X \otimes L \otimes M) \to H^0(X, \Omega^1_X \otimes L \otimes M).$$

(b) Coker $\Phi_{L,M} = \mathrm{Ker}\{H^1(X, N^* \otimes L \otimes M) \to H^1(X, \Omega^1_{\mathbb{P}} \otimes \mathcal{O}_X \otimes L \otimes M)\}$.

(c) Suppose $\Gamma(L) \otimes \Gamma(M) \to \Gamma(L \otimes M)$ is surjective and $H^1(M) = 0$. Then

$$\text{Coker } \Phi_{L,M} \cong H^1(X, N^* \otimes L \otimes M).$$

Proof: We have the exact sequence

$$0 \to N^*(1) \otimes M \to \mathfrak{M}_L \otimes M \to \Omega^1_X(1) \otimes M \to 0,$$

and H^0 of the last map is $\Phi_{L,M}$. Take cohomology of

$$0 \to \mathfrak{M}_L \otimes M \to \Gamma(L) \otimes M \to L \otimes M \to 0,$$

the two hypotheses of (c) imply that $H^1(\mathfrak{M}_L \otimes M) = 0$.

(1.11) Aside from the higher order Gaussians mentioned in the Introduction, there are other generalizations of the map Φ_L. The **k-th associated curve** of $X \subset \mathbb{P}^r$ is the map

$$X \to G(k,r) \subset \mathbb{P}^N,$$

associating to each point of the curve its k-th osculating plane, followed by the Plücker embedding ([G], §9). If $s_0, \ldots, s_r$ form a basis for the sections of $\mathcal{O}_X(1)$, and $F(z) = (s_0(z), \ldots, s_r(z))$ in some local set of coordinates, then the k-th associated curve is given by $F(z) \wedge F'(z) \wedge \ldots \wedge F^{(k)}(z)$. The corresponding natural map analogous to the Gaussian is

$$\wedge^k \Gamma(L) \to \Gamma(K^{\otimes m} \otimes L^{\otimes k}), \qquad m = k(k-1)/2.$$

§2. Surjectivity of $\Phi_{L,M}$ for curves

(2.1) If L and M are line bundles on a curve X, one wishes precise bounds guaranteeing surjectivity of the Gaussian $\Phi_{L,M}$. In [W3], we showed that if $\deg L \geq 2g+2$ and $\deg M \geq 5g+1$, then $\Phi_{L,M}$ is surjective; and, if X is non-hyperelliptic and deg $L \geq 5g+2$, then $\Phi_{K,L}$ is surjective. The method was a variant of the base-point free pencil trick (as used in [Mu1], to study surjectivity of $\mu_{L,M}$). [CHM] proved surjectivity of $\Phi_{L,M}$ for $\deg L$, $\deg M \geq 4g+5$, by proving directly that $H^1(X \times X, p_1^*L \otimes p_2^*M(-2\Delta)) = 0$. The following

Theorem combines the key Lemma 2.3 below (instead of the base-point free pencil trick) with the approach of [Mu1] and [W3].

Theorem 2.2: Let X be a smooth projective curve of genus g.

(a) If $\deg L, \deg M \geq 2g+2$, and $\deg L + \deg M \geq 6g+3$, then $\Phi_{L,M}$ is surjective.

(b)Suppose X is non-hyperelliptic, non-trigonal, and not a plane quintic (i.e., $\mathrm{Cliff}(X) \geq 2$). If $\deg L \geq 4g-3$, then $\Phi_{K,L}$ is surjective.

Proof: We start with the following

Lemma 2.3 (Lazarsfeld): Suppose L is very ample and normally generated, and gives an embedding of X ideal-theoretically defined by quadrics. If $H^1(M \otimes L^{-1}) = 0$, then $\Phi_{L,M}$ is surjective.

Proof: By assumption, the ideal sheaf I_X of X is a quotient $\mathcal{O}(-2)^{\oplus a} \twoheadrightarrow I_X$; restricting to the curve and twisting gives a surjection $(L^{-1} \otimes M)^{\oplus a} \twoheadrightarrow N^* \otimes L \otimes M$. By 1.10, surjectivity of $\Phi_{L,M}$ follows from vanishing of $H^1(L^{-1} \otimes M)$.

Return to (a) of the Theorem. Let D be an effective divisor of degree $d = \deg L - (2g+2)$, and let $L' = L(-D)$. Since $\deg L' = 2g+2$, a theorem of Fujita and St.-Donat implies that L' is very ample, normally generated, and gives an embedding ideal-theoretically defined by quadrics (see, e.g., [L2]). We claim that for general D, $H^1(M \otimes L'^{-1}) = 0$, whence (by Lemma 2.3) $\Phi_{L',M}$ is surjective. But $\deg (M \otimes L'^{-1}) = \deg M - (2g+2) \geq (6g+3) - \deg L - (2g+2) = (2g-1) - d$. Thus, $K \otimes M^{-1} \otimes L'$ has degree $\leq d-1$; as D varies, one gets a d-dimensional family of these divisors. But any family of effective divisors of degree e must have dimension $\leq e$. Therefore, for general D, $K \otimes M^{-1} \otimes L(-D)$ has no sections, and the claim is proved.

Now that $\Phi_{L(-D),M}$ is surjective, the surjectivity of $\Phi_{L,M}$ follows from [W3], (2.9), but we proceed directly. Consider on $X \times X$ the exact sequence

$$(*) \qquad 0 \to (L(-D) \boxtimes M)(-\Delta) \to L(-D) \boxtimes M \to L \otimes M(-D) \to 0.$$

Since

$$\Gamma(L(-D)) \otimes \Gamma(M) \to \Gamma(L \otimes M(-D))$$

is surjective (both degrees are $\geq 2g+1$), and $h^1(L(-D)) = h^1(M) = 0$, we have the vanishing $H^1(X \times X, (L(-D) \boxtimes M)(-\Delta)) = 0$. Twisting (*) by $\mathcal{O}(-\Delta)$, we see the surjectivity of $\Phi L(-D),M$ is equivalent to

$$H^1(X \times X, (L(-D) \boxtimes M)(-2\Delta)) = 0.$$

Since the surjectivity of $\Phi_{L,M}$ follows from the vanishing of $H^1(X \times X, (L \boxtimes M)(-2\Delta))$, it suffices to show

$$H^1(X \times X, p_1^*(L \otimes \mathcal{O}_D) \otimes p_2^*(M) \otimes \mathcal{O}(-2\Delta)) = 0.$$

We use the Leray spectral sequence for $p_1 : X\times X \to X$. Denoting the relevant sheaf by F, note $p_{1*}F$ has finite support on X, so $H^1(X, p_{1*}F) = 0$. We claim $R^1p_{1*}F = 0$. Since $R^2p_{1*}F = 0$, it suffices to show H^1 along the fibres is 0. For $P \in X$, not in the support of D, the fibre is empty. For P in the support of D, we need to consider $H^1(X, M(-2P))$, which is 0 for degree reasons. This gives the desired vanishing and completes the proof of (a) of Theorem 2.2.

As for (b), the hypotheses on X guarantee that the canonical curve is ideal-theoretically defined by quadrics (Petri's Theorem). The result then follows directly from Lemma 2.3.

(2.4) More precise bounds are given in [BEL], using a somewhat different method. In particular, they prove that if X is non-hyperelliptic, then the bound in (a) may be improved to $\deg L + \deg M \geq 6g+2$; and if $\mathrm{Cliff}(X) \geq 3$, and $\deg L \geq 4g+1-3\mathrm{Cliff}(X)$, then $\Phi_{K,L}$ is surjective. One concludes that on a general curve, $\Phi_{K,L}$ is surjective if $\deg L \geq (5/2)g+1$. We suspect that this is not the best bound. We ask the

Question 2.5: On a general curve of genus $g \geq 12$, is $\Phi_{K,L}$ surjective for L very ample, of $\deg L \geq 2g-2$? What is the best bound?

[BEL] also obtains degree bounds for the surjectivity of higher-order Gaussians $\Phi_i : \mathcal{R}_i(L,M) \to \Gamma(K^i \otimes L \otimes M)$.

(2.6) S. Tendian [T] has examined the border-line cases of part (b) of the Theorem. For instance, for X trigonal, he proves $\Phi_{K,L}$ surjective if $\deg L \geq 3g+7$, by computing on the rational normal scroll on which the canonical curve sits. If $\mathrm{Cliff}(X) \geq 2$ and $\deg L = 4g-4$, Lemma 2.3 gives the surjectivity of $\Phi_{K,L}$ unless $L = 2K$; Tendian proves surjectivity of $\Phi_{K,2K}$ in all cases except for bielliptic curves (i.e., elliptic-hyperelliptic), a plane quintic, and some (low genus) conic sections of a del Pezzo surface. For tetragonal curves, he computes using the three-dimensional rational scroll on which the canonical curve sits [Schr]. When $\mathrm{Cliff}(X) \geq 3$, he uses an argument motivated by deformation theory to see that $\Phi_{K,2K}$ is surjective ((3.6.1) below).

§3. Deformations of cones and Gaussian maps

(3.1) Let $X \subset \mathbb{P}^r$ be a smooth and projectively normal subvariety, with normal bundle N_X. We consider the affine cone $C_X \subset \mathbb{A}^{r+1}$, which has an isolated singularity at the vertex. The affine coordinate ring is $R = \oplus\Gamma(X, L^n) = P/I$, where $P = \mathbb{C}[z_0, \ldots, z_r]$, and I is the ideal generated by homogeneous polynomials vanishing on X. The module T^1_R of first-order deformations of R is computed via the exact sequence

$$\mathrm{Der}(P,P)\otimes R \to \mathrm{Hom}_R(I/I^2, R) \to T^1_R \to 0.$$

Since R is graded, so is T^1_R; we write $T^1 = \oplus T^1(i)$, $i \in \mathbf{Z}$. Examining the graded version of the above sequence, one may deduce Schlessinger's formula [Schl]:

$$T^1(i) = \mathrm{Coker}\{\Gamma(\mathbb{P}^r, \Theta_{\mathbb{P}}(i)) \to \Gamma(X, N_X(i))\}.$$

(3.2) To write a first-order deformation of R corresponding to an element of $T^1(i)$, a defining (homogeneous) equation $f \in I$ is perturbed to $f + \varepsilon g$, where g is homogeneous of degree $\deg(f)+i$. In particular, (first-order) deformations of weight 0 are conical; deformations of R of weight ≥0 are equimultiple, since terms of higher degree are added to the equations. On the other hand, first-order deformations of weight ≤ 0 of C_X can be extended to deformations of the projective cone $\overline{C}_X \subset \mathbb{P}^{r+1}$; homogenize the deformed equations using a new variable z, writing $f + \varepsilon g z^j$ for a deformation of weight $-j$. Therefore, one might expect to find that $\overline{C}_X$ has only conical deformations in case $T^1(-i) = 0$, $i > 0$. However, this requires some functorial interpretation, or study of deformations of all orders, just as is needed to show that $T^1 = 0$ implies rigidity.

(3.3) Let us return to the case of a projectively normal curve $X \subset \mathbb{P}^r$, with $L = \mathcal{O}_X(1)$. For the affine cone, the spaces $T^1(i)$ for $i \geq 0$ are not too difficult to compute, e.g., if $\deg L \geq 2g-2$ ([Mu2], [W2]). For the much more delicate negative weight pieces, we have

Theorem 3.4 ([W2]): Let $X \subset \mathbb{P}^r$ be a projectively normal embedding of a smooth curve of genus >0. Then the graded pieces of T^1 of the affine cone satisfy

$$T^1(-i) = (\mathrm{Coker}\ \Phi_{L, K \otimes L^{i-1}})^*, \quad i > 0.$$

Proof: We have by 3.1

$$T^1(-i) = \mathrm{Coker}\{\Gamma(\mathbb{P}^r, \Theta_{\mathbb{P}}(-i)) \to \Gamma(X, \Theta_{\mathbb{P}} \otimes \mathcal{O}_X(-i)) \to \Gamma(X, N_X(-i))\}.$$

We claim the first map is an isomorphism in case i≥1. The Euler sequence gives

$$\begin{array}{ccccccccc} 0 \to & \mathcal{O}_{\mathbb{P}}(-i) & \to & \oplus(\mathcal{O}_{\mathbb{P}}(-(i-1)))^{r+1} & \to & \Theta_{\mathbb{P}}(-i) & \to 0 \\ & \downarrow & & \downarrow & & \downarrow & \\ 0 \to & \mathcal{O}_X(-i) & \to & \oplus \mathcal{O}_X(-(i-1))^{r+1} & \to & \Theta_{\mathbb{P}} \otimes \mathcal{O}_X(-i) & \to 0. \end{array}$$

One must prove the injectivity of

$$H^1(\mathcal{O}_X(-i)) \to \oplus H^1(\mathcal{O}_X(-(i-1)))^r.$$

But, the dual map

$$\oplus H^0(K \otimes L^{i-1})^{r+1} \to H^0(K \otimes L^i)$$

is given by dotting with a basis of $H^0(L)$, hence is really

$$H^0(K \otimes L^{i-1}) \otimes H^0(L) \to H^0(K \otimes L^i).$$

For $i = 1$, the map is surjective since L is very ample (and $g \neq 0$); this plus the normal generation of L implies surjectivity for $i > 1$. Our claim is thus established. Therefore,

$$T^1(-i) = \mathrm{Coker}\{\Gamma(X, \Theta_{\mathbb{P}} \otimes \mathcal{O}_X(-i)) \to \Gamma(X, N_X(-i))\},$$

so

$$T^1(-i)^* = \mathrm{Ker}\{H^1(N^* \otimes L^i \otimes K) \to H^1(\Omega^1_{\mathbb{P}} \otimes \mathcal{O}_X \otimes L^i \otimes K)\}.$$

By Proposition 1.10(a), this last term is the cokernel of $\Phi_{L,K\otimes L^{i-1}}$.

Corollary 3.5: Suppose L defines a projectively normal embedding of the curve X, and $\Phi_{K,L}$ is surjective. Then the affine cone satisfies $T^1(-i)=0$, for all $i\geq 1$.

Proof: We may assume $g\geq 2$. By the Theorem, one must show $\Phi_{L,K\otimes L^{i-1}}$ is surjective, assuming the result for $i=1$. For any line bundles L, M, and N, we have a commutative diagram (where the horizontal maps are Gaussians)

$$\begin{array}{ccc} \mathcal{R}(L,M)\otimes\Gamma(N) & \to & \Gamma(K\otimes L\otimes M)\otimes\Gamma(N) \\ \downarrow & & \downarrow \\ \mathcal{R}(L,M\otimes N) & \to & \Gamma(K\otimes L\otimes M\otimes N). \end{array}$$

Let $M=K$, $N=L^j$. The top row is surjective since $\Phi_{K,L}$ is. The surjectivity of the right vertical map follows from that of

$$\Gamma(K)\otimes\Gamma(L)\otimes\Gamma(K)\otimes\Gamma(L^j)\to\Gamma(K\otimes L\otimes K\otimes L^j),$$

which follows from the two basic results of Petri and Castelnuovo (see the Introduction).

Remarks (3.6.1) If $\deg L\geq 2g+3$, then L defines a projectively normal embedding of X, ideal-theoretically defined by quadrics, and such that the relations among the quadrics are generated by linear ones [Gr1]. It follows from deformation theory that C_X could have no deformations of weight ≤ -2; you can't perturb quadratic equations by constant terms and hope to be able to lift linear relations (e.g., [W1], Cor. 2.8). Therefore, the Gaussians $\Phi_{L,K\otimes L^j}$ are surjective for $j\geq 1$. However, this can also be deduced directly from Theorem 2.2 (a).

(3.6.2) The Theorem also gives a way to compute the corank of a Gaussian $\Phi_{L,K}$, at least when L gives a projectively normal embedding $X\subset\mathbb{P}^r$:

$$\mathrm{Cork}(\Phi_{K,L})=h^0(N_X(-1))-(r+1).$$

(3.7) A key theorem of Pinkham [P] asserts that the $\mathbb{C}^*$-action on C_X extends to the semi-universal deformation. This implies that the base space of this deformation is of the form $\mathbb{C}[[t_1,\ldots,t_\tau]]/J$, where the t_i have weights equal to the negatives of the weights of T^1, and J is generated by weighted homogeneous power series. In particular, if $T^1(-i)=0$, $i>0$, then the defining equations for the total space will have the form

$$F=f+\Sigma t_i g_i+\Sigma t_i t_j g_{ij}+\ldots,$$

where the g_i, g_{ij}, etc, are homogeneous polynomials, of degree $\geq \deg f$. Thus, any adjacency (=nearby singularity) in a deformation has the same multiplicity (and same Hilbert polynomial). In particular, the cone C_X could not be smoothed. More precise results can be proved by giving a functorial definition of equisingular deformation and deformation of weight≥0.

Theorem 3.8 ([W2], 3.2): Suppose L is a very ample, normally generated line bundle on a curve, and $\Phi_{K,L}$ is surjective. Then the affine cone C_X has only equisingular deformations.

(3.9) One may also study deformations of the projective cone $\overline{C}_X \subset \mathbb{P}^{r+1}$. Suppose $Y \subset \mathbb{P}^{r+1}$ is a normal variety, with smooth hyperplane section X, occurring with normal bundle L. If $H^1(\mathcal{O}_Y)=0$, then a construction of Pinkham [P] allows one to degenerate Y to $\overline{C}_X$ in a flat family; take the projective cone over Y, and move a hyperplane section through the vertex, obtaining $\overline{C}_X$ when $t=0$ and Y when $t \neq 0$. (The cohomological condition on Y, which guarantees flatness, is automatic if $X \subset \mathbb{P}^r$ is projectively normal.) Now, the local Hilbert scheme of $\overline{C}_X$ maps smoothly to the deformations of weight ≤ 0 of C_X (Cf. [P], Theorem 5.1); for instance, on the tangent space level

$$H^0(\overline{C}_X, N_{\overline{C}_X}) = \oplus H^0(X, N_X(-i)) \quad (i \geq 0).$$

If $\Phi_{K,L}$ is surjective, then the term on the right contributes only $H^0(X,N_X)$ (from the Hilbert scheme of X in $\mathbb{P}^r$), plus the $r+1$ elements of $H^0(X,N_X(-1))$; these show up in $H^0(\overline{C}_X, N_{\overline{C}})$ as deformations of $\overline{C}_X$ via automorphisms of $\mathbb{P}^{r+1}$ which leave fixed the vertex and a fixed hyperplane section. Thus, all deformations of $\overline{C}_X$ in $\mathbb{P}^{r+1}$ are conical. We summarize by

Theorem 3.10: Let $X \subset \mathbb{P}^r$ be a projectively normal embedding of a curve, with $L=\mathcal{O}_X(1)$. Suppose that $\Phi_{K,L}$ is surjective. Then

(a) The Hilbert scheme of the cone $\overline{C}_X \subset \mathbb{P}^{r+1}$ consists only of conical deformations.

(b) If $Y \subset \mathbb{P}^{r+1}$ is any normal variety, with hyperplane section X for which the normal bundle is L, then $Y \cong \overline{C}_X$.

(3.11) A number of authors ([Ba], [L'v], Zak) use conditions like $T^1(-i)=0$, $i>0$ to imply that any Y as in (b) above must be the cone over (X,L). For instance, the following result was proved by F. Zak, using secant varieties (though L. Ein has given a simpler argument):

Theorem 3.12 (Zak): Let $X \subset \mathbb{P}^r$ be a smooth nondegenerate subvariety of codimension ≥ 2. Suppose $h^0(X,N_X(-1))=r+1$. Then X is not the hyperplane section of any variety Y in $\mathbb{P}^{r+1}$ except for Y the projective cone $\overline{C}_X$.

§4. The Gaussian map Φ_K

(4.1) Let K be the canonical line bundle on a smooth projective curve X of genus $g \geq 2$. We shall study the Gaussian map

$$\begin{array}{ccc} \Phi_K: \wedge^2\Gamma(K) & \rightarrow & \Gamma(K^3) \\ \| & & \| \\ \mathbb{C}^{g(g-1)/2} & \rightarrow & \mathbb{C}^{5g-5}. \end{array}$$

Φ_K was introduced in [W1] to study deformations of the cone over the canonical curve, using the ideas of §3.

Theorem 4.2 ([W1], Thm. 5.9): Suppose X is a smooth curve for which Φ_K is surjective. Then X cannot sit on a (minimal) $K-3$ surface. More generally, if X is a hyperplane section of a normal surface Y with trivial dualizing sheaf, then Y is isomorphic to the cone over the canonical curve of X.

Proof: Using 1.5, the rank of Φ_K for a hyperelliptic curve is $2g-3$, hence this map cannot be surjective. Considering the dimensions of the domain and range of Φ_K, we may assume X is nonhyperelliptic, of genus ≥ 10. Suppose $X \subset Y$, where Y is a $K-3$ surface. Then it is well-known that after possibly blowing down some -2 curves on Y that are disjoint from X, X is a very ample divisor on a $K-3$ surface with rational double points. Necessarily $X \subset Y \subset \mathbb{P}^g$, with $X = Y \cap H$. The corank of $\Phi_{K,K}$ equals that of Φ_K. Thus, by Theorem 3.10, such a Y would have to be the cone over X.

Remarks (4.3.1) Our original proof on the non-existence of the $K-3$ surface Y was slightly different. Via Pinkham's construction, such a Y yields a smoothing of the affine cone C_X; this contradicts Theorem 3.8.

(4.3.2) A completely different proof of Theorem 4.2 is given by Beauville-Mérindol [BM]. Suppose $X = Y \cap H$, with $Y \subset \mathbb{P}^g$ a $K-3$ surface. One has the natural diagram of Gaussians

$$\begin{array}{ccc}
\Lambda^2\Gamma(Y,\mathcal{O}_Y(X)) & \to & \Gamma(\Omega^1_Y(2X)) \\
 & & \searrow \\
\downarrow & \downarrow & \Gamma(\Omega^1_Y(2X)\otimes\mathcal{O}_X) \\
 & & \swarrow \\
\Lambda^2\Gamma(X,\mathcal{O}_X(X)) & \to & \Gamma(X,\Omega^1_X(2X)).
\end{array}$$

The surjectivity of "K implies the surjectivity of

$$\Gamma(\Omega^1_Y(2X)\otimes\mathcal{O}X) \to \Gamma(\Omega^1_X(2X)).$$

From this [BM] deduces that the normal sequence

$$0 \to \Theta_X \to \Theta_Y \otimes \mathcal{O}_X \to N_{X/Y} \to 0$$

splits. Next, observe that a curve X on a $K-3$ surface could have a split normal sequence; for instance, the fixed curve of an involution has this property. The main point of [BM] is to prove that conversely, a split normal sequence for X implies it is the fixed curve of an involution of Y! Since the quotient of Y by such an involution would be a Del Pezzo surface, the genus of X must be ≤ 10. The genus 10 case is ruled out by a separate argument.

(4.4) One easily computes that Φ_K is surjective for a complete intersection curve in $\mathbb{P}^r$, of multidegree $d_1 \leq d_2 \leq \ldots \leq d_{r-1}$, as long as $d_1 + d_2 + \ldots + d_{r-2} > r+1$ ([W1], Thm. 6.2). This condition means that the curve does not sit on a smooth complete intersection surface in $\mathbb{P}^r$

not of general type. Now, Mori-Mukai [MM] and Mukai [Mk] have shown that the general curve of genus g lies on a K-3 surface if and only if $g\leq 9$ or $g=11$. Combined with Theorem 4.2, the following result is natural as well as fundamental:

Theorem 4.5 ([CHM]): For a general curve of genus 10 or ≥ 12, the map Φ_K is surjective.

(4.6) The proof in [CHM] proceeds by writing down an appropriate stable curve X in each of the above genera, and then computing that the Gaussian

$$\Phi_\omega: \Lambda^3\Gamma(X,\omega) \to \Gamma(X, \Omega^1_X \otimes \omega^{\otimes 2})$$

is surjective (here ω is the dualizing sheaf). X is a graph curve--it is a union of rational curves, each intersecting exactly 3 others. It would be interesting to have some explicit smooth curves with Φ_K surjective, especially in genus 10.

(4.7) Another completely different proof of Theorem 4.5 has been given recently by C. Voisin [V]. S. Mukai had pointed out that if the class of a smooth curve X generates Pic of a K-3 surface, then X possesses some pencils L of small degree for which the adjoint line bundle K-L is not projectively normal. Such X on a K-3 are known to be generic in the sense of Brill-Noether-Petri, by [L1]; that is, if $W^r_d = \{L \in \mathrm{Pic}^d(X) \mid h^0(L) \geq r+1\}$, then dim $W^r_d = \rho(g,d,r)$ (the Brill-Noether number) and W^r_d is smooth outside W^r_{d+1}. Voisin then proves the

Theorem 4.7.1 ([V]). Suppose X is a curve of genus $g = 2s+1 \geq 7$, which is generic in the sense of Brill-Noether-Petri. If Φ_K is not surjective, then for every $L \in W^1_{s+2}$,

$$\mu_L: S^2\Gamma(K-L) \to \Gamma(2K-2L)$$

is not surjective.

(There is a similar result in the even genus case). Once one checks that on a general curve, the μ_L's as above are surjective, one has a new proof of Theorem 4.5, as well as a tie-in with Theorem 4.2.

(4.8) It is natural to consider the stratification of the moduli space $\mathfrak{M}_g$ of curves by the corank of the Gaussian. Among the obvious questions to ask:

(a) What are the possible values of the corank?
(b) What are the dimensions of strata?
(c) What is the relationship between this stratification and any other?
(d) What is the generic behavior of K-3 strata?

Voisin proves [V] that for generic curves on K-3's, the corank of Φ_K is ≤ 3 (at least in even genus ≥ 10). We mention other results on the calculation of Φ_K for special curves.

Proposition 4.9 ([W3]). If $g \geq 4$, the corank of Φ_K is $\leq 3g-2$, with equality if and only if X is hyperelliptic.

Proof: Let P be a point of X, with local coordinate t, and let ω be a holomorphic 1-form which is not zero at P. A basis of the 1-forms may be written $f_i\omega = (a_i t^{b_i} + \ldots)\omega$, $1 \leq i \leq g$, $a_i \neq 0$, $0 = b_1 < \ldots < b_g$. Computing $\Phi(f_i\omega \wedge f_j\omega)$ via $f_i df_j - f_j df_i$, whose order of vanishing at P is $b_i + b_j - 1$, we see that $\mathrm{rk}\ \Phi \geq \#\{b_i + b_j \mid 1 \leq i < j \leq g\}$. Because

$$b_1+b_2 < \ldots < b_1+b_g < b_2+b_g < \ldots < b_{g-1}+b_g,$$

the number of such integers is $\geq 2g-3$; an elementary argument shows that if $g \geq 5$, equality occurs iff $b_i = (i-1)b_2 - (i-2)b_1$, all i. If P is a nonhyperelliptic Weierstrass point, then for some s with $3 \leq s \leq g$, one has $h^0(\mathcal{O}(iP)) = 1$, $1 \leq i < s$, and $h^0(\mathcal{O}(sP)) = 2$. By Riemann-Roch, $h^0(K(-iP)) = g-i\ (0 \leq i < s)$, $h^0(K(-sP)) = g-s+1$. Therefore, $b_i = i-1$, $0 \leq i < s$, while $b_s \geq s$; thus one could not have the equality among the b_i's as above, so the rank of Φ is $> 2g-3$. If P is a hyperelliptic Weierstrass point, then $f_i = f_1^i$, so the rank of $\Phi = 2g-3$ (Cf. Prop. 1.5).

Remark (4.10) As pointed out by Arbarello and Ciliberto-Miranda [CM2], this Theorem is implicit in an old result of B. Segre [Se]. Segre considered the smallest possible linear space containing the image of the Gaussian map $X \to \mathrm{Grass}(1,n) \to \mathbb{P}^N$ (the first associated curve); in fact, the above arguments give a proof of Segre's theorem which meets modern standards. Aside from this remark, we know of no classical study of the Gaussian maps Φ_K.

Proposition 4.11: Let X be a nonhyperelliptic curve of genus $g \geq 4$.

(a) [CM2],[Br] If X is trigonal, then $\mathrm{cork}\ \Phi_K = g+5$.

(b) [CM2] If X is elliptic-hyperelliptic ("bielliptic"), $\mathrm{cork}\ \Phi_K = 2g-2$.

(c) [DM] If $X \subset \mathbb{F}_n$, the rational ruled surface over $\mathbb{P}^1$, and $n \geq 5$, then $\mathrm{cork}\ \Phi_K = n+6$.

(d) [CM1] Suppose X is general. For $g \leq 8$, Φ_K is injective; for $g=9$, Φ_K has a one-dimensional kernel; for $g = 11$, Φ_K has a 1-dimensional cokernel.

Proof: (a) is proved for a general trigonal X in [CM2]; the general result is due to J. Brawner, and will be in his UNC Ph. D. thesis. One considers the rational scroll Y in canonical space on which the curve sits. Examining the diagram involving Gaussians on Y (for K_Y+X) and X, it suffices to prove the vanishing

$$H^0(Y, \Omega^1_Y(\log X)(2K+X)) = 0.$$

(c) also requires a calculation on the ruled surface, primarily the surjectivity of the Gaussian on $\mathbb{F}_n$ of the line bundle $\mathcal{O}(K+X)$. (b) is a calculation using the fact that the canonical curve is a quadric section of a cone over an elliptic curve. Finally, the $g \leq 8$ case of (d) follows from consideration of appropriate singular curves; the surprising $g = 9$ and 11 cases follow from a sequence of inequalities.

(4.12) J. Brawner has proved [Br] that tetragonal curves of genus ≥ 15 have cork $\Phi_K \geq 9$. Of course, in this case one tries to compute using the rational normal scroll of dimension 3 on which the canonical curve sits (cf. [Schr]). We had already shown in [W3] that there exist curves of arbitrarily high genus, possessing a g^1_6, for which Φ_K is surjective.

§5. Other questions on Gaussians

(5.1) As mentioned in the Introduction, there exist higher-order Gaussian maps for a pair of line bundles L and M on a variety X. These arise from the filtration

$$\mathcal{R}_i(L,M) = \Gamma(X \times X, I^i \otimes (L \boxtimes M))$$

of $\mathcal{R}_0(L,M) = \Gamma(L) \otimes \Gamma(M)$, with maps

$$\Phi_i : \mathcal{R}_i(L,M) \to \Gamma(X, S^i\Omega^1_X \otimes L \otimes M).$$

We note that when $L = M$, the symmetric product $S^2\Gamma(L)$ is filtered by the $\mathcal{R}_{even}$, while $\wedge^2\Gamma(L)$ is filtered by the $\mathcal{R}_{odd}$.

(5.2) The injectivity of the Gaussians $\Phi_i(L,K-L)$ on a smooth curve X has been investigated by Griffiths ([G]). Suppose L is effective and special, with $d = deg(L)$ and $r+1 = h^0(L)$. Then as in (4.7) L defines a point $L \in W^r_d$. Consider the multiplication map

$$\mu_0 : \Gamma(L) \otimes \Gamma(K-L) \to \Gamma(K).$$

One can show that the tangent space to W^r_d at L is given by the subspace of $T_L Pic^d(X) \cong H^1(X, \mathcal{O}) \cong \Gamma(K)^*$ of functions annihilating the image of μ_0:

$$T_L W^r_d = (\text{image } \mu_0)^{\perp}.$$

The injectivity of μ_0 implies that W^r_d is smooth at L, of dimension $\rho(g,r,d) = g-(r+1)(g-d+r)$ ([ACGH], p. 189). According to Petri's Theorem (e.g., [L1]), on a general curve μ_0 is injective for every L. In general, dim Ker μ_0 is the embedding codimension of the singularity of W^r_d at L; this is examined further in [G], (8.22), via the Gaussian map

$$\mu_1 : \mathcal{R}(L,K-L) = \text{Ker } \mu_0 \to \Gamma(K^2).$$

It is proved there that if μ_1 is injective, then the point $\{L \to X\} \in \mathcal{G}^r_d$ (the space of curves plus line bundles) is smooth, of dimension $3g-3+\rho$. In [G], p. 9.14, Griffiths introduces the higher Gaussians $\mu_i = \Phi_i(L,K-L)$ defined on Ker $\mu_{i-1} = \mathcal{R}_i(L,K-L)$, relating them to osculating i-planes of the image curve in $\mathbb{P}^r$. He proves that μ_r is injective, as is μ_{r-1} "generically."

(5.3) An immersed, non-degenerate curve X in $\mathbb{P}^3$ is called a **contact curve** if the restriction of the contact form ω on $\mathbb{P}^3$ to X is 0. Here, $\omega = X_0dX_1 - X_1dX_0 + X_2dX_3 -$

$X_3dX_2 \in \Gamma(\Omega^1_{\mathbb{P}}(2))$ makes $\mathbb{P}^3$ into a complex contact manifold; on the affine piece $X_0 \neq 0$, ω may be written $dx+ydz-zdy$. In [Bry], R. Bryant describes how to construct every contact curve, and proves

(5.3.1) Every smooth curve embeds into $\mathbb{P}^3$ as a contact curve.

It is easy to see that the twisted rational cubic curve $[s^3,t^3,-3s^2t,st^2]$ is a contact curve (e.g., use (5.4.a)). For any non-degenerate space curve X, let $V \subset \Gamma(\mathcal{O}_X(1))$ be the linear system of hyperplanes, and consider the Gaussian

$$\Phi_X:\Lambda^2V \to \Gamma(X,\Omega^1_X(2)).$$

Proposition 5.4: (a) $X \subset \mathbb{P}^3$ is a contact curve if and only if Φ_X is not injective.

(b) If $\Gamma(X,N^*_X(2))=0$ (N_X=normal bundle of $X \subset \mathbb{P}^3$), then Φ_X is injective.

(c) Other than the twisted cubic, there are no linearly normal contact curves $X \subset \mathbb{P}^3$ satisfying either: X lies on a quadric; X lies on a smooth cubic; or X is a complete intersection.

Proof: A non-0 element of Λ^2V has (as an alternating form) rank 2 or 4, and rank 2 elements cannot be in the kernel of Φ_X (by 1.4). A rank 4 element of $\Lambda^2\Gamma(\mathbb{P},\mathcal{O}_{\mathbb{P}}(1)) \cong \Gamma(\mathbb{P},\Omega^1_{\mathbb{P}}(2))$, where $\mathbb{P}=\mathbb{P}^3$, gives rise after choice of coordinates to the form ω. By Lemma 1.7, the Gaussian Φ_X is equivalent to the restriction map

$$r: \Gamma(\mathbb{P},\Omega^1_{\mathbb{P}}(2)) \to \Gamma(X,\Omega^1_X(2)).$$

This gives (a). The other assertions are straightforward computations, using that r factors through $\Gamma(\Omega^1_{\mathbb{P}}(2)) \to \Gamma(\Omega^1_{\mathbb{P}}(2) \otimes \mathcal{O}_X)$, whose kernel is $H^0(N^*(2))$. For instance, if X lies on a smooth cubic surface S, consider the dual of the sequence of normal bundles for $X \subset S \subset \mathbb{P}$:

$$0 \to \mathcal{O}_X(-3) \to N^* \to N^*_{X/S} \to 0.$$

Thus, $H^0(N^*(2)) \subset H^0(\mathcal{O}_S(2H-X)|_X)$. But using the adjunction formula, the last line bundle on X has degree $X\cdot(2H-X)=d-(2g-2)$; for a linearly normal embedding, this expression is negative, except for a few cases, all of which lie on quadrics.

Question 5.5: (a) Is the twisted cubic the only linearly normal contact curve in $\mathbb{P}^3$?

(b) Let L be a base-point-free complete g^3_d on a curve X. Suppose Φ_L is not injective. Is L equal to three times a g^1_d?

(c) Suppose L is a line bundle of big degree on a curve X. Is $\mathrm{Ker}\ \Phi_L$ generated by alternating forms of rank 4? (Cf. [ACGH], pp. 142 and 255.)

Remarks (5.6) It is not hard to show (calculating as in [W1], Theorem 6.4) that the analogous statement for (c) is true for all line bundles on projective space.

(5.7) If V is an s-dimensional vector space, then the forms of rank ≤ 4 in $\wedge^2 V$ form a subvariety of dimension $4s-10$. Thus, any linear space of codimension $\leq 4s-11$ which contains no rank 2 forms must contains a form of rank 4. We conclude that

(5.7.1) If $\deg L \geq (5/2)g + 5$, then $\mathrm{Ker}\ \Phi_L$ contains a rank 4 element.

(A better bound of degree around $2g+C\sqrt{g}$, $C\sim 5/\sqrt{6}$, can be proved using (5.8) below.) One obtains for each such element a projection to $\mathbb{P}^3$ whose image is a contact curve. One might be able to reprove Bryant's theorem if one could produce an embedding in this way.

(5.8) Because of the relation with self-correspondences on a curve, Beauville asks the following

Question (5.8.1): For a general curve, what is the smallest degree of a line bundle L with Φ_L **not** injective?

One way to produce such an L is to take 3 times a g^1_d, where on a general curve $d = [(g+3)/2]$ is the smallest value; one gets a bound for $\deg$ L of about $(3/2)g$. A better bound is obtained using Brill-Noether theory; find the smallest integer d for which there exists an integer r satisfying

(a) the Brill-Noether number $\rho(g,r,d) = g-(r+1)(r-d+g) \geq 0$

(b) $r(r-1)/2 > 2d+g-1$.

There exists a g^r_d on a general curve by (a), and a dimension count using (b) gives a non-trivial kernel for the Gaussian. Asymptotically in g, the smallest degree obtained in this fashion is $g + \sqrt{g}\ (5/\sqrt{6} + \epsilon)$. We do not know if this bound is best possible.

(5.9) Suppose L is very ample on a smooth projective X, giving $X \subset \mathbb{P}^r$. Let $I_2 = \mathrm{Ker}(S^2\Gamma(L) \rightarrow \Gamma(L^{\otimes 2}))$ denote the space of quadratic polynomials on $\mathbb{P}^r$ vanishing on X. Then the corresponding Gaussian map (as in (5.1)) is

$$\Phi_2 : I_2 \rightarrow \Gamma(X, S^2\Omega^1_X \otimes L^{\otimes 2}).$$

This is easily seen to send

$$\Sigma a_{ij}s_is_j \mapsto \Sigma a_{ij}ds_ids_j,$$

where the $\{s_i\}$ form a basis of $\Gamma(L)$; another formulation, in terms of quadratic forms, is

$$Q(s_1, s_2, \ldots, s_{r+1}) \mapsto Q(ds_1, ds_2, \ldots, ds_{r+1}).$$

In fact, one can prove directly (as described in (1.1)) that these two definitions give a well-defined, natural map. This map should be compared with the "second fundamental form" of Griffiths-Harris ([GH], p. 366); when X is a quadric hypersurface, the image of a generator of I_2 is the symmetric bundle map $S^2\Theta_X \rightarrow N_X = \mathcal{O}_X(2)$, which is the usual second fundamental form of a submanifold.

Question 5.10: Let L be a very ample line bundle on a smooth projective curve X.

(a) What is the geometric significance of $\mathrm{Ker}\ \Phi_L$ being generated by rank 4 elements of $\wedge^2\Gamma(L)$?

(b) What is the geometric significance of the surjectivity of the second Gaussian map $\Phi_2: I_2 \to \Gamma(X, K^{\otimes 2} \otimes L^{\otimes 2})$?

[BEL] provides explicit bounds on deg L which guarantee surjectivity of higher Gaussians.

BIBLIOGRAPHY

[ACGH] E. Arbarello, M. Cornalba, P. Griffiths, and J. Harris, *Geometry of Algebraic Curves*, Vol. 1, Springer, New York, 1985.

[Ba] L. Badescu, "Infinitesimal deformations of negative weights and hyperplane sections," in Algebraic Geometry, Proceedings, L'Aquila 1988, Lecture Notes in Mathematics **1417**, Springer-Verlag (1990), 1-22.

[BM] A. Beauville and J.-Y. Mérindol, "Sections hyperplanes des surfaces K-3," Duke Math. J. **55** (1987), 873-878.

[BEL] A. Bertram, L. Ein, and R. Lazarsfeld, "Surjectivity of Gaussian maps for line bundles of large degree on curves," preprint.

[Br] J. Brawner, UNC Ph. D. thesis, in preparation.

[Bry] R. Bryant, "Conformal and minimal immersions of compact surfaces into the 4-sphere," J. Diff. Geom. **17** (1982), 455-473.

[CHM] C. Ciliberto, J. Harris, and H. P. Miranda, "On the surjectivity of the Wahl map," Duke Math. J. **57** (1988), 829-858.

[CM1] C. Ciliberto and H. P. Miranda, "On the Gaussian map for canonical curves of low genus," Duke Math. J. **61** (1990), 417-443.

[CM2] C. Ciliberto and H. P. Miranda, "Gaussian maps for certain families of canonical curves," preprint.

[DM] J. Duflot and H. P. Miranda, "The Gaussian map for rational ruled surfaces," preprint.

[EKS] D. Eisenbud, J. Koh, and M. Stillman, "Determinantal equations for curves of high degree," Amer. J. Math. **110** (1988), 513-539.

[Gr1] M. Green, "Koszul cohomology and the geometry of projective varieties, I," J. Diff. Geom. **19** (1984), 125-171.

[Gr2] M. Green, "Quadrics of rank four in the ideal of the canonical curve," Invent. Math. **75** (1984), 85-104.

[G] P. Griffiths, "Special divisors on algebraic curves," notes from 1979 Lectures at Regional Algebraic Geometry Conference, Athens, Ga.

[GH] P. Griffiths and J. Harris, "Algebraic geometry and local differential geometry," Ann. Sci. Ecole Norm. Sup. **12** (1979), 355-432.

[K] S. Kumar, "Proof of Wahl's conjecture on surjectivity of the Gaussian map for flag varieties," preprint.

[L1] R. Lazarsfeld, "Brill-Noether-Petri without degenerations," J. Diff. Geom. **23** (1986), 299-307.

[L2] R. Lazarsfeld, "A sampling of vector bundle techniques in the study of linear series," in M. Cornalba et al (eds), *Lectures on Riemann Surfaces*, World Scientific Press (Singapore, 1989), 500-559.

[L'v] S. L'vovsky, "Extensions of projective varieties and deformations," preprint.

[MM] S. Mori and S. Mukai, "The uniruledness of the moduli space of curves of genus 11," in Lecture Notes in Mathematics **1016**, Springer-Verlag (1983), 334-353.

[Mk] S. Mukai, "Curves, K-3 surfaces, and Fano 3-folds of genus 10," in *Algebraic geometry and commutative algebra in honour of M. Nagata* (1987), 357-377.

[Mu1] D. Mumford, "Varieties defined by quadratic equations," on *Questions on algebraic varieties* (C.I.M.E. 1969), Corso, Rome (1970), 30-100.

[Mu2] D. Mumford, "A remark on the paper of M. Schlessinger," in *Complex Analysis*, 1972, Rice University Studies **59** (1973), 113-117.

[P] H. Pinkham, "Deformations of algebraic varieties with $\mathbb{G}m$-actions," Astérisque **20** (1974), 1-131.

[Schl] M. Schlessinger, "On rigid singularities," in *Complex Analysis*, 1972, Rice University Studies **59** (1973), 147-162.

[Schr] F.-O. Schreyer, "Syzygies of canonical curves and special linear series," Math. Ann. **275** (1986), 105-137.

[Se] B. Segre, "Sulle curve algebriche le cui tangenti appartengono al massimo numero di complessi lineari indipendenti," Mem. Accad. Naz. Lincei (6) **2** (1927), 578-592.

[T] S. Tendian, "Deformations of cones over curves of high degree," Ph. D. dissertation (Univ. of N. Carolina, Chapel Hill), July 1990.

[V] C. Voisin, "Sur l'application de Wahl des courbes satisfaisant la condition de Brill-Noether-Petri," preprint.

[W1] J. Wahl, "The Jacobian algebra of a graded Gorenstein singularity," Duke Math. J. **55** (1987), 843-871.

[W2] J. Wahl, "Deformations of quasihomogeneous surfacesingularities," Math. Ann. **280** (1988), 105-128.

[W3] J. Wahl, "Gaussian maps on algebraic curves," J. Diff. Geom. **32** (1990), 77-98.

[W4] J. Wahl, "Gaussian maps and tensor products of irreducible representations," Manuscripta Math **73** (1991), 229-260.

Department of Mathematics
University of North Carolina
Chapel Hill, North Carolina, USA 27599-3250
jw@math.unc.edu

SOME EXAMPLES OF OBSTRUCTED CURVES IN $\mathbb{P}^3$

Charles H. Walter
Department of Mathematics, Rutgers University
New Brunswick NJ 08903, USA

In this paper we compute explicit étale neighborhoods of the points in the Hilbert scheme (the universal family of subschemes of $\mathbb{P}^3$) corresponding to curves in $\mathbb{P}^3$ of a certain type. Most of these curves are obstructed, that is to say, correspond to singularities of the Hilbert scheme. We have several purposes in doing this. One is simply to add to the list of singularities of the Hilbert scheme for which explicit equations have been computed. This is still a very short list which for the most part consists of fairly simple singularities such as two components crossing transversally ([15] 8.6, [14], [7]) or a double structure on a nonsingular variety ([5]). The example of [15] 8.7 seems to be the only explicit singularity known which is more complicated. Our new examples may clarify some of the causes of obstructions, particularly in terms of the cohomology and syzygies of the homogeneous ideal of the curve. A second purpose of the paper is to investigate a question of Sernesi concerning curves of maximal rank. This will be discussed in a moment. But we also simply wish to give an exposition of the deformation theory in terms of which the computation is made so as to make it more accessible to those who study algebraic space curves. We particularly wish to clarify the relationship between the deformations of a space curve (or subscheme) $Y \subset X = \operatorname{Proj} S$, the deformations of the ideal sheaf $\mathcal{J}_Y$ as a coherent sheaf, and the deformations of the homogeneous ideal $I(Y)$ as a homogeneous S–module. In the author's view, this fills a sort of expository gap between theory and examples which exists in most treatments. The problem is that the theory for deformations of subschemes involves computing and lifting infinitesimal deformations of $\mathcal{O}_Y$ as a sheaf of $\mathcal{O}_X$–algebras or of $S(Y)$ as an S–algebra. This leads to the Lichtenbaum–Schlessinger cotangent complex and André–Quillen cohomology, and in principle should lead to computations computing and lifting infinitesimal deformations of say a free differential graded S–algebra resolution of $S(Y)$. Examples, however, are never computed that way because such a resolution is too complicated. In practice, one computes and lifts infinitesimal deformations of a projective resolution of $I(Y)$, effectively computing the deformation theory of $I(Y)$ as a homogeneous S–module. Our perspective is that there are comparison theorems identifying the deformation theories of Y, of $\mathcal{J}_Y$, and of $I(Y)$ in certain circumstances, and so the whole formalism of André–Quillen cohomology may be avoided if one so desires. (We even give an alternate proof of the main comparison theorem which avoids André–Quillen cohomology.) Hence the deformation theory of a space curve may be developed entirely in terms of projective resolutions and Ext, which may not only make it more accessible to some readers, but also perhaps clarify the relationship between deformations and obstructions on the one hand and projective resolutions and cohomology of the original ideal on the other.

The question of Sernesi which our examples settle concerns curves of maximal rank. Roughly speaking, these are curves in $\mathbf{P}^3$ which impose the maximum number of conditions on surfaces of each degree, this maximum being calculated in terms of the degree and genus of the curves and the degree of the surfaces. More precisely, for any n the ideal sheaves $\mathcal{J}$ of these curves satisfies either $H^0(\mathcal{J}(n)) = 0$ or $H^1(\mathcal{J}(n)) = 0$. Such curves are very common, and many large classes of them (*e.g.* projectively normal curves, general nonspecial curves) are unobstructed. Sernesi asked:

Question 0.1. ([1] Question 3) Are all nonsingular curves in $\mathbf{P}^3$ of maximal rank unobstructed?

Example 3.2 will show that the answer is "no". However, the curves of that example, even though they are of maximal rank, do not have *seminatural cohomology*. This is a stronger condition meaning that for each n at most one $H^i(\mathcal{J}(n))$ is nonzero for $i = 0, 1, 2$. So we pose a new question:

Question 0.2. Are all curves in $\mathbf{P}^3$ with seminatural cohomology unobstructed?

The outline of the paper is as follows. In the first section we show how to compute the infinitesimal deformation theory of a finitely generated module over a graded noetherian K–algebra, for K a field. The obstructions to lifting infinitesimal deformations to the next order are certain Massey operations, which we give explicitly. In principle this section is a reprise of say [12], but we include more explicit formulas which are useful for practical computation.

In the second section we compare the different deformation theories relating to deformations of a closed subscheme $Y \subset X = \text{Proj } S$. The different types of deformations are deformations of Y as a subscheme of the fixed X, conical deformations of Y, deformations of the ideal sheaf $\mathcal{J}_Y$ as a sheaf of $\mathcal{O}_X$–modules, and deformations of the homogeneous ideal I as a homogeneous S–module. We show that the Massey product procedure of the first section can be used to compute all of the infinitesimal deformation theories compatibly, provided that one uses appropriate resolutions and cohomology theories. One can then reduce the comparison of deformation theories to comparison of certain global and graded Ext and André–Quillen cohomology groups. Much of our work is substantially a review of known work (especially [9], [10]), but there are some minor improvements in the statements of the theorems. We also give an alternate noncohomological proof of the comparison between say deformations of a subscheme $Y \subset X$ and of the coherent sheaf (of ideals) $\mathcal{J}_Y$ based on the structure theory of exact complexes of Buchsbaum and Eisenbud. This method seems in some ways more natural since it involves finding equations expressing the deformation of an ideal qua module as a deformation of ideals.

In the third section of the paper we apply the comparison theorems of the second section and the procedure of the first section to compute the deformation theory of curves $C \subset \mathbf{P}^3$ satisfying

(0.3) (1) There exists an integer m such that $H^1(\mathcal{J}_C(n)) = 0$ for $n \neq m$,
(2) The homogeneous ideal $I(C)$ has no minimal generators of degree m.

The study of these curves involves three numerical invariants:

$$
\begin{aligned}
a &:= \text{the number of minimal generators of } I(C) \text{ of degree } m+4,\\
b &:= \text{the number of minimal relations of } I(C) \text{ of degree } m+4, \qquad (0.4)\\
c &:= h^1(\mathcal{J}_C(m)).
\end{aligned}
$$

Our result is

Theorem 0.5. *Let C, m, a, b, c be as in (0.3) and (0.4). Then there exists an étale neighborhood of $[C]$ in* Hilb $\mathbf{P}^3$ *which is isomorphic to the subvariety of an affine space* *given by a set of equations* $\Sigma_{j=1}^{b} X_{ij} Y_{jk} = 0$ $(1 \le i \le a,\ 1 \le k \le c)$ *with the* X_{ij} *and* Y_{jk} *linearly independent in* $T^{\dagger}_{[C]}\mathrm{Hilb}\,\mathbf{P}^3$.

In particular, such a curve is obstructed (i.e. the Hilbert scheme is singular) if and only if a, b, c are all positive. We also should remark that the X_{ij} and Y_{jk} do not span $T^{\dagger}_{[C]}\mathrm{Hilb}\,\mathbf{P}^3$. There are other tangent directions in Hilb $\mathbf{P}^3$ along which the curve deforms without encountering obstructions.

In the final section of the paper, we show that curves satisfying (0.3) and (0.4) exist for all triples (a,b,c) of positive integers. The examples given include curves which are both obstructed and of maximal rank (i.e. for each n either $H^0(\mathcal{J}_C(n)) = 0$ or $H^1(\mathcal{J}_C(n)) = 0$). This gives the negative answer to Question 0.1.

Acknowledgements. The negative answer to Question 0.1 has also been produced independently by G. Bolondi, J. Kleppe, and R. Miró–Roig by studying the same examples using a different method. The present paper was prepared with the partial support of a Henry Rutgers Research Fellowship.

Conventions. We reserve k for use as an index, so fields are K. We will denote the graded pieces of a homogeneous module by subscripts on the left, *e.g.* ${}_nE$. This is to avoid confusion with subscripts related to homology. In this paper we are only interested in Hom's and Ext's which preserve degrees, *i.e.* those of degree 0. In order to make this clear we write these as ${}_0\mathrm{Hom}$ and ${}_0\mathrm{Ext}$. Also if $X.$ and $Y.$ are complexes in an abelian category, we write $\mathrm{Hom}^q(X.,Y.)$ for $\Pi_p \mathrm{Hom}(X_{p+q}, Y_p)$. Then $\mathrm{Hom}^{\cdot}(X.,Y.)$ is a complex satisfying $H^q(\mathrm{Hom}^{\cdot}(X.,Y.)) = \mathrm{Ext}^q(X.,Y.)$. Finally, if $\mathcal{F}$ is a quasicoherent sheaf on a projective variety $X = \mathrm{Proj}\ S$, then $\Gamma_*(\mathcal{F})$ denotes the S–module of graded global sections $\oplus_{n\in\mathbb{Z}} \Gamma(\mathcal{F}(n))$. Its right derived functors are $H^i_*(\mathcal{F}) = \oplus_{n\in\mathbb{Z}} H^i(\mathcal{F}(n))$.

1. Computation of the Universal Formal Deformation of a Module

In this section we give a brief review of the computation of the projective hull for deformations of a finitely generated homogeneous module F over a graded Noetherian K–algebra S with $S_0 = K$, where K is a field. This is for deformations over trivially graded Artin local K–algebras A. Here *trivially graded* means that any nonzero $a \in A$ is homogeneous of degree 0. The basics of the method will be needed for the comparison theorems of the second section, while the explicit computations will be used in the proof of the main theorem in the third section. Although this theory is not new, it does not seem to be as well known as it should. This method is sometimes referred to the Massey product method, because the obstruction classes of (1.8) are a form of Massey product. We note that our Massey products are really "commutative" Massey products rather than matric Massey products (as in [13] or [12]) because the multiplication constants $\mu^k_{ij} = \mu^k_{ji}$ appearing in their definition (1.5) are those of a commutative ring. The matric Massey products of [13] have different multiplication constants which are those of a noncommutative ring.

Let K, S and F be as in the previous paragraph. The rings $S\otimes_K A$ inherit the grading of S. We will call *deformations of F* pairs (A,F_A) where A is a trivially graded Artin local K–algebra, and F_A is a homogeneous $(S\otimes_K A)$–module flat over A such that $F_A\otimes_A K \cong F$. (We will not concern ourselves with deformations over more general rings.) Such a deformation is said to be of *order* $\leq r$ if $m_A^{r+1} = 0$. A *universal* r^{th}*–order deformation of F* is a deformation (A,F_A) of F of order $\leq r$ such that for any deformation (B,F_B) of F of order $\leq r$ there exists a unique morphism $A \longrightarrow B$ such that $F_A\otimes_A B \cong F_B$. A *hull* for the deformations of F is a pair $(\hat{H},F_{\hat{H}})$ where $\hat{H}$ is a trivially graded complete local K–algebra, and $F_{\hat{H}}$ is a $(S\otimes_K \hat{H})$–module flat over $\hat{H}$ satisfying $F_{\hat{H}}\hat{\otimes}_{\hat{H}} K \cong F$ such that for all deformations (A,F_A) of F there exists a unique morphism $\hat{H} \longrightarrow A$ such that $F_{\hat{H}}\hat{\otimes}_{\hat{H}} A \cong F_A$. We show how to construct the hull as the inverse limit of the universal r^{th}–order deformations.

Let $(L.,\delta)$ be an S–free resolution of F. Then ${}_0\mathrm{Hom}^{\cdot}_S(L.,L.)$ is a differential graded S–algebra with differential $d(x) = \delta x-(-1)^q x\delta$ for $x \in {}_0\mathrm{Hom}^q_S(L.,L.)$. Its cohomology is the ${}_0\mathrm{Ext}^i_S(F,F)$. If A is an Artin local K–algebra, then ${}_0\mathrm{Hom}^{\cdot}_S(L.,L.)\otimes_K A$ is a differential graded $(S\otimes_K A)$–algebra with differential $d = d\otimes 1$ and with $\delta = \delta\otimes 1 \in R^1\otimes_K A$. Any deformation of F over A has a resolution $(L^{\cdot}\otimes_K A,\delta+\phi)$ with $\phi \in {}_0\mathrm{Hom}^1_S(L.,L.)\otimes_K m_A$ satisfying $(\delta+\phi)^2 = d(\phi)+\phi^2 = 0$. Conversely, any such ϕ defines a flat deformation $\tilde{F}$ of F over A (*cf.* [6] p. 427). Two such ϕ and ψ define the same $\tilde{F}$ if and only if there exists a $u \in {}_0\mathrm{Hom}^0_S(L.,L.)\otimes_K m_A$ such that $\delta+\phi = (1+u)^{-1}(\delta+\psi)(1+u)$. For first order deformations, these conditions become $d(\phi) = 0$ and $\phi = \psi+d(u)$. So first order deformations are defined by classes in ${}_0\mathrm{Ext}^1_S(F,F)\otimes(m_A/m_A^2)$. Let T be the polynomial algebra $K[{}_0\mathrm{Ext}^1_S(F,F)^*]$. The universal first order deformation is defined over T/m_T^2 by the canonical member of ${}_0\mathrm{Ext}^1_S(F,F)\otimes_K(m_T/m_T^2) \cong {}_0\mathrm{Ext}^1_S(F,F)\otimes_K{}_0\mathrm{Ext}^1_S(F,F)^*$.

Now suppose we have a universal r^{th} order deformation of F with resolution $(L.\otimes_K B_r, \delta+\phi)$ with $B_r = T/J_r$ such that $m_T^2 \supseteq J_r \supseteq m_T^{r+1}$. Choose a K–splitting

$$(1.1) \qquad J_r/m_T J_r \underset{\epsilon}{\overset{\pi}{\leftrightarrows}} T/m_T J_r \underset{p}{\overset{i}{\leftrightarrows}} B_r.$$

Then a $\psi \in {}_0\mathrm{Hom}^1_S(L.,L.)\otimes_K(J_r/m_T J_r)$ satisfies

$$(1.2) \qquad (\delta+i(\phi)+\psi)^2 = 0$$

if and only if $d(\psi) = -\pi((i\phi)^2)$. Thus (since we may verify that $\pi((i\phi)^2)$ is a cocycle), the *Massey product* $\overline{\pi((i\phi)^2)} \in {}_0\mathrm{Ext}^2_S(F,F)\otimes_K(J_r/m_T J_r)$ is the obstruction to lifting the deformation to $T/m_T J_r$. It defines an obstruction mapping o_r: ${}_0\mathrm{Ext}^2_S(F,F)^* \longrightarrow J_r/m_T J_r$. We can choose a ψ such that (1.2) is satisfied modulo $\mathrm{im}(o_r)$. Putting $B_{r+1} = T/(\mathrm{im}(o_r)+m_T J_r)$, the resolution of the universal $(r+1)^{st}$ order deformation of F is $(L.\otimes_K B_{r+1}, \delta+i(\phi)+\psi)$.

We now repeat the previous computations in a more explicit form. Choose a system of cocycle representatives $\{\phi_i\}_{i\in\Gamma_1}$ for a basis of ${}_0\mathrm{Ext}^1_S(F,F)$ and a system $\{\theta_t\}_{t\in\Theta}$ of cocycle representatives of a basis of ${}_0\mathrm{Ext}^2_S(F,F)$. In principle it does not matter what bases or what representatives are chosen, but in practice certain choices lead to simpler computations than others. Let $\{f_i\}_{i\in\Gamma_1}$ be a basis of ${}_0\mathrm{Ext}^1_S(F,F)^*$ dual to $\{\overline{\phi}_i\}_{i\in\Gamma_1}$. Then $T = K[f_i]_{i\in\Gamma_1}$, and the universal first order deformation of F has resolution $L.\otimes_K(T/m_T^2)$ with boundary

$$(1.3) \qquad \delta + \sum_{i\in\Gamma_1} \phi_i\otimes f_i.$$

To compute the obstructions to lifting to the second order, we compute the square of (1.3). It is a cocycle involving only the terms quadratic in the f_i. After explicit computation it may be written in the form

$$(1.4) \qquad (\delta + \sum_{i\in\Gamma_1} \phi_i\otimes f_i)^2 = \sum_{t\in\Theta} \theta_t\otimes(\sum_{k\in\Delta_1} a_{kt}f_k) + \sum_{k\in\Delta_1} \beta_k\otimes f_k,$$

where $\{f_k\}_{k\in\Delta_1}$ is the set of quadratic monomials, and the β_k are coboundaries. Choosing cochains ψ_k so that $d(\psi_k) = -\beta_k$, we get

$$(1.5)\qquad (\delta + \sum_{i\in\Gamma_1} \phi_i\bullet f_i + \sum_{k\in\Delta_1} \psi_k\bullet f_k)^2 \equiv \sum_{t\in\Theta} \theta_t\bullet(\sum_{k\in\Delta_1} \alpha_{kt} f_k) \quad (\text{mod } m_T^3).$$

The ideal J_2 is then defined as the the sum of m_T^3 and the span of the $\Sigma_{k\in\Delta_1} \alpha_{kt} f_k$ $(t \in \Theta)$, and the universal second order deformation of F is then defined by the resolution $L.\bullet T/J_2$ with boundary $\delta + \Sigma_{i\in\Gamma_1} \phi_i\bullet f_i + \Sigma_{k\in\Delta_1} \psi_k\bullet f_k$.

To compute the universal third order deformation one repeats the process, computing the left hand side of (1.5) over T. It is a cocycle modulo $m_T J_2$, allowing it to be written in a form analogous to (1.4) modulo $m_T J_2$. One then gets a congruence analogous to (1.5) modulo $m_T J_2$. This process adds third order terms to both sides of (1.5) but leaves the lower order terms alone. One adds yet higher order terms by further iterations. In the limit the $\Sigma_k \alpha_{kt} f_k$ $(t \in \Theta)$ generate an ideal $\hat{J} \subset \hat{T}$, and the analog of (1.5) holds modulo $\hat{m}\hat{J}$. The universal deformation of F is defined over $\hat{T}/\hat{J}$. The *obstruction space* of F is $(\hat{J}/\hat{m}\hat{J})^*$, and it is injected in ${}_0\text{Ext}_S^2(F,F)$ essentially by the righthand side of (1.5).

If at some point the process terminates with equality in the analog of (1.6) (instead of merely congruence), then the universal deformation is algebraizable to $\text{Spec}(T/J)$ where J is the ideal generated by the $\Sigma_k \alpha_{kt} f_k$ $(t \in \Theta)$. This would give an étale open set of some moduli scheme. For examples where this occurs see [15] 8.6–8.7 and [14] Lemma 6 as well as §3 below.

Finally, we should point out what the Massey products are. At the rth stage, let $\{f_i\}_{i\in\Gamma_r}$ denote both a basis of m_T/J_r as lifted to T, and let $\{f_i\}_{i\in\Delta_r}$ be a basis of $J_r/m_T J_r$ lifted to T. If we write $\delta + \Sigma_{i\in\Gamma_r} \phi_i\bullet f_i$ for the differential of the universal r^{th} order deformation of F, then its square modulo $m_T J_r$ is

$$(1.6)\qquad (\delta + \sum_{i\in\Gamma_r} \phi_i\bullet f_i)^2 \equiv \sum_{k\in\Delta_r} (\sum_{ij\in\Gamma_r} \mu_{ij}^k \phi_i\phi_j) \bullet f_k \quad (\text{mod } m_T J_r)$$

where the $\mu_{ij}^k \in K$ are defined by $f_i f_j \equiv \Sigma_k \mu_{ij}^k f_k$ (mod $m_T J_r$). The obstruction to lifting the differential to $T/m_T J_r$ are the Massey products (for $k \in \Delta_r$):

$$(1.7) \qquad \overline{\sum_{ij\in\Gamma_r} \mu^k_{ij}\phi_i\phi_j} = \sum_{t\in\Theta} \alpha_{kt}\overline{\partial}_t \in {}_0\mathrm{Ext}^2_S(F,F).$$

2. Comparison of Deformation Theories

In this section we give comparison theorems for various sorts of deformation theories relating to a closed subscheme $Y \subsetneq X = \mathrm{Proj}\ S$. Much of our work is substantially a review of known work (especially [9], [10]), but there are some minor improvements in the statements of the theorems (such as Theorem 2.3 vs. [10] Theorem 2.2.1). For the most part, the results are proved by comparing certain cohomology groups. But at the end of the section we give an alternate noncohomological proof of the comparison between deformations of Y and deformations of the coherent sheaf $\mathcal{J}_Y$ which uses the structure theory of exact complexes of Buchsbaum and Eisenbud to show explicitly that any deformation of $\mathcal{J}_Y$ is a sheaf of ideals in the appropriate circumstances.

The types of deformations which we compare are:

$$(2.0.1) \qquad \begin{array}{ccc} \text{deformations of } Y & \longrightarrow & \text{deformations of } \mathcal{J}_Y \\ \uparrow & & \uparrow \\ \text{conical deformations of } Y & \longrightarrow & \text{deformations of } I(Y) \end{array}$$

"Deformations of $\mathcal{J}_Y$" (resp. "of $I(Y)$") here means deformations as a sheaf of $\mathcal{O}_X$–modules (resp. as a homogeneous S–module). "Conical deformations of Y" means precisely deformations of Y such that the graded global sections of the deformation of $\mathcal{J}_Y$ give a flat deformation of $I(Y)$. The arrows are natural, functorial maps. The comparison theorems are based on the fact that all the deformation theories can be computed using the method of §1, only with the complexes $L.$ and $\mathrm{Hom}^\cdot(L.,L.)$ modified so that the right type of deformations are computed. This is done as follows. For deformations of $I(Y)$, one proceeds without modification with a projective resolution $L.$ of $I(Y)$ and $\mathrm{Hom}^\cdot(L.,L.)$. For deformations of $\mathcal{J}_Y$, let $\mathfrak{U}$ be an affine open cover of X, and let $C.(\mathfrak{U},\mathcal{O}_X)$ be the Čech homology resolution of $\mathcal{O}_X$ with

$$C_p(\mathfrak{U},\mathcal{O}_X) = \bigoplus_{i_0\cdots i_p} j_{i_0\cdots i_p!}(\mathcal{O}_{U_{i_0}\cap\cdots\cap U_{i_p}})$$

where $j_{i_0\cdots i_p}\colon U_{i_0}\cap\cdots\cap U_{i_p} \to X$ is the inclusion. Then one replaces $L.$ by the total

complex $\mathscr{L}.$ of the double complex of sheaves $\tilde{L}.\bullet C.(\mathfrak{U},\mathcal{O}_X)$, and $\mathrm{Hom}^{\cdot}(L.,L.)$ is replaced by $\mathrm{Hom}^{\cdot}(\mathscr{L}.,\mathscr{L}.)$. For conical deformations of Y, one replaces $L.$ by a free differential graded S–algebra $A.$ resolving $S(Y) = S/I(Y)$. And instead of $\mathrm{Hom}(L.,L.)$ one uses ${}_0\mathrm{Der}^{\cdot}_S(A.,A.)$ because the deformations must preserve the algebra structure. For deformations of Y, one localizes again using a Čech homology complex. Note that if $L.$ is the projective resolution of $I(Y)$ induced by $A.$, then there are natural maps between the $\mathrm{Hom}^{\cdot}$ complexes in the same directions as the arrows of (2.0.1) and these preserve products. Hence any infinitesimal deformations of the type in §1 can be transferred from one theory to the next. We will get isomorphisms of the deformation theories if the maps on the tangent spaces (corresponding to Ext^1) are bijective, and the maps on obstruction spaces (corresponding to Ext^2) are injective. The precise cohomology groups to be compared are

$$\begin{array}{ccc} H^i(Y/X,\mathcal{O}_Y) & \longrightarrow & \mathrm{Ext}^i_{\mathcal{O}_X}(\mathcal{J}_Y,\mathcal{J}_Y) \\ \uparrow & & \uparrow \\ {}_0H^i(S,A,A) & \longrightarrow & {}_0\mathrm{Ext}^i_S(I,I) \end{array} \tag{2.0.2}$$

where $I = I(Y)$, $A = S(Y) = S/I$, and the cohomology in the upper left (resp. lower left) is global André–Quillen cohomol/gy (resp. graded André–Quillen cohomology). (See [11] Chapters 3–5, or [8] Chapters II–IV for global and graded André–Quillen cohomology and its role in deformation theory.)

For the vertical arrows of (2.0.2) there is

Theorem 2.1. (Kleppe) *Suppose Y is a closed subscheme of $X =$* Proj S *for S a graded K–algebra where S is a Noetherian graded ring with irrelevant ideal $\mathfrak{m}$ such that $S/\mathfrak{m} = K$ and $H^0_{\mathfrak{m}}(S) = H^1_{\mathfrak{m}}(S) = 0$. If ${}_0\mathrm{Hom}_S(I,H^1_*(\mathcal{J}_Y)) = 0$, then there is an isomorphism between infinitesimal deformations of Y and infinitesimal conical deformations of Y, and also between infinitesimal deformations of $\mathcal{J}_Y$ and of $I(Y)$.*

Proof. For the first isomorphism, see [9] Remark 3.7. For the second, there is similarly a long exact sequence and a spectral sequence

$${}_0\mathrm{Ext}^i_{\mathfrak{m}}(I,I) \longrightarrow {}_0\mathrm{Ext}^i_S(I,I) \xrightarrow{\eta_i} \mathrm{Ext}^i_{\mathcal{O}_X}(\mathcal{J}_Y,\mathcal{J}_Y) \longrightarrow {}_0\mathrm{Ext}^{i+1}_{\mathfrak{m}}(I,I),$$

$$E_2^{pq} = {}_0\mathrm{Ext}^p_S(I,H^q_{\mathfrak{m}}(I)) \Longrightarrow {}_0\mathrm{Ext}^{p+q}_{\mathfrak{m}}(I,I).$$

Since $H^0_{\mathfrak{m}}(I) = H^1_{\mathfrak{m}}(I) = 0$, and $H^2_{\mathfrak{m}}(I) \cong H^1_*(\mathcal{J}_Y)$, we see that ${}_0\mathrm{Ext}^i_{\mathfrak{m}}(I,I) = 0$ for $i \leq 2$. So η_1 is bijective and η_2 injective, which proves the theorem.

For the horizontal arrows of of (2.0.1), we have the following results. We consider the following situation:

Property 2.2.1. Let $S \longrightarrow S/I = A$ be a surjective morphism of Noetherian rings, and let Z be the system of supports $\{\mathfrak{p} \in \mathrm{Spec}\ S \mid \mathrm{depth}_{\mathfrak{p}}\ S_{\mathfrak{p}} \geq 3\}$. We suppose that $\mathrm{depth}_I\ S \geq 2$, and that $A_{\mathfrak{p}}$ has finite projective dimension over $S_{\mathfrak{p}}$ for $\mathfrak{p} \in V(I) - Z$.

Lemma 2.2. ([10] Lemma 2.2.3) *Suppose $S \longrightarrow S/I = A$ satisfy Property 2.2.1, and let Z be the corresponding system of supports. Then there are isomorphisms* $S \xrightarrow{\sim} \mathrm{Hom}_S(I,I)$ *and* $H^1(S,A,A) \xrightarrow{\sim} \mathrm{Ext}^1_S(I,I)$, *and an injection* $\Gamma_Z(H^2(S,A,A)) \hookrightarrow \Gamma_Z(\mathrm{Ext}^2_S(I,I))$. *If S is graded and I homogeneous, these maps preserve the grading.*

Theorem 2.3. (*a*) *Suppose $Y \subseteq X = \mathrm{Proj}\ S$ is defined by $S \longrightarrow S/I = A$ satisfying Property 2.2.1. Then there is an isomorphism between infinitesimal conical deformations of Y and infinitesimal homogeneous deformations of $I(Y)$.*

(*b*) *Suppose that for the closed subscheme $Y \subseteq X = \mathrm{Proj}\ S$, the stalks of $\mathcal{O}_X \longrightarrow \mathcal{O}_Y$ satisfy Property 2.2.1. Then if $H^1(\mathcal{O}_X) = 0$, there is an isomorphism between infinitesimal deformations of Y and of $\mathcal{J}_Y$.*

Actually, we will show that (b) holds whenever $H^1(Y/X,\mathcal{O}_Y) \cong \mathrm{Ext}^1(\mathcal{J}_Y,\mathcal{J}_Y)$ is finite-dimensional, even if X is not projective.

Proof. (b) We consider the natural maps ζ_i: $H^i(Y/X,\mathcal{O}_Y) \longrightarrow \mathrm{Ext}^i_{\mathcal{O}_X}(\mathcal{J}_Y,\mathcal{J}_Y)$. These ζ_i are the abutment of a morphism of hypercohomology spectral sequences

$$H^p(\xi_q)\colon\ E_2^{pq} = H^p(X,\mathcal{H}^q(Y/X,\mathcal{O}_Y)) \longrightarrow E_2'^{pq} = H^p(X,\mathcal{E}xt^q(\mathcal{J}_Y,\mathcal{J}_Y)),$$

where $\mathcal{H}^q(Y/X,\mathcal{O}_Y)$ denotes the André–Quillen cohomology sheaf. Since $Y \subseteq X$, $\mathcal{H}^0(Y/X,\mathcal{O}_Y) = 0$, while by Lemma 2.2, $\mathcal{H}om(\mathcal{J}_Y,\mathcal{J}_Y) = \mathcal{O}_X$, and ξ_1 is an isomorphism. The morphism of five–term exact sequences yields a commutative diagram with exact rows

$$\begin{array}{ccccccc} 0 \to H^1(Y/X, \mathcal{O}_Y) & \longrightarrow & H^0(X, \mathcal{H}^1(Y/X, \mathcal{O}_Y)) & \longrightarrow & 0 \\ \downarrow \zeta_1 & & \downarrow H^0(\xi_1) & & \downarrow \\ 0 \to \mathrm{Ext}^1(\mathcal{J}_Y, \mathcal{J}_Y) & \longrightarrow & H^0(X, \mathcal{E}xt^1(\mathcal{J}_Y, \mathcal{J}_Y)) & \longrightarrow & H^2(\mathcal{O}_X) \end{array}$$

showing that ζ_1 is an isomorphism.

We now study ζ_2. The edge morphism ϵ: $H^2(Y/X, \mathcal{O}_Y) \to H^0(X, \mathcal{H}^2(Y/X, \mathcal{O}_Y))$ relates global obstructions to local ones. By hypothesis, $\mathcal{O}_{\mathfrak{p},Y}$ has finite projective dimension at all $\mathfrak{p}$ such that $\mathrm{depth}_{\mathfrak{p}}\, \mathcal{O}_{\mathfrak{p},X} \leq 2$. So $\mathcal{J}_{Y,X,\mathfrak{p}}$ has projective dimension 1 at all $\mathfrak{p} \in Y \backslash Z$. By the argument of [6], p. 427, deformations of an ideal of projective dimension 1 are unobstructed. So Y is locally unobstructed at all $\mathfrak{p} \notin Z$. Hence the obstructions for deformations of Y lie in $\epsilon^{-1}(\Gamma_Z(\mathcal{H}^2(Y/X, \mathcal{O}_Y)))$. But by Lemma 2.2, the source of $H^2(\xi_0)$ vanishes, $H^1(\xi_1)$ is an isomorphism, and $\Gamma_Z(\xi_2)$ is injective. So ζ_2 is injective on the obstruction space, completing the proof of (b).

The same argument (with degenerate spectral sequences) proves (a), since the maps derived from Lemma 2.2 preserve the grading.

If $\mathcal{J}_Y$ has finite local projective dimension, we can produce an explicit representation of deformations of an ideal sheaf as an ideal sheaf of a flat deformation. Our method is based on the structure theory of exact complexes of Buchsbaum and Eisenbud. Recall that for ϕ: $F_1 \to F_0$ a morphism of finite free modules over a commutative noetherian ring R, they define $\mathrm{rk}\, \phi$ to be the largest integer r such that $\Lambda^r \phi \neq 0$, and $I(\phi)$ to be the matrix generated by the $r \times r$ minors of ϕ where $r = \mathrm{rk}\, \phi$.

Theorem 2.4. ([2] Theorem 3.1) *Let R be a commutative noetherian ring, and let*

$$0 \to F_n \xrightarrow{\phi_n} F_{n-1} \to \cdots \to F_1 \xrightarrow{\phi_1} F_0$$

be an exact sequence of oriented free R–modules. Let $r_i = \mathrm{rk}\, \phi_i$. Then

(a) *for each i there exists a unique morphism a_i such that $a_{n+1} = 1$ and the diagram*

$$\begin{array}{ccc} \Lambda^{r_i} F_i & \xrightarrow{\Lambda^{r_i}\phi_i} & \Lambda^{r_i} F_{i-1}, \\ & {\scriptstyle a_{i+1}^{\vee}}\searrow \quad \nearrow{\scriptstyle a_i} & \\ & R & \end{array}$$

(*b*) *for all* $i > 1$, Rad $I(a_i) =$ Rad $I(\phi_i)$.

***Second Proof of Theorem* 2.3(*b*) *when* cod $Y \geq 2$ *and* $\mathcal{J}_Y$ *has finite local projective dimension over* $\mathcal{O}_X$.** Let $\mathcal{J}$ be a flat deformation of $\mathcal{J}_Y$ over the spectrum Z of a noetherian local K–algebra B, *i.e.* $\mathcal{J}$ is a coherent $\mathcal{O}_{X\times Z}$ module flat over Z such that the pullback of $\mathcal{J}$ to the closed fiber of $X\times Z \longrightarrow Z$ is isomorphic to $\mathcal{J}_Y$. We show that $\mathcal{J}$ can be embedded in $\mathcal{O}_{X\times Z}$ as the ideal sheaf of a flat deformation of Y.

Assume first that $X = \mathrm{Spec}(A)$ is affine, and that there exists a finite A–free resolution

$$0 \longrightarrow E_n \xrightarrow{\phi_n} E_{n-1} \longrightarrow \cdots \longrightarrow E_2 \xrightarrow{\phi_2} E_1 \xrightarrow{\phi_1} E_0\ (= A)$$

of A/I where $I = \Gamma(X, \mathcal{J}_Y)$. By flatness, we may lift to an $(A\otimes B)$–free resolution

$$0 \longrightarrow F_n \xrightarrow{\psi_n} F_{n-1} \longrightarrow \cdots \longrightarrow F_2 \xrightarrow{\psi_2} F_1$$

of $J = \Gamma(X\times Y, \mathcal{J})$, such that $F_i \otimes_B K = E_i$ and $\psi_i \otimes_B K = \phi_i$. We will construct a lifting of ϕ_1.

Let $e_i = \mathrm{rk}\ E_i = \mathrm{rk}\ F_i$, $r_i = \mathrm{rk}\ \phi_i$, and $r_i' = \mathrm{rk}\ \psi_i$. The Buchsbaum–Eisenbud exactness criterion ([3] Corollary 1) states that $e_i = r_i + r_{i+1} = r_i' + r_{i+1}'$. Hence $r_i = \Sigma_{k=i}^{n} (-1)^{k-i} e_i = r_i'$, and $r_2' = r_2 = e_1 - 1$. Theorem 2.4(a) gives canonical factorizations

$$\begin{array}{ccc} \Lambda^{r_2} F_2 & \xrightarrow{\Lambda^{r_2}\psi_2} & F_1^{\vee} \\ & {\scriptstyle a_2^{\vee}}\searrow \quad \nearrow{\scriptstyle a_1} & \\ & A\otimes B & \end{array} \qquad \begin{array}{ccc} E_1 & \xrightarrow{\phi_1} & A \\ & {\scriptstyle \bar{a}_1^{\vee}}\searrow \quad \nearrow{\scriptstyle \bar{a}_0} & \\ & A & \end{array}$$

where $\bar{a}_1 = a_1 \otimes_B K$. Let $\psi_1 = a_1^{\vee}\cdot(\bar{a}_0 \otimes 1_B)$. Clearly $\psi_1 \otimes K = \phi_1$. Now $a_2 a_1^{\vee} \psi_2$ is the composition

$$F_2 \xrightarrow{\psi_2} F_1 \xrightarrow{(\Lambda^{r_2}\ \psi_2)^{\vee}} (\Lambda^{r_2} F_2)^{\vee},$$

which vanishes since $\Lambda^{r_2+1}\psi_2 = 0$. Thus $I(a_2)\cdot a_1^\vee\psi_2 = 0$. By Theorem 2.4(b) and the exactness criterion [3] Corollary 1, $\operatorname{depth}(I(a_2),A\bullet B) = \operatorname{depth}(I(\psi_2),A\bullet B) \geq 1$, so $I(a_2)$ contains a non-zero divisor, and $a_1^\vee\psi_2 = \psi_1\psi_2 = 0$. Thus ψ_1 induces a morphism $J \longrightarrow A\bullet B$. Since $\psi_1\bullet K = \phi_1$ is injective, ψ_1 is injective and the cokernel is flat by [EGA 0_{III}] (10.2.4). Finally, we note that the map $J \longrightarrow A\bullet B$ induced by $a_1^\vee$ is independent of the choice of resolution ($F.$). It is enough to check this fact localized at every prime of A, and then it follows from the fact that any free resolution of a finitely generated module over a local ring is isomorphic to the direct sum of the (unique up to isomorphism) minimal resolution and an exact complex, and the latter makes no contribution. The choice of $\bar{a}_0\bullet 1_B$ in the definition of ψ_1 is not canonical in the sense that any $a_0 \in A\bullet B$ such that $a_0\bullet_B K = \bar{a}_0$ would do. But if $\mathcal{J}$ defines an ideal of codimension ≥ 2, then $\bar{a}_0 = 1_A$, and $a_0 = 1_{A\bullet B}$ is the only choice (up to multiplication by an appropriate unit) such that ψ_1 induces an injection $J \longrightarrow A\bullet B$ identifying J as an ideal of codimension 2. This completes the proof when X is affine.

For general X, we consider an affine open cover $\{U_i\} = \{\operatorname{Spec} A_i\}$ of X such that each $\Gamma(U_i,\mathcal{J})$ is of finite free dimension. The image of $a_{1i}^\vee\colon F_{1i} \longrightarrow A_i\bullet B$ may be identified canonically as $\bullet_j\,(\Lambda^{f_{ji}} F_{ji})^{(-1)^{j-1}} = \operatorname{Det} J_i$. By the uniqueness assertion in the previous paragraph, the induced morphisms $J_i \longrightarrow \operatorname{Det} J_i$ must agree on intersections, and so piece together to a morphism $\mathcal{J} \longrightarrow \operatorname{Det}(\mathcal{J})$. Since $H^1(\mathcal{O}_X) = 0$, the connected components of Pic X are reduced points. So $\operatorname{Det}(\mathcal{J})$ must be constant as a relative line bundle on $X\times Z/Z$. So $\operatorname{Det}(\mathcal{J}) = p_1^*(\operatorname{Det}(\mathcal{J}))$. Hence we may identify $\mathcal{J}$ as an ideal sheaf via the composition $\mathcal{J} \longrightarrow \operatorname{Det}(\mathcal{J}) \longrightarrow \mathcal{O}_{X\times Z}$ where the second map is p_1^* of the natural map $\operatorname{Det}(\mathcal{J}) \longrightarrow \mathcal{O}_X$. If the codimension of the subscheme defined by $\mathcal{J}$ is at least 2, then $\operatorname{Det}(\mathcal{J}) \cong \mathcal{O}_X$, and an identification $\operatorname{Det}(\mathcal{J}) \xrightarrow{\sim} \mathcal{O}_{X\times Z}$ is the only choice that will identify $\mathcal{J}$ as the ideal sheaf of a closed subscheme of $X\times Z$ of codimension ≥ 2.

3. Proof of the Main Theorem

In this section we use the procedure described in §1 to prove our main theorem.

Proof of Theorem 0.5. Let $S = K[X_0,X_1,X_2,X_3]$ be the homogeneous coordinate ring of $\mathbf{P}^3$, and I the homogeneous ideal of C. Since $H^1_*(\mathcal{J}_C) = K^c(-m)$. The assumption that I has no minimal generators of degree m is equivalent to ${}_0\mathrm{Hom}_S(I,H^1_*(\mathcal{J}_C)) = 0$. So by Theorems 2.1 and 2.3, the deformation theory of C is equivalent to that of the S-module I, which we can now compute using the methods of §1 (without any further reference to cotangent complexes).

First we consider a minimal free resolution of I. By [16] Theorem 2.5 and our definitions of a, b, and c, it is of the form

$$L.: \qquad 0 \longrightarrow S(-m-4)^c \xrightarrow{\delta_1} S(-m-4)^b \oplus S(-m-3)^{4c} \oplus F_1 \xrightarrow{\delta_0} S(-m-4)^a \oplus F_0$$

with F_1 and F_0 are free with no factors of degree $m+4$, and

$$\delta_1 = 0 \oplus ({}^t[X_0\ X_1\ X_2\ X_3] \oplus 1_c) \oplus 0.$$

Since the middle $4c$ rows of the matrix of δ_1 generate $\mathfrak{m}L_2^\vee$,

(*) Any $\beta \in \mathrm{Hom}(L_2,\mathfrak{m}L_1)$ factors as $\alpha\delta_1$ with $\alpha \in \mathrm{Hom}(S(-m-3)^{4c},L_1)$.

To begin the method of §1 we need to choose a set of cocycles in ${}_0\mathrm{Hom}^1(L.,L.)$ whose cohomology classes form a basis of ${}_0\mathrm{Ext}^1_S(I,I)$. Now by minimality, all entries of δ_1 and δ_0 lie in $\mathfrak{m}$. Hence all entries in the matrix of any coboundary in $\mathrm{Hom}^{\cdot}(L.,L.)$ will lie in $\mathfrak{m}$. Therefore there exists a well-defined map

$$\lambda: {}_0\mathrm{Ext}^1_S(I,I) \longrightarrow \mathrm{Hom}(K^c,K^b) \oplus \mathrm{Hom}(K^b,K^a)$$

which sends any member of ${}_0\mathrm{Ext}^1_S(I,I)$ represented by a cocycle in ${}_0\mathrm{Hom}^1_S(L.,L.)$ to the induced map on the factors of the L_i of degree $m+4$. We now choose our set of cocycle representatives of a basis of ${}_0\mathrm{Ext}^1_S(I,I)$ as follows. First we choose a set $\zeta_1, \dots, \zeta_s$ of cocycle representatives of a basis of the kernel of λ. Because the $\overline{\zeta}_p \in \ker\lambda$, each ζ_p maps L_2 into $\mathfrak{m}L_1$. By (*), we may modify the ζ_p by a coboundary so that $\zeta_p|_{L_2} = 0$. Next let ξ_{ij} $(1 \le i \le a,\ 1 \le j \le b)$ be the cochain which sends the jth factor $S(-m-4)$ of L_1 identically onto the ith factor $S(-m-4)$ of L_2, and all other factors of L_2 and L_1 to 0. The ξ_{ij} are cocycles. Similarly let u_{jk} $(1 \le j \le b,\ 1 \le k \le c)$ be the cochain sending the kth factor of L_2 identically onto the jth factor of L_1, and L_1 to 0. Although u_{jk} is not a cocycle, $d(u_{jk}) = \delta_0 u_{jk} \in \mathrm{Hom}(L_2,\mathfrak{m}L_0)$ because all the entries of δ_0 lie in $\mathfrak{m}$. Hence by (*) there exists a $v_{jk} \in \mathrm{Hom}(S(-m-3)^{4c},L_0)$ such that $v_{jk}\delta_1 = -\delta_0 u_{jk}$. Then $\eta_{jk} =$

$u_{jk} \bullet v_{jk} \in {}_0\mathrm{Hom}(L_2,L_1) \bullet {}_0\mathrm{Hom}(L_1,L_0)$ is a cocycle. Note that λ maps the η_{jk} and ξ_{ij} onto a standard basis of $\mathrm{Hom}(K^c,K^b) \bullet \mathrm{Hom}(K^b,K^a)$. Hence the ζ_{ij}, η_{jk}, and ζ_p are a system of cocycle representatives of a basis of ${}_0\mathrm{Ext}^1_S(I,I)$. The members of the dual basis of ${}_0\mathrm{Ext}^1_S(I,I)^* \cong T^*_{[C]}\mathrm{Hilb}\,\mathbf{P}^3$ will be denoted Z_p, Y_{jk}, and X_{ij}. Set $T = K[Z_p, Y_{jk}, X_{ij}]$.

We now consider ${}_0\mathrm{Ext}^2_S(I,I)$. All members of ${}_0\mathrm{Hom}(L_2,L_0)$ are cocycles. Any coboundary lies in ${}_0\mathrm{Hom}(L_2,\mathfrak{m}L_0)$ since the components of the δ_i are in $\mathfrak{m}$, while conversely any member of ${}_0\mathrm{Hom}(L_2,\mathfrak{m}L_0)$ is a coboundary by (*). Since ${}_0\mathrm{Hom}(L_2,\mathfrak{m}L_0) = {}_0\mathrm{Hom}(L_2,F_0)$, we have ${}_0\mathrm{Ext}^2_S(I,I) \cong \mathrm{Hom}(K^c,K^a)$, and the cocycles θ_{ik} sending the kth factor of L_2 identically onto the the ith factor of L_0 are a system of cocycle representatives of a basis of ${}_0\mathrm{Ext}^2_S(I,I)$.

By (1.3) the universal first–order deformation of I has resolution $L.\bullet T/\mathfrak{m}_T^2$ with boundary

$$\delta + \sum X_{ij}\xi_{ij} + \sum Y_{jk}\eta_{jk} + \sum Z_p\zeta_p. \tag{3.1}$$

By (1.4), the obstruction to lifting the boundary to the second order is computed by taking the square of (3.1). It is easily checked that all products of the ξ_{ij}, η_{jk}, and ζ_p vanish except $\xi_{ij}\eta_{jk} = \theta_{ik}$ and $\zeta_p\eta_{jk} \in \mathrm{Hom}(L_2,\mathfrak{m}L_0)$. So (1.4) becomes

$$(\delta + \sum X_{ij}\xi_{ij} + \sum Y_{jk}\eta_{jk} + \sum Z_p\zeta_p)^2 = \sum_{i,k} (\sum_j X_{ij}Y_{jk})\theta_{ik} + \sum Y_{jk}Z_p\zeta_p\eta_{jk}.$$

By (*) the $\zeta_p\eta_{jk}$ are not only coboundaries, but there exist $\psi_{jkp} \in \mathrm{Hom}(S(-m-3)^{4c},L_0)$ such that $d(\psi_{jkp}) = \psi_{jkp}\delta_1 = -\zeta_p\eta_{jk}$. The products of the ψ_{jkp} with the ξ_{ij}, η_{jk}, ζ_p and with each other all vanish. So (1.5) now becomes

$$(\delta + \sum X_{ij}\xi_{ij} + \sum Y_{jk}\eta_{jk} + \sum Z_p\zeta_p + \sum Y_{jk}Z_p\psi_{jkp})^2 = \sum_{i,k} (\sum_j X_{ij}Y_{jk})\theta_{ik}.$$

Since this holds as an equation, not just as a congruence modulo $\mathfrak{m}_T^3$, the process terminates, and by the discussion at the end of §1, the formal moduli of deformations of I are algebraizable over $\mathrm{Spec}(T/J)$ where J is the ideal generated by the ac sums $\Sigma_j X_{ij}Y_{jk}$. This completes the proof of the theorem.

4. Examples

In this section we construct specific examples of curves satisfying (0.3) and (0.4) to which Theorem 0.5 therefore applies. Many of these examples will be both obstructed and of maximal rank, and therefore counterexamples to Question 0.1, as we explained in the introduction. Our basic tool is

Theorem 4.1. ([4] Example 2.1) *Let* $\mathcal{E} = \oplus_{i=1}^{r} \mathcal{O}_{\mathbf{P}^3}(-a_i)$, $\mathcal{F} = \oplus_j \mathcal{O}_{\mathbf{P}^3}(-b_j) \oplus \oplus_k \Omega_{\mathbf{P}^3}(-c_k)$, *and*

$$\mathcal{F}' = \bigoplus_j \mathcal{O}_{\mathbf{P}^3}(-b_j) \oplus \bigoplus_k \mathcal{O}_{\mathbf{P}^3}(-c_k-1)^3 = \bigoplus_{h=1}^{r+1} \mathcal{O}_{\mathbf{P}^3}(-d_h),$$

where $a_1 \le a_2 \le \cdots \le a_r$ *and* $d_1 \le d_2 \le \cdots \le d_{r+1}$. *Suppose* $\phi \in \mathrm{Hom}(\mathcal{E},\mathcal{F})$ *is sufficiently general subject to the condition that* ϕ *does not map any* $\mathcal{O}_{\mathbf{P}^3}(-a_i)$ *isomorphically onto any* $\mathcal{O}_{\mathbf{P}^3}(-b_j)$. *If* $a_i > d_{i+2}$ *for* $1 \le i \le r-1$, *and the characteristic is zero, then* ϕ *is injective, and* $\mathrm{cok}(\phi)$ *is isomorphic to a twist of the ideal sheaf of a nonsingular curve.*

Remark **4.1.1. We have rephrased the example of [4] even beyond the changes due to specializing $\mathbf{P}^n$ to $\mathbf{P}^3$. Namely, the original example defined $\mathcal{F}'$ using $\mathcal{O}_{\mathbf{P}^3}(-c_k-2)^3$ as the replacement for $\Omega_{\mathbf{P}^3}(-c_k)$, and then used the condition $a_i \ge d_{i+2}$ with the assumption that any pairs $a_i = d_j$ will have been canceled before applying the condition. Because we do not wish to make this assumption, so we must use the stronger condition $a_i > d_{i+2}$ in conjunction with the a modified definition of $\mathcal{F}'$. The original proof is adapted by basically shifting the indexing of the filtration of $\mathcal{E}$ by one. So Theorem 4.1 holds whenever the main theorem of [4] holds, *i.e.* in characteristic 0. These particular examples also work in characteristic *p* ([17],) but that is more difficult.**

Example **4.2. Theorem 4.1 may be used to construct many examples of curves satisfying (0.3) and (0.4). For instance, for any triple of positive integers (a,b,c), there exists a curve *C* in $\mathbf{P}^3$, nonsingular (at least if the characteristic is 0 by our last remark), with locally free resolution**

$$0 \longrightarrow \mathcal{O}(-2)^{3c-2} \oplus \mathcal{O}(-4)^b \oplus \mathcal{O}(-5)^{a+1} \xrightarrow{\phi} \Omega^c \oplus \mathcal{O}(-3)^b \oplus \mathcal{O}(-4)^a \longrightarrow \mathcal{I}_C(a+b+2c+1) \longrightarrow 0$$

such that ϕ does not carry any $\mathcal{O}(-4)$ isomorphically onto an $\mathcal{O}(-4)$. Such curves satisfy (0.3) and (0.4) with $m = a+b+2c+1$. The degree and genus of these curves are

$$d = \begin{bmatrix} m+1 \\ 2 \end{bmatrix} + 4a + 3b + 3c + 8,$$
$$g = 2\begin{bmatrix} m+1 \\ 3 \end{bmatrix} + \begin{bmatrix} m \\ 2 \end{bmatrix} + (3a+2b+c+6)m + (3a+b-c+4).$$

(So in the simplest case $(a,b,c) = (1,1,1)$, we have $(d,g) = (33,117)$.) From the long exact sequence of cohomology one may derive that these curves are not contained in any surface of degree $m+1$ or less, so these curves are of maximal rank. But by Theorem 0.5 these curves are obstructed. Thus we get a negative answer to Question 0.1. However, we may compute $h^1(\mathcal{J}_C(m)) = c$ and $h^2(\mathcal{J}_C(m)) = 3a+b+1$. So the curves do not have seminatural cohomology.

References

[1] Ballico, E., Ellia, P.: A Program for Space Curves. Algebraic Varieties of Small Dimension, Torino 1985. Rend. Sem. Mat. Torino, 25–42 (1986).
[2] Buchsbaum, D., Eisenbud, D.: Some Structure Theorems for Finite Free Resolutions. Advances in Math. **12**, 84–139 (1974).
[3] Buchsbaum, D., Eisenbud, D.: What Makes a Complex Exact? J. Algebra **25**, 259–268 (1973).
[4] Chang, M.–C.: A Filtered Bertini–type Theorem, J. reine angew. Math. **397**, 214–219 (1989).
[5] Curtin, D.: Obstructions to Deforming a Space Curve. Trans. Amer. Math. Soc. **267**, 83–94 (1981).
[6] Ellingsrud, G.: Sur le schéma de Hilbert des variétés de codimension 2 dans $\mathbb{P}^e$ à cône de Cohen–Macaulay, Ann. Scient. Ec. Norm. Sup. **8**, 423–432 (1975).
[7] Ellia, P., Fiorentini, M.: Défaut de postulation et singularités du schéma de Hilbert, Ann. Univ. Ferrara, Sez. VII, **30**, 185–198 (1984).
[8] Illusie, L.: Complexe Cotangent et Déformations I. Lect. Notes Math. **239** (1971).
[9] Kleppe, J.: Deformations of Graded Algebras. Math. Scand. **45**, 205–231 (1979).
[10] Kleppe, J.: The Hilbert–Flag Scheme, its Properties and its Connection with the Hilbert Scheme. Applications to Curves in 3–Space. Preprint no. 5, Mat. Inst., Univ. Oslo. (1981)
[11] Laudal, O. A.: Formal Moduli of Algebraic Structures. Lect. Notes Math. **754** (1979).
[12] Laudal, O. A.: Matric Massey Products and Formal Moduli I, in: Algebra, Algebraic Topology and Their Interactions, Lect. Notes Math. **1183**, 218–240 (1986).
[13] May, J. P.: Matric Massey Products. J. Algebra **12**, 533–568 (1969).
[14] Piene, R., Schlessinger, M.: On the Hilbert Scheme Compactification of the Space of Twisted Cubics. Amer. J. Math. **107**, 761–774 (1985).
[15] Pinkham, H.: Deformations of Algebraic Varieties with G_m Action. Astérisque **20** (1974).
[16] Rao, A. P.: *Liaison* Among Curves in $\mathbb{P}^3$. Invent. Math. **50**, 205–217 (1979).
[17] Walter, C.: Some Transversality Theorems in Characteristic p with Applications to Arithmetically Buchsbaum Schemes.

Printed in the United States
By Bookmasters